Nonstoichiometric Oxides

MATERIALS SCIENCE AND TECHNOLOGY

A. S. Nowick and B. S. Berry, ANELASTIC RELAXATION IN CRYSTALLINE SOLIDS, 1972

E. A. Nesbitt and J. H. Wernick, RARE EARTH PERMANENT MAGNETS, 1973

W. E. Wallace, RARE EARTH INTERMETALLICS, 1973

J. C. Phillips, BONDS AND BANDS IN SEMICONDUCTORS, 1973

J. H. Richardson and R. V. Peterson (editors), SYSTEMATIC MATERIALS ANALYSIS, VOLUMES I, II, AND III, 1974; IV, 1978

A. J. Freeman and J. B. Darby, Jr. (editors), THE ACTINIDES: ELECTRONIC STRUCTURE AND RELATED PROPERTIES, VOLUMES I AND II, 1974

A. S. Nowick and J. J. Burton (editors), DIFFUSION IN SOLIDS: RECENT DEVELOPMENTS, 1975

J. W. Matthews (editor), EPITAXIAL GROWTH, PARTS A AND B, 1975

J. M. Blakely (editor), SURFACE PHYSICS OF MATERIALS, VOLUMES I AND II, 1975

G. A. Chadwick and D. A. Smith (editors), GRAIN BOUNDARY STRUCTURE AND PROPERTIES, 1975

John W. Hastie, HIGH TEMPERATURE VAPORS: SCIENCE AND TECHNOLOGY, 1975

John K. Tien and George S. Ansell (editors), ALLOY AND MICROSTRUCTURAL DESIGN, 1976

M. T. Sprackling, THE PLASTIC DEFORMATION OF SIMPLE IONIC CRYSTALS, 1976

James J. Burton and Robert L. Garten (editors), ADVANCED MATERIALS IN CATALYSIS, 1977

Gerald Burns, INTRODUCTION TO GROUP THEORY WITH APPLICATIONS, 1977

L. H. Schwartz and J. B. Cohen, DIFFRACTION FROM MATERIALS, 1977

Zenji Nishiyama, MARTENSITIC TRANSFORMATION, 1978

Paul Hagenmuller and W. van Gool (editors), SOLID ELECTROLYTES: GENERAL PRINCIPLES, CHARACTERIZATION, MATERIALS, APPLICATIONS, 1978

G. G. Libowitz and M. S. Whittingham, MATERIALS SCIENCE IN ENERGY TECHNOLOGY, 1978

Otto Buck, John K. Tien, and Harris L. Marcus (editors), ELECTRON AND POSITRON SPECTROSCOPIES IN MATERIALS SCIENCE AND ENGINEERING, 1979

Lawrence L. Kazmerski (editor), POLYCRYSTALLINE AND AMORPHOUS THIN FILMS AND DEVICES, 1980

Manfred von Heimendahl, ELECTRON MICROSCOPY OF MATERIALS: AN INTRODUCTION, 1980

O. Toft Sørensen (editor), NONSTOICHIOMETRIC OXIDES, 1981

In Preparation

M. Stanley Whittingham and Allan J. Jacobson (editors), INTERCALATION CHEMISTRY

Nonstoichiometric Oxides

Edited by

O. Toft Sørensen

Department of Metallurgy
Risø National Laboratory
Roskilde, Denmark

1981

ACADEMIC PRESS

A Subsidiary of Harcourt Brace Jovanovich, Publishers

New York London Toronto Sydney San Francisco

ACADEMIC PRESS, INC.
111 Fifth Avenue, New York, New York 10003

United Kingdom Edition published by
ACADEMIC PRESS, INC. (LONDON) LTD.
24/28 Oval Road, London NW1 7DX

Library of Congress Cataloging in Publication Data
Main entry under title:

Nonstoichiometric oxides.

(Materials science and technology)
Includes bibliographies and index.
1. Oxides-Addresses, essays, lectures. I. Sørensen, O. Toft II. Series.
QD181.01N66 546'.7212 80-2342
ISBN 0-12-655280-0 AACR2

PRINTED IN THE UNITED STATES OF AMERICA

81 82 83 84 9 8 7 6 5 4 3 2 1

Contents

List of Contributors

Numbers in parentheses indicate the pages on which the authors' contributions begin.

C. R. A. Catlow (61), Department of Chemistry, University College London, London WC1H OAJ, England

A. K. Cheetham (399), Chemical Crystallography Laboratory, University of Oxford, Oxford OX1 3PD, England

LeRoy Eyring (337), Department of Chemistry and The Center for Solid State Science, Arizona State University, Tempe, Arizona 85281

J. A. Kilner (233), Wolfson Unit for Solid State Ionics, Department of Metallurgy and Materials Science, Imperial College, London SW7 2BP, United Kingdom

L. Manes (99), Commission of the European Communities, Joint Research Centre, Karlsruhe Establishment, European Institute for Transuranium Elements, D-7500 Karlsruhe, Federal Republic of Germany

Hj. Matzke (155), Commission of the European Communities, Joint Research Centre, Karlsruhe Establishment, European Institute for Transuranium Elements, D-7500 Karlsruhe, Federal Republic of Germany

O. Toft Sørensen (1), Risø National Laboratory, DK-4000 Roskilde, Denmark

B. C. H. Steele (233), Wolfson Unit for Solid State Ionics, Department of Metallurgy and Materials Science, Imperial College, London SW7 2BP, United Kingdom

H. L. Tuller (271), Department of Materials Science and Engineering, Massachusetts Institute of Technology, Cambridge, Massachusetts 02139

Preface

The properties of many inorganic compounds are greatly dependent on the degree of nonstoichiometry, which can become very extensive in oxide systems, and today nonstoichiometry plays an important role in solid-state research.

Many nonstoichiometric oxides were previously considered to be single-phase materials, extending over a considerable composition range with randomly distributed and noninteracting defects. At the large deviations from stoichiometry observed in many systems, however, this cannot be true, and intuitively the defects must be expected to interact to form either extended clusters (complexes) or large superstructure lattices by long-range ordering. The detailed thermodynamic and structural studies carried out during the past five to ten years have verified that such defect–defect interactions must in fact be taken into account in order to describe the properties and structures of nonstoichiometric oxides.

In the early seventies nonstoichiometric oxides were discussed in several excellent books, but since then many important contributions have improved our understanding of these systems considerably. The aim of this book therefore is to present a detailed and up-to-date account of the present state of the art within this field.

In the first part, the thermodynamic properties of nonstoichiometric oxides are discussed in terms of defect complexes, first in a very general way, using the principles of classical thermodynamics, and then from a statistical thermodynamics point of view. The statistical thermodynamic models and the simple treatment clearly indicate that the nonstoichiometric phase range is much more ordered than was previously believed.

In the second part, transport properties, i.e., diffusion and electrical conductivity, are reviewed. First, the diffusion theories and experimental diffusion coefficients are discussed for several systems and then the electrical properties are discussed for the highly defective ionic and mixed oxide conductors (fast ionic conductors).

Finally, an account is given of the very important results obtained in structural studies, using high-resolution electron microscopy and x-ray and neutron diffraction. These studies, which supplement each other in an excellent manner, have very convincingly confirmed the fact that defect complexes and ordered structures are really formed in nonstoichiometric oxides.

1

Thermodynamics and Defect Structure of Nonstoichiometric Oxides

O. Toft Sørensen
Risø National Laboratory
Roskilde, Denmark

ISBN 0-12-655280-0

I. Introduction

Nonstoichiometric phases are found in many oxide systems at higher temperatures. Especially in oxides in which the cations can exist in several oxidation (valence) states, large deviations from the stoichiometric composition have been observed. The most interesting nonstoichiometric systems are thus found in the transition metal, rare earth, and actinide oxides, which will be considered in this chapter. Previously the thermodynamic properties of these systems were described in terms of randomly distributed and noninteracting point defects. With the large concentration of defects present even at moderate deviations from the stoichiometric composition, however, appreciable interactions between the defects must be expected. Thus clustering, long-range ordering into superstructures, or perhaps even elimination of the defects by a crystallographic shearing mechanism must be taken into account in order to understand the real nature of these nonstoichiometric oxide systems. Structural studies by x-ray and neutron scattering and particularly by high-resolution electron microscopy have also, in accordance with these ideas, confirmed that many nonstoichiometric oxide systems are much more ordered than previously believed. This is described in later chapters, and it is the purpose of the present chapter to discuss the thermodynamic properties of some of the most important nonstoichiometric oxide systems in terms of such defect clusters or in terms of ordered intermediate phases. The approach taken in this analysis is based on classical thermodynamics, whereas a more rigorous statistical thermodynamic approach will be presented for some of the oxide systems in Chapter 3.

Thermodynamic properties and theories of nonstoichiometric oxides have recently been reviewed by Anderson (1970a, b, 1974), Fender (1972), Kofstad (1972), Navrotsky (1974), Mrowec (1978), and Sørensen (1978), and these reviews are recommended for further details.

II. General Thermodynamic Considerations of the Phase Relationships in Nonstoichiometric Oxides

A. Oxygen Activities, μ_{O_2} and $\Delta\overline{G}_{O_2}$

An important thermodynamic quantity to consider in studies on nonstoichiometric oxides is the activity of oxygen in the solid compound or, more exactly, the chemical potential of oxygen μ_{O_2} which is defined as

$$\mu_{O_2} = \mu^\circ_{O_2} + RT \ln P_{O_2} \tag{1}$$

Here $\mu^\circ_{O_2}$ is the chemical potential of oxygen in its standard state (gas, 1 atm) and P_{O_2} is the equilibrium oxygen pressure in the atmosphere surrounding the nonstoichiometric oxide.

Another useful quantity is $\Delta\overline{G}_{O_2}$, the relative partial molar free energy of oxygen, which can be easily calculated from the equilibrium oxygen pressure according to

$$\Delta\overline{G}_{O_2} = \overline{G}_{O_2} - G^\circ_{O_2} = \mu_{O_2} - \mu^\circ_{O_2} = RT \ln P_{O_2} \tag{2}$$

and which will be used throughout this chapter.

From these equations it is clear that μ_{O_2} and $\Delta\overline{G}_{O_2}$ depend on the temperature, but they also depend on the composition, as will be shown later. A nonstoichiometric phase thus shows typical bivariant behavior $[\mu_i = \mu_i(T, x)]$ in contrast to the univariant behavior expected for a two-phase mixture $[\mu_i = \mu_i(T)]$. For a two-component system such as a binary oxide the Gibbs phase rule dictates bivariant behavior for a single phase, and a nonstoichiometric phase can thus be considered a single phase across its composition range. This behavior is illustrated in Fig. 1 which shows a

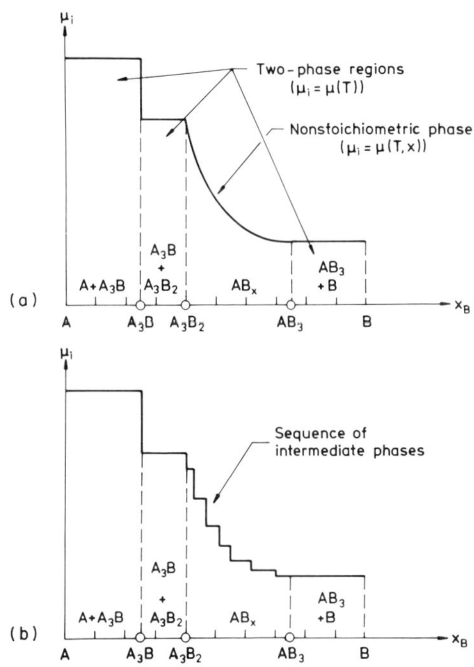

Fig. 1. Hypothetical μ, T, x diagrams for a two-component system at constant temperature showing (a) a nonstoichiometric phase and (b) a sequence of intermediate phases.

μ, T, x diagram at constant temperature for a hypothetical two-component system with the line phases A_3B, A_3B_2, and AB_3 and with a nonstoichiometric phase AB_x between two of the line phases. From the figure it is also clear that a plot of $\Delta\overline{G}_{O_2}$ versus composition, which can be used to evaluate the phase relationships of the system, should show the following characteristic curves for the different types of phases present:

(i) for a two-phase region a horizontal line should be observed;

(ii) for a line phase (ordered compound) a vertical line should be observed; and

(iii) for a nonstoichiometric phase a curve of intermediate slope should be observed. As will be explained later, this curve is expected to show essentially exponential behavior.

In some oxide systems detailed structural studies by x-ray diffraction and high-resolution electron microscopy have shown that apparent nonstoichiometric phases in fact consist of a whole series of line phases (Magnéli phases). This, for instance, is the case for the TiO_x system, where Bursill and Hyde (1972) have shown that crystallographic shear can produce a whole series of ordered phases all following the general formula Ti_nO_{2n-1}. In this case the $\Delta\overline{G}_{O_2}$–composition plot must take the form of a step curve, as is also shown in Fig. 1; however, unless closely spaced and very accurate experimental data are available, this curve can easily be mistaken for a continuous curve. It is, however, doubtful whether sufficient accuracy can be obtained in a thermodynamic study to judge whether a given phase range really is nonstoichiometric or instead consists of a succession of line phases. The real nature of any nonstoichiometric compound can be evaluated only if thermodynamic and structural studies are combined.

B. Free Energy Curves

Although important information about the phase relationships in a nonstoichiometric system can be obtained from a $\Delta\overline{G}_{O_2}$–x plot as described above, the number of possible phases and their width can best be judged from the relative position and shape of the related free energy curves for these phases. If we consider the formation of vacancies in an elemental crystal as an example, the change in free energy ΔG accompanied by the introduction of n vacancies at N lattice sites will be (Swalin, 1972)

$$\Delta G = G - G^* = n\,\Delta H_{\mathrm{v}} - T\left(\Delta S_{\mathrm{c}} + n\,\Delta S_{\mathrm{v}}^{\mathrm{vibr}}\right) \tag{3}$$

where G^* is the free energy of a perfect crystal, ΔH_{v} the enthalpy required to form one vacancy (positive), and ΔS_{c} and ΔS_{v}, respectively, the configurational and vibrational entropies involved in the defect formation.

For N atoms arranged at $N + n$ lattice sites, ΔS_c becomes

$$\Delta S_c = k \ln[(N + n)!/N!n!] \tag{4}$$

and, using Stirling's approximation, gives

$$\Delta S_c = -k\left[N \ln \frac{N}{(N + n)} + n \ln \frac{N}{(N + n)}\right] \tag{5}$$

which substituted in Eq. (3) yields the following expression for the free energy:

$$G = G^* + n(\Delta H_v - T\Delta S_v^{vibr}) + kT\left[N \ln\left(\frac{N}{N + n}\right) + n \ln\left(\frac{n}{N + n}\right)\right] \tag{6}$$

According to Mott and Guerney (1940), ΔS_v can be expressed as

$$\Delta S_{vibr} = xk \ln(\nu/\nu') \tag{7}$$

where x is the number of atoms surrounding the vacancy and ν and ν' are the vibrational frequencies of the perfect lattice and of the surrounding atoms, respectively. Since it must be expected that the vibrational frequencies are slowed down for the atoms surrounding a vacancy, ΔS_v can be expected to be positive.

If the different contributions to G in Eq. (6) are plotted as functions of the number of defects formed, the curves shown qualitatively in Fig. 2 are obtained. The resulting G curve, which also is shown in the figure, has a typical parabolic shape with a minimum below the free energy G^* of the perfect crystal. The introduction of vacancies (defects) thus lowers the free energy of the system until an equilibrium concentration of defects is reached, after which the free energy increases again. When the different contributions are compared, it is also clear that the reason for this behavior is that the positive ΔH_v is more than compensated for by the $T\Delta S_c$ and $T\Delta S_v$ terms. The magnitude of this compensation, however, depends on the temperature and, as also indicated in Fig. 2, the gain in free energy.

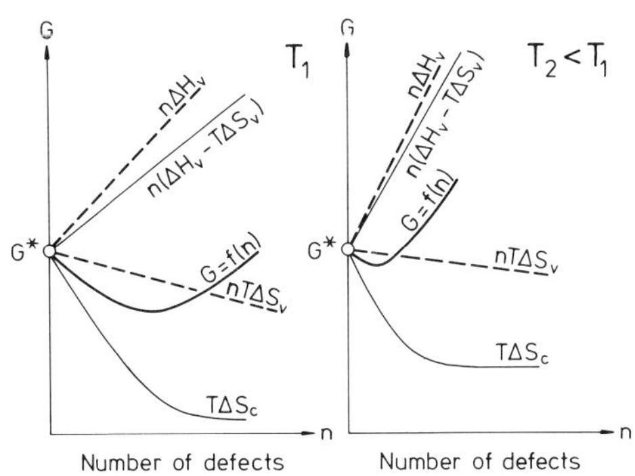

Fig. 2. Schematic representation of the free energy of a crystal as a function of defect concentration.

Thus the number of defects introduced is considerably decreased at lower temperatures, where the shape of the G–x curve essentially is determined by the ΔH_v term. For a line phase (ordered phase), where the formation of defects requires a large expenditure of energy, a sharply pointed G–x curve can thus be expected, whereas a relatively broad, assymmetric G–x curve must be expected for a nonstoichiometric phase in which defects are relatively easily formed.

The defects formed because of the increasing importance of the $T\Delta S$ terms at higher temperatures are termed primary or native defects (internal or thermal disorder) in contrast to those formed by a reaction with the surrounding atmosphere such as are encountered in the nonstoichiometric oxides described in subsequent sections. Primary defects usually occur in pairs, and the simple ratio between the constituents is not changed in this case, e.g., Schottky disorder which consists of an equal concentration of anion and cation vacancies. For some oxide systems, however, the stability of the oxygen and the metal sublattice is not the same, and deviations from the stoichiometric composition can occur as a result of an imbalance

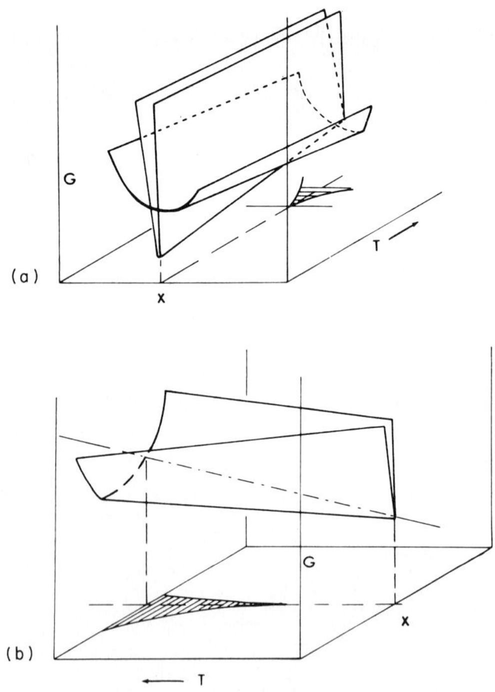

Fig. 3. G–T–x representation for a line and a nonstoichiometric phase. (a) Intersection of the two surfaces making the nonstoichiometric phase more stable at higher temperatures. (b) Progressive change in the shape of the free energy surface of a line phase, which is transformed into a nonstoichiometric phase at higher temperatures. (From Anderson, 1970b.)

among the primary defects. For such systems the sharply pointed G curve for an ordered (line) phase at low temperatures will thus become broader at increasing temperatures, resulting in the formation of a nonstoichiometric phase as shown in Fig. 3 (Anderson, 1970a).

A second possibility for the formation of a nonstoichiometric phase, also shown in Fig. 3, is that the G–T–x surface for a line phase intersects with the surface of an adjacent nonstoichiometric phase, which then becomes the most stable at higher temperatures. In this case the behavior is thus determined from the relative stability of the two types of phases. At intermediate temperatures, however, both phases can coexist as indicated by a common tangent in Fig. 4. In the case of G–x_O^* (mole fraction of oxygen) curves a common tangent shows that the corresponding $\Delta\bar{G}_{O_2}$ value, which can be obtained by extrapolation to $x_O^* = 1$, is the same for both phases. If the system were in equilibrium, the two phases would coexist in a two-phase mixture. However, whether this is observed in a real system depends on how easily the ordered phase can nucleate from a nonstoichiometric phase. If this nucleation is slow or kinetically hindered, then a metastable nonstoichiometric phase could exist up to the line phase,

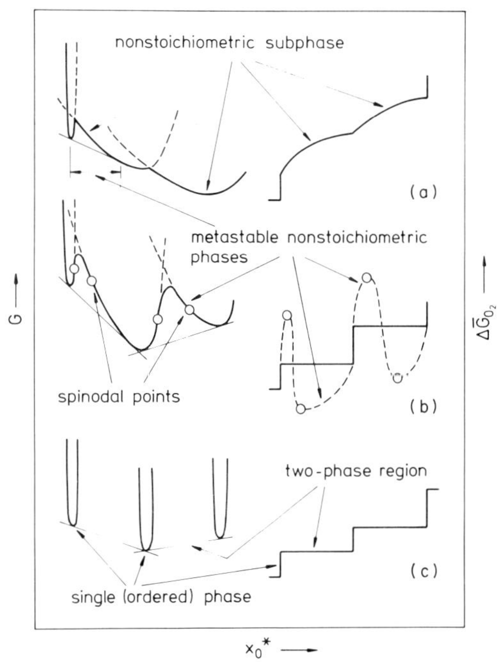

Fig. 4. Shape and relative positions of G–x_O^* curves for line phases and nonstoichiometric phases. Ordered line phases and two-phase regions (c) are formed at intermediate temperatures by spinodal unmixing reactions (b) from nonstoichiometric phases (a). Temperature decreases from (a) to (c).

as indicated in Fig. 4. At still lower temperatures other instabilities can occur. If the overlapping G–x_O^* curves take the spinodal form also shown in this figure, small composition fluctuations will lead to a spontaneous unmixing into a series of line phases and two-phase regions. This unmixing will be irreversible and self-propagating, as it results in a decrease in the total free energy of the system (Cahn, 1961). The series of ordered phases observed at lower temperatures, for instance, in the Ce–O and Pr–O systems, are probably formed in this way, and the observation of unmixing in quenched wüstite ($Fe_{1-y}O$) has also been explained by spinodal decomposition reactions (Manenc *et al.*, 1962; Herai *et al.*, 1964; Greenwood and Howe, 1972).

C. Defect Theories

The primary defects described in the previous section can create only very small deviations from the stoichiometric composition and are usually not considered in studies on nonstoichiometric compounds. More important are the defects formed in reactions between a solid phase and the surrounding atmosphere, which in some oxide systems can create nonstoichiometric phases extending over considerable composition ranges.

A necessary criterion for the formation of a nonstoichiometric phase is that cations can exist in several oxidation (valence) states, and nonstoichiometry is thus primarily observed in transition metal oxides, rare earth oxides, and actinide oxides. Generally, the direction of the change in composition corresponds to a change in the oxidation state of the cations to the next stable state, which results in formation of the following defect types:

1. Oxygen vacancies $V_O^{\cdot\cdot}$: oxygen-deficient oxides in which the cations are easily reduced, e.g., CeO_{2-x}, PrO_{2-x}, PuO_{2-x}.
2. Metal vacancies V_M'': metal-deficient oxides in which the cations are easily oxidized, e.g., $Fe_{1-y}O$, $Mn_{1-y}O$, $Co_{1-y}O$.
3. Oxygen in interstitial lattice positions O_i'': excess oxygen compounds; cations are easily oxidized in these systems, e.g., UO_{2+x}.
4. Metal in interstitial lattice positions $M_i^{\cdot\cdot}$: this type of defect is found in metal-deficient oxides together with V_M'', e.g., in $Fe_{1-y}O$ (Koch and Cohen, 1969).

The notation proposed by Kröger and Vink (1956; see also Kröger, 1964) is used here and throughout this chapter to describe the defects. In this system the type of imperfection is indicated by a major symbol, the location is described by a subscript, and finally the charge of the defect relative to the normal lattice is indicated by a superscript. Examples of the notation of the most important defects are given in Fig. 5.

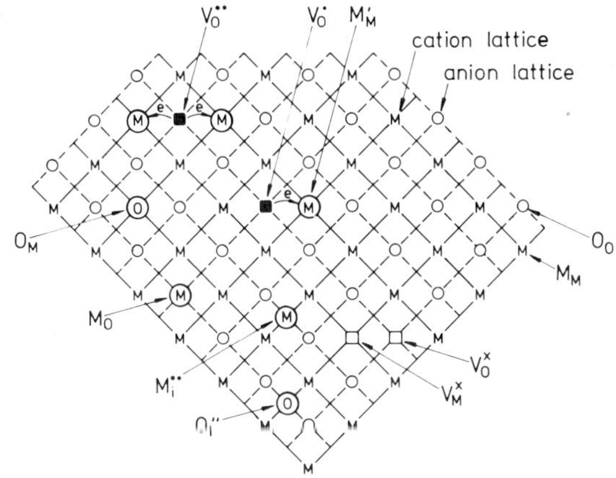

Fig. 5. Defect notation according to Kröger and Vink (1956).

The formation of the different types of defects and the corresponding oxidation or reduction of the cations can be described in a quasi-chemical defect reaction. In the case of oxygen-deficient oxides, for instance, formation and subsequent ionization of the oxygen vacancies can be expressed by

$$O_O \leftrightarrows V_O^x + \tfrac{1}{2}O_2$$

$$V_O^x + M_M^x \leftrightarrows V_O^{\cdot} + M_M'$$

$$V_O^{\cdot} + M_M^x \leftrightarrows V_O^{\cdot\cdot} + M_M'$$

or the total reaction

$$O_O + 2M_M \leftrightarrows V_O^{\cdot\cdot} + 2M_M' + \tfrac{1}{2}O_2 \tag{8}$$

where, as shown in Fig. 5, O_O and M_M represent oxygen and metal atoms at their respective sites in the crystal lattice, V_O^x, $V_O^{\cdot}$, and $V_O^{\cdot\cdot}$ represent a neutral, single, and double positively charged oxygen vacancy, respectively, and M_M' a negatively charged cation (e.g, M^{3+} which is negative relative to the normal cations of M^{4+}) that has taken up one of the electrons liberated during ionization of the neutral oxygen vacancies primarily formed.

For the formation of metal vacancies in a metal-deficient oxide, the following equations apply:

$$\tfrac{1}{2}O_2 \leftrightarrows V_M^x + O_O^x$$

$$V_M^x + e' \leftrightarrows V_M' + h^{\cdot}$$

$$V_M' + e' \leftrightarrows V_M'' + h^{\cdot}$$

$$2Fe_{Fe}^x + 2h^{\cdot} \leftrightarrows 2Fe_{Fe}^{\cdot}$$

or the total reaction

$$\tfrac{1}{2}O_2 \leftrightarrows V_M'' + 2Fe_{Fe}^{\cdot} + O_O^x \tag{9}$$

where V''_M is a double negatively charged metal vacancy and $Fe^{\cdot}_{Fe}$ is an association of $h^{\cdot}$ and Fe^{2+}, which can be considered a Fe^{3+} ion. Here $h^{\cdot}$ is a positive hole in the valence band of the Fe^{2+} ions formed by the uptake of electrons of V^x_M and V'_M.

For both of these compounds the defects are characteristically formed in one of the sublattices, whereas the other is considered to be stable. The defects are furthermore considered to be randomly distributed and noninteracting, which, however, is rather unlikely because of the large defect concentrations observed in some oxide systems.

For $Fe_{0.85}O$, for instance, the apparent vacancy concentration will be 15%. The defect structure for these oxides will therefore be dominated by extended defects in the form of defect clusters, by superstructures, or in some systems by shear structures formed by elimination of the defects by a crystallographic shearing mechanism. The formation of these structures requires interactions between single defects, and the nature of these will be considered briefly here. The forces operating between the defects can be divided into short-range and long-range forces, which can be either attractive or repulsive. In the case of attractive short-range forces, for instance, operating between single defects of opposite charges (e.g., $V^{\cdot\cdot}_O$ and M'_M), energy will be gained when these defects are ordered into defect complexes. Depending on the structure of the host lattice different types of complexes have been observed or postulated for different nonstoichiometric oxides, and in subsequent sections these will be discussed in greater detail for particular oxide systems.

Energy can also be gained in a long-range ordering of the defects or defect complexes, which operates over distances greater than the unit cell. For $FeS_{1.14}$ (pyrrhotite) for instance, Bertaut (1953) has shown by electrostatic lattice energy calculations that considerable energy could be liberated by ordering of the vacancies into alternate cation sheets. For TiO_x Anderson and Burch (1971) have calculated in the same way that the electrostatic energy is lowered by 10–15% for ordered shear structures compared with the energy associated with random distribution of the oxygen vacancies. Long-range ordering thus plays an important role in many nonstoichiometric oxide systems, and they are probably responsible for the formation of the superstructures of large unit cells observed at low temperatures in systems like Ce–O and Pr–O. The nature of the long-range interactions will be considered further in subsequent sections. Together with the short-range forces they will be treated more rigorously by a statistical thermodynamic approach in a later chapter of this book, where the most important theories of defect interactions are also reviewed. A review of some of these theories has also been published recently by Sørenson (1978).

III. Classificatlon of Nonstoichiometric Oxides

Nonstoichiometric oxides can be classified in several ways. First it is important to know the composition range for a given oxide, and as a broad, general classification nonstoichiometric oxide systems can be divided into

(i) essentially stoichiometric oxides with a very narrow composition range,
(ii) nonstoichiometric oxides with a limited composition range, and
(iii) grossly nonstoichiometric oxides with broad composition ranges.

The extent and type of nonstoichiometry also depends on the structure, and for transition metal oxides, rare earth oxides, and actinide oxides nonstoichiometry has been observed in oxides with the following structures:

(a) NaCl structures—nonstoichiometric, e.g., TiO, VO, MnO, FeO;
(b) Fluorite structure—grossly nonstoichiometric, e.g., CeO_2, PrO_2, TbO_2, UO_2, PuO_2;
(c) Rutile structure—essentially stoichiometric, e.g., TiO_2, VO_2;
(d) ReO_3 structure—essentially stoichiometric, e.g., CrO_3, MoO_3, WO_3;
(e) Nb_2O_5 structure—essentially stoichiometric, e.g., Nb_2O_5 and Ta_2O_5.

For rutile—ReO_3 and Nb_2O_5—structures the defect concentrations are considered very small, and for these oxides the nonstoichiometric phases have been shown to consist of a whole series of ordered phases formed by a crystallographic shearing mechanism in which most of the defects are eliminated.

The properties and behavior of a nonstoichiometric system are, however, primarily determined by the type of defect present, and the best classification is therefore according to the nature of these defects:

(1) Oxygen-deficient oxides—e.g., CeO_{2-x} and $(Ca_xZr_{1-x})O_{2-x}$;
(2) Metal-deficient oxides—e.g., $Fe_{1-y}O$;
(3) Oxides showing oxygen and metal deficiencies on both sides of the stoichiometric composition—e.g., $TiO_{i\pm x}$, $VO_{1\pm x}$;
(4) Excess-oxygen oxides—e.g., $UO_{2\pm x}$;
(5) Oxides showing oxygen deficiency on one side of the stoichiometric composition and excess oxygen on the other—e.g., $(U_{1-y}Pu_y)O_{2\pm x}$.

In the case of oxygen-deficient oxides it will be noted that ternary ZrO_2 systems are considered here as nonstoichiometric compounds because of the great number of oxygen vacancies introduced by the substitution. It is, of course, a question of whether these vacancies can be regarded as true defects or as structure elements in a ternary lattice. As will be shown later, interactions between the vacancies appear to play an important role in these systems, and their properties are probably better described in terms of defect complexes than in terms of point defects. Furthermore, their most important properties, high oxygen mobility and high oxygen ionic conductivity, are typical of nonstoichiometric oxides; it is therefore natural also to take these compounds into consideration.

Thermodynamic properties and defect structures for some characteristic nonstoichiometric and grossly nonstoichiometric oxides will be discussed in the following sections.

IV. Thermodynamics and Defect Structure of Oxygen-Deficient Oxides

A. *Thermodynamics, General*

The quasi-chemical equation expressing the formation of doubly charged oxygen vacancies, $V_O^{\cdot\cdot}$, is shown to be

$$O_O + 2M_M \rightleftarrows V_O^{\cdot\cdot} + 2M'_M + \tfrac{1}{2}O_2 \tag{8}$$

Using the law of mass action on this equilibrium gives

$$K_{V_O^{\cdot\cdot}} = [V_O^{\cdot\cdot}] \cdot [M'_M]^2 \cdot P_{O_2}^{(1/2)}$$

which, by introducing the neutrality condition

$$[M'_M] = 2[V_O^{\cdot\cdot}] \tag{10}$$

and by expressing $[V_O^{\cdot\cdot}]$ as the fraction of unoccupied sites in the oxygen lattice—x in MO_{2-x}—i.e., $[V_O^{\cdot\cdot}] = \frac{1}{2}x$ can be converted into

$$K_{V_O^{\cdot\cdot}} = 4[V_O^{\cdot\cdot}]^3 P_{O_2}^{1/2} = \tfrac{1}{2}x^3 P_{O_2}^{1/2} \tag{11}$$

For the formation of $V_O^{\cdot\cdot}$, the composition of the oxide thus depends on the oxygen pressure according to

$$x \propto P_{O_2}^{-1/6} \tag{12}$$

whereas the exponent would become $-\frac{1}{4}$ or $-\frac{1}{2}$ if single charged or neutral vacancies were formed. Other types of defects or defect clusters would give still other exponents, but generally

$$x \propto P_{O_2}^{-1/n} \tag{13}$$

where n is a characteristic number identifying the type of defect.

The important thermodynamic quantity $\Delta\overline{G}_{O_2}$ (relative partial free energy of oxygen) should thus depend on the composition in the following way:

$$\Delta\overline{G}_{O_2} = RT \ln P_{O_2} \propto -nRT \ln x \tag{14}$$

For nonstoichiometric subphases where only one type of defect is predominantly formed, n is constant, and a straight line should be expected in isothermal $\Delta\bar{G}_{O_2}$–$\ln x$ plots. For an analysis of the type of defect present in a nonstoichiometric phase, this plot, which is similar to that used by Brouwer (1954) and Kröger and Vink (1956), is therefore very useful and will be used throughout this chapter.

The basic assumption made in this treatment is that the mass-action law is valid. The conditions to be fulfilled for this law to be applicable are that the defects are randomly distributed and noninteracting, which probably is true only at very small deviations from stoichiometry. As mentioned previously, defect interactions must be expected as the concentration of defects increases but, as pointed out by Kröger (1964), the mass-action law can still be used if the single defects associate into randomly distributed defect complexes that can be treated as a new separate species. This condition is probably still fulfilled at intermediate defect concentrations, but at high concentrations extensive ordering makes any use of the mass-action law doubtful. Information about the phase relationships of an oxide system can, however, still be obtained from a $\Delta\bar{G}_{O_2}$–$\ln x$ plot, and it is worthwhile to consider the whole nonstoichiometric range in this way.

From $\Delta\bar{G}_{O_2}$, the relative partial enthalpy $\Delta\bar{H}_{O_2}$ and the relative partial entropy $\Delta\bar{S}_{O_2}$ can easily be calculated using the standard thermodynamic equations

$$\Delta\bar{S}_{O_2} = -\partial(\Delta\bar{G}_{O_2})/\partial T \tag{15}$$

$$\Delta\bar{H}_{O_2} = \Delta\bar{G}_{O_2} + T\Delta\bar{S}_{O_2} \tag{16}$$

or

$$\Delta\bar{H}_{O_2} = \frac{\partial(\Delta\bar{G}_{O_2}/T)}{\partial(1/T)} \tag{17}$$

The thermodynamic quantity $\Delta\bar{H}_{O_2}$ is especially important, as it can be determined directly by calorimetric measurements. In the following sections experimentally determined $\Delta\bar{H}_{O_2}$ values will be compared with values calculated for some of the most important oxygen-deficient oxides.

Consider again the general equation for the change in free energy ΔG_V accompanied by the introduction of n vacancies, which in Eq. (6) was given as

$$\Delta G_V = n(\Delta H_v - T\Delta S_v^{vibr}) + kT\left[N\ln\left(\frac{N}{N+n}\right) + n\ln\left(\frac{n}{N+n}\right)\right] \tag{18}$$

where ΔS_v^{vibr} is the change in the vibrational entropy involved in the introduction of n vacancies at N lattice sites.

At equilibrium, $\partial G/\partial n = 0$, and differentiation of Eq. (18) gives

$$\partial G/\partial n = \Delta H_v - T\Delta S_v^{vibr} + kT\ln[n/(N+n)] = 0$$

or, upon rearrangement,

$$\frac{n}{N+n} = x_v = \exp\left(\frac{\Delta S_v^{\text{vibr}}}{k}\right)\exp\left(-\frac{\Delta H_v}{kT}\right) = \exp\left(-\frac{\Delta G_v}{kT}\right) \quad (19)$$

The equilibrium constant K_v, for the introduction of vacancies, can be expressed in the same way, assuming that the activity of the crystal is unity. The important conclusion from this very general equation is that neither K_v nor ΔG_v contains the configurational entropy ΔS_v^c.

For the formation of $V_O^{\cdot\cdot}$,

$$\Delta G_{V_O^{\cdot\cdot}}^{\circ} = \Delta H_{V_O^{\cdot\cdot}}^{\circ} - T\Delta S_{V_O^{\cdot\cdot}}^{\text{vibr}} = -RT\ln K_{V_O^{\cdot\cdot}} \quad (20)$$

or, according to Eq. (11),

$$K_{V_O^{\cdot\cdot}} = \tfrac{1}{2}x^3 P_{O_2}^{1/2} = \exp(\Delta S_{V_O^{\cdot\cdot}}^{\text{vibr}}/R)\exp(-\Delta H_{V_O^{\cdot\cdot}}^{\circ}/RT) \quad (21)$$

which by introducing $\Delta\overline{G}_{O_2} = RT\ln P_{O_2}$ and rearrangement gives

$$\Delta\overline{G}_{O_2} = \Delta\overline{H}_{O_2} - T\Delta\overline{S}_{O_2} = -2\Delta H_{V_O^{\cdot\cdot}}^{\circ} + T\left(2\Delta S_{V_O^{\cdot\cdot}}^{\text{vibr}} - 6R\ln x + 2R\ln 2\right) \quad (22)$$

Thus

$$\Delta\overline{H}_{O_2} = -2\Delta H_{V_O^{\cdot\cdot}}^{\circ} \quad (23)$$

$$\Delta\overline{S}_{O_2} = -2\Delta S_{V_O^{\cdot\cdot}}^{\text{vibr}} + 6R\ln x - 2R\ln 2 \quad (24)$$

for the formation of $V_O^{\cdot\cdot}$ in oxygen-deficient oxides, assuming that the defects are randomly distributed and noninteracting as discussed above.

Under these conditions, $\Delta\overline{H}_{O_2}$ should thus be independent of composition and directly proportional to the enthalpy of formation of $V_O^{\cdot\cdot}$. As will be shown later, this is observed only in very restricted composition ranges in practice, indicating that the interactions among the defects are much more pronounced than assumed here.

Assuming that the vibrational entropy is independent of composition, the compositional dependence of $\Delta\overline{S}_{O_2}$ can, according to Eq. (24), be expressed by

$$\partial\Delta\overline{S}_{O_2}/\partial\ln x = 6R \quad (25)$$

A similar expression was derived by Panlener *et al.* (1975), who also proposed the following general expression for cerium interstitials or oxygen divacancies in different ionization states:

$$\partial(\overline{S}_{O_2})/\partial\ln x = (m+1)R \quad (26)$$

where m represents the state of ionization of the defect. The slope of a $\Delta\overline{S}_{O_2}$–$\ln x$ curve thus also depends on the type of defect formed, and straight lines with different slopes should, according to these expressions, be observed for the different nonstoichiometric subphases similar to the $\Delta\overline{G}_{O_2}$–$\ln x$ plots discussed above.

B. Rare Earth Oxides

Oxygen-deficient nonstoichiometric phases existing over a large composition range are found in rare earth oxides such as the Ce–O, Pr–O, and Tb–O systems. As shown in the phase diagrams presented in Fig. 6a, b, the

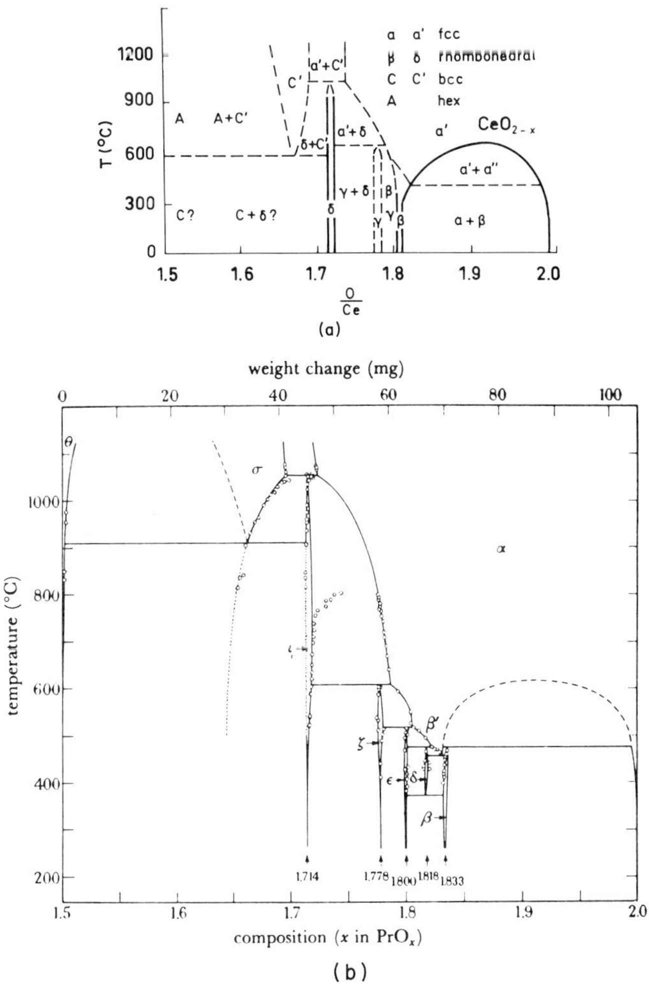

Fig. 6. Phase diagrams for the oxygen-deficient rare earth oxide systems (a) Ce–O and (b) Pr–O. [Ce–O diagram after Blank (1967) and Pr–O diagram after Hyde *et al.* (1966).]

nonstoichiometric phases are formed at higher temperatures, whereas these systems characteristically form a whole series of ordered phases at lower temperatures with compositions that can be described by a common formula, M_MO_{2n-2} (homologous series). It is also interesting to note that this formula apparently also describes the compositions at which phase changes—i.e., changes in the defect structure—take place in the nonstoichiometric phases at higher temperatures, as will be discussed below.

1. Thermodynamic Data for CeO_{2-x}

As clearly demonstrated by Merritt and Hyde (1973) in their study of the TiO_x system, many data points are necessary in order to evaluate the finer details of a nonstoichiometric system. The $\Delta\bar{G}_{O_2}$ data for the CeO_{2-x} phase published by Bevan and Kordis (1964), Panlener *et al.* (1975), and Sørensen (1972, 1976a) are therefore compiled in Fig. 7. When all the data points available for this system are considered, the following details about the CeO_{2-x} phase are revealed:

(i) The straight-line relationship predicted from Eq. (14) is clearly observed, but apparently the slope changes with increasing nonstoichiometry.

(ii) From the slopes of the lines, which depend on the temperature, the corresponding value of n in the expression $x \propto P_{O_2}^{-1/n}$ was calculated and

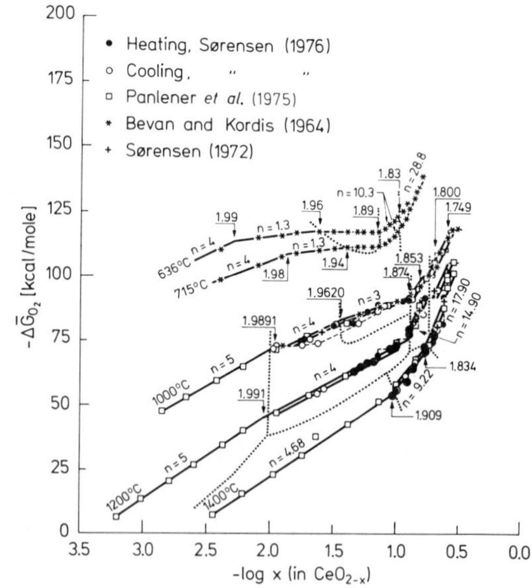

Fig. 7. Relative partial free energies of oxygen, $\Delta\bar{G}_{O_2}$, of CeO_{2-x} as a function of composition ($\log x$).

the values obtained are also given in Fig. 7. According to these values, the nonstoichiometric phase can apparently be divided into subphases which can each be described by a characteristic value of n, indicating that only one type of defect is formed within each subphase.

(iii) The subdivision into subphases can be demonstrated also by plotting the composition at the break in the lines where the slope changes—which can be taken as an indication of the subphase boundaries—as a function of temperature in a normal phase diagram. The diagram for the CeO_{2-x} phase is shown in Fig. 8, and here it is interesting to note that the compositions for the vertical subphase boundaries can be described also by the formula M_nO_{2n-2}.

(iv) From the n values obtained, the following general conclusions about the defect structure of the subphases can be reached:

Subphases with $n \leqslant 6$ can be considered nonstoichiometric with noninteracting defects or defect complexes;

For subphases with $n > 6$, however, substantial ordering of the defects must take place, since such high n values can never be obtained for any type of defect or defect complex. The steep curves should therefore be considered step curves consisting of large vertical steps (single, ordered phases) and small horizontal sections (two-phase regions); see also the

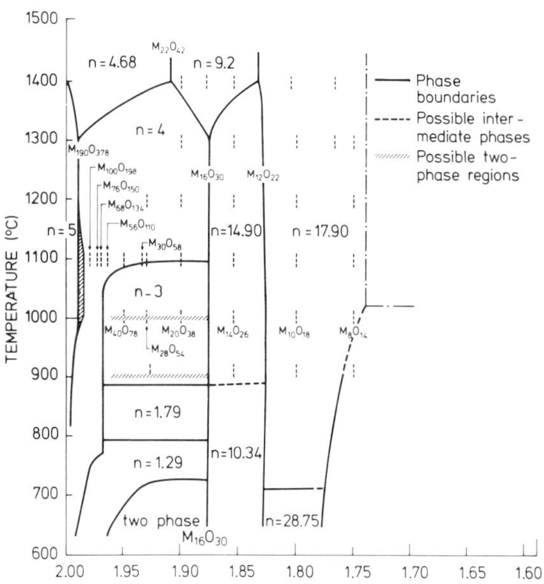

Fig. 8. Diagram of subphases with possible ordered intermediate phases for the CeO_{2-x} phase.

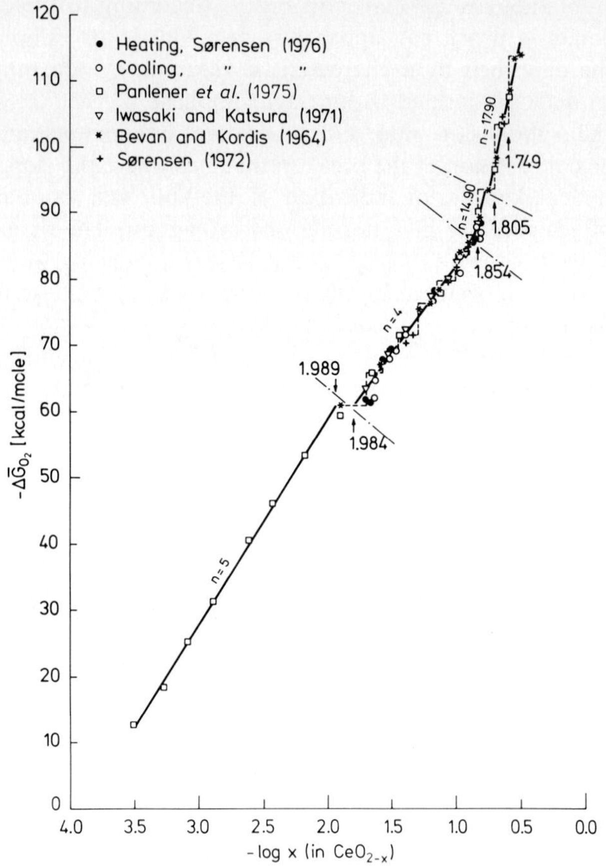

Fig. 9. Relative partial free energies of oxygen $\Delta\bar{G}_{O_2}$ at 1100°C of CeO_{2-x} as a function of composition ($\log x$).

discussion of the TiO_{2-x} system in Section IV.C.1. This is more clearly shown in Fig. 9 where the $\Delta\bar{G}_{O_2}$–$\log x$ lines are drawn on a larger scale.

The $\Delta\bar{S}_{O_2}$ values, calculated according to Eq. (15), are shown in Fig. 10 as a function of composition ($\log x$). In this plot straight lines are again obtained within each subphase and, as for the $\Delta\bar{G}_{O_2}$–$\log x$ plots, the slope of these lines changes from one subphase to the other. A straight-line relationship between $\Delta\bar{S}_{O_2}$ and $\log x$ is also predicted from Eq. (26), which gives the slope in terms of the state of ionization m of single vacancies or divacancies. For a fully ionized divacancy, for instance, $m = +4$ and the corresponding slope should be $5R = 22.9$, whereas the slope for $V_O^{\cdot\cdot}$ should be $6R = 27.45$, which is rather close to the slope obtained for the $n = 4$ region. Other types of defect complexes give different expressions for the

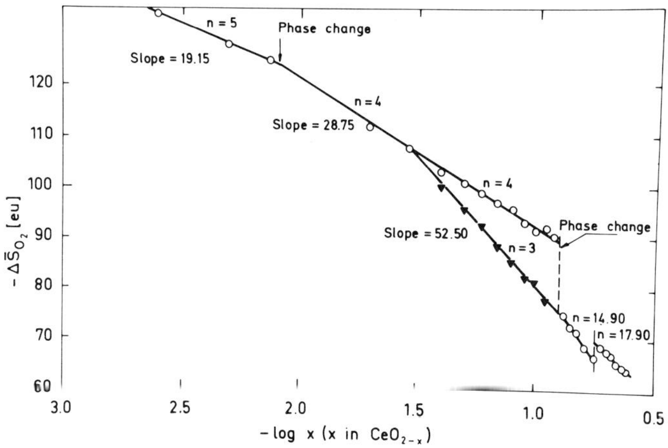

Fig. 10. Relative partial entropy of oxygen $\Delta\bar{S}_{O_2}$ as a function of composition ($\log x$) for the CeO_{2-x} phase.

slope, and it is impossible from this plot to draw firm conclusions about the type of defect predominantly formed within each subphase. The assumption of random and noninteracting defects and of a constant vibrational entropy independent of composition, which is the basis for these expressions, however, seems to be fulfilled for many of the subphases observed.

The $\Delta\bar{H}_{O_2}$ values obtained from Eq. (16) from the $\Delta\bar{G}_{O_2}$ and $\Delta\bar{S}_{O_2}$ data are shown in Fig. 11 as a function of $\log x$, and again straight lines with different slopes are obtained for the different subphases. For randomly distributed and noninteracting defects $\Delta\bar{H}_{O_2}$ should be proportional to the enthalpy of formation of the defects—e.g., $\Delta H_{V_O^{\cdot\cdot}}$—but independent of

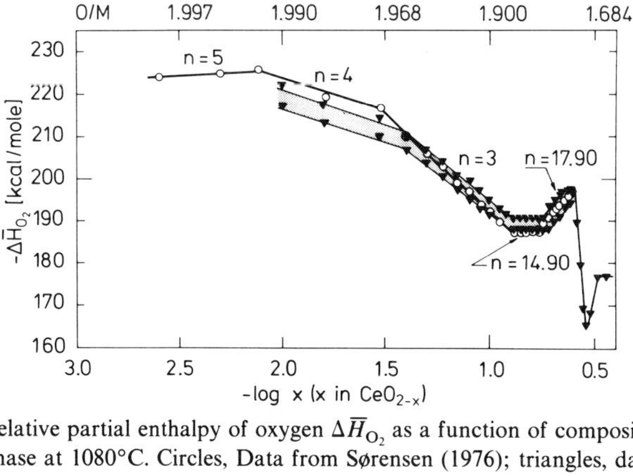

Fig. 11. Relative partial enthalpy of oxygen $\Delta\bar{H}_{O_2}$ as a function of composition ($\log x$) for the CeO_{2-x} phase at 1080°C. Circles, Data from Sørensen (1976); triangles, data reported by Campserveux and Gerdanian (1974).

composition. From Fig. 11 it will be noted that this apparently is the case only for the $n = 5$ and $n = 14.90$ regions, whereas $\Delta\bar{H}_{O_2}$ for the other phases shows linear variations with $\log x$ with substantial slopes. Apparently there is considerable interaction and ordering of the defects in these phases, as will be discussed later. The calculation of $\Delta\bar{H}_{O_2}$ was carried out at 1353 K (1080°C) in order to compare these indirect results with those determined by microcalorimetry by Campserveux and Gerdanian (1974) at this temperature. It is interesting to note that the agreement between the directly measured $\Delta\bar{H}_{O_2}$ values and those determined in a thermogravimetric measurement is excellent, which gives further support to the subdivision of the CeO_{2-x} phase into the subphases shown in the phase diagram in Fig. 8.

2. *Thermodynamic Data for* PrO_{2-x}

Information about the thermodynamic properties of the nonstoichiometric α phase of the Pr–O system (see the phase diagram in Fig. 6b) is surprisingly limited. The reason for this is the very high equilibrium oxygen pressures of these oxides at higher temperatures. In the investigation of Jenkins *et al.* (1970), which hitherto was the most detailed study on this phase, only the composition range $1.830 \geqslant x \geqslant 1.785$ could be covered in the temperature range 400–800°C and at oxygen pressures from 38 to 730 Torr. The $\Delta\bar{G}_{O_2}$ values obtained in this investigation were plotted against temperature, and from changes in the slopes of the resulting straight lines it was concluded that four discrete phases existed in the α phase. The observed slope changes are, however, very small, and experimental evidence for the existence of these subphases is still rather limited. If the $\Delta\bar{G}_{O_2}$ values are plotted as a function of $\log x$, the straight line shown in Fig. 12 is obtained—the data for this plot were obtained from the $\Delta\bar{G}_{O_2}$–T plot of Jenkins *et al.* (1970). From this figure it is clear that the $\Delta\bar{G}_{O_2}$ values vary linearly with $\log x$ in the composition range covered in this investigation, and from this plot there is no evidence for the existence of discrete phases either. The interesting feature is, however, the large value of n obtained in this plot, indicating a high degree of ordering, as is also observed in the CeO_{2-x} phase at large deviations from stoichiometry. There is no evidence for a step curve as in the case of the CeO_{2-x} phase but, as also suggested by Jenkins *et al.*, the ordering can perhaps be explained in terms of a microdomain structure with a very fine texture, i.e., < 100 Å. This idea was originally proposed by Ariya and Popov (1962), who suggested that a nonstoichiometric oxide consisted of coherently intergrown microdomains of the next higher and lower ordered phases. The thermodynamics of microdomains was treated by Ubbelohde (1957, 1966) and Anderson (1970a), and for a further discussion of this concept these references should be consulted.

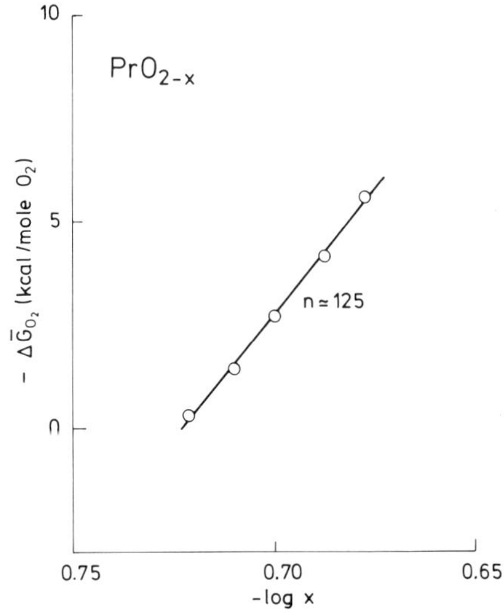

Fig. 12. Relative partial free energies of oxygen, $\Delta\bar{G}_{O_2}$, of PrO_{2-x} as a function of composition ($\log x$) at 612°C. (Data from Jenkins *et al.*, 1970.)

3. Defect Structure of Oxygen-Deficient Rare Earth Oxides

In the early studies on oxygen-deficient oxides, the nonstoichiometry was considered to be due to either interstitial cerium ions or ionized oxygen vacancies. The reason for this was that the pressure dependence of the composition was found to be $x \propto P_{O_2}^{-1/5}$ ($n = 5$) near the stoichiometric composition, which cannot be explained in terms of single oxygen vacancies. In a study on the electrical conductivity of CeO_{2-x} as a function of oxygen pressure Greener *et al.* (1964) also found this pressure dependence, and they concluded that the predominant defects were either quadruply ionized Ce interstitials or completely ionized oxygen vacancy pairs. Blumenthal *et al.* (1971) also interpreted their conductivity data in terms of triply and quadruply ionized Ce interstitials, whereas Kofstad and Hed (1967) suggested a model involving singly and doubly ionized Ce interstitials and electrons localized on Ce ions at normal lattice sites. By assuming that the electrons have only a small probability of occupying the Ce sites nearest the interstitial Ce ions, Kofstad and Hed also introduced a site-blocking effect to explain the increase in n observed at larger deviations from stoichiometry.

In contrast to these suggestions some of the early studies also interpreted the properties of oxygen-deficient phases in terms of oxygen vacancies (see, e.g., Kevane, 1964; Blumenthal and Laubach, 1968). However, from none of these studies was it possible to make a firm conclusion about the defect type, whereas the results obtained recently by Steele and Floyd (1971) in measurements of oxygen self-diffusion in CeO_{2-x} finally seem to support an oxygen vacancy model for these oxides. In recent papers concerning studies on the electronic conductivity of CeO_{2-x} by Blumenthal (1974) and Blumenthal and Sharma (1975) oxygen vacancies are also finally accepted as the predominating defects, and today the role of Ce interstitials is generally considered negligible in oxygen-deficient oxide systems. This is also in accordance with the greater stability of the cation lattice generally observed for these oxides and with the theories of Blank (1974), which suggest that the oxygen ions are actually smaller than the cations and thus more mobile in these fcc oxides.

Several models have been proposed in order to explain the structures of the ordered intermediate phases in the Ce–O and Pr–O systems. Hyde and Eyring (1964), for instance, have proposed that the structural entity that generates the series M_nO_{2n-2} is a linear infinite string of MO_6 octahedra along the $\langle 111 \rangle$ direction surrounded by a contiguous sheath of MO_7 cubes; the fcc structure can be considered to consist of oxygen cubes alternatively containing a metal atom. In M_7O_{12}, M_9O_{16}, and $M_{11}O_{20}$ the strings are parallel and regularly spaced, whereas nonintersecting strings along all four $\langle 111 \rangle$ directions have been proposed for the M_2O_3 phases (type C, cubic) (Sawyer *et al.*, 1965). Thornber *et al.* (1968) subsequently suggested that, as precursors of the defect strings, the vacancies are clustered in pairs at diagonally opposite corners of the cube, which then contains a six-coordinated cation, whereas the surrounding six cubes contain a seven-coordinated cation (see Sørensen, 1978, Fig. 9). Kunzmann and Eyring (1975) proposed that the vacancy pairs were arranged along the $[211]_{fcc}$ direction, which was considered to be the common *a* axis of all the members of the homologous series. Later Summerville *et al.* (1978) applied the same principle in explaining the structure of the β phase ($Pr_{24}O_{44}$, twinned $Pr_{12}O_{22}$ cell). In both investigations the structure proposals were based on high-resolution electron microscopy observations, which will be discussed in more detail in a later chapter.

Still another defect complex was proposed by Martin (1974) and by Hoskins and Martin (1975), who described the structure of intermediate phases of oxygen-deficient oxides by a systematic packing of oxygen octahedra formed by attractive interactions between a central positive vacancy and the surrounding six oxygen ions in the oxygen lattice. These complexes are considered to gather on oblique $\{213\}$ planes, and the structure of the different intermediate phases can be described in terms of the thickness of the defect-free layers separating these planes. One drawback of this model is that it neglects the role of the reduced cations and thus the electronic properties of the crystal. Intuitively, therefore, a model involving either pairing of vacancies around a cation of lower charge or pairing of two reduced cations around an oxygen vacancy, as discussed below, appears to be more correct.

The crystallographic shearing mechanism by which the oxygen vacancies first accumulate on specific crystallographic planes, and then are annihilated by the collapse and shearing of the lattice, does not seem to play as important a role in fluorite oxides as in other systems. A few attempts, however, have been made to introduce this concept for these oxides also to explain the structure of the ordered intermediate phases. Eyring and Holmberg (1963) and Hyde (1971), for instance, showed that the structure of cubic M_2O_3 could be derived from the fcc structure by collapse followed by crystallographic shearing along the $(\bar{1}\bar{1}0)$ planes, whereas Sørensen (1975) proposed shearing along the $(1\bar{2}1)$ planes from an electron microscopy study on CeO_{2-x}. As shown in Fig. 13, crystallographic shearing

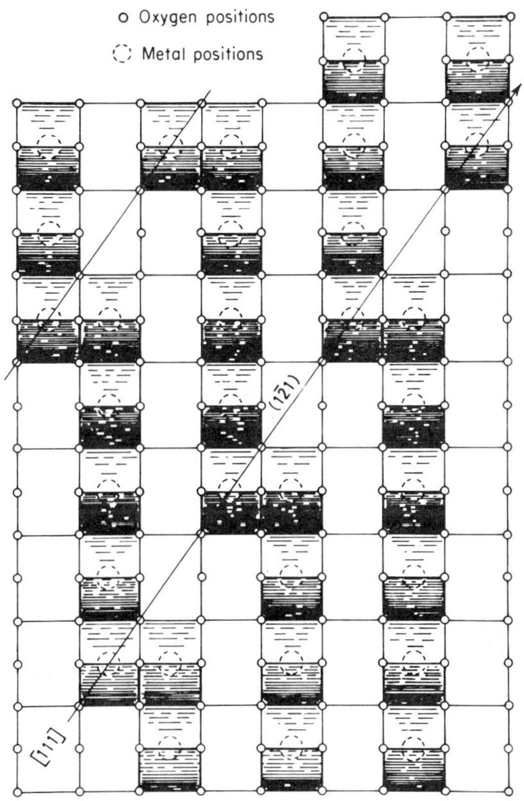

Fig. 13. Bounded projection of oxygen and metal positions on $(10\bar{1})$ showing crystallographic shearing along $(1\bar{2}1)$. Thin slabs of undistorted MO_2 are formed parallel to $(1\bar{2}1)$. (From Sørensen, 1975.)

parallel to the $(1\bar{2}1)$ planes produces a lamellae structure with slabs of undistorted fcc structure and, by varying the thickness of these slabs, which depend on the distance between the $(1\bar{2}1)$ planes participating in the shearing, the compositions of the different members of the homologous series can in principle be obtained. Much more detailed electron microscopy examinations are, however, necessary before the crystallographic shearing mechanism can be really accepted for fluorite oxides.

For less ionic oxygen-deficient oxides, defect complexes consisting of one oxygen vacancy bonded to two adjacent reduced cations were proposed by Schmitz and Marajofsky (1975) and by Manes and Manes-Pozzi (1976). Recently Manes *et al.* (1980) have shown that a systematic packing of these complexes also can describe the structures of all the low-temperature ordered intermediate phases observed in the Ce–O and Pr–O systems, as well as those inferred from thermodynamic studies on the nonstoichiometric CeO_{2-x} phase (Fig. 8). This defect complex is thus a fundamental unit of these oxides, and it will be discussed in some detail below and in a later chapter.

The structure of the MO_2 fcc lattice can be represented in two ways: as oxygen cubes alternatively occupied by cations, or as cation tetrahedra containing oxygen ions. For purely ionic oxides both descriptions apply equally well, but for the oxygen-deficient oxides considered here in which the oxygen sublattice is considered rather unstable, the second description is preferred. The defect complexes consisting of $2(Me^{3+})$, $V_O^{\cdot\cdot}$ are thus naturally formed within cation tetrahedra, and they are therefore termed tetrahedral defects.

When a tetrahedral defect is formed in the fcc lattice, strain is introduced in the surrounding tetrahedra, which increases the lattice energy locally. The magnitude of this strain can be assumed to be inversely proportional to the distance from the central defect, and in the nearest-neighbor tetrahedra it is so large that the formation of new defects is excluded (the formation of vacancies in these positions requires higher energy by the exclusion principle). There are 22 nearest-neighbor tetrahedra in the fluorite structure, and an envelope of 23 tetrahedral positions (including the central tetrahedral defect) can be identified as an extended defect—here designated a T_{23} complex—in the fcc lattice. Figure 14a is a photograph of a model of such a T_{23} complex and Fig. 14b its projection on the (111) planes. In this figure other complex types are also shown, which can be naturally derived from a T_{23} complex in the following way: When the whole lattice has been filled by packing T_{23} complexes in a regular way, the smallest increase in lattice energy caused by the introduction of further defects is observed when the tetrahedra under the smallest strain in the T_{23} complexes—those at the longest distance from the central defect—become a part of a neighboring complex. From the T_{23} complex a T_{19} and a T_7

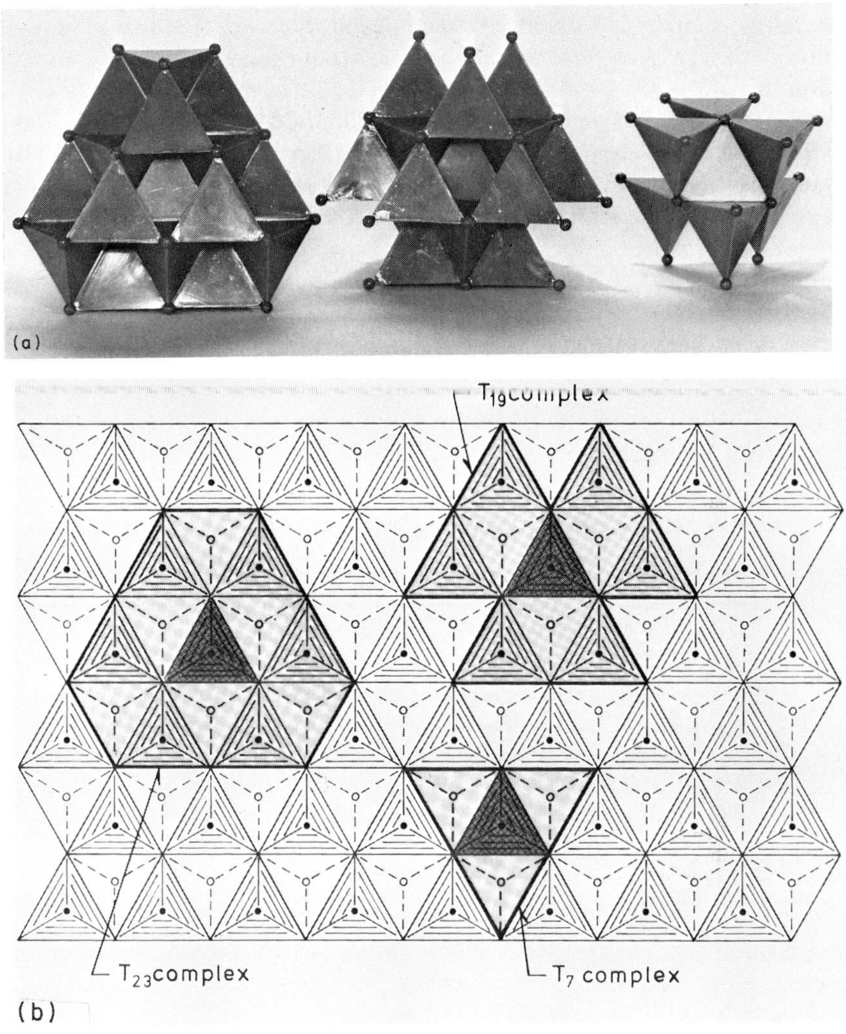

Fig. 14. (a) Photograph of models showing T_{23}, T_{19}, and T_7 complexes and (b) their projection on ($\bar{1}$11). (From Manes *et al.*, 1980; reproduced by permission of IAEA.)

complex, which is formed at larger deviations from stoichiometry, can thus be derived by removing 4 and $4 + (4 \times 3) = 16$ tetrahedra, respectively, as shown in Fig. 14.

The enthalpy changes involved in the formation of a given defect structure can be considered mainly to be determined from two opposing effects:

(i) Enthalpy of formation, which is positive and which increases with decreasing distance between the defects. In subregions where only one type of defect complex is formed, a T_{23} complex, for instance, the enthalpy of formation, can thus be considered constant and independent of composition.

(ii) Enthalpy of interaction, which by Coulomb interactions can be either positive or negative depending on whether there is a repulsion or an attraction between the defect complexes. The tetrahedral defect is neutral—it consists of one $V_O^{\cdot\cdot}$ and two (Me^{3+})—and these interactions are probably very small in these oxides. As explained by Manes and Manes-Pozzi (1976), however, dipoles can be assumed to form in the tetrahedral defects by the charge separation, and dipole–dipole attractions must also be taken into account in these systems.

From the experimental $\Delta\overline{H}_{O_2}$ values shown in Fig. 11 for the CeO_{2-x} phase it will be noted that "lattice stabilization" ($-\Delta\overline{H}_{O_2}$ decreasing with increasing x) apparently takes place in some subregions, whereas "lattice

TABLE I

Packing of Defect Complexes in Subregions Observed for CeO_{2-x}[a]

O/M ratio	$-\Delta\overline{H}_{O_2}$	Packing
2.00–1.99	Constant	Random distribution of T_{23}
1.99–1.96	Decreasing lattice stabilization	Macrocomplexes of four T_{23} (one middle, three upper)
1.96–1.92	Decreasing lattice stabilization	Macrocomplexes of seven T_{23} (one middle, three upper, three lower)
1.92–1.88	Decreasing lattice stabilization	T_{19} formed
1.88		Close-packing T_{19}
1.88–1.83	Constant	T_7 formed
1.83 ($M_{12}O_{22}$)		T_{19} and T_7
1.83–1.82	Increasing lattice destabilization	T_7 formed
1.82 ($M_{11}O_{20}$)		T_{19} and T_7
1.82–1.80	Increasing lattice destabilization	T_7 formed
1.80 ($M_{10}O_{18}$)		T_{19} and T_7
1.80–1.78	Increasing lattice destabilization	T_7 formed
1.78 (M_9O_{16})		T_{19} and T_7
1.78–1.71	Decreasing lattice stabilization	T_7 formed
1.71 (M_7O_{12})		T_7

[a] From Manes *et al.* (1980); reproduced by permission of IAEA.

destabilization" takes place in others. In terms of the two opposing effects mentioned above this must indicate that strong attractive dipole–dipole interactions operate in the former subregions, whereas the larger strain (higher lattice energy) introduced when the distance between the defect complexes becomes smaller produces the $\Delta \overline{H}_{O_2}$ changes observed in the latter subregions.

Based on these ideas the defect structure for the different nonstoichiometric and ordered subphases observed in the CeO_{2-x} system (Fig. 9) can now be constructed by a systematic packing of the T_{23}, T_{19}, and T_7 complexes as summarized in Table I. The structure of the macrocomplexes formed by dipole–dipole attractions consisting of either four T_{23} complexes (one middle and three upper) or seven T_{23} complexes (one middle, three upper, and three lower) is shown in Fig. 15. Figure 16, on the other hand,

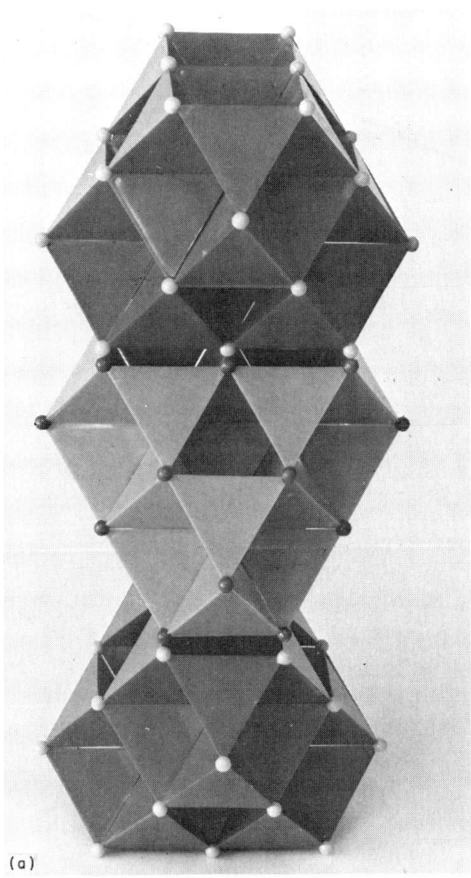

(a)

Fig. 15. (a) Photograph of model showing a macrocomplex consisting of seven T_{23} complexes—three upper (only one is shown), one middle, and three lower (only one is shown).

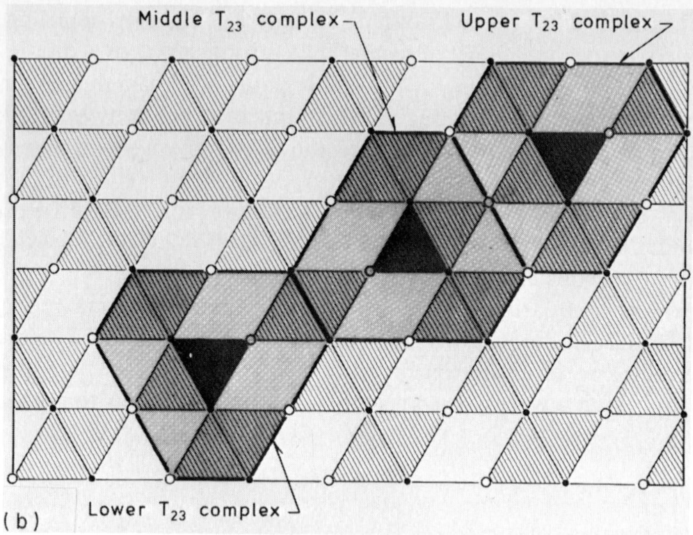

Fig. 15 (*continued*) (b) The projection on (101). (From Manes *et al.*, 1980; reproduced by permission of IAEA.)

shows how the $M_{12}O_{22}$ and M_7O_{12} structures can be constructed by a systematic packing of T_{19} and T_7 complexes. It is interesting to note that the lattice parameters actually measured for these phases (see, e.g., Summerville *et al.*, 1978; Ray and Cox, 1975) are reproduced in these defect structures.

C. Transition Metal Oxides

Previously many of the higher transition metal oxides—TiO_2, V_2O_5, Nb_2O_5, MoO_3, VO_3, etc.—were considered typical oxygen-deficient oxides (TiO_{2-x}, MoO_{3-x}, etc.) which could be treated on the basis of classic defect theories. Numerous studies of this kind have been reported for rutile (Kofstad, 1962; Førland, 1964) and Nb_2O_5 (Blumenthal *et al.*, 1965), but as early as the 1950s the hypothesis had been presented by Magnéli (1953), Wadsley (1955, 1964), and Andersson *et al.* (1957) that the nonstoichiometry in these oxides should rather be explained as a series of ordered phases (Magnéli phases) following a common formula such as Ti_nO_{2n-1}.

This hypothesis was based on extensive x-ray structural studies, and later many detailed electron microscopy studies confirmed that this model is correct (see, e.g., Hyde *et al.*, 1974; Tilley, 1972).

The defects responsible for the deviation from stoichiometry in these

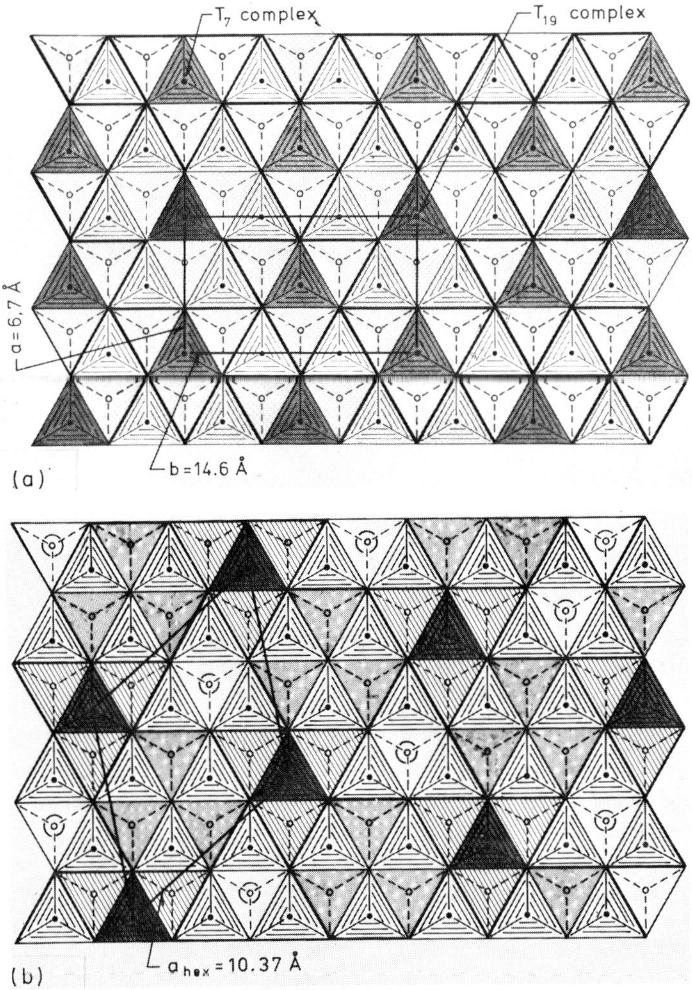

Fig. 16. Projections on (111) of (a) $M_{12}O_{22}$ and (b) M_7O_{12} constructed by a systematic packing of Ti_{19} and T_7 complexes. (From Manes *et al.*, 1980; reproduced by permission of IAEA.)

oxides are crystallographic shear planes (Wadsley defects), which are planar defects formed by a crystallographic shearing mechanism as described in Section IV.B.3.. In this process the original oxygen vacancies are eliminated and, besides the shear planes, the only defects considered present are those inherently formed in the parent lattice between these planes—Anderson (1972) estimated that the concentration of oxygen vacancies in TiO_{2-x} was smaller than 10^{-2} at.% ($x < 10^{-4}$). These oxides therefore cannot be considered "genuine" nonstoichiometric systems, but since they are so well characterized structurally it is worthwhile to consider their thermodynamic properties in some detail.

1. Thermodynamic Data for TiO_{2-x}

One of the most accurate studies on the thermodynamic properties of a nonstoichiometric transition metal oxide of this type is probably the thermogravimetric measurements of Merritt and Hyde (1973) on the TiO_{2-x} system. In this study many close-lying data points with very good precision were obtained, and by plotting these in a $\Delta\bar{G}_{O_2}$–x diagram many ordered phases of the Ti_nO_{2n-1} series were revealed. Generally, the data points obtained for both oxidation and reduction showed that the substoichiometric range $1.80 \leqslant x \leqslant 2.00$ could be divided into several subregions, some of which showed a considerable degree of hysteresis, whereas others showed good reversibility. By comparing these results with the structural information obtained in the detailed electron microscopy examinations reported by Bursill and Hyde (1972), the regions with good reversibility ($1.90 \leqslant x \leqslant 1.93$) were attributed to partly ordered phases in which the shear plane changes from (121) to (132), whereas the regions showing hysteresis, which lie at compositions $x < 1.90$ and $x > 1.93$, respectively, were considered to consist of a whole series of phases formed by (121) or (132) shear planes.

With these structural ideas in mind it is interesting to apply the same type of analysis to the TiO_{2-x} systems as used in the previous sections for the CeO_{2-x} and PrO_{2-x} systems, i.e., to analyze the slopes of a $\Delta\bar{G}_{O_2}$–$\log x$ curve. Unfortunately Merritt and Hyde did not use this type of plot, and their data, which were obtained in a reduction experiment at 1304 K (1030°C) where the composition was changed from $TiO_{1.9990}$ to $TiO_{1.8846}$, are therefore replotted in Fig. 17 in this way. Many of the ordered phases reported in the original curves are also clearly indicated in this figure, but in this context it is interesting to note the steep curve (large n value) in the $x < 1.90$ region, which one would have been tempted to draw if the large number of close-lying data points had not been available. For the region $x > 1.98$ also a very steep curve is obtained, indicating that ordering takes place even at small deviations from stoichiometry and in accordance with the observation of Bursill and Hyde (1971) that oxygen vacancies can be eliminated by a crystallographic shearing process at compositions corresponding to $TiO_{1.999}$. For the curves between the ordered phases in the $1.93 \leqslant x \leqslant 1.98$ region, slopes corresponding to $n = 6$ and $n = 2$, respectively, are observed. In the case of the phase ranges with $n = 6$, oxygen vacancies $V_O^{\cdot\cdot}$ are apparently formed in the parent structure between the

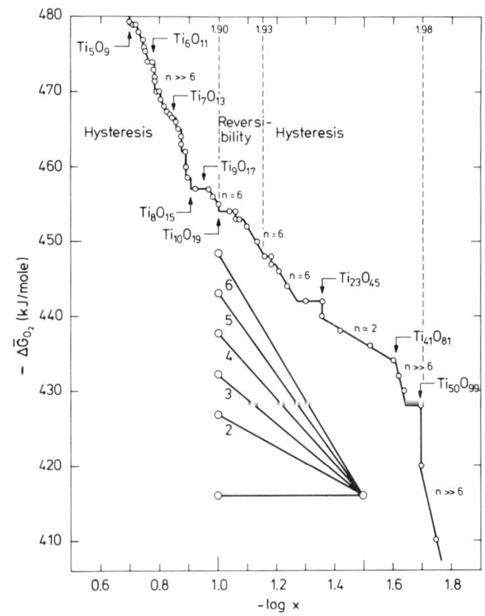

Fig. 17. Relative partial free energy of oxygen, $\Delta\bar{G}_{O_2}$, as a function of composition for the TiO_{2-x} system. (Data from Merritt and Hyde, 1973.)

shear planes, whereas $n \simeq 2$ might indicate the formation of defect complexes.

D. Actinide Oxides

Nonstoichiometry due to oxygen deficiency in the oxides formed by actinide elements is observed mainly in plutonium oxides, PuO_{2-x}, and in transplutonium oxides, AmO_{2-x}, CmO_{2-x}, BkO_{2-x}, and CfO_{2-x}. In spite of the experimental difficulties involved in measurements on these oxides—they are radioactive and thus very toxic and must be handled in airtight and, in some cases, shielded glove boxes—several investigations of their thermodynamic properties and structures have been carried out, and today we have a fairly good knowledge of these systems.

The most important member of this series is PuO_{2-x}, which is fissile and thus an important fuel for power reactors. The properties and structure of these oxides resemble quite closely those of the oxygen-deficient rare earth oxides discussed in previous sections, and the latter have therefore often been used as model substances for PuO_{2-x} oxides. However, when the phase diagram of the Pu–O system shown in Fig. 18 is compared with the

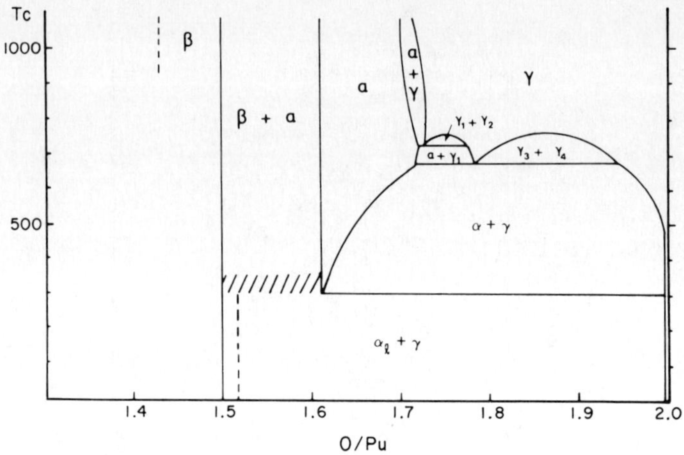

Fig. 18. Phase diagram for the Pu–O system after Boivineau (1976). (By permission of North-Holland Publishing Co., Amsterdam.)

phase diagrams of the Ce–O and Pr–O systems in Fig. 6a, b, it is evident that the ordered intermediate phases formed so abundantly in rare earth oxides apparently do not exist in the Pu–O system. This is also the case for transplutonium oxides, AmO_{2-x} and BkO_{2-x}, whereas ordered phases have been observed at $MO_{1.82\pm x}$ and $MO_{1.714}$ for the Cm–O and Cf–O systems (Chikalla and Eyring, 1969; Turcotte and Haire, 1975). The reason for this difference has been examined by Blank (1967) who concluded that, in the case of the Pu–O system, interaction among the defects at lower temperatures is too weak to produce ordered intermediate phases because of a more pronounced lattice relaxation, although a certain degree of order with respect to the oxygen vacancies cannot be excluded. In a later publication Blank also showed (Blank, 1974, 1975) that properties of actinide and rare earth oxides are better described by a model involving a certain degree of covalent bonding between the cations. The degree of covalency is probably considerably smaller in rare earth oxides, however, and this might also explain the difference between the two systems.

1. Thermodynamic Data for PuO_{2-x}

The $\Delta\bar{G}_{O_2}$ data for the PuO_{2-x} system reported by Atlas and Schlehman (1965, 1966), Markin *et al.* (1964), and Sørensen (1976b) are given in Fig. 19 as a function of $\log x$. Although some of the data of Atlas and Schlehman differ from those reported by the other two authors, it is clear that the predicted linear relationship between $\Delta\bar{G}_{O_2}$ and $\log x$ is also obtained in this system. When the compositions at which the slopes of these lines change are plotted against temperature, the pseudo-phase diagram

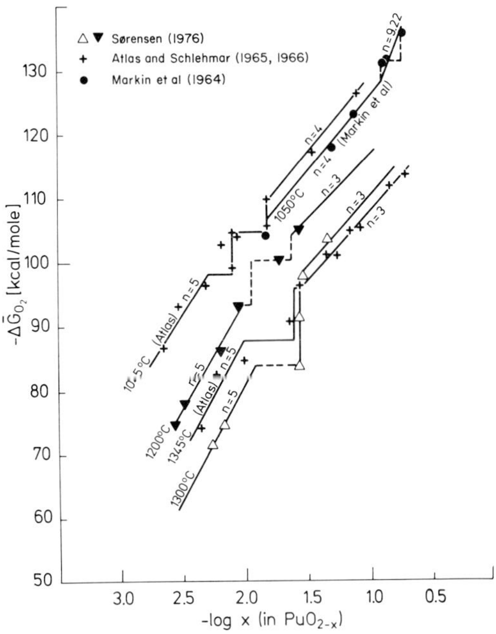

Fig. 19. Relative partial free energies of oxygen, $\Delta\overline{G}_{O_2}$, for PuO_{2-x} as a function of composition ($\log x$).

given in Fig. 20 is obtained. For this oxide system it is also evident that the nonstoichiometric phase range can be divided into subphases each with a characteristic defect being predominantly formed. An interesting feature of this representation is the two-phase regions in the composition range 1.99 $\leqslant$ O/M ratio $\leqslant$ 1.95, which also was observed in the rare earth oxide systems (Fig. 8) and which has been reported for the BkO_{2-x} system, $BkO_{1.91}$–$BkO_{1.93}$ (see Turcotte and Chikalla, 1971).

From the $\Delta\overline{G}_{O_2}$ values plotted in Fig. 19, $\Delta\overline{H}_{O_2}$ and $\Delta\overline{S}_{O_2}$ have been calculated using the standard thermodynamic equations described previously [Eqs. (15) and (17)]. The results obtained are shown in Table II together with $\Delta\overline{H}_{O_2}$ determined by microcalorimetric measurements by Chereau *et al.* (1971). From the table it is clear that the data obtained from Markin's and Sørensen's $\Delta\overline{G}_{O_2}$ values agree very well with the experimental $\Delta\overline{H}_{O_2}$ values reported by Chereau, which further support the idea that several subphases exist in the nonstoichiometric phase range at higher temperatures in the PuO_{2-x} system.

2. *Thermodynamic Data for* AmO_{2-x} *and* BkO_{2-x}

As an example of the thermodynamic properties of nonstoichiometric transplutonium oxides the $\Delta\overline{G}_{O_2}$ values reported for the AmO_{2-x} system by

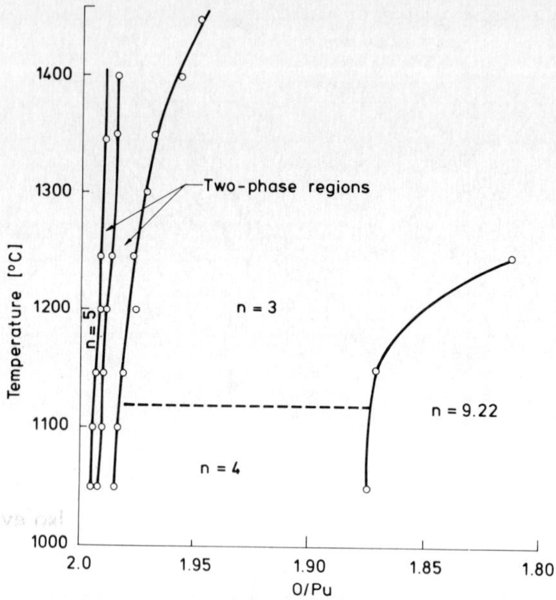

Fig. 20. Diagram of subphase regions in PuO_{2-x}.

TABLE II

$\Delta\overline{H}_{O_2}$ and $\Delta\overline{S}_{O_2}$ for PuO_{2-x} Oxides

Composition $(2-x)$	n	$-\Delta\overline{S}_{O_2}$ (eu)	$-\Delta\overline{H}_{O_2}$ (kcal/mole)	Comments
1.9960	5	121	258[a]	Sørensen (1976b)
1.9960	5	69	158.95[a]	Atlas and Schlehman (1965, 1966)
~1.9960		~125	259	Markin *et al.* (1964)
~1.9960			256	Chereau *et al.* (1971)
1.9498	3	87.5	240.15[a]	Sørensen (1976b)
1.9498	3	72.5	218.65[a]	Atlas and Schlehman (1965, 1966)
~1.950		90	239	Markin *et al.* (1964)
~1.950			241	Chereau *et al.* (1971)

[a] Calculated at 1100°C.

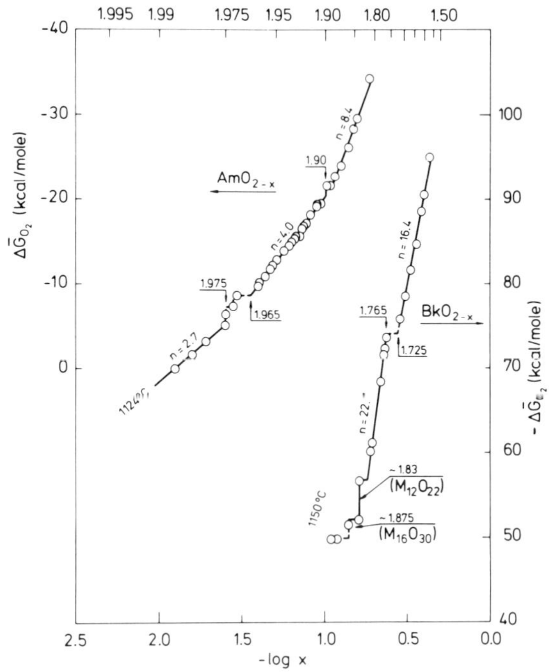

Fig. 21. Relative partial free energies of oxygen, $\Delta\bar{G}_{O_2}$, of AmO_{2-x} and BkO_{2-x} as a function of composition. Data from Chikalla and Eyring (1969) and Turcotte and Chikalla (1971), respectively.

Chikalla and Eyring (1969) and for the BkO_{2-x} system by Turcotte and Chikalla (1971) are plotted in Fig. 21 as a function of composition ($\log x$). Concerning data for CmO_{2-x} and CfO_{2-x} the publications by Chikalla and Eyring (1969), Chikalla and Turcotte (1972), Turcotte *et al.* (1973), and Turcotte and Haire (1975) should be consulted. From the figure it is again clear that the predicted straight-line relationship is obtained both for AmO_{2-x} and BkO_{2-x}, although quite different n values are obtained for the two systems. For AmO_{2-x}, the n values are quite close to those obtained for the PuO_{2-x} system, and it is interesting to note that the pattern of subphases is similar for the two oxides, including the two-phase regions near the stoichiometric composition. Much higher n values are observed for the BkO_{2-x} system for the whole composition range covered in these measurements, and instead of the nonstoichiometric phases characteristic of the other systems considered, BkO_{2-x} apparently consists of a whole series of ordered phases—the high n values correspond to a step curve which, as discussed previously, is formed by ordered phases (vertical curves) with two-phase regions in between (a horizontal curve). Turcotte and Chikalla observed the existence of the phases $BkO_{1.82\pm x}$ and $BkO_{1.92}$, as well as three broad nonstoichiometric phase ranges, whereas the high n values obtained here indicate that the degree of ordering probably is higher than previously anticipated for this oxide system.

3. *Defect Structure of Oxygen-Deficient Actinide Oxides*

The defect structure of the oxygen-deficient rare earth oxides discussed in Section B.3 was based on a tetrahedral description of the fluorite lattice. This description is even more applicable to the fluorite structures of actinide oxides because of their higher degree of covalent bonding between the cations. Therefore the defect types predicted for the CeO_{2-x} system should also be possible for these oxides. From Fig. 22, where the phases observed for the different oxides are compared with those observed for the CeO_{2-x} system, it will be noted that the composition range for the nonstoichiometric subphases ($n \leqslant 6$) is about the same for the three systems PuO_{2-x}, AmO_{2-x}, and CeO_{2-x}. For compositions with O/M ratios smaller than those corresponding to 1.875 ($M_{16}O_{30}$), large n values and thus a high degree of order are found in all systems, and several of the ordered phases such as M_7O_{12}, $M_{16}O_{30}$, and $M_{12}O_{22}$ have been identified for most actinide oxides. In spite of the higher covalency in these systems, which might create a higher degree of lattice relaxation during defect formation as postulated by Blank, the defect structure of actinide oxides at higher

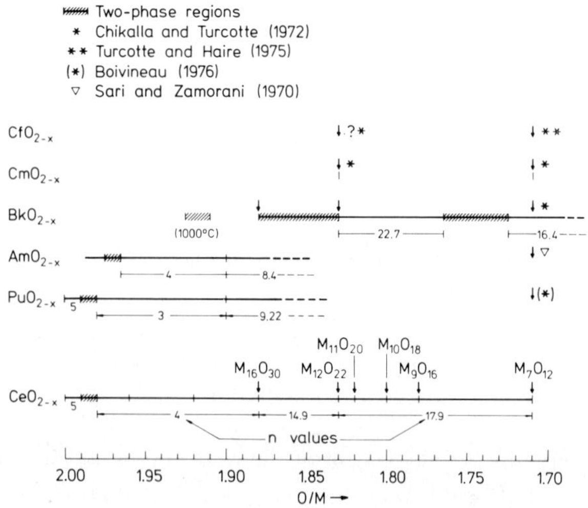

Fig. 22. Subphase regions for nonstoichiometric actinide oxides compared wlth those of the CeO_{2-x} system.

temperatures apparently is the same as that described for rare earth oxides, although more data are necessary for a more rigorous comparison.

V. Thermodynamics and Defect Structure of Metal-Deficient Oxides

A. *Thermodynamics, General*

The quasi-chemical equation describing the formation of double negatively charged metal vacancies, V_M'', was shown in Eq. (9) to be

$$\tfrac{1}{2}O_2 \leftrightarrows V_M'' + 2M_M^{\cdot} + O_O^x \tag{9}$$

Using the law of mass action on this equilibrium and introducing the neutrality condition

$$[M_M^{\cdot}] = 2[V_M''] \tag{27}$$

and expressing $[V_M'']$ as the fraction of unoccupied sites in the cation lattice —y in $M_{1-y}O$—i.e., $[V_M''] = y$, the equilibrium constant for Eq. (9) becomes

$$K_{V_M''} = 4y^3 P_{O_2}^{-1/2} \tag{28}$$

It is clear that the composition of the oxide depends on the oxygen pressure in the following way:

$$y \propto P_{O_2}^{+1/6} \tag{29}$$

and for these oxides the relation between $\Delta\overline{G}_{O_2}$ (relative partial free energy of oxygen) and the composition can thus be generally expressed by

$$\Delta\overline{G}_{O_2} = RT\ln P_{O_2} \propto nRT\ln y \tag{30}$$

where n depends on the type of defects formed. In this case $n = 6$ for the formation of V_M''. As for oxygen-deficient oxides, a straight line should then also be expected in isothermal $\Delta\overline{G}_{O_2}$–ln y plots for these oxides; from the slope of these lines it should in principle be possible to obtain information about the types of defects being formed. For the basic assumptions made in this treatment see the discussion of oxygen-deficient oxides in Section IV.A.

Following again the treatment used for oxygen-deficient oxides,

$$\begin{aligned}\Delta G_{V_M''}^{\circ} &= \Delta H_{V_M''}^{\circ} - T\Delta S_{V_M''}^{\text{vibr}} = -RT\ln K_{V_M''} \\ &= -RT(\ln 4 + 3\ln y - \tfrac{1}{2}P_{O_2})\end{aligned} \tag{31}$$

which by introducing $\Delta\overline{G}_{O_2} = RT \ln P_{O_2}$ gives

$$\Delta\overline{G}_{O_2} = \Delta\overline{H}_{O_2} - T\Delta\overline{S}_{O_2} = 2\Delta H^\circ_{V''_M} - T(2\Delta S^{\text{vibr}}_{V''_M} - 6R\ln y - 2R\ln 4) \tag{32}$$

or

$$\Delta\overline{H}_{O_2} = 2\Delta H^\circ_{V''_M} \tag{33}$$

or

$$\Delta\overline{S}_{O_2} = 2\Delta S^{\text{vibr}}_{V''_M} - 6R\ln y - 2R\ln 4 \tag{34}$$

For randomly distributed and noninteracting defects $\Delta\overline{H}_{O_2}$ should thus be independent of composition. On the other hand, the slope of a $\Delta\overline{S}_{O_2}$–ln y plot should be $-6R$ in the case of V''_M being formed—other slopes will be observed if other defects or defect complexes are formed—if the variation of $\Delta S^{\text{vibr}}_{V''_M}$ with composition is neglected. From Eq. (32) it will finally be noted that $\Delta\overline{S}_{O_2}$ can be determined from the slope of a $\Delta\overline{G}_{O_2}$–T plot, whereas the slope of a $\Delta\overline{G}_{O_2}/T$–$1/T$ plot gives $\Delta\overline{H}_{O_2}$ directly.

The formation of more complicated defect complexes can also be expressed by quasi-chemical equations, and from these similar expressions for $\Delta\overline{G}_{O_2}$, $\Delta\overline{H}_{O_2}$, and $\Delta\overline{S}_{O_2}$ can be derived. Consider, for instance, the Koch–Cohen cluster deduced from x-ray crystallographic examination of ferrous oxide, $Fe_{1-y}O$ (Koch and Cohen, 1969). This complex consists of four interstitial cations, $Fe_i^{\cdot\cdot\cdot}$, in tetrahedral positions—FeO has the rock-salt structure—surrounded by 13 vacant cation sites, $9V'_M$ and $4V''_M$, [see Fig. 8 in review by Sørensen (1978)], and its formation can be expressed by

$$\tfrac{9}{2}O_2 \rightarrow 9V'_M + 4V''_M + 4Fe_i^{\cdot\cdot\cdot} + 5Fe^{\cdot}_{Fe} = C''''' + 5Fe^{\cdot}_{Fe} \tag{35}$$

which gives

$$y^6 \propto P_{O_2}^{9/2} \quad \text{or} \quad y \propto P_{O_2}^{1/1.33} \tag{36}$$

From these equations it is interesting to note that the complex is not neutral and that a compensating atmosphere of Fe^{3+} ions is therefore considered to be formed in the surrounding lattice. The expression for $\Delta\overline{G}_{O_2}$ in the formation of this complex is

$$\Delta\overline{G}_{O_2} = \tfrac{2}{9}\Delta H^\circ_{\text{compl}} - T(\Delta S^{\text{vibr}}_{\text{compl}} - \tfrac{4}{3}R\ln y - \tfrac{2}{9}R\ln C_1) \tag{37}$$

where C_1 is a constant. If this complex is formed, the slope of a $\Delta\overline{S}_{O_2}$–ln y plot should be $-\tfrac{4}{3}R$, which is quite different from the slope of $-6R$ expected when V''_M is formed.

B. Transition Metal Monoxides

The transition metal monoxides showing only metal deficiency and which cover the greatest composition ranges are the $Mn_{1-y}O$ and $Fe_{1-y}O$ oxides; therefore only these compounds will be discussed in the following sections. The $Co_{1-y}O$ and $Ni_{1-y}O$ oxides, however, have interesting electrical and catalytic properties, and for further information the excellent book by Kofstad (1972) is recommended.

1. Thermodynamic Data for $Fe_{1-y}O$

The thermodynamic data discussed below for $Fe_{1-y}O$, and also for $Mn_{1-y}O$, are all based on thermogravimetric equilibration measurements and on electrochemical measurements, both of which can be interpreted in a straightforward manner in contrast to electric conductivity and diffusion measurements as discussed in later chapters.

From the data reported by Vallet and Raccah (1965), Bransky and Hed (1968), Picard and Dodé (1970), and Fender and Riley (1969) the $\Delta\bar{G}_{O_2}$–log y plot shown in Fig. 23 was constructed for the $Fe_{1-y}O$ system. Again it is clear that the data nicely follow the predicted straight-line relationship and that consistent n values can be calculated from the slope of these lines. If the composition at which the slope changes is plotted as a function of temperature, the pseudophase diagram shown in Fig. 24 can be obtained. From this diagram it is evident that the $Fe_{1-y}O$ phase range can be divided into subphases each characterized by a specific n value indicating that each subphase has a characteristic defect structure. Previously Fender and Riley (1969) and Vallet and Carel (1970) proposed that the wüstite region consisted of three regions, and the phase boundaries for these are also drawn in Fig. 24. When these boundaries are compared with those resulting from the present treatment, it is noted that both show the same trend, although a more simplified diagram is obtained by the more rigorous treatment used here.

Fender and Riley proposed that the transition from one region to another was due to order–disorder transitions, and that regions I, II, and III (see Fig. 24) represented increasing order. The boundaries established from the $\Delta\bar{G}_{O_2}$–log y plot, however, are based on defect theories and indicate a change in the defect structure that does not involve an order–disorder transition.

2. Thermodynamic Data for $Mn_{1-y}O$

For the $Mn_{1-y}O$ systems the $\Delta\bar{G}_{O_2}$ values were calculated from data reported by Bransky and Tallan (1971), Hed and Tannhauser (1967),

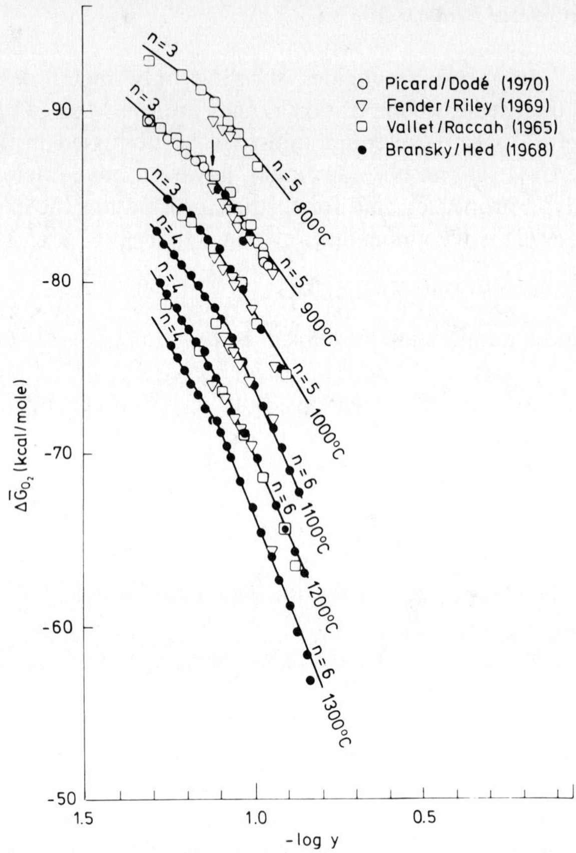

Fig. 23. Relative partial free energies of oxygen, $\Delta\bar{G}_{O_2}$, of $Fe_{1-y}O$ as a function of composition (log y).

Schmahl and Hennings (1969), and Fender and Riley (1970). In Fig. 25 these values are plotted as a function of composition—log y—and it is noted that the predicted straight-line relationship with consistent slopes also is observed for this system, although the data from the different sources do not correspond to each other. The most accurate and consistent data set is believed to be that of Bransky and Tallan, obtained by thermogravimetric equilibration measurements in controlled atmospheres. From these data it will be noted that the slope of the lines changes at specific compositions depending on the temperature. The nonstoichiometric phase range for this system thus also can be divided into subphase regions each with a characteristic defect structure. Previously Fender and Riley proposed a subdivision of the $Mn_{1-y}O$ phase range, and the phase boundaries they observed

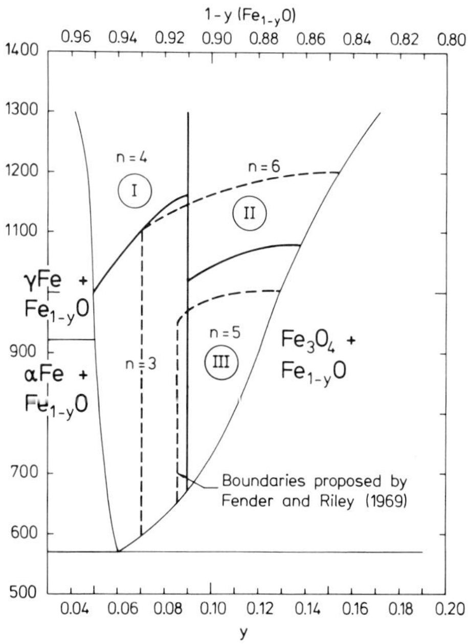

Fig. 24. Diagram of subphases in wüstite, $Fe_{1-y}O$, derived from the $\Delta\bar{G}_{O_2}$–log y plot shown in Fig. 23. Here I, II, and III designate the regions proposed by Fender and Riley (1969).

at ~1170°C are also shown in Fig. 25. The agreement between the two sets of boundaries is not very convincing, but generally it is believed that those derived from Bransky's and Tallan's data are the most correct because they can be explained in terms of defect theories. The Fender–Riley boundaries are, on the other hand, rather vaguely defined, and some of them can be explained by the tetragonal–cubic phase transition in Mn_3O_4 impurities, which takes place at 1162°C (Van Hook and Keith, 1958).

3. *Defect Structure of Metal-Deficient Oxides*

Many different defects or defect complexes have been proposed to explain the thermodynamic and other properties of metal-deficient transition metal monoxides. The most important of these and their corresponding n values are summarized in Table III. The defects can be divided into two categories: single defects, divacancies, and associates of vacancies (negative) and oxidized cations (positive), which assumes that the cationic sublattice is stable and the complexes based on the formation of interstitial cations formed as described in Eq. (35). The former types of defects are rather speculative, whereas the latter are considered to be more realistic

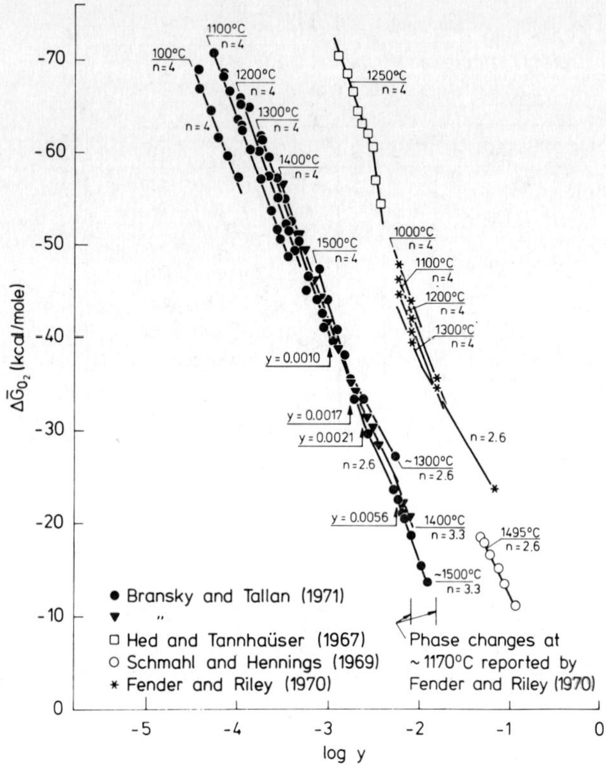

Fig. 25. Relative partial free energies of oxygen, $\Delta\bar{G}_{O_2}$, of $Mn_{1-y}O$ as a function of composition (log y).

since they are based on structure examinations by x-ray and neutron diffraction. As discussed by Catlow and Fender (1975), complexes containing interstitial cations can be built from either corner or edge sharing of the basic complex $[4V_M'' - Fe_i^{\cdot\cdot\cdot}]^{5-}$, which was observed to exist at high temperatures by Cheetham *et al.* (1971). In Fig. 26 the structure of the Koch–Cohen complex and a 6 : 2 cluster formed in this way is illustrated.

When comparing the n values given in Table III for the different possible defects and defect complexes with those observed for the subregions in the $Fe_{1-y}O$ system (Fig. 24), it will be noted that for $(1-y) \geqslant 0.91$ the region with $n = 4$ existing at high temperatures can be tentatively explained by formation of the basic complex which at decreasing temperature associates into a 16 : 5 complex by corner sharing resulting in the $n = 3$ region. The existence of the famous Koch–Cohen complex, which was proposed from an x-ray structure determination on a single crystal with the composition $Fe_{0.902}O$, is thus not supported by the thermodynamic data,

TABLE III

Defect Complexes Proposed for the Metal-Deficient Oxides $Fe_{1-y}O$ and $Mn_{1-y}O$

Defect complex	n	Proposed by
$V^{*}_{m}, V'_{M}, V''_{M}$	2, 4, 6	
$(V_MV_M)''$, $(V_MV_M)''''$	3, 5	Bransky and Tallan (1971)
$(V''_M\,M^{4+}_M)$[a]		Schmahl and Hennings (1969)
$V''_M-Fe_i^{\cdot\cdot\cdot}-V''_M$	4	Roth (1960)
$13V''_M-4Fe_i^{\cdot\cdot\cdot}$[b]	3.33	Koch and Cohen (1969)
$16V''_M-5Fe_i^{\cdot\cdot\cdot}$	3.09	Catlow and Fender (1975)
$4V''_M-Fe_i^{\cdot\cdot\cdot}$	4	Cheetham *et al.* (1971)
$3V'_M-V''_M-Fe_i^{\cdot\cdot\cdot}$	2	
$6V''_M-2Fe_i^{\cdot\cdot\cdot}$	3.5	Catlow and Fender (1975)
$8V''_M-3Fe_i^{\cdot\cdot\cdot}$	2.4	

[a] Proposed for the $Mn_{1-y}O$ system.

[b] This complex is fully oxidized in contrast to the complex discussed in Section V.A [Eq. (35)].

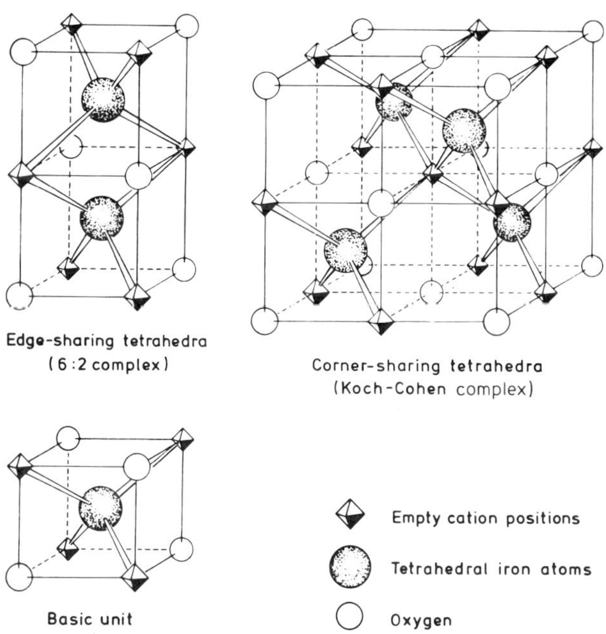

Fig. 26. Structure of defect complexes in metal-deficient oxides formed by corner and edge sharing of a basic tetrahedral defect complex as proposed by Catlow and Fender (1975).

whereas the less compact 16 : 5 complex, which is an element in the inverse spinel structure of Fe_3O_4 (Catlow and Fender, 1975), gives a much better description of the data. The Roth complex, which was inferred from a neutron diffraction study on quenched powders, could also explain the high-temperature $n = 4$ region, but this complex is considered less likely from a structural point of view.

Finally the $n = 6$ and $n = 5$ regions observed for $(1 - y) \leqslant 0.91$ could be explained by the presence of V_M'' and $(V_M V_M)''''$, respectively; however, as the concentration of vacancies increases with decreasing $(1 - y)$, the existence of single independent defects is doubtful, especially as more complicated defect complexes are considered to exist at compositions with higher $(1 - y)$ values. These regions in fact have not been very well examined structurally, but the rapid precipitation of Fe_3O_4 observed by Greenwood and Howe (1972) might indicate that they are highly ordered and that their higher n values are the result of a stepped $\Delta\bar{G}_{O_2}$–log y curve. Kofstad and Hed (1968), who also obtained increasing n values with decreasing $(1 - y)$, explained their data by introducing a blocking effect around the tetrahedral interstitial sites of a Roth complex, which, however, as discussed above, is considered to be less likely.

Comparing the n values of Table III with the values found for the $Mn_{1-y}O$ system (Fig. 25), the three subregions with $n = 4$, 2.6, and 3.3, respectively, can again be explained by formation of the basic tetrahedral complexes ($n = 4$) and their subsequent packing apparently into 8 : 3 complexes by edge sharing followed by the formation of Koch–Cohen complexes by corner sharing in contrast to the $Fe_{1-y}O$ system. These findings should, however, be verified by structural examinations which hitherto have not been performed on this oxide.

VI. Thermodynamics and Defect Structure of Metallic Transition Metal Monoxides

An interesting group of nonstoichiometric oxides are the transition metal monoxides $TiO_{1\pm x}$ and $VO_{1\pm x}$, which at higher temperatures show nonstoichiometry on either side of the stoichiometric composition $MO_{1.0}$. The composition range covered by these oxides is extremely wide and, because both higher and lower oxides of about the same relative stability exist in both oxide systems—Ti_2O_3, V_2O_3, and Ti_2O, $\sim V_2O$ (Kofstad, 1972)—this range is also almost symmetric about $MO_{1.0}$. Another interesting feature of these oxides is that there are a high number of defects in both cationic and anionic sublattices even at the stoichiometric composition. For instance, Banus and Reed (1970) have recently calculated the defect concentrations from x-ray and pycnometric density measurements assuming a random distribution among the lattice sites. The percentage of defects is shown in the accompanying tabulation.

	V_O(%)	V_M(%)	Total (%)
$MO_{1.00}$	15	15	30
$MO_{1.30}$ (upper limit)	0	23	23
$TiO_{0.80}$[a]	24	8	32
$VO_{0.80}$[a]	30	10	40

[a] Lower limit at about $MO_{0.70}$.

At the upper limit of the composition range the oxygen sublattice is thus completely filled, whereas metal vacancies are considered still to exist at the lower limit.

The reason for the large number of defects in the structure of these oxide systems is their specific electronic structure. Both titanium and vanadium ions in these compounds have broad, overlapping *d* orbitals which produce strong metallic bonding between the cations. As pointed out by Anderson (1972), this implies that the excess negative or positive charge involved in defect formation (electrons or electron holes) is completely delocalized and not trapped on the cations. Furthermore, the metallic electrons effectively screen the charged cation and anion vacancies, and the Coulomb interaction between these defects is considerably weakened; this favors the presence of a high concentration of randomly distributed point defects. Because of the high degree of metallic bonding in the oxides they are often termed metallic oxides, although the metallic properties are probably more pronounced for $TiO_{1\pm x}$ than for $VO_{1\pm x}$ (Banus and Reed, 1970). Despite the strong screening of the defects postulated in this model, ordered superlattice structures have, however, been observed at lower temperatures for both oxides, as described in the next section.

Both $TiO_{1\pm x}$ and $VO_{1\pm x}$ have the cubic NaCl structure, and it is interesting to compare the properties of these oxides with those of $NbO_{1\pm x}$, which also has this structure with a large concentration of metal and oxygen vacancies—for $NbO_{1.0}$, $[V_M] \approx [V_O] \approx 25\%$. In contrast to these two oxides $NbO_{1\pm x}$ has a very narrow composition range ($0.98 \leqslant x \leqslant 1.02$) (Banus and Reed, 1970), and the vacancies are highly ordered in this compound with the Nb vacancies at the 000 positions and the oxygen vacancies at the $\frac{1}{2}\frac{1}{2}\frac{1}{2}$ positions (Banus and Reed, 1970). The $NbO_{1\pm x}$ oxides must therefore be the least metallic of this series.

1. Thermodynamic Data for $TiO_{1\pm x}$ *and* $VO_{1\pm x}$

Systematic studies on the thermodynamic properties of transition metal monoxides are very limited. In the few published reports of such studies, the data are generally not detailed enough to allow an analysis of the defects present in these oxides as in the case of the nonstoichiometric systems discussed in the previous sections. Hoch *et al.* (1962), for instance, studied only the two-phase regions bounding the nonstoichiometric phases in their electromotive force measurements on $TiO_{1\pm x}$, $VO_{1\pm x}$, and $NbO_{1\pm x}$ systems. However, based on a statistical model they calculated the defect interaction energies between both like and unlike defects from their data. For TiO, the critical temperature for ordering using these interaction energies was found to agree with experimental data (Hoch, 1963). Another study, which also should be mentioned, is the calorimetric measurements of the enthalpies of formation (ΔH_f°) of both $TiO_{1\pm x}$ and $VO_{1\pm x}$ by Ariya and co-workers (1957; Ariya and Popov, 1962). From their data they concluded that ΔH_f° for any composition could be expressed as a linear function of the enthalpies of the neighboring oxides, and the microdomain concept was suggested to explain this behavior. By this concept, microdomains with the local order of the adjacent phases are considered to coexist in an ideal solution (zero enthalpy of mixing)—for $TiO_{1\pm x}$, for instance, domains with a structure resembling that of Ti_2O, in which the oxygen vacancies are ordered, coexist with Ti_2O_3 which has ordered metal vacancies. However, $TiO_{1\pm x}$ has the NaCl structure, whereas both Ti_2O and Ti_2O_3 are hexagonal close packing; Ariya *et al.* considered that both types of domain had a metastable cubic structure in order to maintain a coherent structure in the nonstoichiometric phase. Although domain structures have been observed for these systems at lower temperatures, as will be discussed in the next section, it is, however, doubtful whether they are of the nature proposed by Ariya *et al.*

Finally, Vasil'eva and Granovskaya (1976) and Vasil'eva *et al.* (1976) have published thermodynamic data—$\Delta\overline{G}_{O_2}$, $\Delta\overline{H}_{O_2}$, and $\Delta\overline{S}_{O_2}$—for $VO_{1\pm x}$ determined by electromotive force measurements in the temperature range 900–1100°C and at room temperature. In this study the lower and upper limits of the nonstoichiometric phase region were determined to be $VO_{0.68}$ and $VO_{1.267}$, respectively, and from the $\Delta\overline{H}_{O_2}$ and $\Delta\overline{S}_{O_2}$ data derived from the experimental $\Delta\overline{G}_{O_2}$ data, it was concluded that the phase range for $VO_{1\pm x}$ apparently consisted of seven subregions with different types of defects responsible for the deviation from the stoichiometric composition. On the assumption that only one type of defect (oxygen or vanadium vacancies) predominates in the regions near the lower and upper phase limits, that vacancies are present in both sublattices simultaneously in the intermediate regions, and that all defects are distributed randomly in all regions, the energies of formation of oxygen and vanadium vacancies and their interaction energies were calculated from a statistical model. These calculations showed that the interaction energy between the oxygen and vanadium vacancies was maximum for the central region near $VO_{1.0}$, indicating that defect complexes and perhaps even an ordered phase is possible at this composition.

2. Defect Structure of Metallic Transition Metal Monoxides

At higher temperatures (above 1000°C) the $MO_{1\pm x}$ phase is considered to be completely disordered and to have diffraction patterns typical of a cubic NaCl-type structure (Andersson *et al.*, 1957). At lower temperatures, however, several ordered intermediate structures have been observed for $TiO_{1\pm x}$ and $VO_{1\pm x}$ by x-ray and electron diffraction studies, which have been described by Watanabe *et al.* (1970) and Bell and Lewis (1971) for the two oxides, respectively. These studies cannot be described in detail in this chapter, but it is interesting to note the type of defect ordering postulated for the two systems. For the $TiO_{1\pm x}$ phase, for instance, Watanabe *et al.* (1970) has suggested that the vacancies are distributed randomly only on every third (110) plane of the original NaCl lattice, and the following ordered superlattice structures were proposed to explain the electron diffraction patterns they observed on specimens annealed at lower temperatures:

$TiO_{0.7-0.9}$ orthorhombic superlattice cell with oxygen vacancies an every third (110) plane;

$TiO_{0.9-1.1}$ monoclinic with both oxygen and titanium vacancies on every third (110) plane; half of the oxygen and titanium atoms are vacant on these planes;

$TiO_{1.20}$ orthorhombic (originally observed by Hilti, 1968), with titanium vacancies on every third (110) plane; half of the titanium atoms are missing randomly from these planes.

Watanabe *et al.* also observed an ordered tetragonal phase with the composition $TiO_{1.25}$. For this phase, however, they proposed that the titanium vacancies were ordered according to the Ni_4Mo structure with the vacancies at the Mo sites, whereas the oxygen sublattice was fully occupied.

A tetragonal superlattice cell has also been observed by electron microscopy for the $VO_{1\pm x}$ system at the composition $VO_{1.23}$ ($V_{52}O_{64}$) by Andersson and Gjønnes (1970). They have proposed that this superstructure, similar to the $Fe_{1-y}O$ system discussed above, consists of an ordered arrangement of defect clusters with one vanadium interstitial at a tetrahedral position surrounded by four vanadium vacancies (basic complex). Other ordered phases have also been observed for the $VO_{1\pm x}$-phase range. Gurevich and Ormont (1957), for instance, found a bcc phase at $VO_{1.30}$ by x-ray measurements, whereas a tetragonal superlattice was observed with electron microscopy by Bell and Lewis (1971) with the composition $VO_{1.31}$ ($V_{244}O_{320}$). In the latter study a superlattice unit cell somewhat larger than that ascribed to $V_{52}O_{64}$ was proposed for $V_{244}O_{320}$, and this expansion was considered to arise from the ordering of additional vanadium vacancies; this cell contains 20 interstitial vanadium atoms, as well as 96 vanadium vacancies.

For both $TiO_{1\pm x}$ and $VO_{1\pm x}$ extensive diffuse scattering has been observed in the electron diffraction patterns. Cowley (1970), who examined $TiO_{1.0}$, has proposed that this scattering originates from a short-range ordering of the metal and oxygen vacancies related to the form of the Fermi surface. This idea was also examined by Castles *et al.* (1971) for TiO, but long-range ordering and the formation of microdomain structures are probably the main cause of this diffuse scattering. A detailed analysis of the diffraction patterns of the scattering observed in the $VO_{1\pm x}$ system, by Bell and Lewis (1971), for instance, indicates that domains with the *c* axis of the tetragonal superlattice are aligned along one of the three possible cubic unit cell axes of the matrix. These are selected randomly, resulting in a uniform distribution of small domains following short anneals, whereas after extended anneals larger domains are believed to arise from a selective coarsening process that favors the growth of neighboring domains with similar *c* axes. Both types of domains have been observed by electron microscopy. Watanabe *et al.* have also observed the formation of domains in the $TiO_{1\pm x}$ system. Near the stoichiometric composition, domains consisting of parallel plates were observed, and this was also the case for the two-phase region $TiO_{(1.1-1.25)}$ in which $TiO_{1.0}$ alternates with $TiO_{1.25}$ domains. In the tetragonal $TiO_{1.25}$ phase, however, only small and randomly distributed domains, but no lamellae structure, were observed, supporting the idea that the defects accumulate on particular crystallographic planes [(110), see above] in $TiO_{1.0}$ but not in $TiO_{1.25}$.

VII. Thermodynamics and Defect Structure of Excess-Oxygen Oxides

Contrary to the situation in transition metal monoxides, the cation sublattice in some oxide systems is so stable that neither cation vacancies nor interstitial cations can form. In systems like the nonstoichiometric uranium oxides, UO_{2+x}, where cations can exist in the +4 and +5 (+6) valence states, deviation from the stoichiometric composition can therefore be achieved only by the introduction of excess oxygen at interstitial oxygen positions. Besides UO_{2+x}, several ternary oxides with uranium as the main constituent also easily form excess-oxygen compounds. The most extensively studied system is, however, the UO_{2+x} oxides and, as an example of the thermodynamic properties and defect structure of nonstoichiometric excess-oxygen compounds, only this oxide system will be considered in this section.

A. Thermodynamics, General

The formation of an interstitial oxygen ion can be expressed by

$$\tfrac{1}{2}O_2(g) \leftrightarrows O_i'' + 2h^{\cdot} \tag{38}$$

In this case, the interstitial oxygen ion is considered to become fully ionized by the removal of two electrons from adjacent cations where two positive holes are formed in the valence band. The cations are thus effectively oxidized to their next stable oxidation state: $U^{4+} \cdot h^{\cdot}$ or U^{5+}. For this reaction the neutrality condition will be

$$[h^{\cdot}] = 2[O_i''] \tag{39}$$

and, using the mass-action law on Eq. (38), x in UO_{2+x} becomes

$$x \propto P_{O_2}^{1/6} \qquad \text{or} \qquad n = 6 \tag{40}$$

If, however, the oxidized cations and the interstitial oxygen ions are not dissociated as assumed above, a neutral complex will be formed according to

$$\tfrac{1}{2}O_2(g) \leftrightarrows [O_i'' \cdot 2h^{\cdot}]^x \tag{41}$$

and in this case

$$x \propto P_{O_2}^{1/2} \qquad \text{or} \qquad n = 2 \tag{42}$$

Much more complicated defect complexes have been proposed to explain the thermodynamic properties and structures of UO_{2+x} oxides. Willis (1964, 1978), for instance, proposed a 2:2:2 complex consisting of four oxygen ions in interstitial positions ($2O_i^a$ and $2O_i^b$) and two oxygen vacancies created by formation of the O_i^a ions [the structure of this complex has been described by the author in a previous review; (see Sørensen, 1978, Fig. 11)]. The formation of the complex can be expresed as

$$2V_i^a + 2V_i^b + 2O_O + O_2(g) = [2(O_i^a \cdot V_O \cdot O_i^b)]'''' + 4h^{\cdot} \tag{43}$$

where V_i^a and V_i^b are two different interstitial sites and O_O is an oxygen atom in the correct oxygen position in the oxygen sublattice. In this case the neutrality condition will be

$$[h^{\cdot}] = 4[\{2(O_i^a \cdot V_O \cdot O_i^b)\}''''] \tag{44}$$

and, using the mass-action law,

$$x \propto P_{O_2}^{1/5} \quad \text{or} \quad n = 5 \tag{45}$$

The oxidized cations can be either dissociated from the complex as assumed in Eq. (43) or attached to the complex in different numbers. The electrons necessary for the formation of O_i^b ions are probably provided by ionization of the oxygen vacancies, and for this complex only the following charges and the corresponding n values are possible:

$$\begin{aligned} &C'; && x \propto P_{O_2}^{1/2} \quad \text{or} && n = 2 \\ &C''; && x \propto P_{O_2}^{1/3}, && n = 3 \\ &C'''; && x \propto P_{O_2}^{1/4}, && n = 4 \\ &C''''; && x \propto P_{O_2}^{1/5}, && n = 5 \end{aligned} \tag{46}$$

where C designates the Willis 2 : 2 : 2 complex $2(O_i^a \cdot V_O \cdot O_i^b)$.

As for the nonstoichiometric oxides considered in the previous sections, $\Delta\overline{G}_{O_2}$ for the excess-oxygen compounds can also be expressed as

$$\Delta\overline{G}_{O_2} = RT \ln P_{O_2} \propto nRT \ln x \tag{47}$$

and for these oxides also a straight line should be obtained in an isothermal $\Delta\overline{G}_{O_2}$–$\log x$ plot provided n is constant. The experimental data for UO_{2+x} plotted in this way are discussed in the next section. Besides the quasi-chemical approach described above, the UO_{2+x} phase has been considered also in terms of statistical thermodynamic models based on the exclusion principle (see, e.g., review in Sørensen, 1978). These models and the application of statistical methods for excess-oxygen compounds are discussed in a later chapter of this book.

B. Nonstoichiometric Uranium Oxides Containing Excess Oxygen

1. Thermodynamic Data for UO_{2+x}

The $\Delta\overline{G}_{O_2}$ values published by Kiukkola (1962), Roberts and Walter (1961), Markin et al. (1968), and Hagemark and Broli (1966) are plotted as a function of $\log x$ in Fig. 27. Of these studies only Hagemark and Broli's were based on the thermogravimetric technique, which is the most accurate, whereas electromotive force measurements (Kiukkola), a gas equilibration technique (Markin *et al.*, 1968), or direct measurements of the equilibrium oxygen pressure were used in the other studies. From this plot it is clear

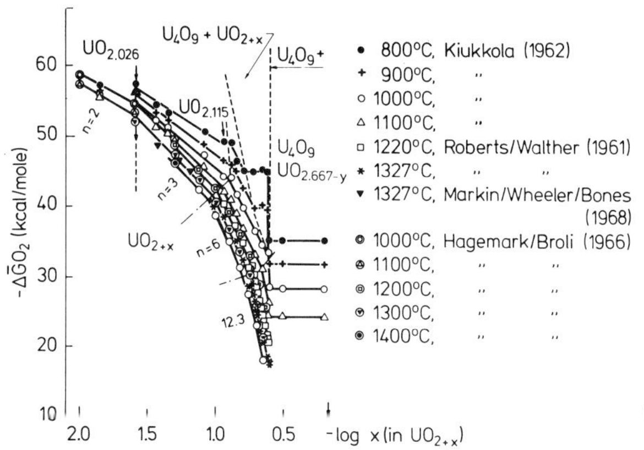

Fig. 27. Relative partial free energies of oxygen, $\Delta\bar{G}_{O_2}$, of UO_{2+x} as a function of composition ($\log x$).

that a straight-line relationship between $\Delta\bar{G}_{O_2}$ and $\log x$ is also obtained for the UO_{2+x} phase and that the slope of the lines changes in a characteristic manner. Obviously, the nonstoichiometric phase range for this oxide can be divided into subregions each with a characteristic defect being predominantly formed. Kiukkola (1962) used a similar plot—$\log P_{O_2}$ versus $\log x$—but found that the slope varied continuously over the nonstoichiometric phase range, indicating a gradual change in the types of defects being formed, which is rather unlikely. In magnitude the slopes obtained in Fig. 27, however, correspond quite closely to those reported by Kiukkola (1962). The pattern of subregions in the UO_{2+x} phase range is also shown in the phase diagram in Fig. 30.

2. *Defect Structure of* UO_{2+x}

An interpretation of the defect structure based on the *n* values shown in Fig. 27 must take the structural findings into account. A systematic study of the structural changes across the UO_{2+x}-phase range still needs to be carried out, and the only structure determination so far is the neutron diffraction work of Willis (1964), who examined a single crystal at 800°C with the composition $UO_{2.12}$. Based on this work the 1 : 2 : 2 defect complex ($1O_i^a$–$2V_O$–and $2O_i^b$; see previous section) was originally proposed, but recently Willis has reported evidence that the most probable defect complex at this composition is of the 2 : 2 : 2 type (Willis, 1978). Catlow and Lidiard (1975) also calculated that the 2 : 2 : 2 complex was the most stable of the two types, but they found in addition that the extended 4 : 3 : 2 complex ($4O_i^a$–$3V_O$–$2O_i^b$) had even higher stability (higher binding energy). The structure of this complex is shown in Fig. 28.

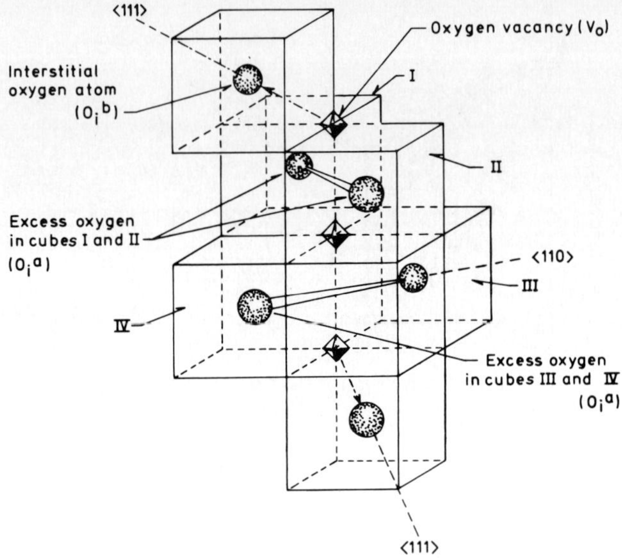

Fig. 28. Structure of 4 : 3 : 2 complex proposed by Catlow and Lidiard (1975) for UO_{2+x}.

With these structural studies and suggestions in mind, the *n* values obtained for the subregions in the UO_{2+x} systems (Fig. 27) can tentatively be interpreted as shown in the accompanying tabulation:

Region	Defect structure
$n = 2$	$\{O_i'' \cdot 2U^{5+}\}^x$
$n = 3$	$\{2(O_i^a \cdot V_O \cdot O_i^b) \cdot 2U^{5+}\}''$ plus two free U^{5+}
$n = 6$	New type of complex but not the 4 : 3 : 2 complex
$n = 12.3$	Highly ordered phase, U_4O_{9-y}

The $n = 6$ region cannot be explained in terms of the 4 : 3 : 2 complex or any other extended complex formed on the same principle, as these all give *n* values smaller than 5. Matsui and Naito (1975) have proposed that the complexes are fully dissociated according to

$$\{2(O_i^a \cdot V_O \cdot O_i^b)\}'''' \leftrightarrows 4O_i'' + 2V_O^{\cdot\cdot} \tag{48}$$

and, using the mass-action law, $n = 6$ is obtained for the formation of O_i'' and $V_O^{\cdot}$ as described above. The coexistence in the lattice of independent O_i'' and $V_O^{\cdot\cdot}$ is, however, doubtful, especially at the high concentration of defects at the large deviation from the stoichiometric composition for this subregion. A new but unidentified complex is thus proposed to exist in the $n = 6$ region.

VIII. Thermodynamics and Defect Structure of Oxides Existing as both Oxygen-Excess and Oxygen-Deficient Compounds

Contrary to the situation in the transition metal monoxides, in which both positive and negative deviations from the stoichiometric composition can be obtained by the formation of metal and oxygen vacancies, respectively, oxygen-excess and oxygen-deficient compounds are formed in fluorite oxide systems by oxygen interstitials and oxygen vacancies, as the cation sublattice in these oxides is considered very stable. To be able to form both an oxygen-excess and oxygen-deficient oxide, however, it is necessary that the cations exist both in valence states higher and lower than those corresponding to the stoichiometric composition. This is not the case for any of the fluorite oxides except under extreme conditions—UO_2 can be reduced to oxygen-deficient compounds only at high temperatures ($> 1800°C$) and under highly reducing conditions—but by partly substituting an oxidizable cation with a reducible cation it is possible to form ternary oxide systems which can show both types of nonstoichiometry.

A typical example of such a ternary oxide is $(U_{1-y}Pu_y)O_{2\pm x}$, which will be considered in this section.

1. Thermodynamic Data for $(U_{1-y}Pu_y)O_{2\pm x}$

The $\Delta\bar{G}_{O_2}$ values published by Woodley (1973), Sørensen (1976b), and Chilton and Edwards (1980) are shown in Fig. 29 as a function of composition ($\log x$). These data have all been determined by thermogravimetric equilibration measurements, but some data obtained by electromotive force measurements, recently published by Woodley and Adamson (1979), are also included in the figure. For these oxides, this plot clearly indicates that $\Delta\bar{G}_{O_2}$ varies linearly with $\log x$ as predicted from the defect theories, and apparently both the hypo- and hyperstoichiometric phase ranges can in this case also be divided into subregions each with a characteristic defect structure as shown in the phase diagram constructed from the $\Delta\bar{G}_{O_2}$–$\log x$ plot in Fig. 30.

2. Defect Structure of $(U_{1-y}Pu_y)O_{2\pm x}$

Detailed structure studies have not yet been reported for these oxides, and the n values found for the subphases shown in Fig. 30 thus cannot be

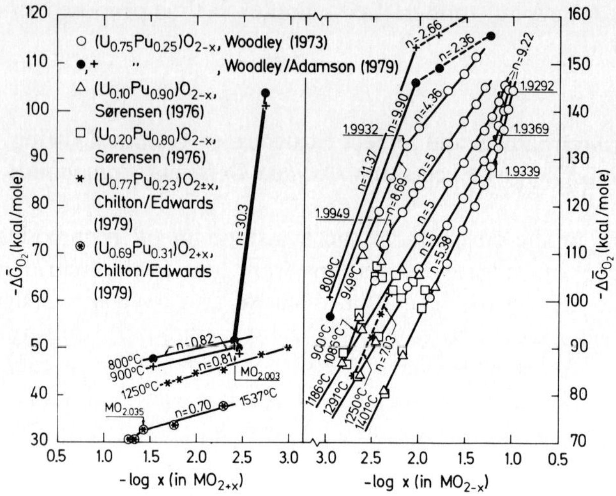

Fig. 29. Relative partial free energies of oxygen, $\Delta\bar{G}_{O_2}$, of mixed U and Pu oxides as a function of composition [$\log x$ in $(U_{1-y}Pu_y)O_{2\pm x}$].

evaluated in terms of defect complexes. Compared to the PuO_{2-x} system (Fig. 20) the two-phase range near the stoichiometric composition is apparently not formed in mixed oxides, indicating that ordering reactions of the defect complexes are much more difficult in these oxide systems. The $n = 5$ region in which single defects or simple defect clusters are probably formed also extends to much lower O/M values for the mixed oxides than for PuO_{2-x}, which again is an indication of the greater stability of the defect structure in mixed oxides. Finally, another difference between the two systems is that the nature of the defect complexes in PuO_{2-x} apparently can change in several steps before the ordered $n = 9.22$ region is reached,

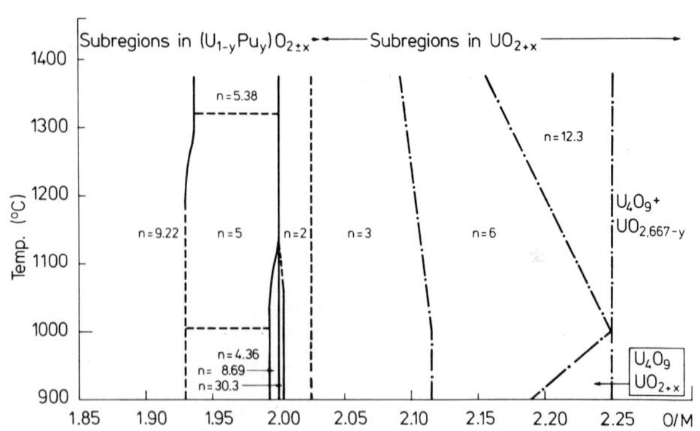

Fig. 30. Diagram of subphases in the nonstoichiometric phase range of $(U_{1-y}, Pu_y)O_{2\pm x}$ and UO_{2+x}.

whereas only one type of defect is observed for the mixed oxide. The PuO_{2-x} oxides are apparently much more reactive, probably because of the greater concentration of Pu ions which play an important role during the formation of defect complexes.

Another interesting feature of the mixed oxides is the high n values found near the stoichiometric composition both in the hypo- and hyperstoichiometric phase range, indicating a high degree of order in these subphases. Corresponding to this observation Tetenbaum and Hunt (1968) reported a minimum in $\Delta\bar{S}_{O_2}$ in the hypostoichiometric region at the composition O/U = 1.98 for UO_{2-x}, which they interpreted as an indication of a partial ordering of lattice defects leading to microstructural changes. For the hyperstoichiometric phase range, however, Markin and Bones (1967) found an entropy maximum from low-temperature electromotive force measurements, which does not correspond to the high degree of ordering indicated here.

Finally, according to the phase diagram in Fig. 30, the n values for the subphase covering the composition range $MO_{2.00}$–$MO_{2.03}$ are apparently different for UO_{2+x} and for the mixed oxide. Contrary to the UO_{2+x} data, the n value for the mixed oxide is rather well defined and, at least for the mixed oxides neutral defect complexes of the type $[2(O_i'' \cdot 2h^{\cdot})]^x$, which would give $n = 1$, seem to be formed near the stoichiometric composition.

IX. Recommendations for Future Work on Nonstoichiometric Oxides

In this chapter, thermodynamics data for the different types of nonstoichiometric oxide systems have been analyzed in a very general way, i.e., the variation in published $\Delta\bar{G}_{O_2}$ values and in some cases $\Delta\bar{H}_{O_2}$ and $\Delta\bar{S}_{O_2}$ data has been examined as a function of $\log x$ or $\log y$, where x or y is the deviation from the stoichiometric composition for $MO_{2\pm x}$ or $M_{1-y}O$, respectively. For the oxides for which extensive and reliable thermodynamic data have been determined, this analysis has shown that the nonstoichiometric phase regions can be divided into subphases; each of these can be described by a characteristic value of n, which is determined from the slope of the linear $\Delta\bar{G}_{O_2}$–$\log x$ plots obtained and which, according to the defect theories, depends on the type of predominating defect present in a given subphase. In some nonstoichiometric oxide systems interpretation of these n values can be based on detailed structural examinations, but for other systems such studies are incomplete or are completely lacking. In order to describe the real defect structure of most nonstoichiometric oxides and thus their properties, systematic structural examinations are therefore badly needed.

The approach used in this chapter assumes that defect complexes are noninteracting and randomly distributed. This is probably true only at small deviations from the stoichiometric composition, and a more sophisticated approach, in which the interactions between the defect complexes are taken into account, is therefore necessary in order to describe the properties of grossly nonstoichiometric oxides properly. Many advanced statistical thermodynamic treatments have been proposed, and some of these will be described in later chapters of this book. For a statistical model to be realistic, however, it is necessary that it be based on the real defect structure and not on a mathematical description, no matter how sophisticated such a description might be. For oxygen-deficient nonstoichiometric fluorite oxides—CeO_{2-x}, PuO_{2-x}, and $(U_{1-y}Pu_y)O_{2-x}$—such a statistical thermodynamic model is being developed (Chapter 3) and, when more nonstoichiometric oxides have been examined structurally, it will be valuable to develop realistic statistical thermodynamic models for these oxides as well.

References

Anderson, J. S. (1970a). *In* "Problems of Non-Stoichiometry" (A. Rabenau, ed.), pp. 1–76. North-Holland Publ., Amsterdam.

Anderson, J. S. (1970b). *In* "The Chemistry of Extended Defects" (L. Eyring and M. O'Keeffe, eds.), pp. 1–20. North-Holland Publ., Amsterdam.

Anderson, J. S. (1972). *Natl. Bur. Stand. (U.S.), Spec. Publ.* No. 364, 295–316.

Anderson, J. S. (1974). *In* "Defects and Transport in Oxides" (M. S. Seltzer and R. I. Jaffee, eds.), pp. 25–48. Plenum, New York.

Anderson, J. S., and Burch, R. (1971). *J. Phys. Chem. Solids* **32**, 923–926.

Andersson, B., and Gjønnes, J. (1970). *Acta Chem. Scand.* **24**, 2250–2252.

Andersson, S., Collén, B., Kuylenstierna, U., and Magnéli, M. (1957). *Acta Chem. Scand.* **11**, 1641–1652.

Ariya, S. M., and Popov, Yu. G. (1962). *J. Gen. Chem. USSR* **32**, 2054–2057.

Ariya, S. M., Morozova, M. P., and Vol'f, E. (1957). *Russ. J. Inorg. Chem.* **2**, 16–30.

Atlas, L. M., and Schlehman, G. J. (1965). *In* "Plutonium 1965" (A. E. Kay and M. B. Waldron, eds.), pp. 838–844. Chapman & Hall, London.

Atlas, L. M., and Schlehman, G. J. (1966). *In* "Thermodynamics," Vol. 2, pp. 407–421. IAEA, Vienna.

Banus, M. D., and Reed, T. B. (1970). *In* "The Chemistry of Extended Defects in Non-Metallic Solids" (L. Eyring and M. O'Keeffe, eds.), pp. 488–521. North-Holland Publ., Amsterdam.

Bell, P. S., and Lewis, M. H. (1971). *Phys. Status Solidi* **7**, 431–439.

Bertaut, E. F. (1953). *Acta Crystallogr.* **6**, 557–561.

Bevan, D. J. M., and Kordis, J. (1964). *J. Inorg. Nucl. Chem.* **26**, 1509–1523.

Blank, H. (1967). "A Comparison of the Pu–O System with the Rare Earth Oxide Systems Ce–O, Pr–O and Tb–O," EUR 3653e. EURATOM, European Atomic Energy Community, Brussels.

Blank, H. (1974). *J. Nucl. Mater.* **51**, 269–270.

Blank, H. (1975). *In* "Thermodynamics of Nuclear Materials 1974," Vol. II, pp. 45–69. IAEA, Vienna.

Blumenthal, J. B., Moser, J. B., and Whitmore, D. H. (1965). *J. Am. Ceram. Soc.* **48**, 617–627.

Blumenthal, R. N. (1975). *J. Solid State Chem.* **12**, 307–318.

Blumenthal, R. N., and Laubach, J. E. (1968). *In* "Anisotropy in Single Crystal Refractory Compounds" (I. W. Vahldick and S. A. Mersol, eds.), pp. 137–150. Plenum, New York.

Blumenthal, R. N., and Sharma, R. K. (1975). *J. Solid State Chem.* **13**, 360–364.

Blumenthal, R. N., Lee, P. W., and Panlener, R. J. (1971). *J. Electrochem. Soc.* **118**, 123–129.

Boivineau, J.-C. (1976). *J. Nucl. Mater.* **60**, 31–38.

Bransky, I., and Hed, A. Z. (1968). *J. Am. Ceram. Soc.* **51**, 231–232.

Bransky, I., and Tallan, N. M. (1971). *J. Electrochem. Soc.* **118**, 788–793.

Brouwer, G. (1954). *Philips Res. Rep.* **9**, 366–376.

Bursill, L. A., and Hyde, B. G. (1971). *Philos. Mag.* **23**, 3.

Bursill, L. A., and Hyde, B. G. (1972). *Prog. Solid State Chem.* **7**, 177–253.

Cahn, J. W. (1961). *Acta Metall.* **9**, 795–801.

Campserveux, J., and Gerdanian, P. (1974). *J. Chem. Thermodyn.* **6**, 795–800.

Castles, J. R., Cowley, J. M., and Spargo, A. E. C. (1971). *Acta Crystallogr., Sect. A* **27**, 376–383.

Catlow, C. R. A., and Fender, B. E. F. (1975). *J. Phys. C* **8**, 3267–3279.

Catlow, C. R. A., and Lidiard, A. B. (1975). *In* "Thermodynamics of Nuclear Materials 1974," Vol. II, pp. 27–43. IAEA, Vienna.

Cheetham, A. K., Fender, B. E. F., and Taylor, R. I. (1971). *J. Phys. C* **4**, 2160–2165.

Chereau, P., Dean, G., and Gerdanian, P. (1971). *C. R. Acad. Sci., Ser. C* **272**, 512–515.

Chikalla, T. D., and Eyring, L. (1969). *J. Inorg. Nucl. Chem.* **31**, 85–93.

Chikalla, T. D., and Turcotte, R. P. (1972). *Natl. Bur. Stand. (U.S.), Spec. Publ.* No. 364, 319–330.

Chilton, G. R., and Edwards, J. (1980). "Thermodynamics of Nuclear Materials 1979," Vol. 1, pp. 357–368. IAEA, Vienna.

Cowley, J. M. (1970). *In* "The Chemistry of Extended Defects in Non-Metallic Solids" (L. Eyring and M. O'Keeffe, eds.), pp. 259–271. North-Holland Publ., Amsterdam.

Eyring, L., and Holmberg, B. (1963). *Adv. Chem. Ser.* No. 39, 46–57.

Fender, B. E. F. (1972). *In* "Solid State Chemistry" (H. J. Emeleus and L. E. J. Roberts, eds.), MTP International Review of Science, Inorganic Chemistry, Series One, Vol. 10, pp. 243–278. Butterworth, London.

Fender, B. E. F., and Riley, F. D. (1969). *J. Phys. Chem. Solids* **30**, 793–798.

Fender, B. E. F., and Riley, F. D. (1970). *In* "The Chemistry of Extended Defects in Non-Metallic Solids" (L. Eyring and M. O'Keeffe, eds.), pp. 54–61. North-Holland Publ., Amsterdam.

Førland, K. S. (1964). *Acta Chem. Scand.* **18**, 1267–1275.

Greener, E. H., Wimmer, J. M., and Hirthe, W. M. (1964). *Proc. Conf. Rare Earth Res., 3rd, 1963* pp. 539–554.

Greenwood, N. N., and Howe, A. T. (1972). *J. C. S. Dalton Trans.* 110–116, 116–121, 122–126.

Gurevich, M. A., and Ormont, B. F. (1957). *J. Inorg. Chem. USSR* **2**(11), 151–165.

Hagemark, K., and Broli, M. (1966). *J. Inorg. Nucl. Chem.* **28**, 2887–2850.

Hed, A. Z., and Tannhauser, D. S. (1967). *J. Chem. Phys.* **47**, 2090–2103.

Herai, T., Thomas, B., Manenc, J., and Bénard, J. (1964). *C. R. Acad. Sci.* **258**, 4528–4530.

Hilti, E. (1968). *Naturwissenschaften* **55**, 130.

Hoch, M. (1963). *J. Phys. Chem. Solids* **24**, 157–159.

Hoch, M., Iyer, A. S., and Nelken, J. (1962). *J. Phys. Chem. Solids* **23**, 1463–1471.

Hoskins, B. F., and Martin, R. L. (1975). *J.C.S. Dalton Trans.* 576–588.

Hyde, B. G. (1971). *Acta Crystallogr., Sect. A* **27**, 617–621.

Hyde, B. G., and Eyring, L. (1964). *Proc. Conf. Rare Earth Res., 4th* pp. 623–644.

Hyde, B. G., Bevan, D. J. M., and Eyring, L. (1966). *Philos. Trans. R. Soc. London, Ser. A* **259**, 583–614.

Hyde, B. G., Bagshaw, A. N., Andersson, S., and O'Keeffe, M. (1974). *Annu. Rev. Mater. Sci.* **4**, 43–92.

Iwasaki, B., and Katsura, T. (1971). *Bull. Chem. Soc. Jpn.* **44**, 1297–1301.

Jenkins, M. S., Turcotte, R. P., and Eyring, L. (1970). *In* "The Chemistry of Extended Defects in Non-Metallic Solids" (L. Eyring and M. O'Keeffe, eds.), pp. 36–53. North-Holland Publ., Amsterdam.

Kevane, C. J. (1964). *Phys. Rev. Sect. 5A* **133**, A1431–A1436.

Kiukkola, K. (1962). *Acta Chem. Scand.* **16**, 327–345.

Koch, F., and Cohen, J. B. (1969). *Acta Crystallogr., Sect. B* **25**, 275–287.

Kofstad, P. (1962). *J. Phys. Chem. Solids* **23**, 1579–1586.

Kofstad, P. (1972). "Non-Stoichiometry, Diffusion and Electrical Conductivity in Binary Metal Oxides," Wiley, New York.

Kofstad, P., and Hed, A. Z. (1967). *J. Am. Ceram. Soc.* **50**, 681–682.

Kofstad, P., and Hed, A. Z. (1968). *J. Electrochem. Soc.* **115**, 102–104.

Kröger, F. A. (1964). "The Chemistry of Imperfect Crystals. North-Holland Publ., Amsterdam; Wiley, New York.

Kröger, F. A., and Vink, H. J. (1956). *Solid State Phys.* **3**, 307–435.

Kunzmann, P., and Eyring, L. (1975). *J. Solid State Chem.* **14**, 229–237.

Magnéli, A. (1953). *Acta Crystallogr.* **6**, 495–500.

Manenc, J., Vagnard, G., and Bénard, J. (1962). *C. R. Acad. Sci.* **254**, 1779.

Manes, L., and Manes-Pozzi, B. (1976). *In* "Plutonium 1975 and Other Actinides" (H. Blank and R. Lindner, eds.), pp. 145–163. North-Holland Publ., Amsterdam.

Manes, L., Sørensen, O. T., Mari, C., and Ray, I. (1980). *In* "Thermodynamics of Nuclear Materials 1979," Vol. 1, pp. 405–426. IAEA, Vienna.

Markin, T. L., and Bones, R. J. (1967). *U.K. At. Energy Auth., Res. Group, Rep.* **AERE-R-5562**.

Markin, T. L., Bones, R. J., and Gardner, E. R. (1964). *U.K. At. Energy Auth., Res. Group, Rep.* **AERE-R-4724**.

Markin, T. L., Wheeler, V. J., and Bones, R. J. (1968). *J. Inorg. Nucl. Chem.* **30**, 807–817.

Martin, R. L. (1974). *J.C.S. Dalton Trans.* 1335–1350.

Matsui, T., and Naito, K. (1975). *J. Nucl. Mater.* **56**, 327–335.

Merritt, R. R., and Hyde, B. G. (1973). *Philos. Trans. R. Soc. London, Ser. A* **274**, 627–661.

Mott, N. F., and Guerney, R. W. (1940). "Electronic Processes in Ionic Crystals." Clarendon Press, Oxford.

Mrowec, S. (1978). *Ceramurgia Int.* **4**, 47–58.

Navrotsky, A. (1974). *In* "Transition Metals" (H. F. Emeleus and D. W. A. Sharp, eds.), MTP International Review of Science, Inorganic Chemistry, Series Two, Vol. 5, pp. 29–70. Butterworth, London.

Panlener, R. J., Blumenthal, R. H., and Garnier, J. E. (1975). *J. Phys. Chem. Solids* **36**, 1213–1222.

Picard, C., and Dodé, M. (1970). *Bull. Soc. Chim. Fr.* **7**, 2486–2487.

Ray, S. P., and Cox, D. E. (1975). *J. Solid State Chem.* **15**, 333–343.

Roberts, L. E. J., and Walter, A. J. (1961). *J. Inorg. Nucl. Chem.* **22**, 213–229.

Roth, W. L. (1960). *Acta Crystallogr.* **13**, 140–149.

Sawyer, J. O., Hyde, B. G., and Eyring, L. (1965). *Bull. Soc. Chim. Fr.* **4**, 1190–1199.

Schmahl, N. G., and Hennings, D. (1969). *Z. Phys. Chem.* **63**, 111–124.

Schmitz, F., and Marajofsky, A. (1975). *In* "Thermodynamics of Nuclear Materials 1974," Vol. I, pp. 457–467. IAEA, Vienna.

Sørensen, O. T. (1972). *Proc. Int. Conf. Therm. Anal., 3rd, Davos, 1971* pp. 31–42.

Sørensen, O. T. (1975). "Rare Earth and Actinide Oxides: Thermodynamic and Electron Microscopy Studies," Risø Rep. No. 331. Risø Natl. Lab., Roskilde, Denmark.

Sørensen, O. T. (1976a). *J. Solid State Chem.* **18**, 217–233.

Sørensen, O. T. (1976b). *In* "Plutonium 1975 and Other Actinides" (H. Blank and R. Lindner, eds.), pp. 123–131. North-Holland Publ., Amsterdam.

Sørensen, O. T. (1978). *In* "Crystal Chemistry, Stoichiometry, Spinodal Decomposition, Properties of Inorganic Phases" (A. M. Alper, ed.), Phase Diagrams: Materials Science and Technology, Vol. V, pp. 75–125. Academic Press, New York.

Steele, B. C. H., and Floyd, J. M. (1971). *Proc. Br. Ceram. Soc.* **19**, 55–71.

Summerville, E., Tuenge, R. T., and Eyring, L. (1978). *J. Solid State Chem.* **24**, 21–31.

Swalin, R. A. (1972). "Thermodynamics of Solids," 2nd ed. Wiley, New York.

Tetenbaum, M., and Hunt, P. D. (1968). *J. Chem. Phys.* **49**, 4739–4744.

Thornber, M. R., Bevan, D. J. M., and Graham, J. (1968). *Acta Crystallogr., Sect. B* **24**, 1183–1190.

Tilley, R. F. D. (1972). *In* "Solid State Chemistry" (H. F. Emeleus and L. E. J. Roberts, eds.), MTP International Review of Science, Inorganic Chemistry, Series One, Vol. 10, pp. 279–313. Butterworth, London.

Turcotte, R. P., and Chikalla, T. D. (1971). *J. Inorg. Nucl. Chem.* **33**, 3749–3763.

Turcotte, R. P., and Haire, R. G. (1975). *Trans. Plutonium Symp., 4th, Baden-Baden* pp. 267–276.

Turcotte, R. P., Chikalla, T. D., and Eyring, L. (1973). *J. Inorg. Nucl. Chem.* **35**, 809–816.

Ubbelohde, A. R. (1957). *Q. Rev. Chem. Soc.* **11**, 246–272.

Ubbelohde, A. R. (1966). *J. Chim. Phys. Phys.-Chim. Biol.* **63**, 33–42.

Vallet, P., and Carel, C. (1970). *Ann. Chim. (Paris)* **5**, 246–249.

Vallett, P., and Raccah, P. (1965). *Mem. Sci. Rev. Metall.* **62**, 1–29.

Van Hook, H. J., and Keith, M. L. (1958). *Am. Mineral.* **43**, 69–83.

Vasil'eva, I. A., and Granovskaya, Z. V. (1976). *Russ. J. Phys. Chem.* **50**, 876–878.

Vasil'eva, I. A., Granovskaya, Z. V., and Gerasimov, Y. I. (1976). *Dokl. Phys. Chem.* **226**, 68–71.

Wadsley, A. D. (1955). *Rev. Pure Appl. Chem.* **5**, 165–193.

Wadsley, A. D. (1964). *In* "Non-Stoichiometric Compounds" (L. Mandelcorn, ed.), pp. 98–209. Academic Press, New York.

Watanabe, D., Terasaki, O., Jostsons, A., and Castles, J. R. (1970). *In* "The Chemistry of Extended Defects in Non-Metallic Solids" (L. Eyring and M. O'Keeffe, eds.), pp. 238–257. North-Holland Publ., Amsterdam.

Willis, B. T. M. (1964). *J. Phys. (Paris)* **25**, 431–439.

Willis, B. T. M. (1978). *Acta Crystallogr., Sect. A* **34**, 88–90.

Woodley, R. E. (1973). *J. Am. Ceram. Soc.* **56**, 116–119.

Woodley, R. E., and Adamson, M. G. (1979). *J. Nucl. Mater.* **82**, 65–75.

2

Defect Clustering in Nonstoichiometric Oxides

C. R. A. Catlow
Department of Chemistry
University College London
London, England

I. Introduction

This chapter is concerned principally with the ways in which large levels of disorder, present in grossly nonstoichiometric oxides, may be stabilized within a single phase. We aim to introduce the reader to this fascinating and diverse area of structural chemistry while pointing out,

ISBN 0-12-655280-0

however, the relevance of our structural considerations to the thermodynamic and transport properties of these phases.

Our understanding of the complex defect structures that form in heavily disordered phases has made considerable advances in recent years. This has been due first to improvements in experimental methods—principally in diffraction and microscopy techniques. Computer-based theoretical techniques have also proved of considerable value. And we draw special attention in this chapter to the role played by such theory in the elucidation of defect structures in oxides.

Our discussion starts, therefore, with a general account of the fundamentals of nonstoichiometry in oxides. The techniques used to investigate defect structure in heavily disordered materials are outlined, and particular systems are then reviewed. First we consider oxides where the nonstoichiometry is incorporated by point defects or defect aggregates. We review our present understanding of rock-salt-structured transition metal oxides, e.g., $Fe_{1-x}O$, and fluorite-structured compounds, e.g., CeO_{2-x} and UO_{2+x}. These two structural types have received the most attention to date. Next we discuss extended defect structures, which form, for example, in TiO_{2-x} and WO_{3-x}; we concentrate on the TiO_2 phase, for which the greatest detail is available.

Our review shall demonstrate that nonstoichiometric oxides may adopt two general modes of stabilizing high levels of disorder; these may be classified as *aggregation* and *elimination*. We shall see that the type of mode adopted is dependent on fundamental properties of the oxide—principally the nature of the oxygen–metal interaction potentials.

II. Nonstoichiometry and Disorder

A. "Real" Nonstoichiometry: Electronic Species and Charge-Compensating Defects

The oxides discussed in this chapter are all strongly ionic materials. In such compounds nonstoichiometry may generally be attributed to variable cation valence. FeO provides a simple example: divalent Fe^{2+} may be readily oxidized to trivalent Fe^{3+}, and a broad nonstoichiometric phase results. In contrast, lattice cation reduction from the tetravalent to the trivalent state is the basic chemical cause of the nonstoichiometry of the CeO_2 and TiO_2 phases. For our present purposes it is acceptable to adopt this simple "chemical" approach to solid-state redox reactions. Thus, employing the examples referred to above, oxidation of FeO may be said to result in distinct *localized* Fe^{3+} ions; reduction of CeO_2 creates localized

Ce^{3+} cations. It is important, however, to note that the electronic species (holes or electrons) created on oxidation or reduction of a compound need not necessarily be localized. A localized species is formed if the electron or hole interacts sufficiently strongly with the lattice for self-trapping to occur. The theory of electron localization, or "polaron" formation, is well developed (see, e.g. Stoneham, 1975; Norgett and Stoneham, 1973). In simple terms localization or self-trapping occurs if the lattice relaxation around the localized state outweighs the accompanying loss of kinetic energy. The high polarity of ionic compounds leads to large relaxation energies favoring localization. Delocalization is, however, quite possible, in which case our simple chemical approach must be modified to include the concepts of band theory. Lattice cation oxidation creates a hole in the valence band; reduction, an electron in the conduction band. The latter concepts will prove to be of use in this chapter, although generally, we shall adopt the simple chemical approach with the implicit assumption of localization of electronic states.

The electronic species created by the redox process perturbs the charge balance in a crystal. This is restored by the creation of *charge-compensating atomistic defects* on whose properties we shall concentrate in this chapter. In the majority of nonstoichiometric compounds the charge compensators are *point defects*, i.e., species confined to one site or a small group of sites. Thus, in $Fe_{1-x}O$, cation vacancies compensate the holes formed on oxidation. The *effective charge* (i.e., the charge relative to that of the species occupying the site in the perfect lattice) of the cation vacancies is negative; they thus neutralize the positively charged holes. The requirement of charge balance also allows holes to be neutralized by anion interstitials. Thus oxygen interstitials are formed on oxidation of UO_2 to give the nonstoichiometric UO_{2+x} phase.

The mode of charge compensation—vacancy or interstitial—can be related to the basic defect thermodynamics of the crystal. Thus in a NaCl-structured compound, vacancy rather than interstitial formation will occur if $g_S < g_F$, where g_S is the free energy of Schottky pair formation and g_F that for the formation of Frenkel pairs. Similar but slightly more complex relations can be derived for fluorite-structured materials, e.g., UO_2. For a discussion of the derivation of such relations we refer the reader to Lidiard (1971, 1974). Here we wish to stress the general point, namely, that there is a close connection between the nature of charge-compensating defects and the fundamental disorder of a crystal.

In reduced crystals, defects with an effective positive charge must be created to balance the negatively charged electrons. In the majority of cases, anion vacancies are formed. CeO_{2-x} provides a good example: a high oxygen vacancy concentration can be accommodated within this fluorite-structured phase. Cation interstitials provide alternative positively charged defects. There is, however, no conclusive proof for the occurrence of these defects as the dominant charge compensators in any electron-excess nonstoichiometric compounds (although their formation in near-stoichiometric TiO_{2-x} has been suggested by some workers; see, e.g., Baumard *et al*, 1977). Simple thermodynamic analyses of the type referred to above show that cation interstitial compensation will occur only if cation Frenkel pairs provide the predominant intrinsic disorder of the crystal. Cation Frenkel energies are, however, generally high for ionic oxides.

Point defect structures predominate in most nonstoichiometric phases. However, in certain compounds the excess charge of the electronic species is compensated by *extended defects*, i.e. defects with one- or two-dimensional periodicity. The best known examples occur in the TiO_{2-x} and WO_{3-x} phases (Tilley, 1972). Reduction of these materials leads to the formation of planar defects known as *shear planes* which have been detected in electron microscopy studies and which are characterized by a change in the mode of linking of the oxygen polyhedra within the plane. Thus the formation of shear planes in the rutile-structured TiO_2 involves a change from edge to face sharing in the TiO_6 octahedra on which the rutile structure is based; in the corner-linked-structured WO_3, corner sharing is replaced by edge sharing of octahedra in the shear plane. This process decreases the oxygen-to-metal (O/M) ratio in the planar defect which is, as a consequence, positively charged; the shear-plane charge can thus compensate the charges of the excess electrons created on reduction.

Further insight into these extended defects is given by the schematic illustration of shear-plane formation in Fig. 1a and b. The change in the

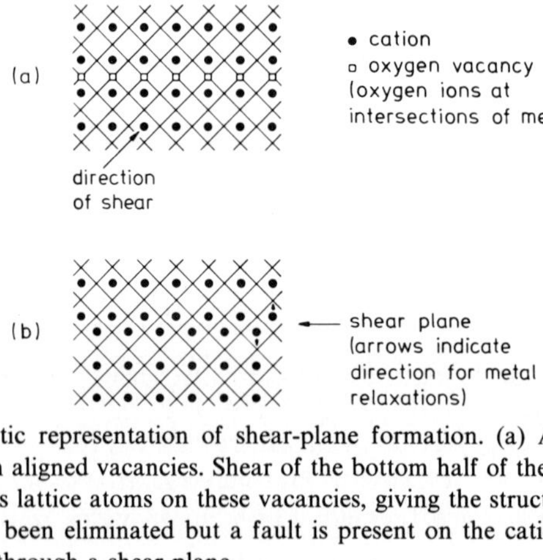

Fig. 1. Schematic representation of shear-plane formation. (a) A hypothetical reduced oxide structure with aligned vacancies. Shear of the bottom half of the crystal in the direction shown superimposes lattice atoms on these vacancies, giving the structure shown in (b) where the vacancies have been eliminated but a fault is present on the cation sublattice. The fault shown is a section through a shear plane.

stacking of the metal sublattice is clearly illustrated, as is the coherence of the oxygen sublattice in the planar fault. The diagrams also show how shear planes can be related to corresponding point defect structures. Indeed we note how formation of the plane can be described in terms of the *elimination of anion vacancies by a process of crystallographic shear*—an important concept to which we shall return later.

Shear-plane formation is a relatively uncommon phenomenon. To date it has been observed only in the nonstoichiometric oxides TiO_{2-x}, VO_{2-x} WO_{3-x} and MoO_{3-x}; although analogous faults giving rise to block structures are found in nonstoichiometric doped Nb_2O_5 phases (Tilley, 1975). The question of the factors controlling the relative stability of shear extended and point defect structures is one of central importance in solid-state chemistry and will be discussed in Section VA.

B. "Simulated" Nonstoichiometry

The previous section has shown how atomistic defects—either point or extended—are created so as to neutralize the charges of the electronic species present in nonstoichiometric crystals. The charge balance may, however, be perturbed by *aliovalent dopant* ions as well as by holes and electrons. An aliovalent dopant is one whose charge differs from that of the host lattice ions, and thus substitution in the host by the dopant ions creates a species with a nonzero effective charge. A charge-compensating defect population is therefore induced exactly as in nonstoichiometric systems.

The relationship between aliovalent doped phases and nonstoichiometric compounds is best illustrated by a comparison of Y^{3+}-doped CaF_2 with UO_{2+x}. High concentrations of Y^{3+} (and other trivalent ions, e.g., the rare earths), may enter the fluorite host at the cation site. The effective positive charge of the dopant ion at this site is compensated by the creation of F^- interstitials. In an analogous fashion, the positive charges of the holes (in effect U^{5+} ions) created on oxidation of UO_{2+x} are compensated by oxygen interstitials. Indeed, the defect structures of Y^{3+}-doped CaF_2 and UO_{2+x} are remarkably similar. And it is for this reason that the doped phase may be said to *simulate* the corresponding nonstoichiometric system.

Other examples of simulated nonstoichiometry will be discussed later in this chapter. The resemblance between real and simulated systems is exact as far as charge balance is concerned. Differences may, however, exist between the two types of defective phases as a result of kinetic factors. Electronic species are mobile. In contrast, dopant ions may have very low mobilities. Thus defect aggregation can be controlled by kinetics in the simulated phase, while thermodynamic control is effective in the corresponding nonstoichiometric compound.

C. Defect Clustering

This topic will be the main theme of this chapter. The origin of aggregation in both real and simulated nonstoichiometric compounds is simple: the favorable Coulomb interactions cause clustering of the electronic species (or dopant ion) and its charge compensator. There is a strong similarity between this clustering and that which occurs between oppositely charged ions in electrolyte solutions. Indeed, the theory of defect interactions in solids has borrowed much of the formalism of electrolyte solution theory. Successful theories of the thermodynamic and transport properties of dilute defect solutions were developed by Lidiard (1957), the models being based on the Debye–Hückel theory with the formation of simple pair clusters. Pair clusters have been shown to be present in defect solutions. Examples are given in Fig. 2; the structures were deduced from electron paramagnetic resonance (EPR) studies. In general, however, simple clusters are replaced by more complex structures at higher defect concentrations (pair clusters are not in general the dominant species at concentrations greater than 1 mole %). The higher levels of disorder in the more grossly nonstoichiometric phases, e.g., $Fe_{1-x}O$ and UO_{2+x}, are stabilized by larger aggregates, often of low symmetry.

Shear-plane formation can also be considered as a special mode of defect aggregation—a point that is evident from Fig. 1a and b where the shear plane is seen as a stabilized vacancy aggregate. The best description is perhaps in terms of aggregation stabilized by *elimination*. Indeed, we believe that, as far as thermodynamics is concerned, this latter description is the most useful way of considering the shear plane. Our discussion of

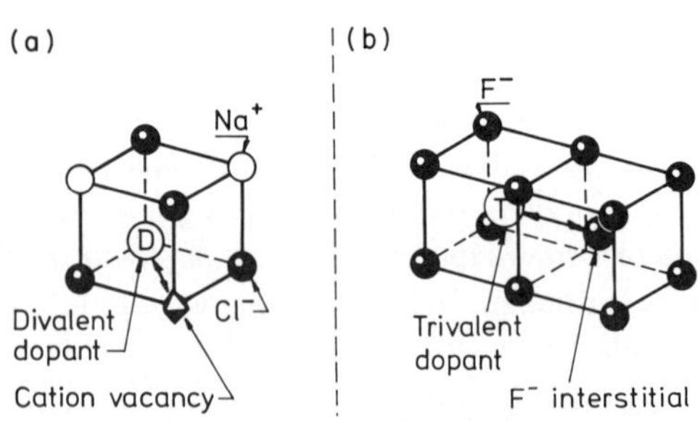

Fig. 2. Simple pair clusters. (a) Dopant vacancy cluster in an alkali halide crystal (NaCl); (b) dopant interstitial cluster in alkaline earth fluoride (CaF_2).

extended defect formation in Section V will therefore concentrate on this aspect of these systems.

Our main concern in the remainder of this chapter is thus with the nature of the complex structures that form in defective oxides. First, however, it is necessary to review the techniques, both experimental and theoretical, for investigating defect cluster structures, including point and extended defect structures.

III. Determination and Characterization of the Defect Structure of Grossly Defective Phases

Three main techniques have been applied in the determination of cluster structures: first, *diffraction* (both x-ray and neutron), second, *high-resolution electron microscopy* (HREM), and third, *computer modeling*. The first two are discussed elsewhere in this book, and our account is therefore brief. More details are given in the case of computer modeling techniques which, as noted, are gaining increasing importance in this field.

A. Diffraction Techniques

For a detailed discussion we refer the reader to Chapter 8 of this volume. Here we wish to stress the following special points, which are of particular importance when diffraction is used to elucidate defect cluster structure:

(a) *Bragg scattering* yields information only on *average* unit cell contents and provides no *direct* information on cluster structure. Bragg intensities are generally analyzed in terms of the *occupation number* of ions at lattice and interstitial sites. However, occupation numbers frequently give clues as to cluster structure, as will be seen in our discussion of the $Fe_{1-x}O$ and UO_{2+x} phases in Sections IV.A.1. and IV.B.1.

(b) *Diffuse scattering* observed in disordered phases can in principle give direct information on individual cluster structures. In practice it has not proved possible to analyze data unambiguously in terms of detailed cluster models for most nonstoichiometric phases. In our discussion of fluorite materials we shall, however, see illustrations of the value of diffuse scattering data.

(c) *Superlattice reflections*, as well as providing knowledge of the ways in which defect or defect aggregates are arranged within an ordered supercell, may also prove of value in the elucidation of the structures of the component individual clusters. The study of $Fe_{1-x}O$ (Section IV.A.1) will illustrate this point.

(d) *Neutron diffraction* has played an especially important role in the study of defective compounds. The advantages of neutron over x-ray methods in such work include the relative ease with which high-temperature studies can be performed, the greater intensity of scattering from light atoms, and the possibility of obtaining more information from powder studies by using the profile refinement technique. The role of neutron studies will receive further attention in our discussion of the defect structure of $Fe_{1-x}O$ and UO_{2+x} in Sections IV.A.1 and IV.B.1.

B. Electron Microscopy

High resolution electron microscopy (HREM), a detailed discussion of which is presented in Chapter 7, is playing an increasingly important role in the study of nonstoichiometry. The main application discussed in this chapter involves the identification and characterization of extended defects, e.g., shear planes in TiO_{2-x}, whose properties are discussed in detail in Section V. A second application concerns the study of ordered arrays of extended defects or point defect clusters, examples of which are considered in Section V.C. If the ordering involves the formation of a defect superlattice, diffraction can in principle yield full information on the structure of the defects and on the way in which they are arranged in the supercell. Analysis of the diffraction data on the resulting large unit cells is, however, often prohibitively difficult. Moreover, if the ordering involves only small groups of extended defects (as in near-stroichiometric TiO_{2-x}) or defect clusters, then only microscopy can yield information on the structure of the groups. A related and fascinating application of the technique concerns the study of faults that develop in the ordered arrays of extended defects. For a discussion of this topic we refer the reader to the review of Anderson (1977).

C. Computer Modeling

Modeling of the disorder in strongly ionic materials provides one of the best examples of the successful use of computer simulation studies. Indeed, this area of computational physics is rapidly becoming an aid that is as valuable in the interpretation of experimental work on disordered ionic solids as molecular dynamic and Monte Carlo simulations are in investigations of liquids. The reason for this arises from the highly complex nature of the system investigated. The power of modern computers is required to model these systems; and guidance as to feasible models is generally necessary if a satisfactory understanding of the experimental data is to be obtained.

Modern computational methods aim at calculating the energetics and structures of point defects and defect aggregates. Extension to the study of two-dimensional defects is an area of active research at present (see Tasker, 1981; Mackrodt and Stewart, 1977), although few results have been reported to date. The calculations are essentially static, i.e., they neglect the effect of the vibrational properties of the crystals. For the application to structural problems discussed in this chapter, this appears to be an unimportant restriction.

The most widely used basis for the calculation of defect structures and energetics is the early work of Mott and Littleton (1938). In essence this method divides the crystal into two regions: an inner one (region I) surrounding the defect, in which all ions are explicitly treated (i.e., their coordinates are adjusted until they are at equilibrium); and an outer region (region II) where a quasicontinuum approach is used in which the total polarization per unit cell is calculated using the measured static dielectric constant and is then divided between the component ions on the basis of their relative displacement polarizabilities.

For a reliable calculation, the inner, explicitly relaxed region must include typically 100 ions (Catlow and Norgett, 1973; Catlow, 1977a). Minimization of the energy of the crystal with respect to the coordinates of all these species, even when the symmetry of the defect and the crystal is exploited, is a major computational exercise. Such calculations have therefore only become feasible because of the development of highly efficient numerical methods for handling the energy minimization of this inner region. The most successful procedures employ Newton–Raphson minimization techniques (Norgett and Fletcher, 1971) in which the second derivatives of the energy of the system with respect to particle coordinates are calculated; such methods are found to converge far more rapidly than steepest-descent or conjugate gradient techniques in which information on first derivatives alone is used to guide the direction of the minimization. Further savings in computer time are effected by a procedure due to Fletcher and Powell (1963) in which the inverse of the matrix of second derivatives is updated in each iteration without recalculation.

A generalized Mott–Littleton procedure, employing the minimization techniques outlined above, has been developed and programmed by Norgett (1974), and is available as the HADES program recently adapted to treat crystal structures of any symmetry by Catlow *et al.* (1981). Calculations using the program for defect energies in alkali halides (Catlow *et al.*, 1979) and alkaline earth fluorides (Catlow *et al.*, 1977a) have proved remarkably successful for these systems; quantitative agreement with experimentally determined defect energies can be obtained from the theoretical calculations. Indeed, these calculations demonstrate that computer simulation methods are essentially exact and that the limitations in the accuracy of the calculated energies arise from limitations in the accuracy of the interatomic potentials used in the simulation of the inner region of the crystals surrounding the defect. For fully ionic crystals such as alkali halides and alkaline earth fluorides reliable interatomic potentials can be developed. Greater difficulties are encountered with oxides where the nature of the bonding is less certain. In this chapter, however, we shall be concerned with a more qualitative application of the techniques. We shall show how the computational studies have played a major role in developing models for defect aggregation in a range of oxide systems, and how they have provided valuable guidance as to the factors that stabilize extended defects in oxides where such structures form. Indeed we shall show in the following section how the *concerted* use of theory and experiment has led to the unraveling of the complex cluster structures of several nonstoichiometric phases.

IV. Point Defect Aggregates

A. Rock-Salt-Structured Transition Metal Oxides

1. Insulating Oxides

In this section we discuss the four insulating rock-salt oxides, MnO, FeO, CoO, and NiO. The divalent oxides of the earlier elements (Ti and V) of the first transition series have a metallic character and will be considered later. Oxidation of all four insulating oxides leads to significant deviations from stoichiometry. For $Mn_{1-x}O$ and $Fe_{1-x}O$ the deviations are large (x may attain 0.15 in $Fe_{1-x}O$). In $Ni_{1-x}O$ and $Co_{1-x}O$, the composition width of the rock-salt-structured phase is much smaller (x is 10^{-4}–10^{-2} at 1000°C in air), although in the Co–O system there is a well-defined spinel phase, Co_3O_4. The explanation of the difference in the composition width of the phases is a challenging problem to which we shall return later.

In the case of MnO and FeO, sufficiently high levels of disorder can be introduced to permit direct determination of the defect structure using diffraction techniques. Such studies (Roth, 1960; Koch and Cohen, 1969; Cheetham *et al.*, 1971b) confirm a defect structure based on the compensation by cation vacancies for the effective charge of the oxidized lattice cations (Section II.A). This would be predicted from the known dominance of Schottky disorder in these crystals. Diffraction studies, however, reveal a more complex defect structure than the simple model suggests, as will be seen below. For $Ni_{1-x}O$ and $Co_{1-x}O$ direct structural investigations are not possible. Transport studies have, however, confirmed that the defect structure for the oxidized phase is again based on cation vacancies.

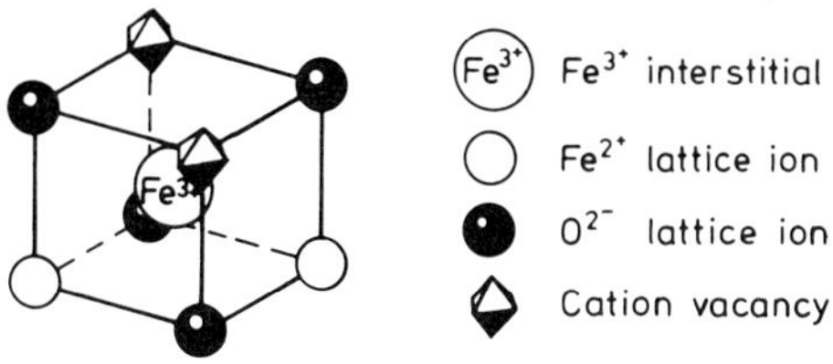

Fig. 3. Interstitial–vacancy cluster in $Fe_{1-x}O$ proposed by Roth (1960).

$Fe_{1-x}O$ shows the greatest deviation from stoichiometry; more details are available therefore on the cluster structures in this oxide than for the other three materials. As noted, several diffraction studies have been reported. The earliest (Roth, 1960) used x-ray techniques and powder samples. The results immediately posed a major problem which has only recently been solved. For in addition to the expected cation vacancy population, Roth found that satisfactory refinements of the powder diffraction data could only be obtained for defect models that included metal interstitials; the results suggested an interstitial concentration roughly half that of the vacancies. This is a most surprising discovery. Simple considerations of the type discussed in Section II.A strongly suggest that one type of defect alone should dominate in a nonstoichiometric phase; and the coexistence of vacancies and interstitials seems to be extremely unlikely. To rationalize this result, Roth proposed that the interstitials were present in a cluster, subsequently known as the Roth cluster, which is illustrated in Fig. 3.

A much more detailed diffraction study using single crystals was subsequently reported by Koch and Cohen (1969). This work confirmed the qualitative feature of Roth's work: Vacancies and interstitials clearly coexist in the $Fe_{1-x}O$ phase. More accurate occupation numbers for the two types of cationic defects were, however, obtained; the refinements gave a vacancy-to-interstitial ratio of three—higher than Roth's value of two—suggesting that the Roth cluster is an oversimplified model of the defect structure of $Fe_{1-x}O$.

In addition, Koch and Cohen (1969) observed superlattice reflections as well as rock-salt Bragg peaks. The defects or defect clusters are evidently ordered. Furthermore, from analysis of the intensities of the superlattice peaks information on the details of the defect structure can be obtained, as was discussed in Section III.A.

Analyses of this type showed that vacancy–interstitial clusters were present in $Fe_{1-x}O$. The data also suggested a high-symmetry structure for the complex. A large high-symmetry complex, with a vacancy–interstitial ratio (R_{VI}) of $\sim$3 was therefore proposed; Koch and Cohen's model is illustrated in Fig. 26 of Chapter 1.

The relative populations of vacancy and interstitial sites, summarized by the ratio R_{VI}, are clearly vital parameters in the elucidation of cluster structure. High accuracy is therefore essential for the defect occupation numbers. Improved accuracy in the measurement of R_{VI} was obtained in the neutron diffraction study of Cheetham *et al.* (1971b). This work also allowed nonstoichiometric compounds to be studied at high temperatures (1400°C) at which $Fe_{1-x}O$ is in thermodynamic equilibrium. In the x-ray work it had been necessary to use quenched samples—a procedure that always leads to doubts as to whether the high-temperature phase has been frozen in during the quench. The main conclusion of the work of Cheetham *et al.* (1971b) was that R_{VI} *varied with stoichiometry*. At low values of x in $Fe_{1-x}O$, i.e., about $x = 0.03$, the ratio is measured as ~4. It falls to ~3 as x is raised to ~0.1. These results are clearly compatible with the Koch–Cohen cluster in the more grossly cation-deficient regions of the phase; a different cluster structure is, however, suggested for nearer-stoichiometric oxides.

The one certain feature emerging from the diffraction work is therefore that extensive and complex modes of defect aggregation occur in these oxides. The experimental results, however, pose two major problems, namely, (i) how does defect aggregation stabilize interstitials, and (ii) how does the cluster structure, as reflected by the ratio R_{VI}, change with stoichiometry. We may also ask more general questions. These concern first a point raised in Section I, namely, how such large levels of disorder are stabilized within a single crystalline phase. A second general problem concerns the relationship between the structure of the nonstoichiometric phase and that of the ordered spinel-structured Fe_3O_4 phase.

All these problems were considered in Catlow and Fender's (1975) detailed study of defect aggregation in $Fe_{1-x}O$, which used the computational techniques outlined in Section IIIC. Interstitial stabilization was the first problem to be examined. A series of calculations was performed designed to test the hypothesis that the interstitials are stabilized essentially by the Coulomb interactions between vacancies and the oppositely charged interstitials. Thus vacancies were introduced into the lattice sites surrounding a tetrahedral Fe^{3+} interstitial (the relevance of the charge of the interstitial will be discussed later). Octahedral lattice Fe^{3+} ions (i.e., holes) were distributed on the surrounding octahedral lattice sites. The binding energy of the resulting clusters was then calculated with respect to component isolated vacancies and holes—the defects into which the clusters will dissociate at infinite dilution.

These calculations have shown that, when all four of the lattice sites surrounding the interstitial are vacant, the cluster becomes bound. The structure of this aggregate, known as the 4:1 cluster, is illustrated in Fig.

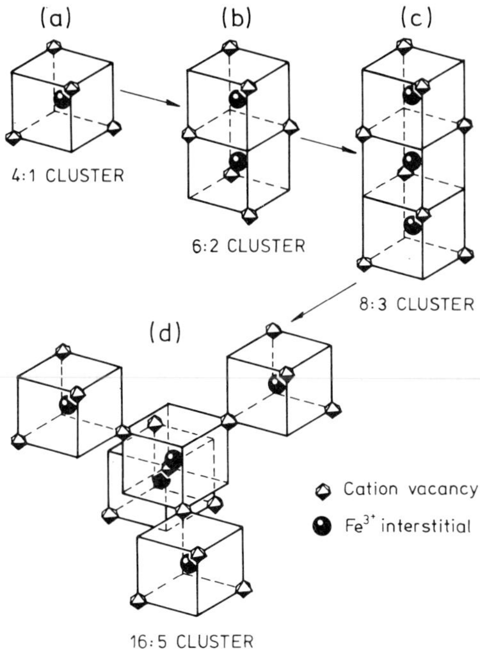

Fig. 4. Insterstitial–vacancy clusters and clustering scheme in $Fe_{1-x}O$ as proposed by Catlow and Fender (1975).

4a. The first problem raised by diffraction studies on $Fe_{1-x}O$ has therefore been solved: *interstitials are stabilized by Coulomb interactions with a surrounding vacancy aggregate*. Conversely, vacancy aggregation—an unusual process involving the clustering of defects of like charge— is made possible by interstitial formation, as the favorable Coulomb interactions with the interstitial stabilize the vacancy aggregate. The binding energy of the 4 : 1 clusters is high (~2 eV per defect). The calculations thus identify this cluster as the basic unit of the defect structure of $Fe_{1-x}O$.

How can further aggregation of 4 : 1 clusters occur? To answer this question we note that the creation of interstitials in an environment of four vacancies is highly favorable energetically. Thus we should look for a mode of aggregation that increases the number of such interstitials, that is, one that lowers the ratio of vacancies to interstitials. This can be achieved by the *sharing of corners or edges between different* 4 : 1 tetrahedral units (face sharing is not compatible with the NaCl structure). The computational calculations found that *small* corner-shared clusters did not show appreciably enhanced stability. The formation of small edge-shared aggregates, however, led to substantial increases in vacancy-binding energies. The

TABLE I

Binding Energies for Clusters in $Fe_{1-x}O$[a]

Cluster	Binding energy per vacancy (eV)
4 : 1	1.98
6 : 2	2.42
8 : 3	2.52
Koch–Cohen	2.1
16 : 5	2.38

[a]After Catlow and Fender (1975).

simplest of such aggregates comprises two 4 : 1 clusters sharing a common edge, giving the 6 : 2 cluster illustrated in Fig. 4b. The calculations suggested that, in addition, larger 8 : 3 clusters would form (Fig. 4c); results are summarized in Table I. Further cluster growth by edge sharing does not seem to be possible. The calculations showed that the vacancy-binding energies in larger edge-shared clusters were not appreciably greater than those in the 4 : 1 clusters—an effect we attribute to the increasing importance of repulsive interactions between interstitial Fe^{3+} ions.

The initial mode of aggregation of 4 : 1 clusters appears therefore to involve the formation of small edge-shared aggregates—principally 6 : 2 and 8 : 3 clusters. For further growth, however, a different cluster morphology must be adopted. The computer simulation studies showed that for large aggregates corner sharing could become favored. Two types of corner-sharing clusters were considered. First, the Koch–Cohen cluster in which a central *vacancy* is shared by four vacancy tetrahedra, and second, the 16 : 5 cluster where a central *4 : 1 cluster* shares each of its corners with a second 4 : 1 cluster (Fig. 4d). Both clusters show vacancy binding energies competitive with those of the edge-shared aggregates, as will be noted from Table I; that of the 16 : 5 aggregate is, however, larger, suggesting that this aggregate will form in preference to the Koch–Cohen cluster. It is of significance to note therefore that the *16 : 5 cluster is an element of the inverse spinel structure* adopted by Fe_3O_4 and could indeed be considered a nucleus of this ordered structure.

Initial aggregation by edge sharing is thus replaced by the formation of larger corner-shared clusters with possible spinel nucleation. These models are summarized in Fig. 4. They are in accordance with and indeed explain the experimental data on $Fe_{1-x}O$. In particular the observation by Cheetham *et al.* (1971b) of a reduction with increasing x in the ratio R_{VI} from 4 to 3 accords with the formation of 4 : 1 clusters at low deviations from stoichiometry, and with further aggregation into 6 : 2 and 8 : 3 clusters as the vacancy concentration is raised. The formation of 16 : 5 spinel nuclei is obviously compatible with the ultimate precipitation of the ordered spinel phase. Further support for the models has been claimed recently by Battle and Cheetham (1979), who find that the magnetic structure of $Fe_{1-x}O$, as revealed by polarized neutron-scattering studies is consistent with the proposed cluster structures.

A detailed picture of defect clustering in $Fe_{1-x}O$ has therefore emerged from a combination of theory and experimental investigations. Do such structures, however, form in the other insulating nonstoichiometric oxides? And if so, how can we explain the marked differences, referred to earlier, between the phase diagrams of the four oxides?

The dominant factor determining the cluster morphologies discussed above is the Coulomb interaction between oppositely charged defects. Thus we might expect similar cluster structures in the four oxides, as all of them have similar lattice parameters and ionicities. Calculations confirmed this prediction. In all four materials similar cluster-binding energies were obtained. One crucial factor is, however, influenced by the nature of the metal cation. This is the charge of the interstitial cation which in the computational studies discussed above was assumed to be 3. Calculations for the 4 : 1 cluster with divalent interstitials result, however, in similar binding energies. The choice of trivalent interstitials for the case of FeO can still, however, be justified, as the higher ligand field energies of Fe^{2+} favor the retention of these ions at the octahedral lattice sites. From similar arguments, based on ligand field terms, we conclude that in CoO and MnO the *divalent* ions will be favored at the interstitial sites. Calculations (Catlow *et al.*, 1977b) showed, however, that edge-shared clusters containing divalent interstitials were not bound appreciably more strongly than the 4 : 1 cluster, but increased binding was calculated for the 16 : 5 aggregates. This suggests that, when divalent interstitials are present, cluster growth occurs *directly* from the 4 : 1 to the large corner-shared clusters, and edge-shared aggregates are not formed. An explanation of the smaller stoichiometry range in MnO and CoO may now be proposed. In these oxides extensive disorder cannot be stabilized in the edge-shared clusters that form in $Fe_{1-x}O$. Thus, when the defect concentration is sufficiently high for aggregation to occur beyond the 4 : 1 clusters, the 16 : 5 spinel-like clusters form with precipitation of the spinel phase. A narrower composition range is observed, as large vacancy concentrations cannot be accommodated within the nonstoichiometric phase by clusters that are morphologically different from the related ordered phase.

We should note that the models discussed above differ from those commonly assumed in treatments of the thermodynamic and transport properties of these oxides (Kofstad, 1972). Analyses of these data have generally considered only the association between single vacancies and holes, i.e., vacancy aggregation has been ignored. The work discussed in this section shows that such models are inadequate for the more grossly nonstoichiometric regions of the phase. Moreover, Catlow *et al.* (1977b) have shown that the binding energies of vacancy aggregates are considerably greater than those of simple vacancy–hole aggregates; the difference is such as to suggest that the simple clusters have a very limited composition range, i.e., when the vacancy concentration reaches a value sufficient for the defects to aggregate, this occurs directly to the 4 : 1 cluster without the formation of significant concentrations of single vacancy–hole complexes. This argument has been given quantitative demonstration by a mass-action treatment using the calculated cluster binding energies (Catlow *et al.*, 1977b). The conclusion clearly has major consequences for the analysis of diffusion data as well as thermodynamic properties.

Defect clustering in insulating transition metal oxides is evidently therefore a more complex process than the simplicity of the structure of these materials suggests. A similar degree of complexity is observed in the metallic divalent oxides discussed in the next section.

2. *Metallic Oxides—TiO and VO*

The rock-salt-structured oxides TiO and VO have good metallic conductivities. These interesting compounds also show high ranges of composition. Thus, in Ti_xO, x varies from 0.65 to 1.25, while in V_xO, the composition of the phase ranges from $x = 0.85$ to $x = 1.25$. In addition, high levels of disorder are present even in the stoichiometric oxides. Thus in TiO, 20% of both metal and oxygen sites are vacant. The reason for these remarkably high levels of disorder are uncertain, but the most plausible explanation, discussed by Goodenough (1974), relates the defect and electronic properties: the creation of defects reduces the lattice parameter, thus increasing the intercationic overlap of the 3d orbitals and hence lowering the conduction band energies.

Several studies have demonstrated the existence of extensive short-range order of the vacancies in these oxides. Details of the complex defect structures have not, however, been elucidated. The most detailed investigations were reported by Morinaga and Cohen (1976). In both compounds they found evidence for metal interstitials, although the concentrations were less than those found in $Fe_{1-x}O$. The cation-deficient phases may therefore contain clusters similar to those discussed in the previous section for the insulating oxides. The cluster structure in these oxides is, however, far less certain than in the case of the insulating materials.

B. Fluorite-Structured Oxides

The fluorite structure, adopted by several oxides (e.g., UO_2, PrO_2, ThO_2, and CeO_2) shows exceptional adaptability, tolerating high levels of disorder which may be introduced either by dopants or by redox processes. The structure of the fluorite lattice is illustrated in Fig. 5. We draw attention to the large interstitial sites whose presence results in interstitial disorder dominating in this lattice. And since the lower Madelung term results in the binding of the anions to their lattice sites being less strong than that of the cations, anion Frenkel pairs are the dominant intrinsic defects in fluorite compounds. Aliovalent dopants or redox reactions thus lead to the creation of extrinsic concentrations of anionic defects: vacancies are introduced by low-valence cation dopants or by reduction; interstitials are present in the anion-excess systems created on oxidation or by doping with high-valence cations.

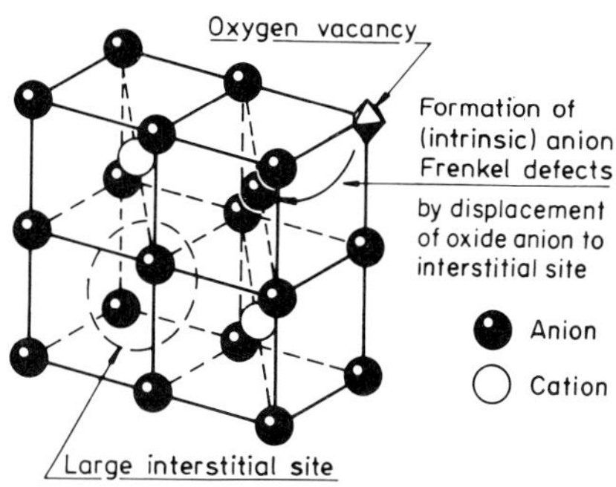

Fig. 5. The fluorite structure.

1. Anion-Excess Oxides

The only widely studied example here is provided by the nonstoichiometric oxide UO_{2+x}. Anion-excess phases can also be prepared by doping fluorite oxides with pentavalent ions, e.g., Nb^{5+}; these systems have, however, received relatively little attention.

Willis (1963, 1964) reported an investigation of the defect structure of UO_{2+x} using Bragg neutron-scattering techniques; he also studied the corresponding ordered U_4O_9 phase. The results of his study on $UO_{2.13}$ were at first sight puzzling. Interstitials were detected not at the body centers of the cubic interstitial sites marked in Fig. 5 but at sites considerably displaced from this symmetric position. Two types of interstitials were found—one displaced along the $\langle 110 \rangle$ axis from the center of the intersti-

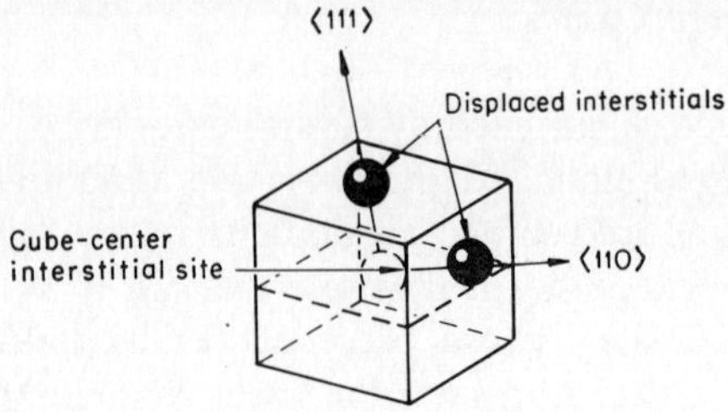

Fig. 6. ⟨110⟩ and ⟨111⟩ displaced interstitials in the fluorite lattice.

tial site and the second along the ⟨111⟩ axis; the displacements are illustrated in Fig. 6. In addition vacancies were detected at the lattice anion sites; the concentrations of the vacancies and of the two types of interstitial were roughly equal.

These results, obtained from a crystal with a high interstitial concentration, contrast with those available from EPR studies of other fluorite-structured systems with low levels of interstitial disorder. (EPR studies of the defect structure of UO_2 are not possible, owing to the complex magnetic structure of the material.) Rare earth doped alkaline earth fluorides (in which the positive effective charge of the dopant is compensated by F^- interstitials) have been extensively studied by EPR and ENDOR.† The work of Baker *et al.* (1968) has conclusively shown that at low interstitial concentrations (1 mole % or less) the F^- ions occupy the symmetric body center interstitial site. However, neutron diffraction studies (Cheetham *et al.*, 1971a) on more heavily doped crystals containing 5 mole % or higher interstitial concentrations have revealed a defect structure for the rare earth doped fluoride close to that found by Willis for UO_{2+x}.

The latter observations suggest that the low-symmetry defect structure is a general feature of anion-excess fluorites; moreover it leads to an explanation of the origin of this unusual structure. Distorted interstitials appear to be present only at higher defect concentrations; defect clustering may therefore be invoked to explain their formation. Willis proposed the interstitial aggregate shown in Fig. 7a, known as the 2 : 2 : 2 cluster. A similar structure was suggested by Cheetham *et al.* (1971a) to explain the neutron data on 5 mole % Y^{3+}-doped CaF_2. For higher interstitial concentrations more complex clusters were needed to account for the vacancy and interstital occupation numbers obtained from Bragg scattering studies; formation of the 4 : 3 : 2 cluster, illustrated in Fig. 7b, was proposed. Willis's data on U_4O_9 led to defect occupation numbers compatible with a structure based on an ordered array of 4 : 3 : 2 clusters, although the details of the

†The ENDOR technique (electron nuclear double resonance) leads to greatly enhanced resolution compared with conventional EPR, and allows complex hyperfine structure to be resolved.

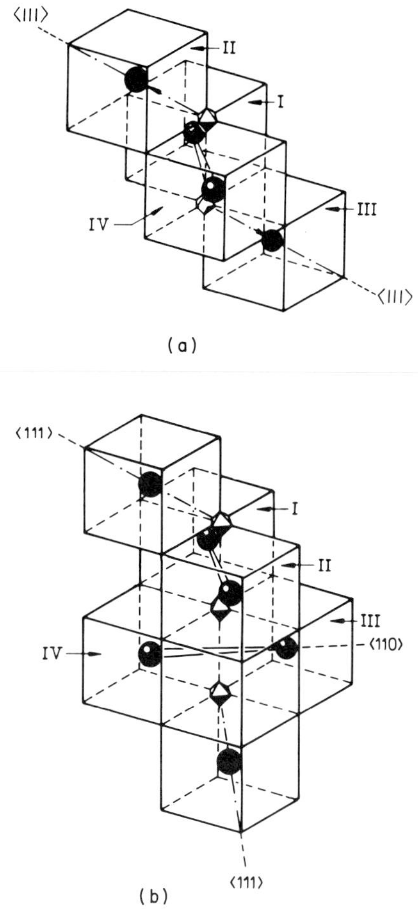

Fig. 7. (a) 2 : 2 : 2 clusters in UO_{2+x}. ⟨110⟩ interstitials in cubes I and IV. ⟨111⟩ interstitials in cubes II and III. (b) 4 : 3 : 2 clusters in UO_{2+x}. ⟨110⟩ interstitials in cubes I–IV. Remaining two cubes contain ⟨111⟩ interstitials. ●, oxygen interstitial; ○, oxygen vacancy.

structure of this ordered compound are far from certain. Diffuse scattering studies have also been reported for doped CaF_2 (Childs *et al.*, 1972). No conclusive results emerge, but analyses of the data are consistent with the formation of 2 : 2 : 2 and related clusters. In addition, the data suggest some association of the dopants with the interstitial cluster. This significant point will be discussed further in the light of the models we have proposed to explain the formation of these unusual clusters.

The problem of the stability of the clusters discussed above has again been illuminated by the computational techniques described in Section

IIIC. Indeed, the theoretical studies have led to a simple mechanism for the formation of the 2 : 2 : 2 and related aggregates (Catlow, 1973). The 2 : 2 : 2 cluster is essentially an *interstitial* dimer stabilized by a coupled interstitial–lattice relaxation mechanism. These models require the species of effective positive charge (the dopant ions or the holes) to be localized at the lattice cation sites nearest the interstitial cluster. The two compensating interstitials then move inward along the $\langle 110 \rangle$ axis as shown in Fig. 7a; the driving force for this relaxation is the electrostatic interaction with the dopant ions. This inward $\langle 110 \rangle$ interstitial relaxation is accompanied by $\langle 111 \rangle$ relaxation of the neighboring lattice ions toward interstitial sites. Calculations showed that such relaxations were essential for the stability of the cluster, as they remove the stress acting on these lattice ions due to the presence of neighboring interstitials. Similar models can be proposed for the 4 : 3 : 2 cluster which is essentially an interstitial trimer stabilized by the same type of coupled interstitial–lattice relaxation mechanism. We should note that recent studies (see, e.g., Gettman and Greiss, 1978) have suggested that, at higher defect concentrations, larger dopant aggregates may form, although the extent to which they are compatible with structural data at lower concentrations has not yet been demonstrated.

The combination of neutron scattering and computer simulation has therefore once more revealed complex cluster mechanisms for stabilizing high levels of disorder. Similar complexity is found in the anion-deficient oxides discussed in the next section, although here there is less understanding of the factors determining the nature of the defect aggregation.

2. *Anion-Deficient Fluorite Oxides*

Reduction of the fluorite-structured oxides UO_2, PuO_2, PrO_2, and CeO_2 leads to anion-deficient phases of which CeO_{2-x} is the most widely studied example. Analogous phases may be created by doping these and other oxides (e.g., ThO_2 and HfO_2) with trivalent or divalent ion cations, e.g., Y^{3+} and Ca^{2+}. An interesting and important example of the latter system is provided by ZrO_2, which when pure has a monoclinic lattice at low temperatures, but is stabilized in the cubic fluorite structure by the introduction of Ca^{2+} or Y^{3+}.

Interest in anion-deficient fluorites has been focused on their transport and thermodynamic properties. These defective compounds show very high anion mobilities—sufficient in some cases for the materials to be classified as *superionic conductors*. Such materials have electrical conductivities comparable with those observed in molten salts; they have attracted a great deal of attention in recent years (Van Gool, 1972; Mahon and Roth, 1976). Thermodynamic studies have tended to concentrate on the oxides UO_{2+x}

and CeO_{2-x} and the mixed oxides of which $(U, Pu)O_{2-x}$ is important owing to its use as a reactor fuel. Structural studies are, however, reported. We review these below concentrating on those aspects that are most relevant to the transport and thermodynamics of the materials discussed subsequently.

a. Structural Data. The presence of anion vacancies in both reduced and doped fluorite oxides has been confirmed by both x-ray and neutron studies (Faber *et al.*, 1976). Diffraction work has also shown that a series of ordered phases is formed in several phases. CeO_{2-x} provides a good example illustrating the complexity of these systems; the phase diagram is shown in Fig. 6 of Chapter 1. Similar complexity is found in PrO_{2-x} where there exists a distinct homologous series of oxides of general composition Pr_nO_{2n-2}. Few data are available on UO_{2-x} which is only stable at high temperatures. Ordered phases are also found in doped systems; an example is provided by the doped ZrO_2 phases discussed later in this section.

Unfortunately, despite the obviously fascinating structural chemistry of these systems, many of the structures of ordered anion-deficient systems remain unsolved. Problems arise from two sources: first, the difficulties in obtaining single crystals, and second, the low scattering power for x-rays of the oxygen relative to the metal sublattice; details of the oxygen sublattice structure may therefore be inaccessible. Both problems can in principle be overcome in neutron studies in which the metal sublattice no longer dominates the scattering and in which detailed structural information may be obtained from powders using the profile refinement technique. The value of powder neutron studies has indeed been shown by the study of Steele and Fender (1974) on Y^{3+}-doped ZrO_2 and ThO_2, where detailed information on the lattice relaxation around vacancies was obtained.

Despite the lack of detailed structures for many of the reduced fluorite oxides, it has been established that their defect structure is, in several cases, dominated by a cluster comprising two oxygen vacancies at the opposite ends of a body diagonal in one of the anion cubes of the fluorite structure. This vacancy dimer, illustrated in Fig. 8, is presumably stabilized by neighboring trivalent ions. It is probable, although not certain, that such clusters do not exist in isolation but only as ordered defect superstructures. Different spacings in these ordered arrays give rise to the different phases of the homologous series of reduced oxides present in the PrO_{2-x} system. These structures have been extensively investigated by diffraction and microscopy techniques. Eyring (Chapter 7) presents a review of this complex and fascinating field.

For doped materials, our knowledge of the detailed defect structure is less than that for the nonstoichiometric fluorite oxides. The study of doped

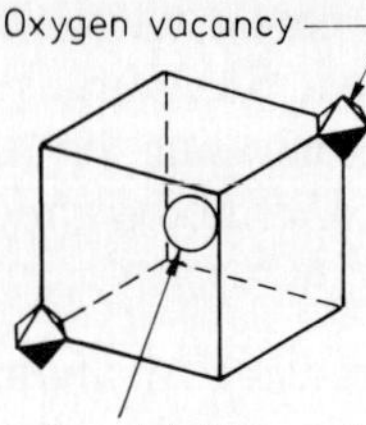

Fig. 8. ⟨111⟩ vacancy dimer in fluorite oxides.

systems is complicated by the low mobility of the dopant ion referred to in Section II.B. Only at high temperatures is kinetic control of the defect structure assured. Both divalent- and trivalent-doped oxides have been studied. In some cases (e.g., Ca–ZrO_2) ordered phases (e.g., $Ca_2Zr_4O_9$) may be precipitated, although the precise nature of the structure of these phases is controversial (Allpress and Rossell, 1975, Morinaga and Cohen, 1980). There is ample evidence for short-range order [see, e.g., the studies of Dexpert-Ghys *et al.* (1977) on ZrO_2–Y_2O_3]. Again, however, the details of the ordering are generally not known in any detail. More work is clearly required. In particular, the nature of the cation distribution in doped oxides is of considerable importance, as we shall see in the following section.

To date, little theoretical work has been reported on the structure of these systems. Their complexity has precluded application of the computer simulation techniques discussed in Section III.C, although recent developments in the field may bring these systems within the range of these methods. As discussed in Chapter 1, Sørensen has developed models of vacancy ordering based on the dominance of a simple cluster of one vacancy and two reduced or dopant cations—a structural principle different from the one discussed in the previous paragraph. (The formation of this cluster, described in Chapter 1, also forms the basis of a detailed statistical mechanical theory of these phases developed by Manes and co-workers (Manes, 1976; Manes and Manes-Pozzi, 1976), which is further discussed in Chapter 3.) Sørensen's treatment is based on ordering of the basic cluster with the assumption that the presence of a cluster at a site prevents other clusters from occupying adjacent sites in the lattice—the so-called site exclusion principle. His models appear to be compatible with the structures of certain of the reduced fluorite oxides. The formation of ⟨111⟩ vacancy dimers does not, however, naturally follow from the treatment.

The structural properties of anion-deficient fluorites will clearly remain a challenging field for experimentalists and theoreticians for some years. Similar uncertainties in theoretical interpretation are found for the transport properties discussed in the next section, although here we find a greater wealth of experimental data.

b. Transport Studies. In all fluorite-structured materials the mobilities of the anions are several orders of magnitude greater than those of the cations. Particularly high anion diffusion coefficients are observed in the anion-deficient phases—an effect that is unquestionably attributable to the low activation energies for anion vacancy migration. Low values (in the region 0.3–0.5 eV) have indeed been calculated for the oxygen vacancy activation energies in fluorite oxides (Catlow, 1977a). Values derived from analysis of the experimental data discussed below are again low, although they seem, in general, to be higher by ~0.2 eV than the calculated energies.

The conductivities[†] of anion-deficient fluorites have been extensively studied owing to the use of the materials as solid-state electrolytes. Conductivity measurements have been made as a function of temperature and of the composition of the oxides. The measured variation of the conductivity σ is usually represented as an Arrhenius plot, i.e., a plot of $\ln(\sigma T)$ versus T^{-1}, the slope of which gives the Arrhenius energy (in units of the gas constant) for the conduction process. In general it is found for fluorite oxides doped with trivalent ions that at higher temperatures (typically 800°C) the Arrhenius energy is ~0.6 eV while at lower temperatures (~400°C) the value decreases to ~0.8 eV. The interpretation of this behavior is simple: at low temperatures extensive dopant–vacancy clustering occurs, and the Arrhenius energy has a substantial contribution from the energy required to dissociate the vacancy from the cluster. At higher temperatures the vacancies are at least partially dissociated, giving an Arrhenius energy closer to the vacancy activation energy. For a detailed discussion we refer the reader to Chapter 5 (Kilner and Steele).

The effect of defect clustering on the conductivity of the fluorite oxides should be stressed. Indeed, in our opinion, attempts to enhance the conductivities of these materials should be based on considerations of this factor, since according to the calculations (Catlow and Butler, 1981), the free vacancy mobility does not vary strongly with the nature of the oxide. As we have seen, detailed understanding of defect aggregation in these oxides has not yet been achieved. Two important and related observations can, however, be made regarding the effect of vacancy–dopant interactions on conductivity. These concern:

(i) *Differences between the behavior of reduced and doped oxides.* Striking differences are often observed between the anion transport in the two

[†]In all fluorites ionic conductivity is entirely effected by the anions; i.e., the anion transport number is unity.

types of anion-deficient phases. The work of Steele and Floyd (1971) on CeO_{2-x} and Y^{3+}-doped CeO_2 provides a good example, and the results are summarized in Table II. In general, larger variations in activation energy with composition are found for the former compared with the latter systems.

TABLE II

Arrhenius Energies for Anion Transport in Y^{3+}-Doped CeO_2 and Reduced CeO_2[a]

Crystal	Arrhenius energy (eV)
5% Y^{3+}-doped CeO_2	0.85
10% Y^{3+}-doped CeO_2	0.9
20% Y^{3+}-doped CeO_2	0.9
$CeO_{1.95}$	0.3
$CeO_{1.9}$	0.25

[a] After Steele and Floyd (1971). All experiments performed in the temperature range 600–1100°C.

This difference between the two systems is at first puzzling since, as regards valence, dopant ions resemble reduced cations. However, a radical difference between the two systems arises from the contrast between the high mobilities of the conduction-band electrons present in the reduced phases and the low mobilities of the dopant cations. Low cation mobility is a general property of the fluorite lattice (Lidiard, 1974) that is particularly pronounced in fluorite oxides. An equilibrium distribution on the cation sublattice is therefore not to be expected except at the highest temperatures. It is probable therefore that the most thermodynamically stable clusters do not form in doped oxides. In contrast, clustering in the reduced phases will always be thermodynamically controlled, since here migration of only the highly mobile oxygen vacancies and electrons is required.

The contrast between kinetic and thermodynamic control of the cation sublattices in the two systems is probably responsible for the difference in their transport properties. The variation in cluster structure with composition will be more pronounced for the reduced than for the doped system because the low dopant mobility may prevent formation of the more thermodynamically favored cluster in the latter. The large variation of activation energy with composition obtained for reduced compared with doped crystals may then, in a general sense, be rationalized. The low mobility of the dopant ions in fluorite oxides is, we believe, also responsible for a second puzzling feature of the transport properties of these materials, which is discussed below.

(ii) *Anomalous variation of conductivity with dopant concentration in reduced materials.* Plots of σ versus dopant concentration are given in Fig. 9 for a Y^{3+}-doped CeO_2. We note that a *maximum* is observed, in the region of ~8 mole % of dopant ions. Such behavior *cannot* be predicted for any defective systems in full thermodynamic equilibrium; the concentration of the mobile dissociated vacancies must increase monotonically with the total vacancy concentration (up to a concentration of 50%) regardless of the nature of the clustering, providing there exists an equilibrium distribution of clusters. Partial kinetic control of the defect structure of the oxides seems therefore to be necessary to explain the behavior illustrated in Fig. 9.

No detailed theories of this effect have yet been developed. We can, however, see qualitatively how dopant immobility could lead to the observed behavior. Let us assume that there is a random distribution of dopant ions on the cation sublattice. At low dopant concentrations, such a distribution will result in most dopant cations being isolated (i.e., all the nearest-neighbor cation sites surrounding the majority of dopant ions will be occupied by lattice cations). In this case, the maximum extent of vacancy–dopant aggregation will be the formation of simple clusters involving single vacancies bound to the isolated dopant cations. However, as the concentration of dopant ions is raised, an increasing number of clusters of two, three, or more dopant ions must occur, given a random distribution of

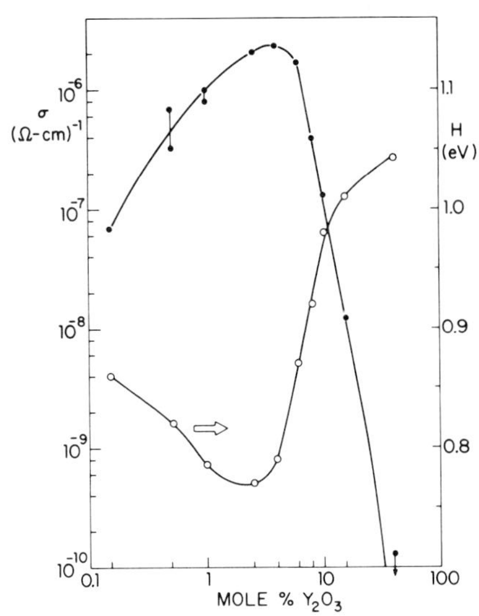

Fig. 9. Plot of conductivity versus dopant concentration for Y^{3+}-doped CeO_2. (After Nowick *et al.*, 1979.) Also shown is the variation of the Arrhenius energy for the conductivity. •, conductivity (σ); ∘, Arrhenius energy (H).

the dopant. These will provide deeper traps for the vacancies, resulting in a reduction in vacancy mobility. Thus, although increasing the dopant concentration will raise the total number of vacancies, vacancy mobility, however, will be decreased. At sufficiently high concentrations the latter factor begins to dominate, leading to the observed reduction in mobility.

Kinetic control of the cation distribution in doped fluorite oxides can therefore clearly have a major influence on oxygen transport in their systems. We conclude this section with a discussion of how the same factor may also have a marked effect on the thermodynamics of these systems.

c. Equilibrium Oxygen Partial Pressures. A detailed discussion of this field is given by Manes in Chapter 3. However, in the context of our present discussion of kinetic control of the properties of doped oxides, we should consider an observation of Schmitz (1975), who noted that the thermodynamic data on the mixed oxide $(U_{1-y}Pu_y)O_{2-x}$ could be described by the simple relationship

$$P_{O_2}^{-1/4} = Ky/(y-2x) \tag{1}$$

where P_{O_2} is the equilibrium oxygen partial pressure and K is a constant. It has been shown (Catlow, 1977b) that this equation can be predicted by a simple mass-action treatment, assuming the formation of a cluster of one dopant ion and one vacancy.† The predominance of such a cluster, at least in the region of low dopant concentrations, is predicted from the model of a random dopant cation distribution as argued in the previous section.

The claim that the observation of Schmitz supports the random cation model has been questioned (Matzke and Manes, 1978). It is argued that simple mass-action treatments are not appropriate for these systems and that purely statistical arguments would lead to significant concentrations of larger clusters at the Pu concentrations to which many of the thermodynamic measurements refer. This field is clearly controversial (Catlow, 1978), and more detailed theories based on the random dopant model are required.

In summary, our knowledge of defect aggregation in reduced and doped fluorite oxides is incomplete. Further theoretical and experimental studies are needed to elucidate the nature of defect clustering in these materials, as this evidently has a major influence on the widely studied thermodynamic and transport properties of the systems.

This completes our account of point defect systems. The remainder of the chapter is concerned with oxides containing extended defects.

† An earlier treatment (Schmitz, 1975) claiming to derive the equation, assuming the formation of a cluster of one vacancy and two dopant ions, has been shown to be incorrect (Catlow, 1977b).

V. Extended Defect Structures

We recall from Section II that extensive disorder may be stabilized not only by point defect aggregation but by the elimination of point defects with the formation of extended planar defects. Shear-plane formation in the oxides TiO_{2-x} and WO_{3-x} was discussed with the aid of a simple schematic diagram (Fig. 1a and b). Detailed structural information, obtained from microscopy and diffraction techniques, is now available (see Tilley, 1972; Anderson, 1973). In this section we concentrate on a number of fundamental problems raised by the observation of extended defects. We consider first the stability of these structures and their relation to point defect models. We pay special attention to the role that theoretical studies have played in explaining why shear-plane formation is confined to relatively few nonstoichiometric compounds—a problem raised in Section II. We then discuss the factors that control the ordering of extended defects, which is observed in a large number of materials. Finally, we consider briefly the various mechanisms proposed for shear-plane nucleation.

A. Shear-Plane Structures and Energetics

Figure 1a and b, discussed in Section II, showed how the elimination of vacancies with accompanying crystallographic shear results in the formation of a shear plane—a defect that may be described in terms of a change, within the plane, of the mode of linking of the MO_6 octahedra. We should note, however, that shear planes can be related to other point defect structures. Thus Fig. 10b illustrates a conservative extended defect (i.e., one whose formation does not lead to a change in the stoichiometry of the crystal) known as an antiphase boundary (APB). Addition of titanium ions to the interstitial sites, adjacent to the APB, that are marked in Fig. 10b, generates the shear phase illustrated in Fig. 10c. APBs may exist in stoichiometric TiO_2. Shear-plane formation in reduced TiO_{2-x} could therefore occur by metal interstitial capture by preexisting APBs.

We return later to the implications regarding the mechanism of shear-plane formation of this alternative description of the extended defect. The point to be stressed here is that shear-plane formation eliminates point defects; indeed, *shear planes may be considered as a special form of point defect aggregate*. This immediately raises the question of why this mode of defect aggregation is not more widespread. In particular, it is at first puzzling why vacancy elimination by crystallographic shear should not in general be a favored process. Vacancies are energetically expensive species (their formation energies in, for example, rutile and fluorite oxide, are

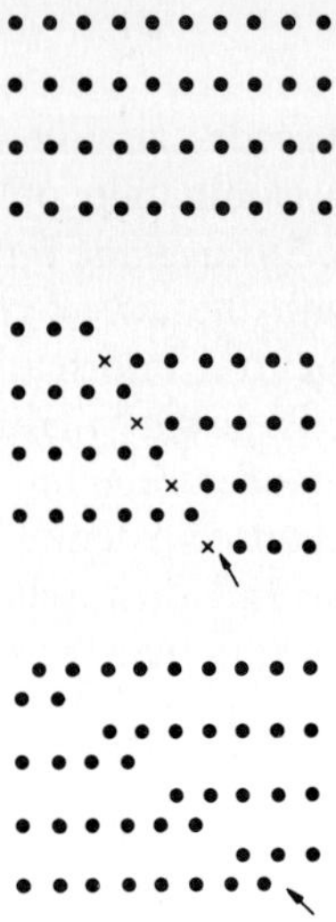

Fig. 10. The relationship of shear-plane to ABP structures. (Only cations in the ⟨100⟩ projection of rutile are shown.)

typically in the range 15–20 eV (Catlow, 1977a; Catlow *et al.*, 1981)); vacancy elimination should therefore be energetically favored. This argument, however, ignores one essential structural feature of shear planes. Metal–metal distances in the plane are shorter when compared with those in the perfect crystal. The formation of shear planes is therefore opposed by the consequent increased cation–cation repulsions. And we suggest that shear planes are not generally formed in nonstoichiometric compounds because in the majority of structures the increased cation–cation repulsions that accompany the formation of these defects outweigh the favorable energy associated with the vacancy elimination.

Does any special mechanism operate to reduce cation–cation repulsions in the plane in materials where shear planes are observed? We have suggested that metal atom relaxations play an essential role here (Catlow and James, 1978, 1981; Catlow, 1980). Relaxation of metal ions away from the planes clearly reduces the unfavorable repulsion energy. Detailed diffraction studies on Ti_4O_7 (Marezio *et al.*, 1973)—a compound containing ordered arrays of shear planes—have shown that there are indeed large metal relaxations of ~0.3Å. We propose that these are vital in stabilizing the extended defects.

This proposal has received support from two sources. First, theoretical lattice energy calculations of shear-plane energetics have recently been reported. These calculations resemble the computationally based theories of defect energetics discussed in Section III.C in that they employ a Born model of the ionic solid using simple interionic pair potentials. They are, however, performed on a perfect lattice (for which a unit cell structure must be specified) rather than a defect configuration. [A general efficient lattice energy program is now available—PLUTO, written by Catlow and Norgett (1978).] In the case of the shear plane, calculations were performed on the compounds Ti_4O_7 and Ti_5O_9 which contain ordered arrays of planar defects. From the calculated lattice energies a shear-plane formation energy may be deduced. The calculated value contains a term arising from the interactions of the shear planes; this term is not significant in our present discussion, although it is obviously of vital importance in discussing shear-plane ordering—a topic raised in the following section.

To test our suggestion that relaxations are a major factor in stabilizing shear planes, we performed two sets of calculations. First we calculated shear-plane energies for the *observed* structures (including, of course, the metal relaxations) in Ti_4O_7; next we calculated the energy of an *ideal* shear-plane structure in which the metal atoms have been returned to the centers of their octahedra. The calculations showed a large difference of ~10 eV† between the energies of the observed and ideal structures; ion relaxation therefore strongly stabilizes shear planes. Moreover, when shear-plane formation energies are compared with those of the possible point defect structures for reduced rutile (a topic discussed below), it is found that the energy of the latter species is higher by only ~2 eV. Thus, without the stabilizing relaxation energy, shear-plane formation would be prohibitively endothermic when compared with the creation of point defects. The relaxations are thus *essential* for the stability of the extended defect.

The second source of support for our relaxation model is in an empirical observation originally made by Tilley (1977), who noted that shear planes were formed in compounds with exceptionally high values of the static dielectric constant, ϵ_0. Thus the quoted values for ϵ_0 in TiO_2 and WO_3 are ~150 and ~300, respectively. In contrast, in the rutile-structured SnO_2, where reduction does apparently not result in shear-plane structures, ϵ_0 has a more normal value of ~15. The relaxation model immediately explains this correlation. Compounds with high values of ϵ_0 have high cation displacement polarizabilities; such structures should therefore permit large stabilizing cation relaxations around the shear plane. The large static dielectric constants are not therefore the *direct* cause of shear-plane stability; shear-plane formation and high values of ϵ_0 are both related to the same factor, namely, the large cation displacement polarizabilities. Both

†Energies are expressed as shear-plane energy per eliminated oxygen ion. This permits ready comparison with the energetics of point defect structures.

have the same fundamental cause, i.e., the softness of the metal–oxygen potential.

The above discussion thus, once more, demonstrates the power of computationally based theory when used in conjunction with experimental data. The next section continues our account of shear-plane stability by considering in detail the equilibrium between point and extended defect structures.

B. Equilibria between Point and Extended Defect Structures

Two problems have been debated in recent years in this controversial field. The first has concerned the basic question of whether point defects can exist in compounds where shear planes are known to form. The second problem assumes the existence of point defects but is concerned with their nature: In TiO_{2-x}, for example, both anion vacancy and cation interstitial models have been proposed.

The question of the existence of point defects is, we believe, now settled. Considerable evidence has now accumulated in support of the idea that a point defect population is present in equilibrium with shear planes. Indeed, it is difficult to avoid this conclusion when the statistical mechanics of shear structures is considered. As already discussed, shear planes can be considered as specially stabilized aggregates of point defects—either vacancies or interstitials. We can therefore write down equilibria of the type

$$\text{shear plane} \rightleftharpoons \text{point defects} \tag{2}$$

At sufficiently low defect concentrations, the higher configurational entropy of the point defect structures must lead to shear-plane dissociation; although if the point defect structures are very much higher in energy than the shear planes, dissociation may not occur at measurable defect concentrations.

The problem has been investigated theoretically (James, 1979; Catlow and James, 1981; Catlow, 1981). Shear-plane energies in TiO_{2-x}, calculated as discussed in Section V.A, have been compared with the energies of corresponding point defect structures. Calculations of defect energetics using the Mott–Littleton procedure described in Section III.C strongly favor anion vacancy models for reduced rutile; reduction with cation interstitial formation is calculated to be much more endothermic. The nature of the point defects in reduced rutile is, as noted, a controversial one, with several workers favoring the interstitial model. Nevertheless, if for the present we assume our calculations are correct, we can write for the extended-point defect equilibrium:

$$(\text{s.p.}) \rightleftharpoons V_A + 2e^- \tag{3}$$

where by (s.p.) we indicate a shear plane, V_A is an anion vacancy, and e^- a conduction band electron. These equilibria imply that the electrons created on reduction of TiO_2 are trapped by the positively charged shear plane and liberated when this defect dissociates.

Our calculations predicted an energy of ~2 eV for reaction Eq. (3). We have seen in the previous section how this calculated energy plays an important role in our arguments concerning shear-plane stabilities. This value can also be used in a simple mass-action treatment of the extended–point defect equilibrium. This theory, we note, ignores all entropy terms other than the configurational entropy of the point defects. (It can readily be shown that the configurational entropy associated with larger structural units, such as shear planes, is negligible.) The resulting prediction for crystals at 1000 K is for shear-plane dissociation at $x \sim 2 \times 10^{-3}$; i.e., anion vacancies aggregate into shear planes when the deviation from stoichiometry exceeds this value. A significant range of defect concentrations thus exists in which point defect structures predominate.

These conclusions are strongly supported by the conductivity studies of Baumard *et al.* (1977). Their results may be summarized in the form of plots of $\log(\sigma)$ versus $\log P_{O_2}$ illustrated in Fig. 11. The data were obtained at a variety of temperatures. It is seen that a sharp change in slope occurs at a composition that varies with temperature. Such a change is indicative of extensive clustering. Aggregation of anion vacancies into shear planes provides the obvious explanation. We note that the work of Baumard *et al.* (1977) predicts that aggregation into shear planes occurs at deviations from stoichiometry higher than suggested by our calculations. The calculated

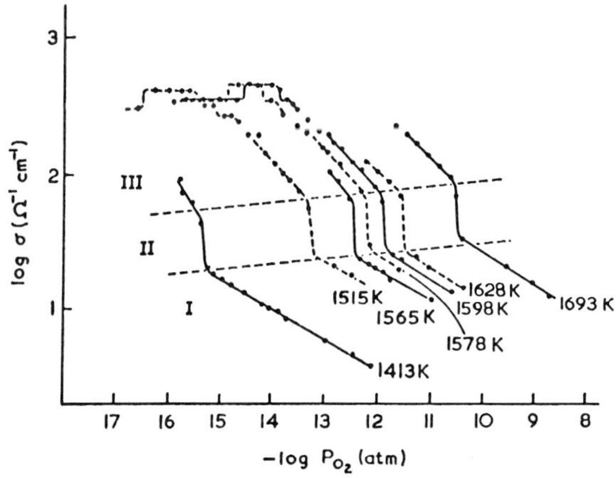

Fig. 11. The variation of conductivity with oxygen partial pressure for TiO_{2-x} (after Baumard *et al.*, 1977). Clustering of point defects into shear planes is postulated for region II.

energy of dissociation of shear planes into point defects appears therefore to be somewhat high, although only a small change in our calculated energy is needed in order to give accurate agreement with the data of Baumard and co-workers.

We believe that the work of Baumard *et al.* (1977), with the support provided by the calculations, has established the existence of point defect–shear plane equilibria. Nevertheless, as noted, the nature of the point defects formed is controversial. Analysis of the conductivity data of Baumard *et al.* (1977) supports the cation interstitial model, although the evidence is indirect, being based on the variation of conductivity with oxygen partial pressure. In contrast, the calculations strongly support the vacancy model. This controversy, moreover, relates to another area of uncertainty in the solid-state chemistry of these systems, that is, the mechanism of shear-plane formation. We shall return to this kinetic problem after considering the thermodynamics of a further important feature of shear-plane systems, namely, the long-range ordering of the extended defects that occurs in several systems.

C. Shear-Plane Ordering

Shear planes show a remarkable tendency to form ordered structures—a feature immediately revealed by microscopy studies. Indeed, it is now well established that there exists an homologous series of oxides, Ti_nO_{2n-1}, with long-range order based on regularly spaced shear planes, whose spacing increases with n. For higher values of n ($\sim$15–18) the interplane separations can be surprisingly large, $\sim$200–300 Å. At higher temperatures long-range order is no longer observed in near-stoichiometric TiO_{2-x}. Groups of regularly spaced planes are, however, seen in microscopy studies (Bursill *et al.*, 1971).

Similar behavior is noted for WO_{3-x}. The phenomenon appears to be general in solids containing extended defects. Anderson (1973) discusses a range of solid solutions, described as adaptive structures, of which Ta_2O_5–WO_3 provides an example, where the solute is incorporated as a planar defect; ordering of the planar defects occurs with a continuous range of interplanar spacings, resulting in a continuous range of compositions. A more exotic example is provided by the observation of the void lattice in irradiated metals. Spherical voids of diameter 50–100 Å are observed in several metals under irradiation; ordering to give a superlattice with a void separation of $\sim$100 Å has been observed in microscopy studies of several metals, e.g., Mo. For detailed discussions of this remarkable phenomenon we refer to the work of Bullough, Stoneham, and co-workers (Tewary and Bullough, 1972; Stoneham, 1971).

Large-scale periodicity involving extended defect structures appears therefore to be a general phenomenon. Details of ordered shear-plane compounds are available in several reviews (Tilley, 1972; Anderson, 1973). Here we wish to comment on our present understanding of the fundamentals of the phenomenon. Clearly, the ordering of extended defects implies appreciable defect interactions at the separations at which ordering is observed.† Theories of extended defect interactions have been developed, following the lead given by Bullough, Stoneham, and co-workers in their successful treatment of the void lattice. It has generally been assumed that elastic interactions provide the dominant term. This is unquestionably true for the case of void interactions in metals. It is more questionable for shear-plane interactions in ionic oxides, although Stoneham and Durham (1973) have have shown that the electrostatic interaction between these defects is negligible at the observed spacings provided the shear-plane charge is locally neutralized by the conduction-band electrons.

Stoneham and Durham (1973) developed a theory of the elastic interactions between a hypothetical ⟨100⟩ shear plane in the ReO_3 structure (a distorted version of which is adopted by WO_3). Essentially, they use a continuum elasticity approach to calculate the energy arising from the defect forces exerted by one plane on the elastic strain field of a second plane. Simple assumptions are made as to the nature of the defect forces. The theory successfully predicts appreciable interactions between planes at the observed separations. In addition, an interaction energy function is calculated for infinite arrays of shear planes as illustrated in Fig. 12. A minimum in the interaction energy is calculated for spacings of ~80 Å—a value close to that observed in the shear-plane groups in near-stoichiometric TiO_{2-x}. The theory predicts a more complex oscillatory function when the interaction of isolated pairs of shear planes is considered.

Stoneham and Durham's approach has been developed by Iguchi and Tilley (1977) who considered real shear-plane systems in WO_{3-x}; their theory also makes use of microscopy data to evaluate the magnitude of the defect forces. However, the most promising development in this field is, we believe, the application of the atomistic calculations discussed in Section V.A for the case of Ti_4O_7 and Ti_5O_9. No assumptions as to the nature of the extended defect interactions are required here; and a set of calculations for the Ti_nO_{2n-1} series could in principle provide a detailed understanding of shear-plane interactions.

In summary, therefore, a considerable quantity of data is available on ordered shear-plane structures. A general understanding of the phenome-

†This assumes of course that thermodynamic control is operative in these systems. This appears to be the case in TiO_{2-x} where shear plane mobility is known to be appreciable.

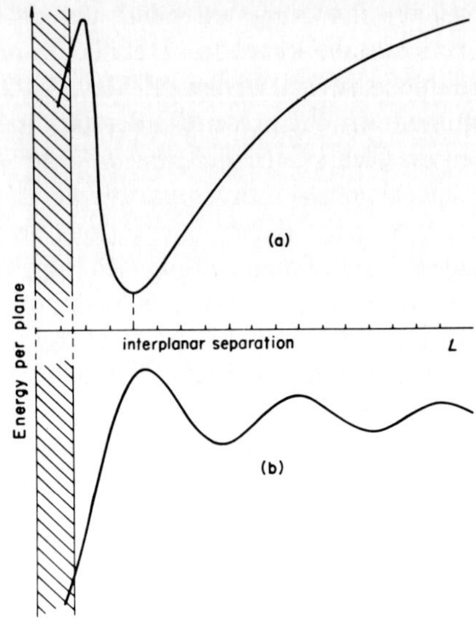

Fig. 12. Variation of shear-plane interaction energies with separation as calculated by Stoneham and Durham (1973), for (a) shear-plane superlattices and (b) pairs of shear planes.

non is beginning to emerge from theoretical approaches to the problem of shear-plane interactions. Theories based on continuum elasticity have proved successful in providing a basic framework. Atomistic calculations are required for detailed predictions.

D. Mechanisms of Shear-Plane Formation

Our account of this controversial field will be brief. Both homogeneous and heterogeneous nucleation mechanisms have been proposed, including a special case of the latter involving shear-plane nucleation at the surface followed by migration of the plane into the bulk of the crystal. In the brief account that follows we emphasize the role of recent computed modeling techniques in this area.

1. Homogeneous Mechanisms

A simple example of this class of mechanism was proposed by Anderson and Hyde (1967). They suggest that vacancies aggregate into a disk; the disk then collapses (in very much the way as was discussed in our schematic account of shear-plane formation illustrated in Fig. 1a and b) to give a shear-plane nucleus bounded by a dislocation loop. Growth of the plane occurs by vacancy trapping by the loop. Such models assume of course that vacancies are the dominant point defects. As remarked, this point is controversial, with several workers suggesting cation interstitials as the major point defect species. Homogeneous nucleation mechanisms based on cation interstitial aggregation can be proposed (James, 1979). They are, however, more complex than the vacancy mechanisms. The initial formation of the interstitial cluster must be followed by rearrangement of the cation sublattice.

However, the essential feature of both mechanisms, on which we now concentrate, is the initial formation of point defect aggregates which act as precursors for the planar defects. We can investigate the stability of such aggregates using the computer simulation techniques. Several aggregates have been examined, involving vacancies and interstitials (the clusters must of course be surrounded by conduction-band electrons—Ti^{3+} ions in our model—to achieve electroneutrality). In no case was an appreciable binding energy obtained. The result is hardly surprising in view of the high static dielectric constant of TiO_2 ($\epsilon_0 \sim 150$) which must drastically reduce the Coulomb interactions between the point defects and electrons; this is the term we would expect to provide most of the binding energy.

This inability to form stable point defect precursors strongly militates against the homogeneous nucleation mechanisms. Our discussion thus continues with the alternative models based on heterogeneous nucleation.

2. *Heterogeneous Formation Mechanisms*

The simplest of such mechanisms is again suggested by the schematic discussion of shear-plane formation given in Section V.A. We recall that shear planes can be generated by the addition of interstitials to APBs (Fig. 10). An obvious mechanism for shear-plane formation then involves interstitial capture by preexisting APBs. It is possible to elaborate this mechanism by proposing that interstitials are created by reduction at the surface. They are then trapped at the point at which the APB intersects the surface; rapid interstitial migration occurs down the APB and thus generates the shear plane. Calculations have provided some support for this mechanism (James, 1979), as it has been shown that the lowest-energy APB has an $\langle 011 \rangle$ orientation which on capture of a cation interstitial is converted into a $\langle 121 \rangle$ shear plane—the observed orientation of the planes in TiO_2.

Indeed, we believe that the APB mechanism is the most plausible of those proposed to date. Objections can be raised to alternative, heterogeneous mechanisms involving shear-plane nucleation at the surface followed by migration of the plane into the bulk; these are summarized by Bursill and Hyde (1972). A feature to note regarding the APB mechanism is the implied correlation between the rate of reduction of TiO_2 and the concentration of preexisting APBs. No clear evidence is available as yet on this question, whose further investigation would be of interest.

VI. Summary and Conclusions

This chapter has looked in detail at the two ways that may be adopted for stabilizing extensive disorder in nonstoichiometric solids. We have seen how large levels of disorder may be stabilized by point defect *aggregation*; alternatively *elimination* of point defects occurs with extended defect formation. In both cases the defect structures generated have a surprising degree of complexity. Elucidation of the structural and kinetic problems associated with these structures requires a combination of sophisticated theoretical and experimental techniques. Indeed, the complementary use of theory and experiment is, we believe, one of the most important features of this area of study, and the one that promises considerable further advancement in the near future.

Acknowledgments

This chapter has profited considerably from discussions with my colleagues R. James, B. E. F. Fender, A. M. Stoneham, A. B. Lidiard, and J. M. Thomas.

References

Allpress, J. G., and Rossell, H. J. (1975). *J. Solid State Chem.* **15**, 68.
Anderson, J. S. (1973). *J.C.S. Dalton Trans.* 1107–1115.
Anderson, J. S. (1977). *J. Phys.* (Paris) **C7**, 17–27.
Anderson, J. S. and Hyde, B. G. (1967). *J. Phys. Chem. Solids* **32**, 923–926.
Baker, J. M., Davies, E. R., and Hurrell, J. P. (1968). *Proc. R. Soc. London, Ser. A* **308**, 403–412.
Battle, P. D. and Cheetham, A. K. (1979). *J. Phys. C* **12**, 337–345.
Baumard, J. F., Panis, D., and Anthony, A. M. (1977). *J. Solid State Chem.* **20**, 43–51.
Bursill, L. A., Hyde, B. G., and Philp, D. K. (1971). *Philos. Mag.* **23**, 1503–1513.
Bursill, L. A., and Hyde, B. G. (1972). *Prog. Solid State Chem.* **7**, 177–253.
Catlow, C. R. A. (1973). *J. Phys. C* **6**, L64–L70.
Catlow, C. R. A. (1977a). *Proc. R. Soc. London, Ser. A* **353**, 533–561.
Catlow, C. R. A. (1977b). *J. Nucl. Mater.* **67**, 236–238.
Catlow, C. R. A. (1978). *J. Nucl. Mater.* **74**, 172–173.
Catlow, C. R. A. (1981). *In* "The Physics and Chemistry of Refractory Oxides," Proc. NATO Summer School. To be published.
Catlow, C. R. A., and Butler (1981). To be published.
Catlow, C. R. A., and Fender, B. E. F. (1975). *J. Phys. C* **8**, 3267–3279.
Catlow, C. R. A., and James, R. (1978). *Nature (London)* **272**, 603–605.
Catlow, C. R. A., and James, R. (1981). To be published.

Catlow, C. R. A., and Norgett, M. J. (1973). *J. Phys. C* **6**, 1325–1339.
Catlow, C. R. A., Norgett, M. J., and Ross, T. A. (1977a). *J. Phys. C* **10**, 1627–1640.
Catlow, C. R. A., and Norgett, M. J. (1978). *U. K. At. Energy Res. Establ., Memo.* **AERE-M 2763**.
Catlow, C. R. A., Fender, B. E. F., and Muxworthy, D. G. (1977b). *J. Phys. (Paris)* **C7**, 67–71.
Catlow, C. R. A., Corish, J., Diller, K. M., Jacobs, P. W. M., and Norgett, M. J. (1979). *J. Phys. C* **12**, 451–464.
Catlow, C. R. A., James, R., Mackrodt, W. C., and Stewart, R. (1981). *Phys. Rev.* (in press).
Cheetham, A. K., Cooper, M. J., and Fender, B. E. F. (1971a). *J. Phys. C* **4**, 3107–3121.
Cheetham, A. K., Fender, B. E. F., and Taylor, R. I. (1971b). *J. Phys. C* **4**, 2160–2165.
Childs, P. E., Steele, D., and Fender, B. E. F. (1972). *J. Phys. C* **5**, 2677–2688.
Dexpert-Ghys, J., Faucher, M., Dexpert, H. and Coro, P. (1977). *J. Phys. (Paris)* **C7**, 95–106.
Faber, J., Jr., Seitz, M. A., and Mueller, M. H. (1976). *J. Phys. Chem Solids* **37**, 903–907, 909–915.
Fletcher, R., and Powell, M. J. D. (1963). *Comput. J.* **6**, 163–168.
Gettman, P., and Greiss, O. (1978). *J. Solid State Chem.* **26**, 255–263.
Goodenough, J. B. (1974). *In* "Solid State Chemistry" (C. N. R. Rao, ed.), p. 215. Dekker, New York.
Iguchi, E., and Tilley, R. J. D. (1977). *Philos. Trans. R. Soc. London* **286**, 55–85.
James, R. (1979). Ph. D. Thesis, Univ. of London, London.
James, R., and Catlow, C. R. A. (1977). *J. Phys. (Paris)* **C7**, 32–35.
Koch, F., and Cohen, J. B. (1969). *Acta Crystallogr., Sect. B* **25**, 275–287.
Kofstad, P. (1972). "Non-Stoichiometry, Diffusion and Electrical Conductivity in Binary Metal Oxides," Wiley (Interscience), New York.
Lidiard, A. B. (1957). *In* "Handbuch der Physik" (S. Flügge, ed.), Vol. XX, pp. 246–348, Springer-Verlag, Berlin and New York.
Lidiard, A. B. (1971). "The Theory of Imperfect Crystals in Solids," I.A.E.A., Vienna.
Lidiard, A. B. (1974). *In* "Crystals with the Fluorite Structure: Electronic, Vibrational and Defect Properties" (W. Hayes, ed.), pp. 101–184. Oxford Univ. Press (Clarendon), London and New York.
Mackrodt, W. C., and Stewart, R. F. (1977). *J. Phys. C* **10**, 1431–1446.
Mahon, G. D., and Roth, W. L. (1976). "Superionic Conductors." Plenum, New York.
Manes, L. (1976). *In* "Plutonium 1975 and Other Actinides" (H. Blank and R. Lindner, eds.), pp. 182–183, North-Holland Publ., Amsterdam.
Manes, L. and Manes-Pozzi, B. (1976). *In* "Plutonium 1975 and Other Actinides" (H. Blank and R. Lindner, eds.), pp. 145–163. North-Holland Publ., Amsterdam.
Marezio, M., Dernier, P. D., McWhan, D. B., and Reimika, J. P. (1973). *J. Solid State Chem.* **6**, 213–221.
Matzke, H., and Manes, L. (1978). *J Nucl. Mater.* **74**, 167–171.
Morinaga, M., and Cohen, J. B. (1976). *Acta Cryst.* **A32**, 387.
Morinaga, M., and Cohen, J. B. (1980). *Acta Cryst.* **A36**, 520.
Mott, N. F., and Littleton, M. J. (1938). *Trans. Faraday Soc.* **34**, 485–499.
Norgett, M. J. (1974). *U. K. At. Energy Auth., Res. Group, Rep.* **AERE-R 7650**.
Norgett, M. J., and Fletcher, R. (1971). *J. Phys. C* **3**, L190–L192.
Norgett, M. J., and Lidiard, A. B. (1972). *In* "Computational Solid State Physics" (F. Hermann, W. W. Dalton, and T. R. Koehler, eds.), pp. 385–412, Plenum, New York.
Norgett, M. J., and Stoneham, A. M. (1973). *J. Phys. C* **6**, 229–246.
Nowick, A. S., Wang, D. Y., Park, D. S., and Griffith, J. (1979). *In* "Fast Ion Transport in Solids" (P. Vashishta, J. N. Mundy, and G. K. Shenoy, eds.), pp. 673–679, Elsevier/North-Holland Publ., Amsterdam.

Roth, W. L. (1960). *Acta Crystallogr.* **13**, 40.

Schmitz, F. (1975). *J. Nucl. Mater.* **58**, 357–360.

Steele, B. C. H., and Floyd, J. M. (1971). *Proc. Br. Ceram. Soc.* **19**, 55–76.

Steele, D., and Fender, B. E. F. (1974). *J. Phys. C* **7**, 1–11.

Stoneham, A. M. (1971). *J. Phys. F* **1**, 778–784.

Stoneham, A. M. (1975). "Theory of Defects in Solids." Oxford Univ. Press (Clarendon), London and New York.

Stoneham, A. M., and Durham, P. (1973). *J. Phys. Chem. Solids* **34**, 2127–2135.

Tasker, P. W. (1981). *Philos. Mag.* (in press).

Tewary, V. K., and Bullough, R. (1972). *J. Phys. F* **2**, L69.

Tilley, R. J. D. (1972). *In* "Solid State Chemistry" (H. F. Emeleus and L. E. J. Roberts, eds.), MTP International Review of Science, Inorganic Chemistry, Series One, Vol. 10, pp. 279–313. Butterworth, London.

Tilley, R. J. D. (1975). *In* "Solid State Chemistry" (H. F. Emeleus and L. E. J. Roberts, eds.), MTP International Review of Science, Inorganic Chemistry, Series Two, Vol 10, p. 73. Butterworth, London.

Tilley, R. J. D. (1977). *Nature (London)* **269**, 229–230.

VanGool, W. (1972). "Fast Ion Transport in Solids," Proc. NATO Summer School. North-Holland Publ., Amsterdam.

Willis, B. T. M. (1963). *Proc. Br. Ceram. Soc.* **1**, 9–19.

Willis, B. T. M. (1964). *J. Phys. (Paris)* **25**, 431–439.

3

A New Method of Statistical Thermodynamics and Its Application to Oxides of the Lanthanide and Actinide Series

L. Manes

Commission of the European Communities, Joint Research Centre
Karlsruhe Establishment, European Institute for Transuranium Elements
Karlsruhe, Federal Republic of Germany

ISBN 0-12-655280-0

I. Statistical Thermodynamic Concepts for Grossly Nonstoichiometric Compounds

A. Introduction

Statistical treatments of randomly distributed, noninteracting point defects fail in grossly nonstoichiometric solids, where concentrations up to one-eighth of the atoms of the lattice may be reached (e.g., $UO_{2.25}$ in a UO_{2+x} lattice) before the phase limits are attained. It is intuitive that, through the interplay of interactions, the defects become ordered, constituting clusters or "seeds" of the new phase already within the disordered phase. Alternatively, one can say that they are assimilated within microdomains or regions of the lattice in which they disappear as defects and in which the constituting atoms form the derived structure.

The disordered structure of a grossly nonstoichiometric compound may be represented therefore as a collection (or ensemble) of lattice atoms partially ordered over some lattice volume, as well as a collection (or ensemble) of interacting defects assimilated (or ordered) over the same lattice volume. These lattice volumes (or microdomains) are continuously being formed and destroyed by thermal motion: therefore the phase shows fluctuations in order in time and space. These fluctuations are also fluctuations in local density and local symmetry of the lattice. A good statistical description, in this representation, would be given by a thermodynamic assessment of the equilibrium number and dimensions of structural microdomains (Anderson, 1970), as well as by a thermodynamic assessment of the equilibrium number of defects being ordered by some particular ordering law in the lattice. A fluctuation may extend over many interatomic distances and grows continuously into another fluctuation, determining phase decomposition when it occupies a considerable portion of the lattice. In turn, at sufficiently high temperatures, fluctuations will be reduced at only a few atomic distances (eventually, one), and the concept of microdomains will lose its meaning; the random distribution of weakly interacting defects may then be the best statistical way to represent the system. Within a fluctuation, a defect possesses a certain mean energy which is a function of its position; if the defect is positioned well within the fluctuation, it may be considered a completely assimilated defect and its mean energy will more or less resemble the potential energy an assimilated defect has in a given structure (cohesive energy per assimilated defect); if on the contrary it is positioned in a region of intergrowth between microdomains, its mean energy will be essentially determined by the interaction between the two adjacent regions or fluctuations.

To represent correctly by statistical thermodynamics this system of fluctuations in order of the defects, the most attention, when writing the partition functions, has been given to the form of the interaction, i.e., to the enthalpy of the system. But it is intuitively evident that as great a role is played by the way the fluctuations are distributed in the lattice, i.e., by the configurational entropy of the system.

As yet, it has been difficult to extend the description of the interaction beyond the hypothesis of nearest-neighbor interaction (attraction or repulsion). As for the configurational degeneracy determining the configurational entropy, either the combinational form typical of random distribution or, in the case of repulsion, what is called the site exclusion principle, has been employed. It has been difficult, however, to represent correctly what is a central observation concerning the thermodynamics of these systems, i.e., the strong correlation between their enthalpy and entropy functions. This strong correlation must be connected to the way the intergrowth of fluctuations occurs and ought to involve, conceptually, some entropic reason that distinguishes in a fundamental way a random distribution of independent or weakly interacting point defects, as observed for a slight departure from stoichiometry, from a distribution of microdomains or regions of order of the defects. An obvious difference is indicated by the fact that each fluctuation in order occupies some volume of the lattice. It is likely that, in the homogeneous space constituting the lattice, each fluctuation tends to maximize its own volume and tries to impose its own "ordering law" on the whole lattice, being hindered in this process by the presence of other fluctuations in order. This introduces an entropic competition between the fluctuations, which is absent from the random (i.e., casual) distribution concept. This also introduces naturally the concept of a correlation between *enthalpy* (i.e., the ordering law) and *entropy* in the thermodynamic properties of the system.

In the treatment of nonstoichiometric systems, another point to which the thermodynamicist should pay attention is the presence of spinodals in the high-temperature thermodynamic curves of the disordered phase and the fact that they should be considered residues of the phase diagram in high-temperature thermodynamics. Spinodals of a given order are defined as zeroes, i.e., extrema, in some order of the derivatives of the partition function (or of the free energy G or of the chemical potential μ). Although the concept of an order–disorder transition should be better defined in the case of nonstoichiometric solids, spinodals clearly recall this concept. They are related to the enthalpy as well as to the (configurational) entropy of the system; whereas the former determines essentially the temperature range in which a spinodal of some order exists and the critical temperature at which it disappears, the latter determines essentially the compositions at which a spinodal occurs. This composition is not random; it represents, for instance, a composition at which the lattice is (almost) saturated with a particular form of order. If one thinks of the entropic competition discussed above, the position of the spinodals describing the point in a G–T–composition diagram at which a certain arrangement of the defects or of the atoms, hence a certain type of fluctuation, becomes critical is a joint manifestation of the form of the enthalpy and of the entropy of the system.

Spacing statistics, which are the subject of this chapter, are essentially based on the concept of fluctuations in the order of the defects. They do not attempt to provide a detailed description of the interaction. Rather, they assign to each defect an average cohesive energy typical of the particular order in which it is assimilated. The first challenging problem for spacing statistics is therefore to assess these mean energies and to connect them with the type of order of the fluctuation.

On the other hand, spacing statistics use an extension of the site exclusion principle for configurational degeneracy in order to describe the entropic competition between the different fluctuations in order and to describe their intergrowth.

B. *The Thermodynamics of Nonstoichiometric Phases*

1. *G–T–x and μ–T–x Diagrams*

A nonstoichiometric phase is fully characterized by its G–T–x and μ–T–x diagrams, where G is the free energy, μ the equilibrium chemical potential, x the departure from stoichiometry, e.g., in an oxide of general formula $MO_{n\pm x}$ [where n is the oxygen-to-metal (O/M) ratio in the stoichiometric oxide] and T the temperature. These diagrams have been amply discussed in Chapter 1 of this book.

When more than one phase is present at low temperatures in the composition region for which, at a high temperature, a nonstoichiometric disordered phase exists, inflections may occur (Thorn, 1970), which are characterized by a zero value for the first derivative $\mu' = (\partial\mu/\partial x)_T$ of the chemical potential μ and, since they form an horizontal tangent, by $\mu'' = (\partial^2\mu/\partial x^2)_T > 0$ (minima in μ'). In Fig. 1 (Thorn, 1970), they are marked as inflections I and said to be of degenerate homogeneity. It is shown (Thorn, 1970) that these inflections may be separated by other types of inflections (type J in Fig. 1) characterized by a vertical tangent ($\mu' = 0$, $\mu'' < 0$ in μ'), which are said to be of intrinsic homogeneity. They correspond to single-phase regions at low temperatures.

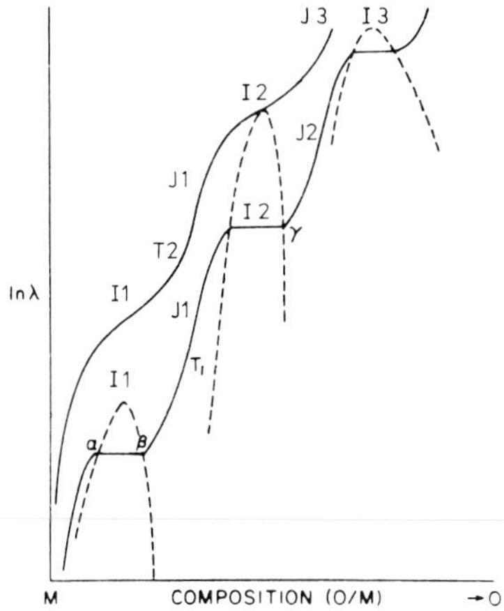

Fig. 1. Activity–composition isotherms illustrating three diphasic regions at T_1 and their envelopes and homogeneous regions at T_2 ($T_2 > T_1$). Inflections for T_2 at I_1, I_2, etc., are residues of diphasic regions, hence are identified as degenerate regions of homogeneity. Inflections at J_1, J_2, J_3, etc., remain when the temperature is lowered at T_1, hence are identified as regions of intrinsic homogeneity. (From Thorn, 1970, reproduced by permission of North-Holland Publishing Co., Amsterdam.)

2. *Spinodal Instabilities*

An inflection of type I found for the high-temperature thermodynamic functions of the disordered phase of nonstoichiometric compounds can be assumed to be a high-temperature residue of the spinodal instability of the free energy G of the system of defects causing nonstoichiometry (Anderson, 1970). The spinodal instability can be defined as the presence, in the chemical potential, of two neighboring extrema or spinodals (maximum and minimum; $\mu' = 0$, $\mu'' < 0$, $\mu'' > 0$, respectively) between which a composition range exists in which $\mu' < 0$. Any composition between the spinodals is absolutely unstable, since any spontaneous fluctuation in composition occurring in the system is irreversible and self-propagating (Cahn, 1961). Therefore, spinodal instability denotes the existence of a diphasic region. At the closure of a diphasic region in the phase diagram, the two spinodals merge with the inflection point I, and the μ-versus-x curve still shows just an inflection point with an horizontal tangent. This inflection point may be preserved at $T > T_c$ and thus be considered a residue of the diphasic region.

The two phases present at $T < T_c$ are perhaps generated from one another by the assimilation of point defects as structural elements (Anderson, 1970). Also, in a nonstoichiometric compound, the symmetry properties of the two phases are not very far apart (Anderson, 1970; Thorn, 1970). The transition is often represented as an order–disorder transition of the defects.

This transition is related to the presence of zeros in μ'. And we have mentioned the disappearance of zeros in μ' (i.e., of the spinodals) and the appearance of the inflection I when going from $T < T_c$ to $T > T_c$, which recalls the diphasic region and is represented by the condition $\mu'' = 0$.

Let us call the inflection I a spinodal of second order. One may safely assume that the situation met when studying the thermodynamic functions of a nonstoichiometric compound with increasing temperature is an increase in the order of the derivatives of the free energy G (or of the partition function Z). This can also be interpreted as an increase in the order of the order–disorder transitions of the defects (Thorn, 1970).

3. *The Relationship between the Enthalpy and Entropy Functions in a Nonstoichiometric Phase*

A central observation concerning the thermodynamics of nonstoichiometric phases is the strong relation (Thorn, 1970) between the enthalpy and the (configurational) entropy functions of the system, especially between the partial molar quantities $\overline{H}$ and $\overline{S}$. In Chapter 1 of this book, this connection is shown for the Ce–O system.

This relation does *not appear* in all thermodynamic models in which the nonstoichiometry is explained on the basis of a random distribution of independent defects. However, if the *clustering* or *segregation* of defects is considered, both being devices by which defects order themselves in the lattice by means of an attractive or repulsive interaction, this connection appears to be inevitable. In fact, the two processes make the positions around an (isolated) defect *more* or *less* favorable for the introduction of a second defect. If this is true, the random hypothesis of distribution of the defects can no longer be sustained, since the defects are no longer independent, and both the configurational entropy and the enthalpy are therefore affected.

C. *Thermodynamic Models with Nearest-Neighbor Interactions between Defects*

1. *Introduction*

An approach to the problem of relating the low-temperature phase diagrams of nonstoichiometric systems, showing phase separation, to the high-temperature thermodynamic functions of the disordered phase is to consider attractive or repulsive interaction between nearest-neighbor defects. The resulting models have been reviewed many times (Anderson, 1970; Sørensen, 1978). The essential feature of these models is the ability to generate spinodals at low temperatures and to define a critical temperature T_c as a function of the interaction energy parameters.

2. Nearest-Neighbor Attraction—The Case of UO_{2+x}

The nearest-neighbor attraction model is usually an application of the Ising–Langmuir thermodynamic functions (Anderson, 1970; Ziman, 1972) in which the attractive interaction is defined by a single parameter ϵ. The N defects are first considered to be randomly distributed over B sites; the interaction energy ϵ operates between those that happen to be nearest neighbors. The partition function is therefore of the type

$$Z = a(T)\binom{B}{N}\exp\left[\frac{-p(N^2/2B)\epsilon}{KT}\right] \tag{1}$$

where $N^2/2B$ represents the *probability* that a pair will be found in the lattice, p the number of nearest neighbors at a defect site, and all terms that do not depend on the interaction are grouped in $a(T)$. Notice that in this approach the configurational entropy $\binom{B}{N}$ is written as if the defects were independent.

An application of the model of nearest-neighbor interaction to $UO_{2\pm x}$ was made by Thorn and Winslow (1966).

The defects considered responsible for nonstoichiometry are (a) for UO_{2-x}, oxygen vacancies, and (b) for UO_{2+x}, oxygen interstitials. The presence of electronic disorder, i.e., of heterocharged cations in the metal sublattice $(U^{+3})^{z'}$, $z' < 0$ in UO_{2-x}; $(U^{+5})^{z'}$ or $(U^{+6})^{z'}$, $z' > 0$ in UO_{2+x} (where z' is the effective charge with respect to the lattice) is not explicitly taken into account.

Consistent with the formalism of nearest-neighbor interaction, interaction terms E_{VV} between vacancies and E_{II} between interstitials (but not E_{VI}) are introduced. The formalism generates spinodals, as is appropriate for this kind of treatment, and phase limits can be therefore derived for the disordered phase of $UO_{2\pm x}$.

This gives satisfactory results for the phase limit of UO_{2-x}, showing that the hypothesis of interacting vacancies can be considered valid for the substoichiometric phase of UO_2.

In terms of expression of the isotherm, modified by introduction of the defect-depending vibrational contribution for vacancies, a critical temperature T_c is derived for the substoichiometric phase and is given by

$$2.7326T_c = E_{VV}/K \tag{2}$$

where E_{VV} is derived from the lattice dynamics of the fluorite structure (defining a Debye temperature for the vacancies) as well as from thermodynamic data for the UO_{2-x} phase. The two values are found to be in reasonably good agreement.

In UO_{2+x}, the number of sites available for interstitials is, in the first instance, dictated by the geometry of the fluorite lattice: $B_{int} = \alpha B$, where B is the total number of cation sites. The value $\alpha = 1$ (one interstitial per U atom) suggested by the structure is in reasonably good agreement with experimental thermodynamic curves up to $UO_{2.08}$ only.

For higher stoichiometries agreement was found when α was reduced to $\frac{1}{4}$ (one interstitial per four U atoms), a fact that points to ordering of the defects, with exclusion of interstitial lattice sites when positioning one interstitial.

This theoretical observation opens the way to consideration of another type of model of interacting defects, which is the subject of the next subsection.

3. *Nearest-Neighbor Repulsion*

The effect of an essentially repulsive interaction is segregation of defects. Ordering may be achieved by trying to minimize the repulsion by keeping defects as far apart from one another as possible.

An essential statistical tool for this kind of treatment is the *site exclusion principle*, which directly involves the configurational entropy. This principle states that the presence of a defect at a site precludes the occupation of a specified number of adjacent available sites by the same kind of defect. It is clear that, by this principle. the configurational entropy is decreased. Furthermore, the lattice is saturated by a well-defined number of defects; introduction of more than this number tends to increase the repulsive energy to a limit that the crystal can no longer bear.

Speiser and Spretnak (1955) have used this treatment in investigating interstitial defects in a nonstoichiometric crystal.

In this treatment, the simple combinatorial degeneracy $\binom{\alpha B}{N}$ for a system of interstitial defects is replaced by

$$\Omega = \frac{C^N[(\alpha/C)B!]}{N![(\alpha/C)B - N]!} \tag{3}$$

where N is the number of interstitials, B the number of lattice sites, αB the number of interstitial lattice sites available, and C the number of lattice sites *excluded* because one interstitial lattice site is occupied by one interstitial defect. It is seen that, if $C = 1$, i.e., no interstitial lattice site is excluded other than the one occupied, expression (3) reduces to random configurational degeneracy. In Eq. (3) the term C^N describes a residual freedom of disposition of the interstitials in the C positions available.

In Eq. (3), C was varied parametrically. Hagemark (1966), in explaining the equilibrium data of UO_{2+x}, used a variant of this model where oxygen interstitials in the UO_{2+x} fluorite lattice excluded, on the average, 12 adjacent sites. The fluorite lattice of UO_{2+x} is in this case saturated with interstitial oxygens at $UO_{2.25}$ (U_4O_9).

The site exclusion principle modifies the simple random picture for the distribution of the point defects, creating an envelope C within which no other defect can be placed. In this way it transforms the defect from a point defect to a more extended object having volume C and a degree of freedom within this volume. The configurational degeneracy is reduced in Eq. (3) with respect to the random distribution of independent point defects.

An extension of the site exclusion principle as first formulated is at the basis of the statistical method (spacing statistics) that is the subject of this chapter.

D. The Equilibrium of Microdomains

1. The Microdomain Concept

We have seen that the effect of interactions between point defects in a nonstoichiometric phase is to create regions of order in the lattice by clustering and segregation. It is possible to depict this effect as a process that passes gradually through different steps:

(a) The defects cluster or segregate at preferential sites determined by the type of interaction.

(b) The extension of this process creates in the lattice regions of ordering of the defects with respect to the normal (or host or parent) lattice, which, in their local symmetry and interdefect (and interatomic) distances, resemble a crystal structure other than the host structure: true seeds, for instance, of the crystal structures, which are generated at lower temperatures in the phase diagram of the compound.

(c) Finally, the order extends throughout the whole lattice, with the formation of new crystals and well-defined surfaces between them, hence of new phases, and a diphasic region is obtained.

One can picture this gradual process also by saying that, in the nonstoichiometric phase at a temperature higher than T_c thermal fluctuations exist in which the lattice assumes different configurations of order, which vary from the parent structure to the derived structure. These thermal fluctuations are evidently thermal fluctuations in order of the point defects, but they are also thermal fluctuations in the atoms of the lattice, hence thermal fluctuations in the structure of the lattice.

The microdomain concept views the lattice, at a given temperature $T < T_c$, as the locus at which thermal fluctuations in structure occur. At thermal equilibrium, it is equivalent to describe the lattice by giving the equilibrium number of thermal fluctuations at which the lattice assumes a certain structure, or to picture the lattice as a thermal assembly of regions or microdomains each having a well-defined crystal structure: the host structure or one (or many) structures, which may be stable only at lower temperatures. Note that the stability of the microdomain structure at lower temperatures is not a necessary condition: All possible arrangements of the lattice atoms may be represented in the microdomain population even though these structures may be disproportionate in stabler structures at lower temperatures. In the second description, however, not only the number of microdomains of definite structure but also their size characterizes the thermal distribution. The microdomain concept completely abandons the point defect description of the nonstoichiometric phase. A microdomain is an extended perturbation of the host lattice, hence is an extended defect. It may also be understood as the locus at which point defects become ordered.

2. *The Phase Rule*

The thermal stability of the fluctuations in structure is given by $\Theta_{\text{fluct}} = E_{\text{fluct}}/kT$, where E_{fluct} is the mean energy per atom (or per assimilated point defect) within the fluctuation. The microdomain population is then characterized by a series of thermal stability terms $\Theta_{\text{micr}}(D, \text{structure})$, which are a function of the dimension D and of the type of structure represented in the microdomain. There is a difficulty, however, in defining the energies E_{micr} associated with Θ_{micr}.

Consider two adjacent microdomains in a lattice and, for simplicity, let there be two microdomains, one representative of the structure of the host lattice and the other representative of some derived structure.

There is no net separation of the two regions by way of a well-defined structural discontinuity or surface. On the contrary, the two regions grow gradually into one another in a coherent way, and the region of this coherent intergrowth will be a disordered region, extending some atomic shells, in which one progresses gradually from some typical value of energy per atom (or per assimilated defect) E_0 of the host structure to another value E_1 for the derived structure. The presence of these regions of intergrowth (or surfaces) complicates the thermodynamic treatment of microdomains. It is a situation resembling that in a liquid, in which the order of the crystalline structure of the solid has been destroyed (Anderson, 1970).

Notice that, as a consequence, the two microdomains are not energetically independent. Throughout the region of intergrowth, interaction exists between the atoms of the two microdomains (or between the lattice defects). In order to minimize the total energy of the lattice, and given a certain freedom of movement of the atoms in the intermediate surface region, the two microdomains may tend to (a) change their dimensions in an optimal way, or (b) orient themselves in an optimal way.

At sufficiently low temperatures, well-defined surfaces between the two regions will be formed and a diphasic region will occur in the phase diagram. All this has consequences from a thermodynamic point of view. Ubbelohde (1957, 1966) has pointed out that, although the disordered phase of a nonstoichiometric solid displays divariant behavior with respect to the phase rule, this may be due mainly to the fact that, in describing a microdomain the phase rule should be modified into

$$F = c - 2 + P + \sum \pi \tag{4}$$

where $\sum \pi$ describes the additional degrees of freedom of the microdomains with respect to one another. Hence the divariance of the equilibrium between microdomains might be due to this additional term, and the phase diagram might contain, on the G or μ curve, regions of pseudodiphasic equilibrium, which are divariant only because of the additional term $\sum \pi$ in the phase rule.

3. Integral Enthalpies

The concept of microdomains is supported by the thermochemical observation that in most grossly nonstoichiometric compounds the formation enthalpy ΔH_f of a compound of intermediate composition is a *linear function of the composition* having as coefficients the formation enthalpies of the stoichiometric compounds between which it is situated. For instance, for an oxide of general composition MO_{1+y} $(0 < y < 0.5)$, which is situated between MO and M_2O_3 ($MO_{1.5}$), one has (Ariya and Mozozowa, 1958; Ariya and Popov, 1962)

$$\Delta H_f^\circ(MO_{1+y}) = (1 - 2y)\,\Delta H_f^\circ(MO) + 2y\,\Delta H_f^\circ(MO_{1.5}) \tag{5}$$

Although Eq. (5) allows the hypothesis of ideal mixing, at the level of the partial molar enthalpies and entropies it is felt that the hypothesis is not sufficient. Sørensen (1978) points out that because of the difference in lattice parameters considerable local strain would be imposed on a UO_{2+x} lattice, if one considers the formation of U_4O_9 (the next oxide in the phase diagram) domains in a U_4O_8 (or UO_2) matrix as attempted by Dequesnoy and Marion (1964). In reality, at the level of partial molar quantities, not only can ideal behavior no longer be sustained, but $\overline{H}$ and $\overline{S}$ appear to be correlated.

4. Nonideal Models of Microdomains

The departure from ideality in microdomain equilibria is stressed by the very general formalism proposed by Thorn (1970) for a system in which one microdomain [hereafter indicated in all related quantities by the superscript (2)] is continuously growing in a matrix with a different structure [hereafter indicated by the superscript (1)]. In the system AB_y, N_A is the total number of atoms A, N_B the total number of atoms B (supposed constant), $N_A^{(2)}$ and $N_B^{(2)}$ the number of atoms A and B ordered in the microdomain, and $N_A^{(1)}$ and $N_B^{(1)}$ the number of atoms A and B in the matrix. Suppose that $E_1^{N_A}$ and $E_2^{N_B}$ are energies (considered dependent on the variable total number of A atoms N_A) and g_1 and g_2 degeneracies for the matrix and microdomain, respectively. An appropriate partition function for the system is

$$Z = \int \cdots \int \left[g_1 \exp\left(\frac{-E_1}{kT} \right) \right]^{\rho} \left[g_2 \exp\left(\frac{-E_2}{kT} \right) \right]^{\rho} \lambda_A^{N_A} \lambda_B^{N_B} \, dE_1 \, dE_2 \, dN_A \, dN_B \tag{6}$$

(or the equivalent expressed as a sum), where $\lambda_i = \exp(\mu_i / kT)$ and μ_i is the chemical potential of atom i ($i = \mathrm{A, B}$).

The function ρ in Eq. (6) is called the correlation function and has the form

$$\rho = (N_A - N_A^{(2)})/(N_A^{(1)} - N_A^{(2)}) \tag{7}$$

with

$$N_A^{(1)} + (1 - \rho) N_A^{(2)} = N_A \tag{8}$$

At $N_A = N_A^{(1)}$, $\rho = 1$; at $N_A = N_A^{(2)}$, $\rho = 0$; this implies an ordering of the atoms N_A completely in the matrix and in the microdomain, respectively.

At a given composition AB_y, i.e., at a fixed number N_A of atoms A, we can vary the relative number $N_A^{(2)}$ with respect to $N_A^{(1)}$ (keeping in mind that $N_A^{(1)} + N_A^{(2)} = N_A$), for instance, by varying the temperature. In this way we displace the thermal equilibrium between the matrix and the microdomain in the solid. It is easy to see from Eq. (7) that ρ will have a discontinuity at $N_A^{(1)} = N_A^{(2)} = \frac{1}{2} N_A$, passing from $+\infty$ to $-\infty$. In the first range of ρ (1 to $+\infty$) the microdomain is growing into the matrix; in the second range, the matrix is growing into the microdomain, and we can obtain a correlation function having a positive sign simply by interchanging the meaning of the superscripts (1) and (2) in all the equations.

Suppose that N_A (i.e., y in AB_y) is kept constant; the function ρ (or, better, its reciprocal $s = 1/\rho$) is similar to the order parameters as defined in order–disorder theories, e.g., for metallic alloys (see Fowler and Guggenheim, 1939).

From the partition function, the chemical potentials μ_i as well as their derivatives of any order can be worked out, thus allowing analysis of the experimental thermodynamic functions. This treatment implies essentially that the coherent intergrowth of the microdomain and the matrix can be represented by an ordering of the constituting atoms into one or the other structure, the correlation function ρ acting as the ordering parameter.

The energy of a microdomain (e.g., E_1 in the above treatment) might be conveniently split into δ, representing the cohesive stabilizing energy of the new structure per atom, and $\sigma(D)$, representing the energy of the boundary zone of coherent growth (surface), which is a function of the size of the microdomain. Here D is the size or dimension of the microdomain. In $\sigma(D)$ are included the strains produced by the microdomain on the matrix host lattice; $\sigma(D)$ may depend also on the mutual orientation of the microdomains with respect to one another and to the matrix. It is important to note that, at lower temperatures and close to the diphasic regions, the energy $\sigma(D)$ has a predominant role in determining the thermodynamics of the system. Also, the mutual orientation of the microdomains can give rise to long-range forces, inducing the long-range order necessary for the creation of a new structure (Thorn, 1970).

Including $\sigma(D)$ in the partition function, Anderson developed a statistics of microdomains. The average size $\bar{D}$ at temperature T is given, in these statistics, by

$$\bar{D} - D_0 + 1 = [1 - X(1 - b\lambda^{ex})]^{-1}\exp[-\sigma(D)/kT] \qquad (9)$$

Here D_0 is the minimum size allowed for the microdomain, X its molar fraction, b a constant given by the molecular formula of the compound the microdomain represents, $\lambda^{ex} = \exp(\mu^{ex}/kT)$, where μ^{ex} is the difference between the chemical potential of the microdomain in a disordered structure and in a low-temperature ordered structure (the microdomain system is stable only when $\mu^{ex} < 0$).

The energy δ, on the other hand, prevails in determining the thermodynamics of the system at higher temperatures (far from the diphasic region). It is important to note that, to the extent that a microdomain can also be represented as a crystal region in which point defects have become ordered, creating the seeds of a new structure, this quantity is related to interaction energies contributing to the formation of defect clusters. We see here the relation between microdomain statistics and the statistics of interacting defects discussed previously (Thorn, 1970).

II. Basic Concepts and Formalism of Spacing Statistics

A. *The Basic Ideas of Spacing Statistics*

1. Fluctuations in the Short-Range Order of the Defects

As seen in Section I, a disordered phase of a nonstoichiometric compound shows local fluctuations in the order of the point defects responsible for nonstoichiometry. Also, the disordered phase may be pictured as an assembly of microdomains of different lattice structures where defects have been assimilated, assuming therefore local order.

Spacing statistics are essentially based on this concept of fluctuations in the short-range order of the point defects. It is important to visualize the defective lattice in order to understand the formalism and the applications of spacing statistics.

A random state for N defects would correspond to an average distance $\bar{R}$ between them. A more ordered state, for instance, that of a region in which a new crystal structure is forming, would have an average interdefect distance R different from $\bar{R}$. Let C be a volume, supposedly spherical, surrounding a defect situated in the center, which has a radius of $R/2$. This volume defines a region in which no other defect is present and is a function of the order achieved among the defects in the structure. To it corresponds an average density $1/C$ for the ordered population of defects.

Suppose that we are able to sample a large volume of sample volumes V_s out of the lattice and to determine for this volume V_s the average spacing C between the defects. If an interaction exists favoring a certain C_i, the number n_i of samplings in which this C_i is the average spacing of the defects will be increased with respect to the number of samplings showing the random spacing. Let us call the situations in each V_s local thermal fluctuations in order between the defects. Then n_i is the population of thermal fluctuations i in the lattice characterized by an average spacing C_i.

2. Short-Range and Long-Range Order

Suppose now a situation in the lattice in which an interaction causes ordering of the defects over a limited range (short-range order) characteristic of a certain fluctuation i. Local sampling as described above reveals a thermal distribution in which the population of thermal fluctuations n_i and the average spacing C_i are most frequent. We say that our system is characterized by this type of short-range order and that the average spacing C_i and the local density $1/C_i$ are a measure of it. But some longer-range interactions might be also present, by which all the microdomains of the short-range order considered previously are in some way ordered with respect to each other. This means that, over a very large region of the crystal, two short-range situations, characterized by very close values C_j and C_i, are in some way related by this longer-range interaction. The sampling procedure performed to detect the short-range order might not be sufficient to detect this longer-range order.

Thus a given sampling procedure provides information about a certain type of order characterized by a certain range or local pattern, leaving out information about longer-range or less local patterns. The set $[C_i]$ of spacings of the defects and the corresponding set $[n_i]$ of thermal fluctuations characterize the thermal equilibrium with respect to some local or short-range type of order, leaving out information about longer-range order that our sampling was unable to detect.

3. *Thermal Stability of a Fluctuation i*

In saying that a fluctuation i was caused by an interaction, we meant that, because of this interaction, each defect possessed a certain mean energy E_i within the fluctuation i. It is this mean energy E_i (with respect to the energy $\bar{E}$ that the defect would possess in a random state) that makes the fluctuation i thermally stable. It is clear that thermal motion tends to bring the defects into the random state. It is better therefore to define the *thermal stability* of the fluctuation i as $\theta_i = E_i/kT$. The situation in the lattice, relative to some short-range order, is therefore characterized by a set $[E_i]$ of average energies per defect and by a corresponding set $[\theta_i]$.

A cluster of point defects can be considered a special type of very stable short-range order where the average spacing is characterized by the dimensions of the cluster and by the average interaction of the cluster with other clusters and the lattice. In this way the cluster concept may be introduced in this representation of the disordered lattice.

It is clear that the mean energy E_i characterizes *only* the interaction that causes the short-range order represented in the fluctuation i. If two fluctuations i and j are related, as in the preceding paragraph, by some other longer-range interaction, this will add to the mean energies E_i and E_j another term E_{ij}.

4. *The Partition Function of Spacing Statistics*

From now on we assume that the sampling procedure described above has provided a set $E_{fi}(C_i)$ of energies per C_i. This is equivalent to saying that, in a lattice composed of B sites (lattice B), we have been able to solve a Hamiltonian $\mathcal{H}$ for all possible *interactions* between N defects by expanding it as

$$\mathcal{H} = \sum_{i=0,L} n_i(C_i)\mathcal{H}_i(C_i) + \mathcal{H}_{\mathrm{res}} \tag{10}$$

$$\sum_{i=0,L} n_i(C_i) = N \tag{11}$$

where each $\mathcal{H}_i$ represents the Hamiltonian of one defect interacting with other defects *external* to a definite, defect-free envelope or spacing C_i. The $E_{\mathrm{f}i}(C_i)$ can be thought of as being the eigenvalues of the one-defect Hamiltonians $\mathcal{H}_i(C_i)$. The term $\mathcal{H}_{\mathrm{res}}$ represents the interactions *not* included in the set of $\mathcal{H}_i$'s (e.g., longer-range interactions) and can be thought of as being a perturbative term.

The total energy E_{A} of an ensemble constituted by a set $[n_i^A]$ of defects can therefore be written as

$$E_A = \sum_{i=0,L} n_i^{\mathrm{A}} E_{\mathrm{f}i} + \epsilon \tag{12}$$

where ϵ derives eventually from the perturbative term $\mathcal{H}_{\mathrm{res}}$ and

$$E_{\mathrm{f}i} = E_{\mathrm{f}} + E_i(C_i) \tag{13}$$

where E_{f} is the formation energy of one defect in the lattice.

The expansion Eq. (10) without any perturbative term $\mathcal{H}_{\mathrm{res}}$ [i.e., $\epsilon = 0$ in Eq. (12)] defines objects, which may be called blocks C_i, that are *independent* and have a well-defined volume C_i. In this case we will speak of pure spacing statistics.

If we consider a canonical ensemble (Fowler and Guggenheim, 1939) of independent blocks C_i, the partition function Z_{s} of this ensemble is given by

$$Z_s = \sum_{\mathrm{A}} \Omega_{\mathrm{A}} \exp(-E_{\mathrm{A}}/kT) \tag{14}$$

with the conditions: $\sum_{i=0,L} n_i^{\mathrm{A}} = N$, $\sum_{i=0,L} n_i^{\mathrm{A}} E_{\mathrm{f}i} = \bar{E}$, where $\bar{E}$ is the thermal mean energy of the canonical ensemble. All thermodynamic functions and the equilibrium set $[n_i]$ can be calculated by extracting from Eq. (14) the most probable term and maximizing it with respect to the set $[n_i]$ (e.g., by the usual method of Lagrangian undetermined multipliers).

If $\epsilon \neq 0$, then Eq. (5) is modified to

$$Z = Z_{\mathrm{s}} Z(\epsilon); \qquad Z(\epsilon) = \sum_{\mathrm{A}} \exp(-\epsilon/kT) \tag{15}$$

B. The Configurational Degeneracy Ω_{A} for Pure Statistics

The functional form of the degeneracy Ω_{A} in Eq. (14) is written by generalizing the site exclusion principle employed in the case of pairwise repulsion between nearest-neighbor defects (Section I, C). However, this generalization can be shown *to be independent* of any particular nature of the interaction (whether repulsive or attractive) and to depend *only* on the fact that we have transformed, by the expansion Eq. (11), the system of point defects in a system of noninteracting blocks C_i having a characteristic spacing (or envelope) C_i associated with them.

In order to find the functional form of Ω_A and to demonstrate this assumption, we describe a conceptual experiment, which may be called the *interdiffusion* or *intergrowth* of two structures each characterized by a single species of block C_i. For simplicity we restrict ourselves to a set $[C_i] = [C_0 > C_1]$ $(L = 1)$, i.e., to only two types of short-range order.

(a) Suppose first that lattice B is separated by a barrier into two parts, B_0 and B_1, each containing *only* the block having the same subscript and such that $B_0 + B_1 = B$. The barrier prevents interdiffusion of the two species, so that this system does not maximize the configurational entropy S of the two populations. The entropy S_{sep} is in this case, for a given B_0 and B_1,

$$S_{\text{sep}} = S_0 + S_1 = k(\ln g_0 + \ln g_1) \tag{16}$$

where g_0 and g_1 represent the configurational degeneracies of the two populations n_0 and n_1 in the two separated volumes B_0 and B_1. The two degeneracies g_0 and g_1 are functions of the choice of the two volumes B_0 and B_1 and are given by Eq. (3), since we describe the randomization of one block in its proper volume as in this equation. Suppose that S_{sep} has a maximum for two particular volumes B_0^* and B_1^*, which we define as $B_0^* = C_0 n_0 + K$ and $B_1^* = B - C_0 n_0 - K$, where K is a certain number of sites. The $g_{\max} = g_0^* g_1^*$ is given by

$$g_{\max} = \binom{B_0^*/C_0}{n_0}\binom{B_1^*/C_1}{n_1} C_0^{n_0} C_1^{n_1} \tag{17}$$

The system so defined has a total energy given by $\bar{E} = n_0 E_{f0} + n_1 E_{f1}$ (by which we indicate that the barrier does not modify the two eigenvalues E_{f0} and E_{f1} and that the total energy equals the average energy $\bar{E}$ of the canonical ensemble).

(b) We now lift the barrier, and interdiffusion of the two populations of blocks in lattice B occurs *without a change* in the total energy $\bar{E}$.

Let us isolate an envelope of C_0 sites, containing a C_1 block at its center, anywhere in the lattice (and preferably in B_1). Since we have assumed that the blocks are noninteracting, dissipation of energy from this volume will be zero (or extremely slow). Any process occurring within this volume will be an *adiabatic process* and thus spontaneous if coupled with an increase in entropy. Thus the process

$$d(C_1)(P - 1)\ (\text{lattice sites}) \rightarrow d(C_0) \tag{18}$$

which represents the tendency of a defect in block C_1 $[d(C_1)]$ to move through the whole envelope C_0, thus becoming a defect in block C_0 $[d(C_0)]$, by diffusing into the $(P-1)$ lattice sites $(P = C_0/C_1)$, will be spontaneous since it involves an increase in entropy:

$$S_{(0)} - S_{(1)} = K(\ln C_0 - \ln C_1) = K \ln P \tag{19}$$

It follows that, in the interdiffusion of the two populations of blocks when the barrier is lifted, block C_0 can grow at sites where a C_1 is situated, thus expanding into the whole of the volume B_1^*, but not vice versa. [The presence of a "thermal bath," which defines the canonical ensemble (Fowler and Guggenheim, 1939), insures, however, the reconstitution of the two populations n_0 and n_1 in lattice B at equilibrium.] The final state will have the same total energy $\bar{E}$ but *a degeneracy* Ω *given by*

$$\Omega = [w_0!/(w_0 - n_0)!\, n_0!][w_1!/(w_1 - n_1)!\, n_1!]C_0^{n_0}C_1^{n_1} \tag{20}$$

where $w_0 = B/C_0$, $w_1 = (B - C_0 n_0)/C_1$, and $\Omega > g_{\max}$ of Eq. (17).

In this experiment, the functional form is found for Ω_A, and it is shown that this form ensures maximization of configurational entropy for the two populations of blocks.

The functional form Ω_A given by Eq. (20) can be extended to a set $[C_i]$ with $0 \leqslant i \leqslant L$. The irreversible process of Eq. (18) holds, in fact, for any couple $C_i > C_{i+1}$, and we can generalize Eq. (20) as (Atlas, 1970)

$$\Omega_A = \prod_{i=0,L} \Omega_i \tag{21}$$

$$\Omega_i = [w_i!/(w_i - n_i)!\, n_i!]C_i^{n_i} \tag{22}$$

$$w_i = (B + C_i n_i - \sum_{r=0,i} C_r n_r)/C_i \tag{23}$$

Equation (23), which gives the number of available positions for block C_i, is obtained by assuming a particular method of filling lattice B with the populations $n_0, \ldots, n_L$. This procedure consists of first placing the largest block C_0, at the sites left free of C_0, then the next smaller block C_1, and so on down to the smallest block C_L. Block C_i prevents the formation of the smaller block $C_{i\pm 1}$ at sites, *but not vice versa*; Eq. (23) is therefore derived by a multiple application of the site exclusion principle.

We have seen earlier that the site exclusion principle has been introduced (Speiser and Spretnak, 1955) to describe the effect of repulsion between defects. For the same purpose, it has been used in applications of spacing statistics (Atlas, 1968a, b; Manes and Manes-Pozzi, 1976; Manes *et al.*, 1979) that we will discuss later. In another presentation of these statistics (Atlas, 1970), it is employed without any special assumption as to whether repulsion or attraction is acting between defects. It is in reality only a consequence of the fact that we are maximizing the occupation of lattice B with blocks having their own volume C_i; it represents the correct configurational degeneracy for the disposition of extended defects in a lattice of B points.

C. *The Spacing Rule in Spacing Statistics*

The expansion Eq. (10) is governed by the kind of short-range-order situations considered, since they determine the choice of the individual Hamiltonians $\mathcal{H}_i(C_i)$. In practice, the possibility of defining particular clusters of defects dictates the particular choice of the set $[C_i]$. We call the rule by which the set $[C_i]$ is generated the spacing rule.

In a system of defects in a nonstoichiometric phase, all interactions and all types of order may be thought to be described when decreasing the spacing C_i of a lattice unit step I (e.g., a lattice vector) that is dependent on the particular structure in which the defects are positioned. The spacing rule may therefore be expressed as (Atlas, 1968a, b, 1970)

$$C_i = C_0 - iI \tag{24}$$

The corresponding set $[E_{fi}]$ represents in this case a spatial discretization of the interaction energy ϵ, as well as of the different types of order by which the interaction between defects is systematically varied in steps of the lattice vector.

Other spacing rules can, however, be employed, as we will see in the following.

D. *The Thermodynamic Functions for Pure Statistics* ($\epsilon = 0$)

1. *Integral Thermodynamic Functions*

The free energy G is given by

$$G = -kT\ln Z = \sum_{i=0,L} (n_i E_{fi}) - T(k\ln(w_i^{w_i} C_i^{n_i}/(w_i - n_i)^{(w_i - n_i)} n_i^{n_i})) \tag{25}$$

where the first term represents the enthalpy H and the second term the entropy S of the system, for a distribution $[n_i]$ of blocks at thermal equilibrium.

The enthalpy is that expected for an *ideal* solution of blocks in lattice B: When two structures, containing the separated populations of blocks (as in the conceptual experiment described above) are mixed, no change in enthalpy occurs; i.e., *no enthalpy of mixing is present* in the pure statistics. This agrees with the observations of Ariya (Ariya and Popov, 1962) discussed in Section I.D. This result, however, supposes that the expansion Eq. (10), hence the correspondence assumed between the sets $[C_i]$ and $[E_{fi}]$, validly describes interaction *at least up to a certain minimum range* C_L. Any enthalpy of mixing is in this way represented only by one perturbative ϵ of Eq. (12).

On the other hand, the dissolution of the two structures increases the entropy considerably. The formation of nonstoichiometric phases should be governed mostly by an *entropy of mixing*, as often suggested (Anderson, 1970; Thorn, 1970).

Finally, entropy and enthalpy are closely related because of the existence of a correspondence between the sets $[C_i]$ and $[E_{fi}]$ and of a spacing rule. The observation of such a relationship is central to the problem of nonstoichiometric phases, as discussed in Section I, B.

2. *The Partial Molar Thermodynamic Functions*

The chemical potential μ_i for the block C_i is obtained, as usual, by differentiating Eq. (25) with respect to the corresponding population n_i.

We start by considering the two cases $L = 0$ ($[C_i] = [C_0]$) and $L = 1$ ($[C_i] = [C_0 > C_1]$).

In the case $L = 0$, only one block is present and is distributed randomly in lattice B. The equation is ($n_0 = N$)

$$\mu = \mu^\circ + kT \ln[N/(w_0 - N)] = \mu^\circ - T[k \ln(a_{\text{un}}/a)] \quad (26a)$$

$$\mu^\circ = E_f - kT \ln C_0 \quad (26b)$$

where a_{un} ($a_{\text{unused sites}}$) and a are, respectively, the fraction of lattice B not occupied by blocks $[a_{\text{un}} = C_0(w_0 - N)/B]$ and the fraction occupied by blocks ($a = C_0 N$). The quantity $w_0 = B/C_0$ is given by Eq. (23) when $L = 0$. Equations (26a) and (26b) are what is expected from the random distribution of a volumetric defect or block when the site exclusion principle is employed (Speiser and Spretnak, 1955): Notice the entropic term $kT \ln C_0$ in μ^*, which expresses the intrinsic entropy of one block. The μ° is the standard chemical potential as usually defined in thermodynamics.

In the case $L = 1$, simple algebra shows the following expressions for the chemical potentials μ_0 and μ_1 of the blocks C_0 and C_1:

$$\mu_0 = \mu_0^\circ - T[k \ln(a_{\text{un}}/a_0)] \quad (27a)$$

$$\mu_1 = \mu_1^\circ - T[k \ln(a_{\text{un}}/a_1)] \quad (27b)$$

$$a_{\text{un}} = (B - C_0 n_0 - C_1 n_1)/B \quad (27c)$$

$$a_0 = C_0 n_0/B; \qquad a_1 = C_1 n_1/B \quad (27d)$$

$$\mu_0^\circ = E_0 - T[k(\ln C_0 - \ln F^{(P-1)})] = E_0 - T(k \ln C_0 + \bar{S}_{\text{intgr}}) \quad (27e)$$

$$\mu_1^\circ = E_1 - kT \ln C_1 \quad (27f)$$

where $\bar{S}_{\text{intgr}}$ stands for $\bar{S}_{\text{intergrowth}}$. Here we have maintained the same definition as in Eq. (26a) for the standard chemical potentials μ_i°. It is seen, however, that μ_0° ($L = 1$) differs from Eqs. (26a) and (26b) by the term

$$\bar{S}_{\text{intgr}} = -k \ln F^{P-1} = k \ln(a_{\text{un}}/a_{\text{intgr}})^{P-1} \tag{28}$$

where a_{un} has been already defined, $P = C_0/C_1$, and

$$a_{\text{intgr}} = (B - C_0 n_0)/B \tag{29}$$

represents *the fraction of lattice B in which further intergrowth can occur* between the two blocks C_0 and C_1. This is the volume in which the two populations n_0 and n_1 *compete* when the total number $N = n_0 + n_1$ of defects is varied.

The quantity $F^{-1} = a_{\text{un}}/a_{\text{intgr}}$ represents the concentration of unused lattice sites that can take part in the process of intergrowth described by Eq. (18). The relationship of Eq. (28) to the process [Eq. (18)] is evident.

The quantity $\bar{S}_{\text{intgr}}$ governs the whole statistics. At thermal equilibrium ($\mu_0 = \mu_1$), it is easy to show the following relations for the thermal populations n_0 and n_1 (normalized now according to the rule $y_i = n_i/w_0$, $y = \sum_{i=0,L} y_i$) and for F:

$$M_0 = a_{\text{intgr}}/a_0 = (1/C_0)\exp\{\theta_0[P/(P-1)]\bar{S}_{\text{intgr}}/k - \mu/kT\} \tag{30a}$$

$$M_1 = a_{\text{intgr}}/a_1 = (1/C_1)\exp(\theta_1 - \mu/kT) \tag{30b}$$

$$F = a_{\text{intgr}}/a_{\text{un}} = (1 + M_1)/M_1 \tag{30c}$$

$$y_0 = (1 + M_0)^{-1} \tag{30d}$$

$$y_1/y_0 = \exp[-(\theta + \bar{S}_{\text{intgr}}/k)] \tag{30e}$$

Here we used the thermal stabilities ($\theta_i = E_{fi}/kT$) defined above, and we have defined the *difference in thermal stabilities* between the two blocks: $\theta = \theta_1 - \theta_0$.

An advantage of Eqs. (30a)–(30f) is to give

$$y = y(\mu/kT, [\theta_i], [C_i]) \tag{31}$$

after Eqs. (30a)–(30f) have been used to eliminate concentration variables. This gives the chemical potential μ as a function only of the total concentration variable y, which is the total number of defects (e.g., of vacancies in a MO_{2-x} oxide), hence the only concentration *observable*, and of the results of the expansion Eq. (10) (or the assumptions taken for the two sets $[C_i]$ and $[E_{fi}]$).

Simple algebra extends these results to any number of blocks L. The set of equations ruling the statistics is given by

$$M_i = (1/C_i)\exp(\theta_i - \mu/kT) \prod_{t=(i+1),L} F_t^{P_{it}} \tag{32a}$$

$$M_L = (1/C_L)\exp(\theta_L - \mu/kT) \tag{32b}$$

$$y_i = (1/M_i) \prod_{s=0,i} F_s^{-1} \tag{32c}$$

$$F_s = (1 + M_s)/M_s \tag{32d}$$

The advantage of these relations is once again to give the chemical potential μ (although in implicit form) as a function of y and of the sets $[C_i]$ and $[E_{fi}]$ governing the expansion Eq. (10).

The quantities F_s are also in this case related, through Eq. (18), to the intergrowth of block C_i into block C_{i+1}.

To summarize:

(a) In the formalism of spacing statistics, the entropic intergrowth of blocks C_i governs the thermal equilibrium.

(b) The chemical potential μ is a function of the observable total concentration y *and* of the entropic changes due to this intergrowth.

3. *The Derivative μ' of the Chemical Potential: Spinodal Points*

From Eq. (31) the derivative $\mu' = (\partial\mu/\partial N)_T$ can be calculated. For $L > 1$, it is easier to take into account the inverse function

$$\rho = [\partial N/\partial(\mu/kT)]_T \tag{33}$$

In Fig. 2, μ and μ' are given at different θ for $L = 1$. We summarize the information in Fig. 2 as follows:

(a) The chemical potential μ for the system described in pure spacing statistics has second-order spinodal points [inflections of type I (Thorn, 1970)] when $\theta > 0$. These inflections disappear for $\theta \to 0$, i.e., for $\Delta = (E_{f1} - E_{f0}) > 0$, when $T \to \infty$. By the arguments of Thorn (1970), and Section I, B, these inflections can be interpreted as residues, in the single disordered phase, of lower-temperature diphasic regions. For $\theta < 0$, no second-order spinodal points are found, indicating great stability of the block C_1.

(b) The chemical potential μ of pure spacing statistics *does not* contain first-order spinodal points, hence the presence of a diphasic region. Therefore, pure spacing statistics describe a monophasic system. Nevertheless, they may store information (as second-order spinodal points) on the residues of lower-temperature diphasic regions for characteristic values of y. To describe the system completely, the total partition function Z of Eq. (15)

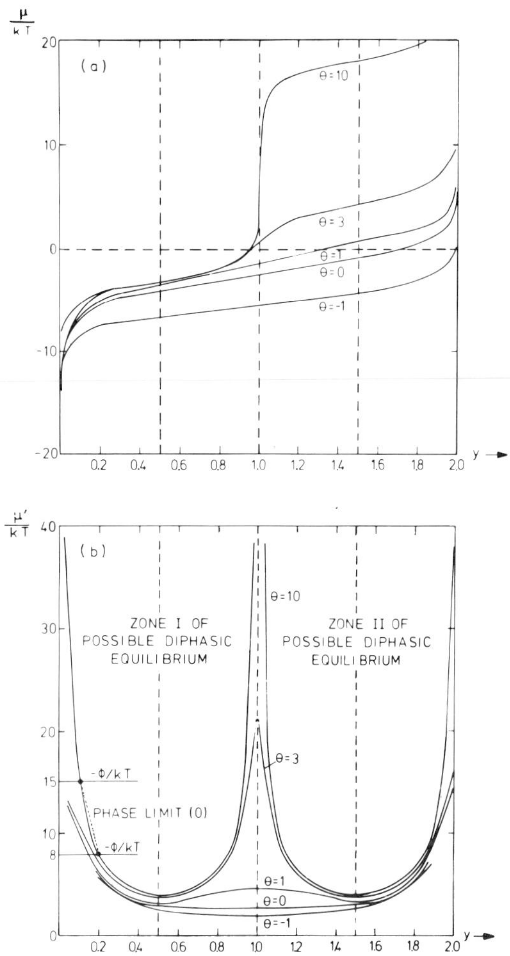

Fig. 2. (a) The chemical potential μ_s/kT as a function of the composition y and of the difference θ in thermal stability between two structures (0) and (1) (spacing rule: $C_1 = C_0/2$, $P = 2$). (b) The first derivative of the chemical potential μ_s' ($\epsilon = 0$): μ_s'/kT, as a function of the composition y and of the difference θ in thermal stability between two structures (0) and (1) (spacing rule: $C_1 = C_0/2$, $P = 2$). Conditions for first-order spinodals at two temperatures, T_1 and T_2, and a phase limit for structure (0) are indicated (the phase limit has been incorrectly situated at the position of the spinodals).

must be used, thus allowing an interblock interaction ϵ to appear. Nevertheless the variable Z_s in Eq. (15) can be considered a "storage" partition function for the intergrowth of two blocks C_0 and C_1 in lattice B.

Figure 3 shows the same calculations for $L = 4$ with a special hypothesis for the sets $[C_i]$ and $[E_{fi}]$, which will be discussed later when the application

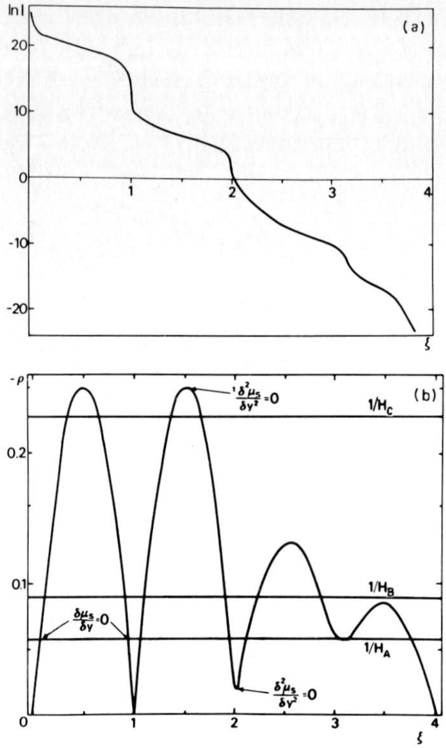

Fig. 3. (a) μ/kT versus composition y for sets $[C_i = C_0/i]$, $[\theta_i]$ and $L = 4$. (The curve has been used to calculate the phase diagram of CeO_{2-x}; see Section III). (b) The function $\rho = dy/d(\mu/kT)$ for the same sets. First-order spinodal points are indicated as intersections with $-kT/\phi = 1/H$.

of spacing statistics to the substoichiometric fluorite phases MO_{2-x} is considered.

A series of inflections of type I, separated by inflections of type J, are generated, as is appropriate for a description of the disordered high-temperature single phase of a nonstoichiometric system, which at lower temperature separates into a series of new phases (Thorn, 1970). An example may be given, once again, by the separation at low temperatures of a MO_{2-x} fluorite lattice into the M_nO_{2n-2} series.

E. Spacing Statistics for Two Strongly Interacting Defect Systems

Thus far spacing statistics have been used to describe a system composed of one type of defect that achieves local short-range order through interaction. Atlas (1970) has extended the statistics to the description of a system composed of two interacting subsystems of defects of different types. For instance, in lanthanide and actinide $MO_{2\pm x}$ oxides having the fluorite structure, the presence of oxygen interstitials or vacancies in the oxygen sublattice induces the presence of oxidized or reduced cations in the metal sublattice.

In this case, the statistics are defined by two sets of spacings $[C_i^{\mathrm{A}}]$ and $[C_j^{\mathrm{B}}]$ and two spectra of eigenvalues $[E_{ij}^{\mathrm{A}}]$ and $[E_{ji}^{\mathrm{B}}]$, where the presence of the two subscripts i, j is justified by the fact that one has to take into account the interaction between point defects of the same type (e.g., repulsive electrostatic interaction) and the interaction between point defects of different types (e.g., attractive electrostatic interaction). The basic concepts are the same as those already presented for the system of defects of one type, and the reader is referred to (Atlas, 1970) for the development of a formal approach.

The chemical potential μ can also in this case be expressed *only* as a function of the total concentration of defects y and the characteristic sets $[C_i^{\mathrm{A}}]$, $[C_j^{\mathrm{B}}]$, $[E_{ij}^{\mathrm{A}}]$, $[E_{ji}^{\mathrm{B}}]$.

F. Spacing Statistics as Statistics for Microdomain Formation

Thus far we have discussed spacing statistics as a tool for describing the progressive short-range order of point defects. We now introduce some concepts for use of the statistics to describe microdomain formation (Section I).

We may picture the formation of a microdomain as a progressive process whereby the assimilation of point defects begins with their association into a stable cluster. Clusters then aggregate and become ordered in larger and larger regions. The presence of a cluster perturbs the surrounding host lattice and influences the neighboring clusters.

Spacing statistics are employed in the case of clusters to connect the spacing rule, generating the sets $[C_i]$ and $[E_{\mathrm{f}i}]$, with conditions of formation and aggregation of the clusters. The presence of one cluster, in the lattice for instance, can be characterized by

(a) a stable spatial configuration of the point defects involved in the cluster and a bonding energy E_{b} of the cluster;

(b) A perturbation field $\Delta(\bar{r})$ extending radially around the cluster in the lattice (e.g., a strain field);

(c) an interaction field $\mathcal{E}(\bar{r}_{ij})$ between the clusters i and j, depending on their mutual distance r_{ij} (e.g., electrostatic interaction, if the clusters possess a charge with respect to the lattice or a dipole moment).

The aggregation of clusters may result in

(d) a variation in the perturbation field Δ (e.g., a greater strain field when a pair of clusters is present);
(e) a variation in the interaction field $\mathcal{E}$ between clusters, due, for instance, to complete or partial compensation by charges or dipoles.

The method of introducing conditions (d) and (e) in a spacing rule is of course not unique, but in principle the knowledge (or assumption) of the spatial configuration of the cluster, of the energy E_b, and of the fields Δ and $\mathcal{E}$ may lead to evaluation of

(a) the set $[C_i]$, which may be composed of the characteristic envelopes of host lattice containing one cluster (or a pair of clusters) and a certain shell of host lattice perturbed by the cluster;
(b) the set $[E_{fi}]$, which would be characterized by the energy E_b and some value of the perturbation Δ in the shell of the associated host lattice.

The consideration of $\mathcal{E}$ might finally provide the term ϵ not contained in the set $[E_{fi}]$.

This type of reasoning considers a microdomain as locus at which defects become ordered via suitable clustering. The set $[E_{fi}]$, which is defined for the defects, can be constructed in such a way as to take care of the difference between the two energies δ and σ discussed in Section I. A defect may perhaps be present at the surface between two microdomains or in the bulk of the microdomains; it will possess, in the two cases, different E_{fi} values. The sampling volume V_s for the statistics must of course be chosen to be sensitive to this variation in E_{fi}.

G. First-Order Spinodal Points: The role of ϵ in Z.

Suppose now that we use the full partition function Z of Eq. (15) to describe our system of defects or clusters.

From Eq. (15) it is seen that $\mu' = 0$, i.e., first-order spinodal points occur when

$$\mu_s' = -\mu'(\epsilon) \tag{34}$$

The reduction of the interaction energy between defects or clusters into a set of spacing levels $[E_{fi}]$ was achieved by averaging this interaction energy within a local pattern of order. Longer-range, interblock interaction is therefore not represented in the set $[E_{fi}]$. Hereafter the interblock interaction not reduced into $[E_{fi}]$ will be called the long-range interaction ϵ.

A knowledge of ϵ is required in searching for first-order spinodal points. This is certainly no trivial problem, since it requires a very detailed modeling of the system of defects in lattice B. Nevertheless, the presence of the storage partition function Z_s may simplify the problem, since the interaction must be known only for distances larger than the spacings $[C_i]$.

Suppose, as in the case discussed in the preceding paragraph, that just one type of defect or cluster can assume the spacings of the set $[C_i]$. In the long-range interaction ϵ, we may think that, over some larger interdefect distance, the defects no longer "see" one another as belonging to classes i. Hence ϵ can be given the form

$$\epsilon = N^2\Phi \tag{35}$$

(if only the interaction between pairs of defects is taken into consideration), where Φ is some average potential per pair of defects. We will now examine the effect of this form of ϵ on the partition function Z.

Equation (34) becomes

$$\mu_s' = -\Phi \tag{36}$$

and extrema in μ are found if $\Phi < 0$, i.e., if Φ represents an attractive interaction.

In Fig. 2b $-\Phi/kT$ is represented by a horizontal line, and the intersections with μ_s'/kT represent extrema in μ versus y, hence diphasic regions. It is seen that there may be one or two diphasic regions depending on the position of $-\Phi/kT$. Both $y = \frac{1}{2}$ and $y = (P + 1)/2$ represent points I on the μ curve.

From the approximate values of μ_s'/kT at $y = \frac{1}{2}$ and $y = (P + 1)/2$ closure of the diphasic zones I and II in Fig. 2 occurs at

$$T_c(\text{I}) = -4k/\Phi \tag{37}$$

and

$$T_c(\text{II}) = -4k/\Phi(P - 1)^2 \tag{38}$$

which are attained by raising the temperature T.

The analysis of the effect of this simple form of ϵ given by Eq. (35) illustrates the meaning of the storage partition function Z_s. Given a well-defined spacing rule generating the sets $[C_i]$ and $[E_{fi}]$, the storage partition function Z_s individuates the possible location of diphasic regions. These regions appear at lower temperatures, and a phase diagram of the system can be drawn.

Above the critical temperatures T_c, on μ versus y curves, inflections of type I, separated by inflections of type J at the characteristic y's marking the saturation concentrations $y = p^{(i+1)i}$ (where lattice B is saturated by the

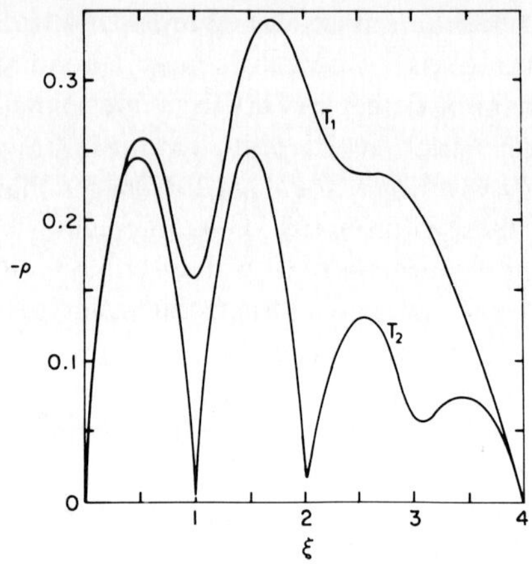

Fig. 4. The function $\rho = dy/d(\mu/kT)$ for the same sets as in Fig. 3 but at different temperatures ($T_1 > T_2$). Notice that some of the extrema in ρ transform into inflections when the temperature is raised.

different blocks C_i), persist in a certain temperature range and disappear at very high temperatures. Figure 4, however, shows that even for high temperatures the characteristic points $y = p^{(i+1)i}$, as well as $y = \frac{1}{2}(p^{i(i+1)} - p^{i(i-1)})$, remain as loci, where the μ' curve undergoes a change, i.e., as zero in the $\mu^n = (\partial^n \mu/\partial y^n)_T$ derivatives. The storage partition function Z_s appears therefore as a device that connects the low-temperature and high-temperature thermodynamics through the configurational mechanism; this has been illustrated in the preceding paragraphs.

H. Spacing Statistics as a Code for Reading the Experimental Thermodynamic Curves

The characteristic oscillations in the derivatives of the storage partition function Z_s suggest possible utilization of spacing statistics which may permit information about the size of point defect clusters and the general form of the interaction energy to be extracted from the high-temperature thermodynamics. In fact the formalism of spacing statistics can be used to interpret the thermodynamic functions at a given temperature with respect to the total number of defects N (for instance, x in $MO_{2 \pm x}$). It has been shown elsewhere in this book that the μ-versus-$\ln x$ as well as the $\overline{H}$-versus-$\ln x$ curves experimentally determined show characteristic values of x where the slope changes. The pseudophase diagrams (Sørensen, 1978) constructed from these curves show striking similarities for grossly non-stoichiometric compounds of the same type. Assume therefore that the concentrations at which this occurs are significant as being correlated with some of the concentrations y in which Z_s shows oscillations, hence with particular values of the spacing set $[C_i]$. Spacing statistics used as a code therefore open up the possibility of gaining information about the size and extension of characteristic clusters of point defects in the high-temperature thermodynamics of these compounds. Once these clusters are identified, it is possible to use this information to construct lower-temperature-derived structures. This analysis in the case of substoichiometric fluorite phases will be discussed in the following section.

I. Conclusions

In the most general partition function of spacing statistics for a system of interacting defects or clusters, information about the nature and the saturation concentration of the different species and of their short-range interactions can be stored in a storage partition function Z_s, and other longer-range interactions can therefore be separated by factorization. This allows description of the disordered phase of nonstoichiometric compounds and prediction of their phase diagrams. Spacing statistics act as a simplified statistics of microdomains.

Spacing statistics include some of the features required for a realistic description of the intergrowth of microdomains of closely related structures:

(a) In integral thermodynamic quantities, pure spacing statistics permit the enthalpy to be reduced to that of an ideal solution of the two structures.

(b) The intergrowth is, however, governed by the entropy of the mixture, and spacing statistics, even in their pure form, include information about spinodal compositions; this is the meaning of the parameter of intergrowth defined above.

(c) At any step, the partial quantities $\overline{H}$ and $\overline{S}$ are strongly correlated through the functional relation of the two sets $[C_i]$ and $[E_{fi}]$.

(d) Any residual enthalpy of mixing is included in the longer-range interaction ϵ not included in $[E_{fi}]$; this interaction determines unmixing and the appearance of diphasic phases.

The application of spacing statistics is of course subject to the spacing rule establishing the set $[C_i]$ and to the method employed to define the set $[E_{fi}]$ of spacing levels.

In the following part, some examples of the application of spacing statistics will be given. Interest will be focused on the spacing rule adopted and on the methods employed to define the set $[E_{fi}]$.

III. Applications of Spacing Statistics

A. *Introduction*

The formalism of spacing statistics illustrated in Section II has been applied to the description of the nonstoichiometric fluorite phases of binary compounds $MO_{2\pm x}$ (M = Ce, U, Pu) and of ternary compounds (U, Pu) O_{2-x}. These applications are reviewed in this section.

The first application, by Atlas (1968a, b, 1970), is a general attempt to describe the interaction between defects in the fluorite lattice over a wide range of compositions and temperatures. The oxides considered are UO_{2+x} and CeO_{2-x}. In both oxides, the system considered is composed of two subsystems A and B:

(a) In CeO_{2-x}, oxygen vacancies V^z (subsystem A) and reduced cations $M^{z'}$ (subsystem B) in which z and z' represent ionization effective charges of the point defects with respect to the lattice (z and z' can also be fractional; they are taken as $z = 1$, $z' = -1$);

(b) In UO_{2+x}, Willis clusters (Willis, 1964) with an effective charge $z = -1$ (subsystem A) and U^{+5} ions with an effective charge $z' = 1$ (subsystem B).

The application of spacing statistics by Atlas passed through two different stages, characterized by different physical models and mathematical devices for generating the spacing set $[E_{fi}]$:

(a) In the first stage (Atlas, 1968a, b), the two subsystems are considered independent; furthermore, the progressive ordering of the defects is considered to be governed essentially by electrostatic repulsion between equally charged defects.

(b) In the second stage (Atlas, 1970), the two subsystems are considered to be strongly interacting and their progressive ordering to be due to both repulsive *and* attractive interaction. Therefore this stage represents a general attempt to treat charge–charge Coulomb interaction between defects in a lattice and consequently will be very carefully examined.

The second application (Manes *et al.*, 1979) was specially intended for the construction of phase diagrams of substoichiometric fluorite phases of binary and ternary oxides. It implies the assumption of a fundamental cluster (the tetrahedral defect), which also acts as a building block in the construction of low-temperature stable structures. It generates the sets $[C_i]$ and $[E_{fi}]$ of spacing statistics by formulating hypotheses on the spatial extension of the clusters and of the strain field generated by them in the fluorite lattice. It implies, finally, a long-range interaction energy between these clusters.

B. The General Interaction Statistics of Atlas

1. Spacing Statistics for Two Independent Subsystems

Consider the two independent subsystems defined above. In the lattice, each defect A is surrounded by an unknown number of other defects A and defects B at various distances.

Consider a pair of equally charged defects A separated by a distance d_i. When one of them is taken as the center, the second can be located at any lattice site lying on a sphere of diameter d_i. They will exert on each other a repulsion that may be called E_i^{A}, and the sphere of diameter d_i may be called the interaction volume for this repulsion. Alternatively, for each defect A of the pair, an envelope or spacing C_i^{A} can be defined, containing C_i^{A} possible positions for defects A and having a radius $r_i = \frac{1}{2} d_i$. Likewise, a spacing C_j^{B} and a repulsive energy E_j^{B} can be defined for a defect B at a distance d_j from another equally charged defect B.

Suppose now that defects A of spacing C_i^{A} are in a local pattern of order for defects A as defined in Part II. The spacing C_i^{A} is a measure of this order, and its reciprocal is the local density for defects A. By varying C_i^{A} with a spacing rule, e.g.,

$$C_i^{\mathrm{A}} = C_0^{\mathrm{A}} - i$$

from some value C_0^{A} to some lower limit C_L^{A}, the set $[C_i^{\mathrm{A}}]$ can be generated and, correspondingly, the set $[E_i^{\mathrm{A}}]$ of repulsive spacing levels.

If the subsystems A and B are independent, the local patterns of order for A are not influenced by those for B. Hence another set $[C_j^{\mathrm{B}}]$ and another set $[E_j^{\mathrm{B}}]$ of repulsive levels can be generated for the subsystem B independently of A.

In reality, the envelope or spacing C_i^{A} for defect A also contains, in the real lattice, defects B [e.g., if A are the oxygen vacancies V in CeO_{2-x}, the volume C_i^{A} also contains, in the sublattice, defects $(\mathrm{Ce})^{z'}$]. Between the two species, an attractive interaction will occur. The net energy of defect A at the center of C_i^{A} will be the resultant of the repulsion *and* attraction components. In order to overcome this difficulty and assign a well-defined value E_i^{A} to defect A in the spacing C_i^{A}, the idea of independence of the two subsystems is utilized in a second way.

It is considered that in the spacing C_i^{A} defects B carrying the charge z' are distributed uniformally over all the possible B sites in the envelope C_i^{A}. Conversely, for spacing C_j^{B}, the same is true of defects A. This means that defect A experiences *constant attraction* from defects B, and vice versa. The independence of the two systems is used to transform the distribution of one of the subsystems into a uniform distribution when treating the repulsive interaction of the other subsystem.

The total energy E of the crystal is then

$$E = \sum_{i=0,L^A} n_i^A E_{fi}^A + \sum_{j=0,L^B} n_j^B E_{fj}^B + N_{AB}(N) E_{AB} \tag{39}$$

where $[E_{fi}], [E_{fj}]$ are obtained from E_i^A, E_i^B by the addition of constant formation energies, as in Eq. 13, n_i^A and n_j^B are thermal populations, N_{AB} represents the total number of pairs A–B present in the lattice (a function of the total population N^A of one of the defects (since $N^B = (z'/z)\, N^A$ for electroneutrality), and $\bar{E}_{AB}$ is some average attractive potential. The total degeneracy of the system is then given by

$$\Omega = \Omega_S^A \Omega_S^B \tag{40}$$

where Ω_S^A and Ω_S^B have the form given by Eq. (21).

Atlas (1968a, b) gave expressions for E_i^A and E_j^B in terms of the dimensions of the spacings considered spherical, for instance,

$$E_i^A = \frac{z^2 e^2}{2Ka}\left(\frac{4\pi}{3C_i^A}\right)^{1/2}\left(\frac{7C_0^A + i}{C_i}\right) \tag{41}$$

where e is the electron charge, K a dielectric constant (see later), and a the lattice constant of the host fluorite structure.

This way of treating the attractive interaction is equivalent to (a) considering the ordering of defects as being guided mostly by interdefect repulsion, and (b) separating completely the fluctuations in order of subsystems A and B. The last assumption can hardly be true. In fact, the general process of phase formation with decreasing temperature is caused by a simultaneous ordering of both subsystems in lattice microdomains. In CeO_{2-x}, for instance, the formation of another phase (e.g., one of the Ce_nO_{2n-2} low-temperature phases) implies a progressive ordering of both oxygen vacancies and reduced cations in the lattice.

The attraction term is given by the total number of A–B pairs in the crystal. It has the form

$$\epsilon_{\text{attraction}} = N^2 \bar{E}_{AB} \tag{42}$$

This form of attraction might be understood as an average attractive field that the reference defect experiences besides the repulsion. In this sense, it plays a role in the statistics that is similar to that of the term ϵ, as discussed in Section II.

2. *Energy Levels for Spacing Statistics for Two Strongly Interacting Subsystems*

In a later paper Atlas (1970) corrected, for UO_{2+x}, the physical concept and the mathematical device by which the set of spacings and the set of spacing levels are generated. The correction involves points (a) and (b) of the preceding section. The fluctuations in order in sublattices A and B are no longer considered independent. Hence the statistical treatment of the configurational degeneracy is modified (Section II.E). The two defect subsystems are strongly correlated in a degeneracy term Ω^{AB} (Atlas, 1970).

The configurational degeneracy Ω^{AB}, which is derived by using the site exclusion principle, has the same form when the lattice is filled first with subsystem A and then with subsystem B, or vice versa. This stresses the strong interdependence of the two subsystems.

In this case both repulsive *and* attractive electrostatic interactions between the two subsystems guide the ordering process and are included in the two sets $[F_{ij}^{A}]$ and $[F_{ji}^{B}]$ of Section II.E.

To calculate the spacing levels, consider that

(1) A fluctuation in order of two strongly interacting subsystems A and B in the host lattice is described in a sample volume V_s by two local densities, one for subsystem A equal to $1/C_i^A$, and one for subsystem B equal to $1/C_j^B$. The reasoning appears immediately correct if one considers this fluctuation to represent a well-defined crystal structure, as suggested in Section II. Suppose the host lattice to be the fluorite lattice of UO_{2+x} and the two subsystems to be the Willis clusters and U^{5+} cations. Suppose the fluctuations to represent the U_4O_9 structure. One can describe the U_4O_9 structure by ordering the Willis clusters and the U^{5+} cations in a definite way, thus generating two local densities $1/C^{A}_{(UO_2)}$ and $1/C^{B}_{(U_4O_9)}$. Any fluctuation in V_s showing a departure from this particular type of order, intermediate between the random distribution and the U_4O_9 structure, is due to a deficit of one or the other or both species in V_s; hence it gives rise to other values of $1/C_i^A$ and $1/C_j^B$.

(2) Consider now a particular defect A in a particular fluctuation *ij*. Its average energy is due to the local density $1/C_i^A$ of the surrounding defects A *and* to the local density $1/C_j^B$ of the surrounding defects B. This average energy therefore can be called E_{ij}^A. Conversely, the average energy of a defect B in the same fluctuation would be E_{ji}^B.

(3) We come now to the generation of set $[E_{ij}]$. Imagine for this purpose a fluctuation in order *ij*. All lattice sites for subsystem A will have a probability $1/C_i^A$ of containing a defect A, and all lattice sites for subsystem B a probability $1/C_j^B$ of containing a defect B in the fluctuation. Thus, the electrostatic repulsion from other defects A, *as well as* the electrostatic attraction from defects B, will act on a defect A. The resultant interaction E_{ij}^A on A may be calculated by means of a Madelung-type lattice summation

$$E_{ij}^{A} = \mathfrak{M}_{ij}^{A}\phi \tag{43}$$

where $\mathfrak{M}_{ij}^{\mathrm{A}}$ is a Madelung constant and in which all lattice points are weighted by the two probabilities $1/C_i^{\mathrm{A}}$ and $1/C_j^{\mathrm{B}}$. The expression obtained is

$$E_{ij}^{\mathrm{A}} = \frac{zz'e^2}{2Ka}\left[\frac{-1}{C_j^{\mathrm{B}}}\sum_{i=1,l_1}\frac{P_k^{\mathrm{B}}}{\lambda_k r_k^{\mathrm{AB}}} + \frac{z^2}{zz'}\frac{1}{C_i^{\mathrm{A}}}\sum_{k=1,l_2}\frac{P_k^{\mathrm{A}}}{\lambda_k r_k^{\mathrm{AA}}}\right]$$
$$= E_{ij}^{\mathrm{A}}(\text{attraction}) + E_{ij}^{\mathrm{A}}(\text{repulsion}) \tag{44}$$

The bracketed expression is an adaptation of Evijen's (1932) version of the Madelung series. Limits l_1 and l_2 determine how many shells of A and B sites are to be incorporated in the series to ensure convergence. The constant λ_k accounts for the sharing of boundary lattice sites with adjacent shells. The constants P_k^{B} and P_k^{A} represent the number of B and A lattice points in the shell k, and r_k^{AB} and r_k^{AA} the distance of this shell from the reference defect A.

If one compares this description with the description employed by Atlas and discussed in the preceding subsection, one notices the disappearance of the term ϵ from the formalism which was due, in the case of independent subsystems, to averaging of the electrostatic attraction. Hence the new treatment of the two subsystems (and of the thermodynamics of UO_2) cannot generate first-order spinodals whereas the first treatment can. We will see that this is indeed the case when examining, in the following discussion, the results of the two treatments.

Equation (44) is quite general for the description of charge–charge electrostatic interactions between defects in a fluctuation ij. It generates the average interaction energy E_{ij}^{A} (and in a similar way E_{ji}^{B}) for the defects in any local fluctuation. Possibly, it can be extended to the interaction of other poles by a suitable definition of the local density for dipoles. Hence it represents a considerable advance in treating the interaction of charged defects, since it allows the statistics of these systems to be extended to very large concentrations of defects.

Notice, however, that the great generality of the method may not be adequate for the full representation of the disordered phase. We will examine two cases:

a. Interblock Interaction. As discussed in Part II, the sets of spacings and of spacing levels are essentially constructed based on the consideration of local patterns or local fluctuations in short-range order of the defects. Since the short-range order is defined by the two spacings C_i^{A} and C_j^{B}, Eq. (44) describes the short-range order established in a fluctuation ij. Longer-range order, however, might connect the fluctuation ij with some other fluctuation lm: for instance, by some interaction orienting the patterns with respect to one another in the lattice. General analysis of the interaction energy might give terms E_{iljm}, to be calculated differently from Eq. (44) and which might be called long-range or interblock interaction and coincide with ϵ of Section II.G.

b. Clustering. The formation of clusters can be included in the degeneracy term Ω^{AB} of Atlas (1970) as a particular condition simplifying Ω^{AB}, e.g., local (i.e., within one or more particular spacings) compensation of charges (which differs from the general electroneutrality condition throughout the crystal). But, if so, Eq. (44) ought to be corrected, in principle, by an extra bonding term—the energy counterpart of the configurational condition—ensuring local compensation within the spacings chosen.

In general, Eq. (44) describes electrostatic interactions between point defects and thus produces an accurate description of the disordered phase as long as clusters or seeds of new structures are not formed. Also, the eventual long-range interaction between these clusters (which finally determines phase separation) is not represented by Eq. (44).

Given, however, its great generality, a new field of theoretical research is opened up by Eq. (44), which should consider expansions of the interaction energy between defects more general than the charge–charge interaction described by this equation.

3. *Results of the Spacing Statistics of Atlas*

One of the advantages of the Atlas model in both versions discussed above is that in principle only one parameter (i.e., E_f) is required to reproduce the thermodynamic functions of the compounds to which it is applied.

In fact, in the spacing sets, C_0 is a counting tool and C_L can be chosen with regard to the geometry of the lattice and of the defects considered. Equations (41) and (44) for the sets of spacing levels do not need any parameters once the dielectric constant k is known. The energy $\overline{E}_{AB}$ can also in principle be calculated.

As for similar solid-state properties, a problem is created by the value chosen for the dielectric constant k which appears as a scaling factor in the Coulomb energies and which should be a function of the concentration and distance of the defects as well as of the temperature (Atlas, 1970).

However, the model is governed only by the computational effort required.

We will now examine the results obtained by the Atlas thermodynamic model.

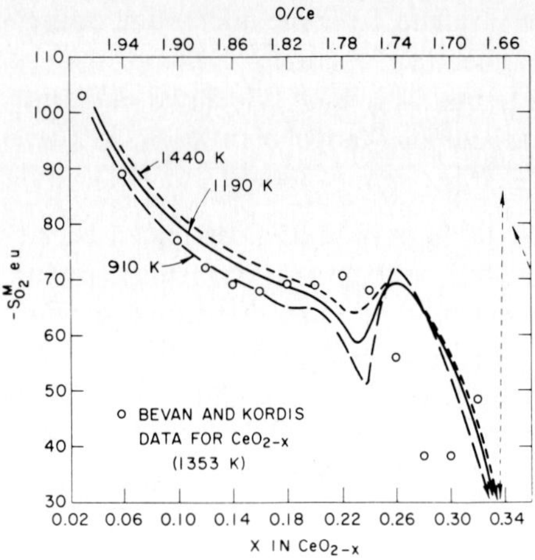

Fig. 5. Computed relative partial molar entropy of oxygen in CeO_{2-x}. (From Atlas, 1968a; reproduced by permission of Pergamon Press. Data from Bevan and Kordis, 1964.)

a. CeO_{2-x}. Figures 5–7 show $\overline{S}(O_2)$, $\overline{H}(O_2)$, and $\overline{F}(O_2)$ for CeO_{2-x} at various temperatures, which compare fairly well with experimental results. The range of x, which is determined by the value chosen for C_0^A and C_L^A, is $0.03 \leqslant x \leqslant 0.5$.

For all the thermodynamic quantities, inflections in the calculated curves occur at compositions such that they can be correlated with the diphasic regions of CeO_{2-x}. The reversions, however, seem to be stable at higher temperatures than in the experimental phase diagram. The occurrence of these reversions is, according to the arguments of Section II,

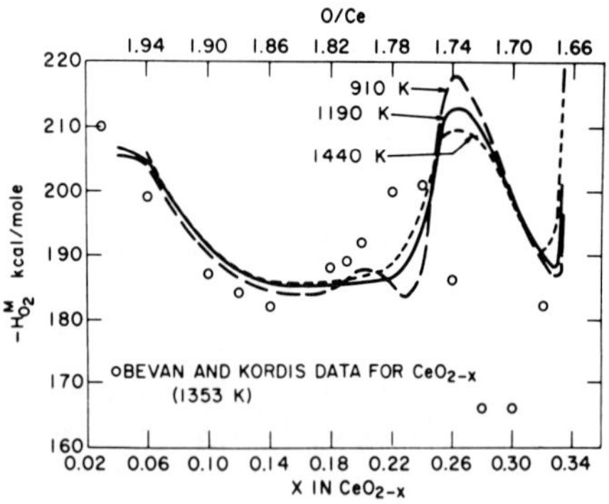

Fig. 6. Computed relative partial molar enthalpy of oxygen in CeO_{2-x}. (From Atlas, 1968a; reproduced by permission of Pergamon Press. Data from Bevan and Kordis, 1964.)

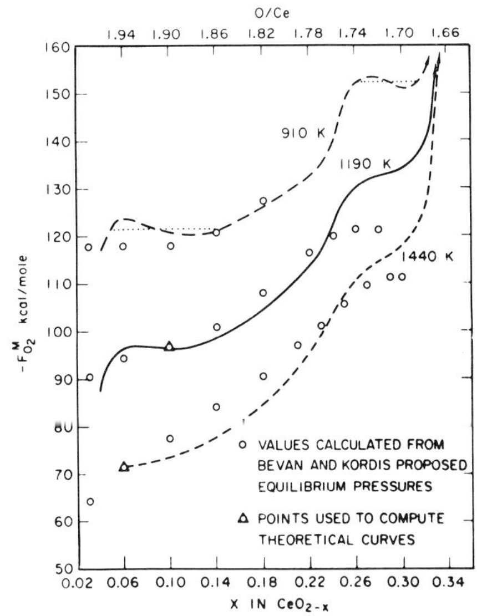

Fig. 7. Computed relative partial molar free energy of oxygen in CeO_{2-x}. (From Atlas, 1968a; reproduced by permission of Pergamon Press. Data from Bevan and Kordis, 1964.)

related to the presence of an average interaction energy ϵ and to the chosen spacing set $[C_i]$.

It can be shown that, in Eq. (41), $E_{i+1}^{A} - E_i^{A} = \Delta_{(i+1)i}$ is not very large. Therefore, the Atlas model resembles the $L = 1$ model illustrated in Section II, with $\theta_{L,0} = \theta$, and at relatively high temperatures the two reversions shown correspond to the predictions of this model. It is evident that at lower temperatures more reversions are revealed because of the increase in the differences in thermal stability $\Delta_{(i+1)i}$. The occurrence of these reversions in the composition of diphasic regions in the Ce–O diagram suggests that the model can describe the thermodynamics of CeO_{2-x}, although with rather crude assumptions.

b. UO_{2+x}. For this oxide both forms of spacing statistics illustrated in this section have been employed. It is interesting therefore to compare the results of the two treatments. In both cases, based on the values of the limits of the spacing sets, the interval of compositions explored is $0.04 \leqslant x \leqslant 0.5$.

In Figs. 8a and 9a, $\bar{S}(O_2)$ are calculated with the statistics of independent subsystems. The agreement with the experimental points is reasonably good. Reversions (due to the presence of a term ϵ) are developed, as in the former case, for $x > 0.2$ and for $0.06 < x < 0.1$ in $\bar{F}(O_2)$ (Fig. 10a). The

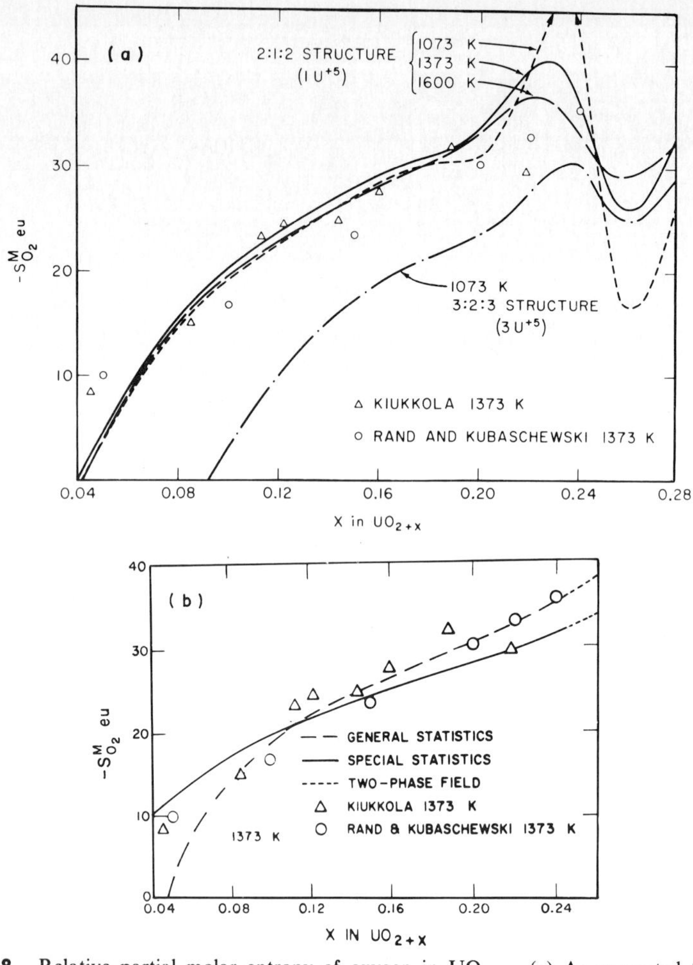

Fig. 8. Relative partial molar entropy of oxygen in UO_{2+x}. (a) As computed from a model of two independent subsystems A and B (Atlas, 1968b; Kiukkoka, 1962; Rand and Kubaschewski, 1963). (b) As computed from a model of strongly interacting subsystems A and B (Atlas, 1970; Kiukkola, 1962; Rand and Kubaschewski, 1963). [Reproduced by permission of Pergamon Press (Atlas, 1968b) and North-Holland Publ. (Atlas, 1970).]

former is related to the UO_{2+x}–U_4O_{9-y} diphasic phase, and the latter, which is not present in the phase diagram, to observed anomalies in the lattice constant-versus-x curve (Atlas, 1968b) of UO_{2+x} at $UO_{2.15}$.

In Figs. 8b, 9b, and 10b, the same range of compositions is explored with the statistics of strongly dependent subsystems. The agreement with

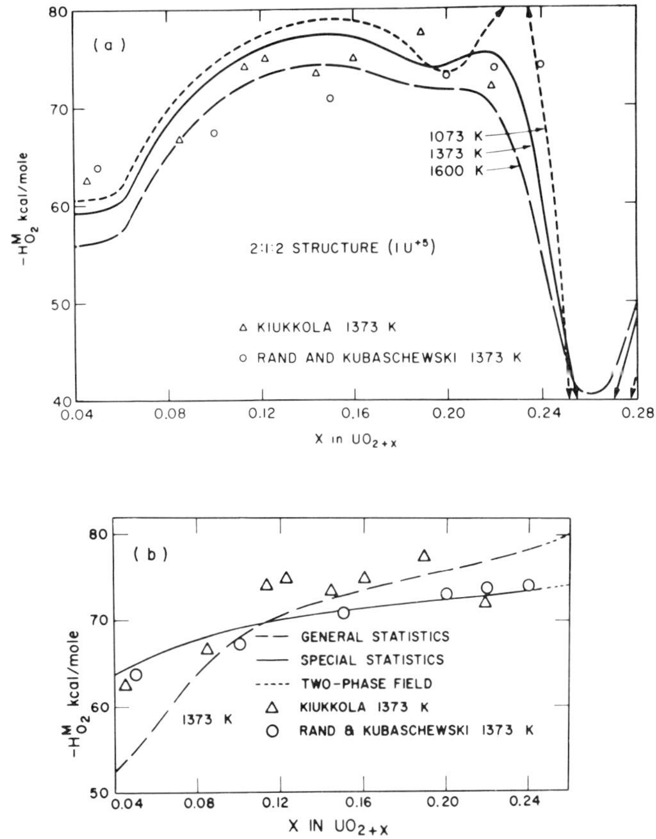

Fig. 9. Relative partial molar enthalpy of oxygen in UO_{2+x}. (a) As computed from a model of two independent subsystems A and B (Atlas, 1968b; Kiukkola, 1962; Rand and Kubaschewski, 1963). (b) As computed from a model of strongly interacting subsystems A and B (Atlas, 1970; Kiukkola, 1962; Rand and Kubaschewski, 1963). [Reproduced by permission of Pergamon Press (Atlas, 1968b) and North-Holland Publ. Co. (Atlas, 1970).]

the experimental points is definitely improved, but the reversions disappear, as expected, since no ϵ term is contained in the partition function Z. This illustrates the point we have made above. Since more information about the stability of short-range fluctuations in the order of the defects has been stored in the storage partition function Z_s, the description of the disordered phase improves but the spinodals contained in it are no longer first-order. To investigate more closely the region of diphasic equilibria and its residues, an expression for longer-range interaction is needed.

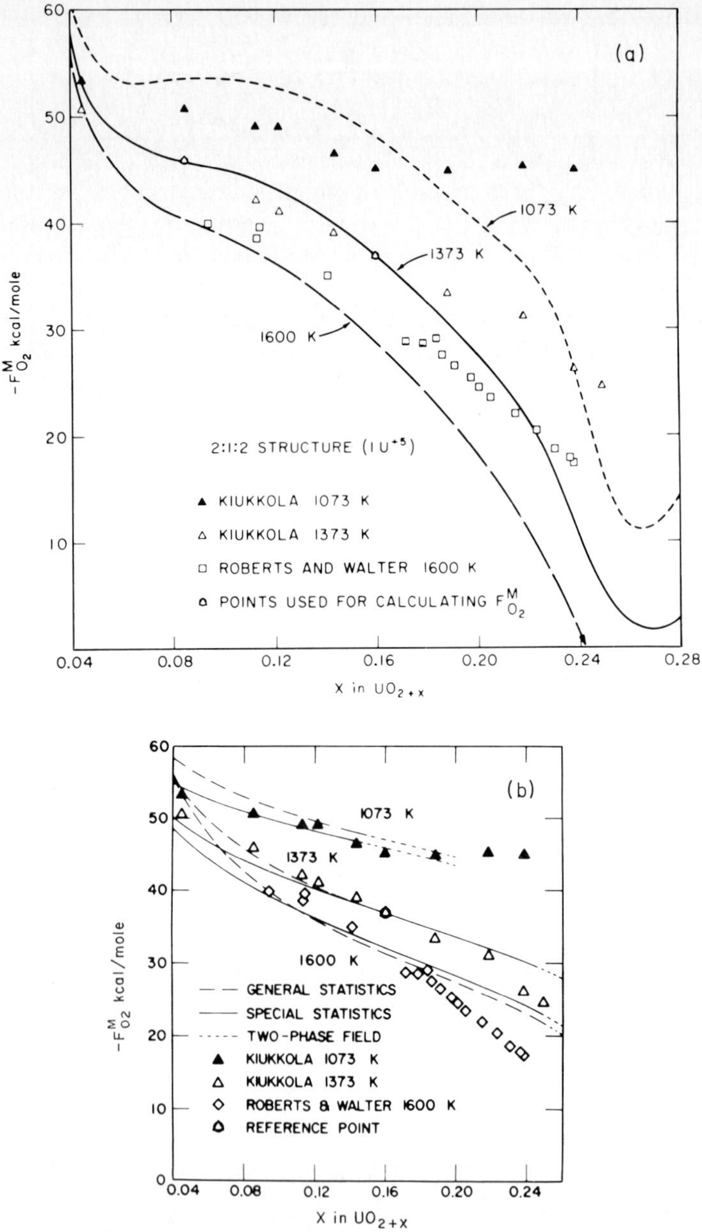

Fig. 10. Relative partial molar free energy of oxygen in UO_{2+x}. (a) As computed from a model of two independent subsystems A and B (Atlas, 1968b; Kiukkola, 1962; Roberts and Walter, 1961). (b) As computed from a model of strongly interacting subsystems A and B (Atlas, 1970; Kiukkola, 1962; Roberts and Walter, 1961). [Reproduced by permission of Pergamon Press (Atlas, 1968b) and North-Holland Publ. Co. (Atlas, 1970).]

C. Spacing Statistics of Tetrahedral Defects

1. The Definition of the Tetrahedral Defect

The tetrahedral defect has been introduced (Manes and Manes-Pozzi, 1976; Manes *et al.*, 1979) in an attempt to relate the high-temperature thermodynamics of the disordered substoichiometric phases of some MO_{2-x} systems [namely, CeO_{2-x}, PuO_{2-x}, ternary systems $(U, Pu)O_{2-x}$] to their phase diagrams. Although not an observed entity like, for instance, the Willis cluster (Willis, 1964), it was described by Manes and Sørensen (Manes *et al.*, 1979) on the basis of crystallographic arguments as being the building block of the low-temperature Mo_nO_{2n-2} structures found in many lanthanide (Pr, Ce, Tb) and actinide (Am, Cm, Bk, Cf) oxides. Saturation concentrations by tetrahedral defects of the fluorite lattice in the high-temperature disordered phases of these oxides can also be correlated (Manes *et al.*, 1979) with the characteristic pseudo-phase lines seen in Sørensen's diagrams (see Sørensen, 1978). These arguments are discussed in detail in Chapter 1 of this book.

As commonly accepted, the basic point defects of the substoichiometric phases of MO_{2-x} oxides are oxygen vacancies ($V_O^{\cdot\cdot}$) and reduced cations (M'_M). Here we define the tetrahedral defect as a *local* bond characterized by a bonding energy, $E_b < 0$, within an oxygen coordination tetrahedron in the fluorite structure, between two reduced cations and one oxygen vacancy. Such a unit ($2M'_M V_O^{\cdot\cdot}$)

(a) is no longer a point defect but has its own volume in the lattice (the tetrahedral volume);

(b) constitutes a neutral entity with respect to the lattice but has a dipole moment, since the centers of positive and negative charge do not coincide, if a charge transfer has taken place between the vacancy and the cations (the center of the positive charge being in the vacancy and that of the negative charge being on one of the tetrahedron's sides, halfway between the reduced cations);

(c) represents local short-range ordering with respect to the "free" defects and thus has an important effect on the total energy and entropy of the crystal; and

(d) may be packed in different ways, thus extending the range of ordering of the defect population.

Having defined a fundamental cluster, we proceed, in the application of spacing statistics, as suggested in Section II:

(a) by defining a sampling volume V_s, in which we observe a set of spacings $[C_j]$ (notice that hereafter the first spacing will be called C_1 instead of C_0 as elsewhere in this chapter);

(b) by generating a strain field Δ introduced in the lattice by the presence of tetrahedral defects variously packed in the sampling volume V_s; this field generates both the sets $[C_j]$ and $[E_{fj}]$ and imposes a spacing rule; and

(c) by taking into account a residual longer-range ϵ due to the dipolar attraction between tetrahedral defects.

In the model, a further detail is introduced—dissociation of the tetrahedral defects in their constituents:

$$(2M'_M V_O^{\cdot\cdot}) \rightleftharpoons 2M'_M + V_O^{\cdot\cdot} \tag{45}$$

Hereafter, n_ξ represents the total number of tetrahedral defects, and n_0 the total number of free vacancies $V_O^{\cdot\cdot}$ ($2n_0$ is the total number of M'_M).

Since the two sets $[C_j]$ and $[E_{fj}]$ generate a set $[n_j]$ of thermal fluctuation populations for the tetrahedral defects in different spacings, we have

$$\sum_{j=1,L} n_j = n_\xi \tag{46}$$

2. *The Sampling Volume* V_s

It is assumed that an isolated tetrahedral defect in the fluorite lattice would create a strain field Δ perturbing this lattice *at least* to an (arbitrary) spacing equal to the volume enclosing the nearest-neighbor tetrahedra. This permits the definition of a spacing C_1 (the first of the set C_j), which by inspection of the fluorite structure contains $C_1 = 23$ nearest-neighbor tetrahedra or, equivalently, 23 oxygen sites. Any other tetrahedral defect being formed within this envelope will increase the strain field, since the perturbation of the lattice will be much greater; this provides a repulsive lattice energy tending to destabilize the lattice, which is the source of the set $[E_{fj}]$ of the spacing statistics.

We may consider this envelope C_1 the sampling volume V_s on which to construct the statistics. The isolated tetrahedral defect has in this volume the average spacing $C_1 = V_s$. Successive packings of tetrahedral defects are characterized by spacings C_j $(j > 1)$.

Therefore the fluctuations in order (and the blocks $[C_j]$) are generated. Notice that they are of a segregative nature, since the strain field Δ is repulsive.

3. *The Set* $[E_{fj}]$

For the formation energy E_{f1} of an isolated tetrahedral defect, the following relation holds:

$$E_{f1} = E_f + E_b + \Delta = E_f + \delta, \qquad \delta = E_f + \Delta \tag{47}$$

where E_f is the formation energy of one $V^{\cdot\cdot}$ and two M' in their unassociated state, E_b the boundary energy of the point defects in a tetrahedral defect, and Δ the strain parameter of the isolated tetrahedral defect.

The bonding energy E_b is dependent on the type of bond established within the tetrahedron between the $V_O^{\cdot\cdot}$ and the two M'_M; it must essentially involve some kind of local redistribution of the electronic charge left free in the oxygen vacancy. This implies a departure from the purely ionic point charge model, hence some covalency or some molecularity. [As an indication of E_b (obviously difficult to assess) it was taken in PuO_{2-x} (Manes and Manes-Pozzi, 1976) to be the difference in formation enthalpy $\Delta H_f(MO_2)$–$\Delta H_f(MO_{1.5})$ between the dioxide and the sesquioxide—thus the tetrahedral defect was considered a molecular seed of the Me_2O_3 structure.]

The strain parameter Δ is, on the contrary, considered to be related to the change in volume of the original tetrahedron when transformed in a tetrahedral defect. The increase in the ionic radius of the cation when going from M^{4+} to M^{3+} (even though partially compensated for by a collapse of the hole created by the oxygen vacancy) suggested, as a reasonable picture, an increase in the volume of the tetrahedron, hence an elastic strain on the surrounding lattice (it would be interesting, though, to examine the opposite hypothesis).

Consider now that the change in volume due to the formation of a tetrahedral defect is *isotropic*, i.e., does not involve deformation and/or reorientation of the tetrahedra within C_1. The change in lattice strain *within* C_1 can in this case be expressed as

$$\Delta_j = \alpha_s(C_j^{-1} - C_1^{-1}) = \Delta(p^{-1} - 1) \tag{48}$$

where α_s is an appropriate strain factor, $\Delta = \alpha_s C_1^{-1}$, and $p = C_j/C_1$ the fraction of the volume C_1 per tetrahedral defect. The set $[E_{fj}]$ is thus generated as

$$E_{fj} = E_f + E_b + \Delta + \Delta_j = E_{f1} + \Delta_j \tag{49}$$

4. *The Spacing Rule*

Evidently, p depends on the detailed geometry and symmetry of the volume C_1 defined above. For instance, not all tetrahedral positions in the shell of nearest-neighbor tetrahedra are equivalent: Some of them (see Chapter 1) are closer, and some of them are farther from the central tetrahedron. Therefore the set $[E_{fj}]$ depends on the exact definition of p. Furthermore, one might assume a sampling volume larger than C_1 so as to include larger associations of tetrahedral defects and a better definition of the strain field given by Eq. (48) (e.g., considering derformations and/or reorientations along special crystallographic directions).

TABLE I

C_j and n for Some Lanthanide Oxide Systems[a]

n	Formula	C_j	Systems	O/M from M_nO_{2n-2}	O/M from Eq. (50)	Experimental stability range of the phase
7	M_uO_{12}		Pr–O, Tb–O, Ce–O	1.714		1.713–1.719
		$C_3 = 7.7–8$			1.740	
9	M_9O_{16}		Pr–O, Ce–O	1.778		1.776–1.778
10	$M_{10}O_{18}$		Pr–O	1.800		1.799–1.801
11	$M_{11}O_{20}$	$C_3 = 11.5–12$	Pr–O, Ce–O, Tb–O	1.818	1.826	1.817–1.820
12	$M_{12}O_{22}$		Pr–O, Ce–O	1.833		1.833–1.836
24	$M_{24}O_{46}$	$C_1 = 23–24$			1.913	
∞	MO_2	$C_0 = \infty$	All systems	2.00	2.00	2.00

[a]Characteristic envelopes C_j for spacing statistics of the tetrahedral defects and comparison with n in M_nO_{2n-2} subphases of some lanthanide oxide systems.

However, in the first application of spacing statistics, the simplifying assumption was made that the envelope C_1 was roughly *equally partitioned* among the tetrahedral defects contained at a certain level of packing. This gives the relation

$$C_j = C_1/j \tag{50}$$

which can be assumed to be the spacing rule for this application of the statistics. This leads to a definition of the set $[E_{fj}]$ as

$$E_{fj} = (j-1)\Delta \tag{51}$$

Notice that the rule of spacing given by Eq. (50) leads to an interesting consequence when the saturation concentrations of the fluorite lattice are computed with respect to each type j of packing. The O/M ratio for a packing j is given by $(O/M)_j = 2(1 - 1/C_j)$, to be compared with $(O/M)_n = 2(1 - 1/n)$ which results from the formula M_nO_{2n-2} for the low-temperature phases in a M–O substoichiometric fluorite system (e.g., M = Ce; see Chapter 1). The comparison between C_j and n is given in Table I. The C_j calculated by Eq. (50) represents fairly well an average value of the observed n in M_nO_{2n-2} structures.

5. The Dipolar Long-Range Interaction Energy ϵ

Evaluation of the dipolar interaction energy ϵ has been obtained (Manes and Manes-Pozzi, 1976) by performing a Boltzmann average over the mutual orientations of the dipoles (well defined by the symmetry of the fluorite structure) formed within the tetrahedral defects, and over a mutual dipole–dipole distance. This gives

$$\epsilon = (4\pi z^2 e^2/3Ka)I(T)n^2 = n^2\Phi \tag{52}$$

where ze is the effective charge creating the dipole, a the lattice constant, K a dielectric constant (taken as a computing parameter between K_∞ and K_0 in the system considered). For a discussion of the difficulty involved in choosing K in this kind of treatment, see (Atlas, 1970; Manes and Manes-Pozzi, 1976). Here $I(T)$ is a slowly varying Langevin-type function. Notice that ϵ is calculated from a given value of K for *each temperature* in this treatment.

6. The Mutual Dependence of the Energy Parameters

Even with the simplifying assumptions employed, all the energy parameters are strongly interdependent. In fact, suppose that a certain type of bond is formed in the tetrahedral defect consistent with a considerable amount of covalency. This will give a very negative E_b. However, the change in volume will in this case probably be small, since the ionic radii of the cations will not vary greatly. Hence E_b and Δ in Eq. (47) (Blank, 1975) will be correlated. The resulting electronic distribution within the tetrahedral defect will give a lowered effective charge ze on the vacancy and a lowered dipole moment of the tetrahedral defect. This imposes a correlation among E_b, Δ, and ϵ.

Therefore the model employed here seems to be especially adapted for comparing the thermodynamics of different fluorite phases (e.g., that of CeO_{2-x} and of PuO_{2-x}) and obtaining information on the bond in these systems by comparison of the parameters E_b, Δ, and ϵ that best fit their thermodynamic functions.

7. The Partition Function

The partition function Z takes care of dissociation of the tetrahedral defects into the constituent point defects. This was accomplished by randomizing the n_0 and $2n_0$ populations of vacancies and reduced cations of lattice sites *not occupied* by tetrahedral defects:

$$\Omega_{\text{vacancies}} = \Omega_v = (2N_M - n_\xi)!/(2N_M - n_\xi - n_0)!\, n! \tag{53a}$$

$$\Omega_{\text{electrons}} = \Omega_e = (N_M - 2n_\xi)!/[N_M - 2(n_\xi + n_0)]!\,(2n_0)! \tag{53b}$$

The partition function Z therefore (only the most probable term is shown) is

$$Z = [\Omega_v \Omega_e \exp(-n_0 E_f/kT)]\left\{\left(\prod_{j=1,L} \Omega_j\right)\exp\left[\sum_{j=1,L} (n_j E_{fj}/kT)\right]\right\} \times [\exp(-\epsilon/kT)] = Z_0 Z_s Z(\epsilon) \quad (54)$$

subject to

$$n_0 + \sum_{j=1,L} n_j = n_0 + n_\xi = N_v \qquad \text{(total number of vacancies)} \quad (55)$$

The chemical potential μ is given by (Manes *et al.*, 1979)

$$\mu = \mu_0(n_0, n_\xi) + \mu_{\text{short range}}(n_\xi) + \mu_{\text{long range}}(n_\xi, \epsilon) \quad (56)$$

which is dependent on only two concentration variables n_0 and n_ξ. These two variables may be considered as representing disorder and short-range order due to the tetrahedral defects, respectively.

Finally, Eq. (56) gives conditions for first-order spinodal points, as discussed in Part II. In particular, for T very low, where it can be shown that $N_v \cong n_\xi$, the first-order spinodal points are determined by the matching of the short-range and long-range parts of the chemical potential μ; i.e., the repulsion potential due to the strain field is counterbalanced by the attraction potential due to dipolar interaction (see Chapter 1).

8. *Some Results of the Spacing Statistics of Tetrahedral Defects for Substoichiometric* CeO_{2-x}, PuO_{2-x} *Oxides*

Figure 11 shows the $\overline{H}(O_2)$-versus-O/M curve calculated for the CeO_{2-x} system. It is seen that the characteristic minima and maxima found experimentally (Campserveux and Gerdanian, 1974) are due to saturation of the lattice, at different O/M ratios, with tetrahedral defects—note the correspondence with the curve for $D_\xi = d\xi/dx$. It is worth recalling that the presence and the meaning of these maxima and minima have not been explained previously (Campserveux and Gerdanian, 1978). The same curve calculated for the PuO_{2-x} system, which shows no experimental minima, can be related to the lower strain introduced into the fluorite lattice by tetrahedral defects, perhaps as a result of the lower ionicity of the bond (Blank, 1975) (see Fig. 12) (Chereau, 1972).

When expressing μ as in Eq. (56), the presence of diphasic regions in the phase diagram can be easily deduced by differentiation with respect to x and searching for extrema and/or inflection points. When this is done, the Sørensen diagrams discussed in Chapter 1 are fairly well reproduced at rather high temperatures (Manes *et al.*, 1979).

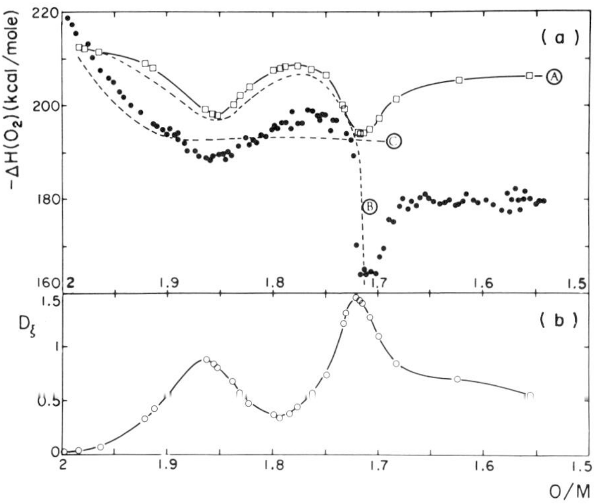

Fig. 11. Calculated $\Delta\overline{H}(O_2)$ and D_ξ (derivatives of the concentration of tetrahedral defects) versus the O/M ratio for CeO_{2-x}. Experimental data (solid dots) (Campserveux and Gerdanian, 1974). Curve A: $\delta = -4$, $\Delta = 10$, and $\phi = -48$ kcal/mole^2; $L = 4$. Curve B: same parameters, but $L = 3$. Curve C: fitting of model presented in (Campserveux and Gerdanian, 1978). (Reproduced by permission of IAEA.)

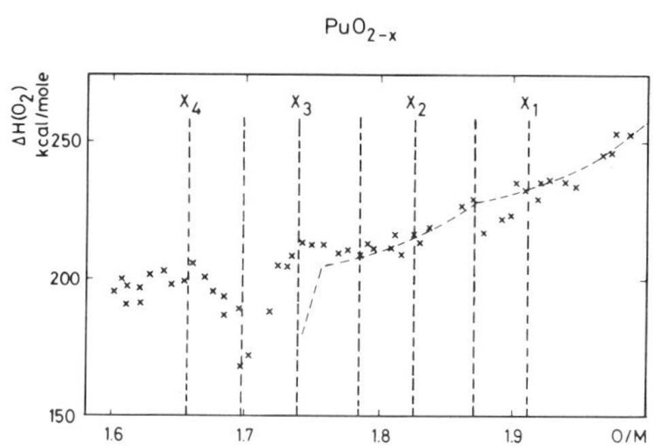

Fig. 12. Calculated $\Delta\overline{H}(O_2)$ versus O/M ratio for PuO_{2-x} (Manes, 1976). Experimental data from Chereau (1972). (Reproduced by permission of North-Holland Publ. Co., Amsterdam.)

Figure 13 shows the agreement between the m values $[m = d\ln p_{O_2}/d\ln x = (\mu' x)/kT]$ used to build the pseudophase diagram for CeO_{2-x} and the equivalent m_c calculated with the help of the present theory (Section II.F. and H, Fig. 2).

Figure 14 represents an attempt to calculate the phase diagram of CeO_{2-x} with the energy parameters used in the calculation of $\Delta\overline{H}(O_2)$ (Fig. 11). In the region of validity of the model, which is the fluorite structure region (region A), the diphasic bell and the 1–5 phase boundary are

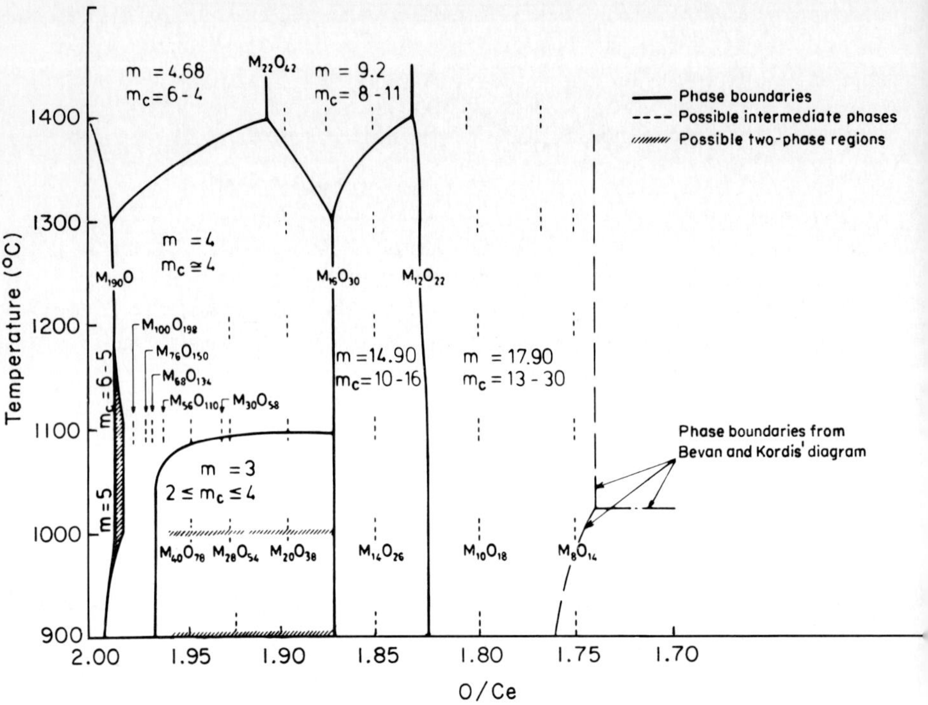

Fig. 13. Pseudophase diagram for the CeO_{2-x} system showing subregions with possible ordered intermediate phases. The m_c values have been calculated from the thermodynamic model (Manes *et al.*, 1979). (Reproduced by permission of IAEA.)

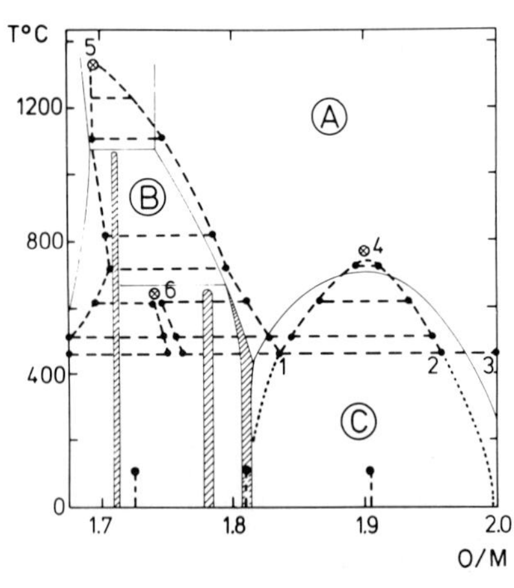

Fig. 14. A phase diagram calculated with the present model compared with the Ce–O phase diagram. Same parameters as in Fig. 11. Dotted line: calculated diagram (Manes *et al.*, 1979). (Reproduced by permission of IAEA.)

reproduced fairly well. In regions B and C the agreement is not as good, but interesting trends are discovered—note, for instance, the monophasic region terminating at 6.

9. The Extension of Spacing Statistics to Ternary $U_{1-z}Pu_zO_{2-x}$ *Systems*

Spacing statistics were applied (Manes and Manes-Pozzi, 1976) to ternary oxides $U_{1-z}Pu_zO_{2-x}$ with the fluorite structure. This extension is possible because the spacing statistics are able to expand the total energy E of the defective crystal as a function of local short-range-order energy parameters. In $U_{1-z}Pu_zO_{2-x}$ ternary oxides, the thermodynamic observation of Markin *et al.* (1967) that the oxygen potentials $\Delta\bar{G}(O_2)$ were not functions of the two independent composition variables x and z characterizing these oxides, but rather of a variable, called the plutonium valence, which is defined as

$$V_{Pu} = 2(1 - x/z) \tag{57}$$

suggested local ordering to these authors.

The plutonium valence rule has been found to be satisfied only in a limited range of Pu compositions z, i.e., $0.05 \leqslant z \leqslant 0.30$, and not to be valid at rather high temperatures. Also, a similar rule does not seem to apply for the analogous $(U_{1-z}Ce_z)O_{2-x}$ system. Hence it must be considered only an indication of a connection between the concentration of oxygen vacancies and Pu^{3+} reduced cations, probably suggesting local short-range order between the two subsystems of defects.

The probable existence of short-range order between Pu^{3+} and oxygen vacancies has inspired a series of other statistical or thermodynamic cluster models, which we will not review here in detail, differing in the type of cluster or clusters assumed (Schmitz and Marajofsky, 1975; Schmitz, 1975; De Franco and Gatesoupe, 1976; Catlow, 1977). Here we are more interested in showing how the formalism and ideas of spacing statistics may be applied to ternary systems. It is clear from much of this chapter that the application of spacing statistics to another type of basic cluster, or to a series of progressive aggregations of point defects, is the way to improve the description of the thermodynamic functions by introducing other descriptions of short-range order. Therefore application of spacing statistics to any of the cluster systems described in the above-cited references would be a useful approach in the study of these ternary oxides.

10. The Sets $[C_j]$ *and* $[E_{fj}]$ *for Ternary Systems*

The ideas discussed in the preceding subsection for CeO_{2-x} have been used in the description of $U_{1-z}Pu_zO_{2-x}$ oxides. Only the hypothesis of a tetrahedral defect has been widened to consider short-range order situations in which, in a coordinated tetrahedron of an oxygen site, oxygen vacancies are bound with *at least two Pu ions*. Since the distribution of Pu ions in the U–Pu–O fluorite lattice is considered random, this condition imposes a limit on the C_1 and C_L values of the set C_i used in the statistics and makes C_1 a function $C_1(z)$ of the plutonium composition z.

The condition that a coordination tetrahedron should contain *at least* two Pu ions (randomly distributed in the lattice) gives rise to configurations not exactly equal to the tetrahedral defect discussed above. In fact, the coordination tetrahedron may also contain a *pair* or a *triplet* of Pu ions. For all these situations, however, the same bonding energy E_b and the same strain parameters Δ are considered to hold. A coordination tetrahedron containing a pair, a triplet, or a quadruplet of Pu ions is called a reference defect. This reference defect can be written in all cases as $2\,Pu'V^{\cdot\cdot}$, as in Section C.1.

The spacing C_1 is different for a pair, a triple, or a quadruplet of Pu ions if we consider only the nearest-neighbor tetrahedra connected with the Pu ions constituting the reference defect; this is a restriction of the rule given above for C_1 in tetrahedral defects, which has an intuitive basis in the fact that the volume occupied by uranium ions cannot contain the $2\,Pu'V^{\cdot\cdot}$ defects since these are formed only in regions where Pu ions are present.

Furthermore, given a certain Pu composition z in a $U_{1-z}Pu_zO_{2-x}$ ternary oxide, there is a well-defined probability, determined by z, that the C_1 envelope will be occupied by other Pu ions, and different configurations of occupation of the envelope C_1 have to be considered (see, for example, Fig. 15).

By a somewhat lengthy but rigorous counting procedure, the three characteristic envelopes C_1 (pair), C_1 (triplet), C_1 (quadruplet) can be

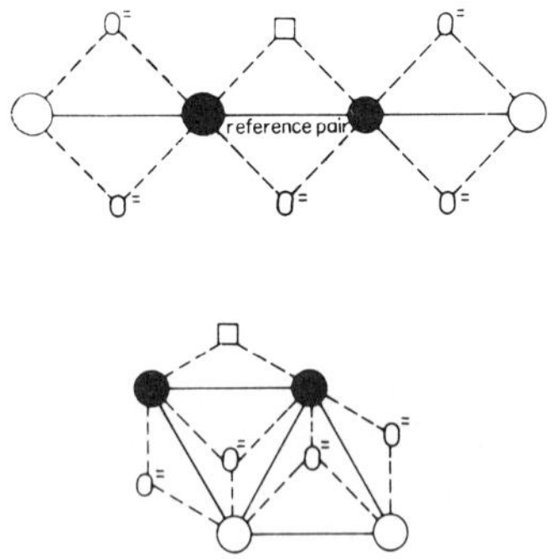

Fig. 15. Two possible situations for occupation of the external shell of a reference pair by Pu atoms (Manes, 1976). (Reproduced by permission of North-Holland Publ. Co., Amsterdam.)

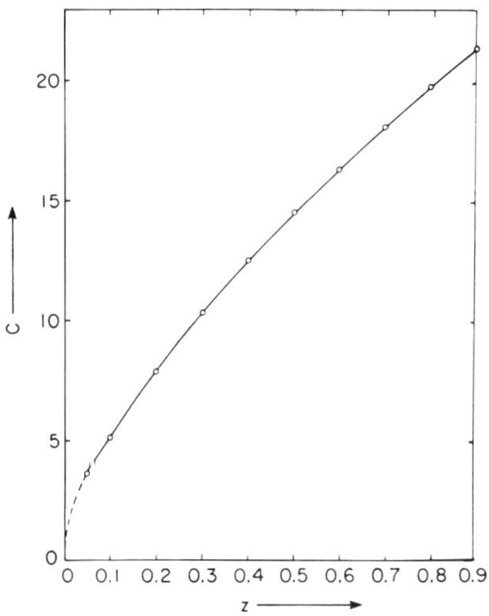

Fig. 16. The spacing $C_1(z)$, in the set $[C_j]$ $(1 \leqslant j \leqslant L)$, as a function of the Pu content z.

determined as a function of z (Manes and Manes-Pozzi, 1976). Then, since the three possible situations are considered equal from an energy point of view, an average $\overline{C}_1(z)$ is calculated.

This procedure consists essentially in excluding the volume of the lattice occupied by uranium ions from the statistics when Pu ions are randomly distributed. The sampling volume V_s is therefore computed in a probabilistic way from the volume randomly occupied by Pu ions only.

The sets $[C_j]$ and $[E_{fj}]$ are then computed by employing the same rules given by Eqs. (50) and (51). Figure 16 shows $\overline{C}_1(z)$ as a function of the Pu composition z.

11. The Saturation Concentrations

For the same reason, the number of tetrahedra in which a reference defect can be formed is also restricted. If N_M is the number of cation sites in $U_{1-z}Pu_zO_{2-x}$ and $2N_M$ the number of oxygen sites (and of coordination tetrahedra), lattice B of Section II will be constituted only by

$$B = N_{O'} = 2N_M p_t(z) \tag{58}$$

where $N_{O'}$ is the number of coordination tetrahedra with *at least* two Pu ions (called tetrahedra O′) and p_t is the probability that a tetrahedron O′ will exist at a given composition z. This probability is given by

$$p_t(z) = p_2 + p_3 + p_4 = \sum_{i=2,3,4} \binom{4}{i} (1-z)^{4-i} z^i \tag{59}$$

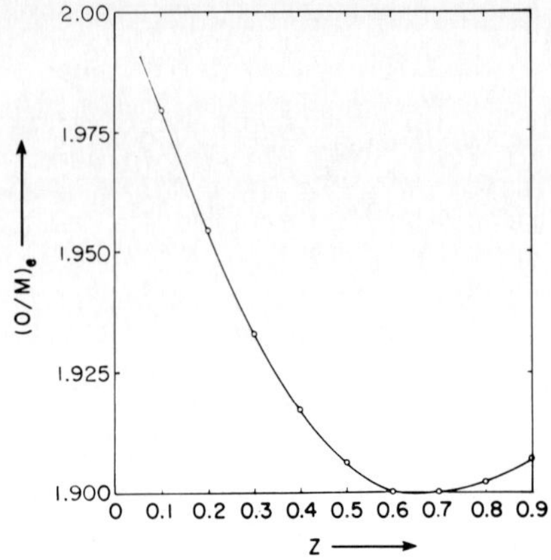

Fig. 17. The $(O/M)_{s1}$ saturation stoichiometries for spacing $C_1(z)$ as a function of the Pu content z (Manes, 1976). (Reproduced by permission of North-Holland Publ. Co., Amsterdam.)

The saturation oxygen deficiencies $x_j(z)$ in $U_{1-z}Pu_zO_{2-x}$ for different spacings C_j of the reference defects therefore are

$$x_j(z) = 2p_t(z)/C_j(z) \tag{60}$$

This formula is analogous to the one discussed in Section C.4 for binary compounds (Fig. 17). At approximately $x_j(z)$, spinodal points may be expected, i.e., diphasic separation or inflections in μ, and changes in slopes in $\overline{H}$ and $\overline{S}$, as discussed in Section II.

12. Results of the Application of Spacing Statistics to $U_{1-z}Pu_zO_{2-x}$

The concepts discussed previously lead to some interesting results in the application of spacing statistics to the $U_{1-z}Pu_zO_{2-x}$ system. The full discussion of these results can be found in Manes and Manes-Pozzi (1976). Here we summarize them briefly.

a. Phase Diagram and General Properties of the $U_{1-z}Pu_zO_{2-x}$ *System.* A prediction of the phase diagram of the $U_{1-z}Pu_2O_{2-x}$ system is obtained *only* from a consideration of the saturation oxygen deficiencies x_j and not from the complete application of Eq. (34) of Section II.G.

The agreement of this prediction with the ternary phase diagram of $U_{1-z}Pu_zO_{2-x}$ of Sari *et al.* (1970) and with other experimental observations is satisfactory.

In Chapter 4 (Matzke), an observation is reported on the plutonium diffusion coefficients as a function of oxygen deficiency, which show a

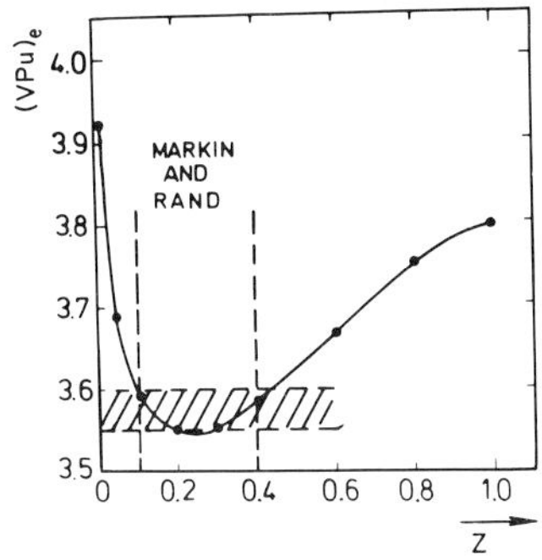

Fig. 18. $V(\mathrm{Pu})_{s1} = 4 - 2x_{s1}(z)/z$ as a function of the Pu content z. (Manes, 1976). (Reproduced by permission of North-Holland Publ. Co., Amsterdam.)

break (indicating saturation of the carriers, presumably Pu^{3+} ions) at about the saturation concentration x_1 in $U_{0.8}Pu_{0.2}O_{2-x}$.

Also, Norris (1979) was able to describe the characteristic variation with stoichiometry of thermal diffusion coefficients in oxygen redistribution in $U_{1-z}Pu_zO_{2-x}$ oxides by employing the present statistical treatment.

b. The Plutonium Valence Rule. The plutonium composition's range of validity for the plutonium valence rule is explained by the statistics as that in which the ratio $x_1(z)/z$ is approximately constant. In this case, the thermodynamics of the system can be written as a function of a single variable, $V_{\mathrm{Pu}} = 4 - 2(x/x_1)$, as required by the rule. In Fig. 18, this is shown to occur, because of the very wide minimum, in the range $0.05 \leqslant z \leqslant 0.3$.

c. The Dipolar Interaction Energy. An observation based on the calorimetrically determined $\overline{H}(O_2)$-versus-O/M curves by Chereau (1972) in a series of $U_{1-z}Pu_zO_{2-x}$ systems was that, for $z = 0.50$, the $\Delta\overline{H}(O_2)$-versus-O/M curve seemed to be stoichiometry-independent (Figs. 19 and 20). This feature is well reproduced when deriving the dipolar attraction $\epsilon(z)$ as in Section C.5, which is in the present treatment a function of the plutonium composition z [in fact, the spacing $C_i(z)$ is a function of z].

d. The Thermodynamic Functions of $U_{1-z}Pu_zO_{2-x}$. In Figs. 19 and 20, the calculated $\overline{H}(O_2)$-versus-O/M curves for different values of the plutonium composition z are presented and compared with the results of Chereau (1972). Spacing statistics predict the characteristic peaks shown in these curves.

In Manes and Manes-Pozzi (1976) the calculated $\overline{S}(O_2)$ for $z = 1$ (PuO_{2-x}) and $z = 0.25$ compare fairly well with the experimental values of Chereau (1972) and Woodley (1972), respectively.

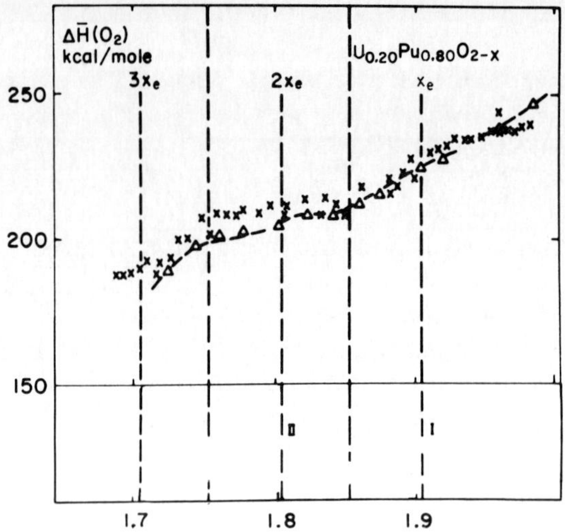

Fig. 19. Experimental $\Delta\overline{H}(O_2)$ values (Chereau, 1972) (×) are compared with calculated (△) for $U_{0.20}Pu_{0.80}O_{2-x}$ (Manes, 1976). (Reproduced by permission of North-Holland Publ. Co., Amsterdam.)

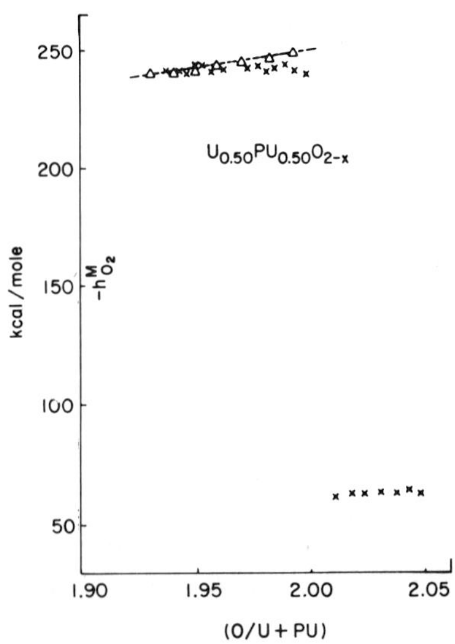

Fig. 20. Experimental $\Delta\overline{H}(O_2)$ values (Chereau, 1972) (×) are compared with calculated (△) for $U_{0.50}Pu_{0.50}O_{2-x}$ (Manes, 1976). (Reproduced by permission of North-Holland Publ. Co., Amsterdam).

IV. Conclusions and Recommendations for Future Work

The conceptual basis of the statistics discussed in this chapter is the description of order fluctuations in defects in nonstoichiometric phases by their average spacing set $[C_i]$ and reduction in the interaction between defects, at least up to some range, in a set $[E_{fi}]$. When this is done, the mixing of order fluctuations or of structural microdomains in a disordered phase is viewed as an *ideal* mixing, with respect to enthalpy, in pure spacing statistics. The configurational description of their intergrowth, dominated by what we called entropic competition in Section I.A, contributes, however, a very important mixing entropy which governs all the thermodynamic functions and determines, for a given choice of the $[C_i]$ and $[E_{fi}]$ sets, the spinodal compositions. Unmixing of the disordered phases in diphasic regions of the phase diagram is left to the part of the interaction not included in $[E_{fi}]$: the longer-range interaction between the microdomains, possibly the formation of surfaces between them.

This opens two lines of development:

(a) Improvement in the evaluation of the mean cohesive energies per defect $[E_{fi}]$ within a fluctuation, perhaps by extending the Atlas (1970) treatment of Eq. (44);

(b) Improvement in the evaluation of the longer-range interaction ϵ, for which a more detailed description of the microdomains is needed.

In order to develop the theory along these lines, spacing statistics, used to interpret experimental thermodynamic functions as suggested in Section II.H, may be a useful tool, especially if combined with structural observations. In Chapter 1 of this book, accurate study of the chemical potential of a large number of nonstoichiometric oxides shows that subphases can be recognized in the monophasic region of their phase diagrams. The importance of the pseudo-phase diagrams that can be constructed in this way (Sørensen, 1978) must be stressed, as they can be easily read with the help of the formalism of spacing statistics. It is to be hoped that from them, as well as from structural information, the nature of the basic seeds of low-temperature structures (or, equivalently, of the true extended defects causing nonstoichiometry) can be inferred.

References

Anderson, J. S. (1970). *In* "Problems of Nonstoichiometry" (E. Rabenau, ed.), pp. 1–76. North-Holland Publ., Amsterdam.

Ariya, S. M., and Morozowa, M. P. (1958). *J. Gen. Chem. USSR* **28**, 2647–2652.

Ariya, S. M., and Popov, Y. G. (1962). *J. Gen. Chem. USSR* **32**, 2054–2057.

Atlas, L. M. (1968a). *J. Phys. Chem. Solids* **29**, 91–100.

Atlas, L. M. (1968b). *J. Phys. Chem. Solids* **29**, 1349–1358.

Atlas, L. M. (1970). *In* "Extended Defects in Non-Metallic Solids" (L. Eyring and M. O'Keeffe, eds.), pp. 425–443. North-Holland Publ., Amsterdam.

Bevan, D. J. M., and Kordis, I. (1964). *J. Inorg. Nucl. Chem.* **26**, 1509–1523.

Blank, H. (1975). *In* "Thermodynamics of Nuclear Materials 1974," Vol. II, pp. 46–69. IAEA, Vienna.

Cahn, J. W. (1961). *Acta Metall.* **9**, 795–801.

Campserveux, J., and Gerdanian, P. (1974). *J. Chem. Thermodyn.* **6**, 795–800.

Campserveux, J., and Gerdanian, P. (1978). *J. Solid State Chem.* **23**, 73–82.

Catlow, C. R. A. (1977). *J. Nucl. Mater.* **67**, 236–238.

Chereau, P. (1972). Ph.D. Thesis, Rep. CEA-R-4402. Univ. de Paris-Sud, Orsay.

De Franco, A., and Gatesoupe, J. P. (1976). *In* "Plutonium 1975 and Other Actinides" (H. Blank and R. Lindner, eds.), pp. 133–143. North-Holland Publ., Amsterdam.

Dequesnoy, A., and Marion, F. (1964). *C. R. Acad. Sci.* **258**, 4550–4552.

Evijen, H. M. (1932). *Phys. Rev.* **39**, 675–687.

Fowler, R. H., and Guggenheim, E. A. (1939). "Statistical Thermodynamics." Cambridge Univ. Press, London.

Hagemark, K., and Broli, M. (1966). *J. Inorg. Nucl. Chem.* **28**, 2837–2850.

Kiukkola, K. (1962). *Acta Chem. Scand.* **16**, 327–345.

Manes, L., and Manes-Pozzi, B. (1976). *In* "Plutonium 1975 and Other Actinides" (H. Blank and R. Lindner, eds.), pp. 145–163. North-Holland Publ., Amsterdam.

Manes, L., Sørensen, O. T., Mari, C. M., and Ray, I. (1980). *In* "Thermodynamics of Nuclear Materials, 1979," Vol. 1, p. 405. IAEA, Vienna.

Markin, T. L., Bones, R. J., and Gardner, E. R. (1967). *IAEA Tech. Rep., Ser.* No. 79.

Norris, D. I. R. (1979). *J. Nucl. Mater.* **79**, 118–127.

Rand, M. H., and Kubaschewski, O. (1963). "The Thermochemical Properties of Uranium Compounds." Oliver & Boyd, Edinburgh.

Roberts, L. E. J., and Walter, A. J. (1961). *J. Inorg. Nucl. Chem.* **22**, 213–229.

Sari, C., Benedict, U., and Blank, H. (1970). *J. Nucl. Mater.* **35**, 267–277.

Schmitz, F. (1975). *J. Nucl. Mater.* **58**, 357–360.

Schmitz, F., and Marajofsky, A. (1975). *In* "Thermodynamics of Nuclear Materials 1974," Vol. I, pp. 457–468. IAEA, Vienna.

Sørensen, O. T. (1978). *In* "High Temperature Oxides" (A. M. Alper, ed.), Part V, pp. 75–125. Academic Press, New York.

Speiser, R., and Spretnak, G. W. (1955). *Trans. Am. Soc. Met.* **47**, 493–502.

Thorn, R. J. (1970). *In* "Extended Defects in Non-Metallic Solids" (L. Eyring and M. O'Keeffe, eds.), pp. 395–424. North-Holland Publ., Amsterdam.

Thorn, R. J., and Winslow, G. H. (1966). *J. Chem. Phys.* **44**, 2032–2048.

Ubbelohde, A. R. (1957). *Q. Rev. Chem. Soc.* **11**, 246–272.

Ubbelohde, A. R. (1966). *J. Chim. Phys. Phys.-Chim. Biol.* **62**, 33–42.

Willis, B. T. M. (1964). *J. Phys. (Paris)* **25**, 431–439.

Woodley, R. W. (1972). Rep. HEDL-TME 72-89. Hanford Engineering Development Laboratory, Westinghouse Hanford Co., Richland, Washington.

Ziman, J. M. (1972). "Principles of the Theory of Solids," 2nd Ed. Cambridge Univ. Press, London and New York.

4

Diffusion in Nonstoichiometric Oxides

Hj. Matzke

Commission of the European Communities, Joint Research Centre
Karlsruhe Establishment, European Institute for Transuranium Elements
Karlsruhe, Federal Republic of Germany

ISBN 0-12-655280-0

I. Introduction

Diffusion processes in oxides have been studied extensively during the past three decades for two reasons: to better understand the defect structure and to determine the species and the basic material transport phenomena that are rate controlling for high-temperature kinetic processes of technological interest.

Experience has shown that the physical and chemical properties of transition metal oxides are mainly related to their deviation from stoichiometry, which itself depends on two important parameters, namely, the temperature and the equilibrium oxygen partial pressure. Electric conductivity and diffusion are among the properties that react most dramatically to the oxygen-to-metal (O/M) ratio. The related processes of matter transport, such as creep, sintering, grain growth, and pore mobility, also depend strongly on the deviation from stoichiometry. This chapter will deal mainly with self-diffusion as measured by tracer techniques, but the above related processes will also be considered to some extent. Electric, and in this context predominantly ionic, conductivity will only be mentioned briefly whenever necessary, since it is the topic of another chapter in this book.

The concept of isolated point defects has become increasingly less attractive in the past decade, and it is being replaced by microdomain and cluster models, at least for given temperature ranges and for larger deviations from stoichiometry. The virtually complete elimination of point defects by the formation of shear structures is also briefly considered in this chapter. These topics are elaborated on in more detail in other chapters of this book. Finally, since diffusion-controlled processes often have to compete with matter transport by evaporation–condensation, in particular in the presence of the driving force of a thermal gradient, questions connected with vapor pressure and evaporation are also dealt with to some extent.

The substances considered have been selected on the basis of their degree of nonstoichiometry and the availability of reliable diffusion data (Table I). They are:

(i) *Oxides of the fluorite structure.* These include ZrO_2, a common ceramic in research and industry (e.g., for use in solid-state galvanic cells), CeO_2, and the actinide oxides ThO_2, UO_2, and PuO_2. The series of actinide dioxides is of great interest in nuclear applications. The present generation of nuclear reactors uses UO_2 as nuclear fuel. Fast breeder reactors at present

TABLE I

Homogeneity Range and Availability of Diffusion Data for Nonstoichiometric Oxides[a]

	Stability range			Diffusion data available[b]	
Oxide	$(O/M)_{min}$	$(O/M)_{max}$	$\Delta(O/M)$	Oxygen	Metal
TiO	0.65	1.25	0.60	−	+
Ti_2O_3	1.501	1.512	0.011	−	−
TiO_2	1.992	2.000	0.008	+ +	+ +
VO	0.80	1.30	0.50	−	−
MnO	1.00	1.18	0.18	−	+
FeO	1.045	1.200	0.155	−	+ +
Fe_3O_4	1.336	1.381	0.045	−	+ +
CoO	1.000	1.012	0.012	+	+ +
NiO	1.000	1.001	0.001	+	+ +
ZrO_2 [c]	1.67	2.004	0.33	+	+
CeO_2 [c]	1.78	2.00	0.22	+	−
ThO_2 [c]	1.9	2.00	0.1[c]	+ +	+ +
UO_2	1.65[d]	2.25	~0.60[d]	+ +	+ +
PuO_2	1.61	2.00	0.39	(+)	+ +
$(U_{0.8}Pu_{0.2})O_2$	1.90	2.25	0.35	(+)	+ +
PrO_2	1.75	2.00	0.25	−	−
Nb_2O_5	2.35	2.5	⩽ 0.15	+	−

[a]The data in this table for the transition metal oxides are those published by Reed (1970); for ZrO_2 by Ruh and Garrett (1967); for CeO_2 by Bevan and Kordia (1964); for UO_2, $(U, Pu)O_2$, and PuO_2 by the International Atomic Energy Agency (1967); for PrO_2 by Hyde *et al.* (1966); and for Nb_2O_5 by Kofstad (1972).

[b]+ +, extensive data; +, some reliable work; (+) or −, no quantitative data.

[c]Or chemical doping with, for example, CaO or Y_2O_3 to typically $x \sim 0.15$.

[d]Only near the melting point (~2500°C).

employ mixed $(U, Pu)O_2$, and may in the future use $(U, Th)O_2$. In addition, these three actinide oxides constitute an interesting series of isomorphic substances, since ThO_2 is practically always stoichiometric, UO_2 easily dissolves large amounts of interstitial oxygen to form anion-excess compositions UO_{2+x}, and PuO_2 [or $(U, Pu)O_2$] can accommodate large concentrations of oxygen vacancies in the reduced state of anion-deficient PuO_{2-x}.

(ii) *Oxides of the rock-salt structure.* Here the most studied and technologically important oxides, FeO, NiO, and CoO, will be considered, though other materials will also briefly be mentioned (mainly MnO). No relevant diffusion data are so far available for interesting substances such as VO.

(iii) *The rutile structure, hence* TiO_2. Diffusive matter transport in TiO_{2-x} has been studied to some extent. Also, TiO_{2-x} is of great scientific interest because of the ease of formation of shear planes and the Magnéli phases Ti_nO_{2n-1}.

(iv) *More complex oxides.* Some limited results are available for more complex nonstoichiometric oxides, such as Nb_2O_5. These will be briefly summarized, though an exhaustive discussion of their diffusional properties is not yet possible.

II. Crystal Structures Considered

A. *Fluorite*

In the oxides of the fluorite or CaF_2 structure, MO_2, each metal atom M is surrounded by eight equivalent nearest-neighbor O atoms each of which in turn is surrounded by a tetrahedron of four equivalent M atoms (Fig. 1). A typical feature of the fluorite structure are the large $(\frac{1}{2}, \frac{1}{2}, \frac{1}{2})$ octahedral interstitial holes in which interstitial ions (e.g., excess oxygen in UO_{2+x}) can easily be accommodated.

The fluorite oxides to be discussed are UO_2, PuO_2 [and the mixed oxide $(U, Pu)O_2$], CeO_2, and ZrO_2 (at $T \geqslant 1400°C$ or when stabilized with, for example, CaO). Comparison with other isomorphic substances, predominantly ThO_2 but also fluorides such as CaF_2 and BaF_2, is often useful. For instance, a melting point relation can be helpful in deciding which of a range of experimental results for MO_2-type substances is reliable. These isomorphic materials are stoichiometric unless deviations from stoichiometry are produced chemically, e.g., by adding CaO or Y_2O_3 to produce oxygen-deficient thoria, $(Th, Ca)O_{2-x}$.

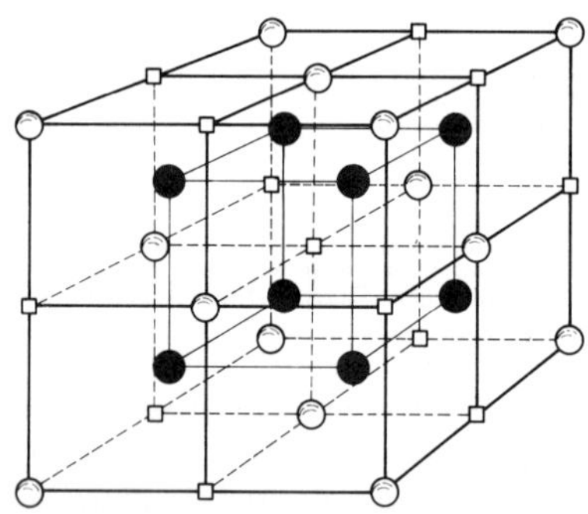

Fig. 1. The fluorite (CaF_2) structure.

B. *Rock Salt*

The NaCl or rock-salt structure (Fig. 2) can be envisaged as being a close-packed cubic array of large oxygen ions with smaller metal ions (here Fe, Ni, Co, Ti, and Mn, but also Nb, V) occupying the octahedral

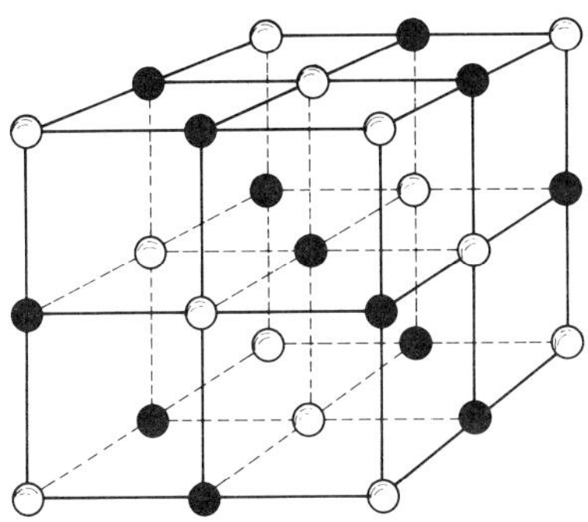

Fig. 2. The rock-salt (NaCl) structure.

interstices. Normally, each cation (anion) is surrounded by six anions (cations). However, some of these oxides, in particular FeO, MnO, TiO, VO, and NbO, can be highly nonstoichiometric (Table I), such that their rock-salt structures become highly defective.

C. Rutile

The crystal structure of rutile, the high-temperature form of TiO_2, is more complex than the basic types of cubic structure described above, since it is noncubic and thus leads to anisotropy in defect properties. Figure 3 shows the rutile structure. The sublattice of Ti^{4+} ions is body-centered tetragonal, the c dimension being much smaller than the a dimension, $c/a = 0.644$ (Baur, 1956). The structure can also be regarded as consisting of TiO_6 octahedra sharing edges and corners in such a way that each oxygen ion belongs to three neighboring octahedra. The oxygen octahedra are slightly distorted. When this structure is observed parallel to the (short) c-axis, open channels become apparent, which allow fast diffusion of, for example, small interstitial ions (Fig. 4), but also cause anisotropy in any diffusion process. Consequently, a complete description of the diffusion of any species in the rutile structure requires two principal diffusion coeffi-

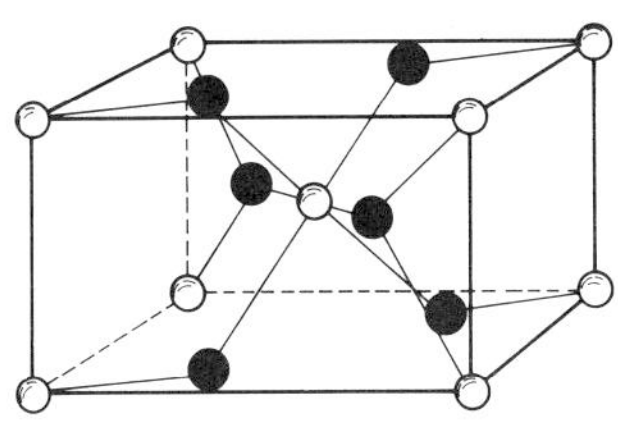

Fig. 3. The rutile (TiO_2) structure.

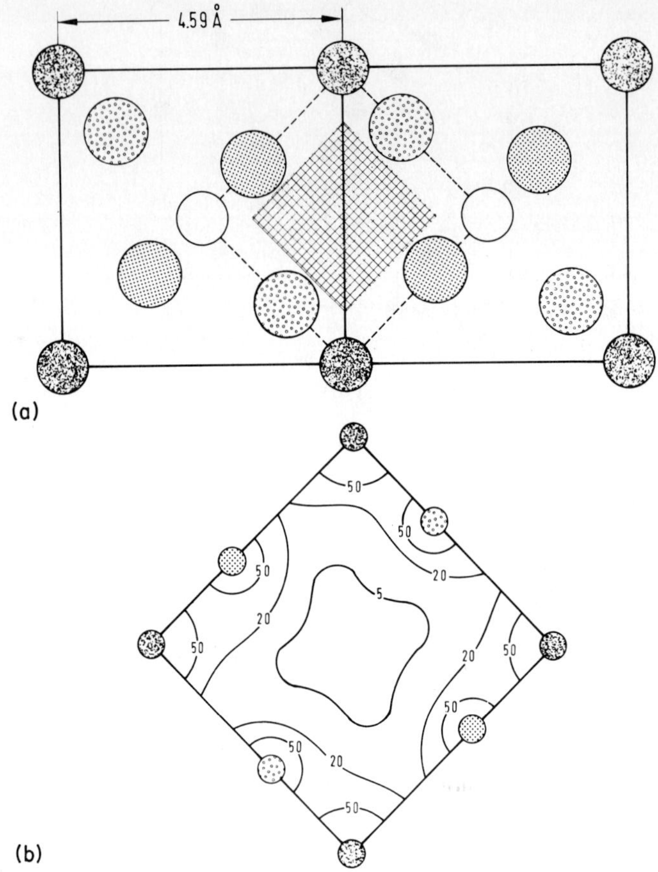

Fig. 4. The open channels along the c axis in the rutile structure, indicated by the dashed lines and the hatched square in (a) and by the low energy area in (b) representing potential energy contours (in eV) calculated using the static continuum form of the Molière function.

cients, one parallel (D_{zz}) and one perpendicular (D_{xx}) to the z axis; to this purpose, the c axis is oriented such that it coincides with the z axis of a Cartesian coordinate system. A diffusion coefficient D in any direction with the angle ψ between this diffusion direction and the z axis is then given by the relation

$$D = D_{zz} \cos^2 \psi + D_{xx} \sin^2 \psi.$$

For rutile, deviations from stoichiometry occur only in the substoichiometric phase, TiO_{2-x}. The stability range is between about 1.99 and 2.00 (Fig. 16; see also Blumenthal and Whitmore, 1963). Two types of defects are generally considered for substoichiometric rutile, either titanium interstitials or oxygen vacancies. Various associations of oxygen vacancies with Ti^{3+} ions are possible. Of particular interest also are diffusion phenomena in the phases formed during further reduction, i.e., between $TiO_{1.66}$ and $TiO_{1.99}$, where point defects are eliminated by the formation of planar faults known as *shear planes* (see, e.g., Anderson, 1972; see also Section III.D).

III. Point Defects and Clustering

Deviations from stoichiometry produce point defects, most likely oxygen vacancies or metal interstitials in substoichiometric (hypostoichiometric) compounds and oxygen interstitials or metal vacancies in overstoichiometric (hyperstoichiometric) compounds. Similar effects can be achieved by chemical doping, e.g., by adding CaO to ThO_2. Such a material contains one oxygen vacancy for each added impurity atom, resulting in charge neutrality. Finally, point defects are also created thermally in any material provided the temperature is high enough.

These defects can exist as single, isolated point defects at low enough concentrations (about 0.5% down to 10^{-2}%, depending on structure and binding conditions). At higher concentrations, the defects will aggregate into clusters, will become ordered, or will be eliminated by the formation of two- or three-dimensional defects (dislocation loops, shear planes, voids). As a matter of fact, previously unsuspected order has been revealed in almost every grossly nonstoichiometric system such that point defects in these ordered structures may play only a nonessential role in kinetic behavior. The different possibilities of clustering and order are described below.

A. Point Defects

Point defects include vacancies and interstitial atoms in the different sublattices of the oxide MO. To describe these defects, the symbols used by Kröger and Vink (1956) and Kröger (1964) will be adopted. This system is introduced in Fig. 5 in Chapter 1 (Sørensen). Briefly, the imperfection is designated by a capital letter and its site by a subscript. Thus

V_M = metal or cation vacancy
M_i = metal or cation interstitial
V_O = oxygen or anion vacancy
O_i = oxygen or anion interstitial
V_i = vacant interstitial site
Mf_M = foreign metal atom (impurity) on metal site
Mf_i = foreign metal atom (impurity) on interstitial site.

Point defects can be neutral or charged relative to the perfect crystal.

Such charged defects are indicated by a superscript; e.g., $V_O^{\cdot\cdot}$ is a double positively charged oxygen vacancy, and O_i'' is a negatively charged oxygen interstitial. Because of their charges, point defects tend to cluster at higher concentrations as a result of Coulomb interaction.

There are different ways of achieving charge equilibrium in a stoichiometric crystal:

(i) *Frenkel disorder* in which equal concentrations of metal vacancies, V_M'', and metal interstitials, $M_i^{\cdot\cdot}$, are created.

(ii) If the same process occurs in the oxygen sublattice, the oxygen Frenkel disorder creating equal concentrations of $V_O^{\cdot\cdot}$ and O_i'' is often referred to as *anti-Frenkel disorder*.

(iii) *Schottky disorder* is the last (existing) possibility. Charge neutrality is achieved by removing equal concentrations of metal and oxygen atoms from the lattice to the surface, hence by creating equal concentrations of $V_O^{\cdot\cdot}$ and V_M''. These defects are called Schottky pairs in MO and Schottky trios in MO_2, since again a neutral molecule, this time consisting of two oxygen and one metal atom, must be removed from the crystal.

In the intrinsic temperature range, usually above $0.5T_m$, thermally created defects dominate. Their concentration is determined by the molar free energy of formation, ΔG_F or ΔG_S, for formation of a *pair* of Frenkel defects or a Schottky pair or trio. Thus, if

(N_{V_O}) = equilibrium fraction of oxygen vacancies

(N_{O_i}) = equilibrium fraction of interstitial oxygen ions

the temperature dependence of the concentration of oxygen Frenkel defects will be

$$(N_{V_O})(N_{O_i}) = \exp(-\Delta G_{FO}/RT) \quad (1)$$

In the extrinsic (impurity-controlled) range, usually at $T \leqslant 0.5T_m$, and in nonstoichiometric oxides, the defect concentration is constant and determined by the concentration of impurities or the deviation from stoichiometry. For compounds like oxides, however, the dominating defect species (e.g., oxygen vacancies and interstitials in fluorite-type oxides) are often not rate controlling for diffusion-dependent processes. Rather, the slower-moving minority defects (e.g., metal defects in fluorite-type oxides) are rate determining since, to achieve true mass transport, both species of the oxide have to be transported. As a consequence, a delicate interplay of the different defect equilibria occurs. As an example, the case of $UO_{2\pm x}$ is treated in Table II. UO_2 has a wide range of hypo- and hyperstoichiometric compositions (Table I). The resultant equations of Table II are obtained by combining Eq. (1) for oxygen defects with the remaining two possibilities: defects of the cation-Frenkel type, FM, and the Schottky type, S, of disorder. The first type involves metal vacancies and interstitials, (N_{V_M}) and

TABLE II

Defect Equilibria in the Fluorite Structure $MO_{2\pm x}$

Material	Defect concentration[a]	
MO_2	$(N_{O_i}) \approx 2(N_{V_O}) = \sqrt{2}\ \exp(-\Delta G_{FO}/2kT)$	(2)
	$(N_{V_M}) \approx 2\exp[-(\Delta G_S - \Delta G_{FO})/kT]$	(3)
	$(N_{M_i}) \approx \frac{1}{2}\exp[-(\Delta G_{FM} + \Delta G_{FO} - \Delta G_S)/kT]$	(4)
MO_{2+x}	$(N_{O_i}) \approx x$, assuming $x \gg (N_{V_O})_{\text{thermal}}$	(5)
	$(N_{V_O}) \approx (1/x)\exp(-\Delta G_{FO}/kT)$	(6)
	$(N_{V_M}) \approx x^2 \exp[-(\Delta G_S - 2\,\Delta G_{FO})/kT]$	(7)
	$(N_{M_i}) \approx (1/x^2)\exp[-(\Delta G_{FM} + 2\,\Delta G_{FO} - \Delta G_S)/kT]$	(8)
MO_{2-y}	$(N_{V_O}) \approx y/2$, assuming $y/2 \gg (N_{O_i})_{\text{thermal}}$	(9)
	$(N_{O_i}) \approx (2/y)\exp(-\Delta G_{FO}/kT)$	(10)
	$(N_{V_M}) \approx (4/y^2)\exp(-\Delta G_S/kT)$	(11)
	$(N_{M_i}) \approx (y^2/4)\exp[-(\Delta G_{FM} - \Delta G_S)/kT]$	(12)

[a] Equilibrium fraction of oxygen vacancies (N_{V_O}), of oxygen interstitials (N_{O_i}), of metal vacancies (N_{V_M}), and of metal interstitials (N_{M_i}).

(N_{M_i}), and the second vacancies in both the metal and the oxygen sublattices. Thus

$$(N_{V_M})(N_{M_i}) = \exp(-\Delta G_{FM}/RT) \tag{13}$$

$$(N_{V_O})^2(N_{V_M}) = \exp(-\Delta G_S/RT) \tag{14}$$

Two basic assumptions are made in this treatment: Neither association of defects nor interaction of defects with impurities occurs.

Table II shows that the crucial quantity for most defect concentrations is ΔG_{FO}, the formation energy of an oxygen Frenkel defect. Also, most *effective* formation energies are composite quantities. The defect concentrations for $MO_{2\pm x}$ calculated with these equations are shown in Fig. 11 (Section V.A). The most drastic change in defect concentration occurs near the stoichiometric composition. Relations similar to those in Table II can also be set up for other oxides, provided the bonding is predominantly ionic, hence charge equilibrium is reached by the formation of point defects. The above considerations hold for small defect concentrations. Hence they are true for thermally created defects in stoichiometric compounds where typical defect fractions at the melting point are of the order of 10^{-4}. For deviations from stoichiometry exceeding a fraction of the order of 0.005 or 0.01, and at low enough temperatures, defects interact. This interaction may be in the form of small-scale clustering in microdomains, as in long-range order, or as virtual elimination of point defects by the formation of shear planes.

B. Point Defect Clusters

The Schottky–Wagner treatment of randomly distributed defects fails at higher defect concentrations. Defect clustering is extensively discussed in other chapters of this book and will therefore be only briefly dealt with here in connection with the substances relevant to diffusion results, such as UO_{2+x}, $(U,Pu)O_{2-x}$, and $Fe_{1-x}O$. Neutron diffraction has shown that UO_{2+x} has the same space group (Fm3m) as UO_2, thus precluding any long-range ordering as reported for the higher oxide U_4O_9 (see below). The neutron diffraction work of Willis (1963, 1964a, b, 1965) proved short-range clustering of oxygen at large deviations from stoichiometry. UO_{2+x} is formed by the incorporation of interstitial oxygen, O_i, rather than by forming uranium vacancies, V_U, as was shown in Belle (1961) by measurements of lattice parameter and density. The excess oxygen does not occupy the center of the large $(\frac{1}{2}, \frac{1}{2}, \frac{1}{2})$ interstitial sites (Fig. 1), but rather two other sites are populated: the O′ site located ~1 Å along the ⟨110⟩ axis, and the O″ site displaced along the ⟨111⟩ axis from the $(\frac{1}{2}, \frac{1}{2}, \frac{1}{2})$ position, roughly midway between this site and a normal oxygen site. In addition to the excess oxygen, a significant number (about one to two) of adjacent normal oxygen atoms are displaced, thus creating an identical number of oxygen vacancies. This arrangement of defects in clusters is understandable, since the distances from O′ and O″ sites to the nearest normal oxygen atoms are only ~1.5 Å. This is prohibitively small in comparison to the normal O–O interionic distances of ~2.6 Å.

Figure 5a shows a schematic simplified picture of clustering in MO_{2+x}. In UO_{2+x}, the clusters consist of more atoms, e.g., in an array of 2:2:2 at the two sites and the connected vacancies. The work of Cheetham *et al.* (1970; Cheetham, 1973) on CaF_2 plus 10 mole % YF_3 simulating MF_{2+x} confirmed the existence of such 2:2:2 or 3:4:2 clusters even in the fluoride, indicating that they are an intrinsic feature of anion-excess fluorite

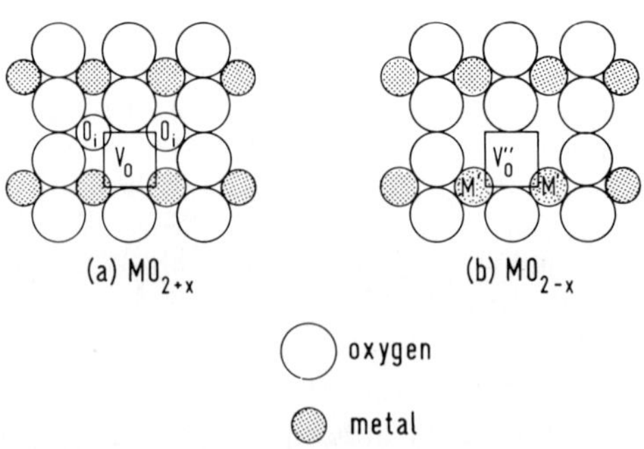

Fig. 5. Schematic presentation of defect cluster formation in a dioxide: (a) MO_{2+x} (b) MO_{2-x}.

compounds. A direct confirmation of the oxygen displacements was obtained by channeling techniques (Matzke *et al.*, 1971). The scattering of a deuteron beam off the heavy U atoms was largely independent of crystal orientation. Deep minima for the aligned crystal for all three directions indicated a near-perfect U sublattice. (A more careful analysis shows that the U atoms are displaced somewhat, by $\lesssim 0.25$ Å, from their normal positions.) In contrast, the same deuteron beam interacting with the oxygen sublattice via a $^{16}O(d, p)^{17}O$ reaction encountered largely blocked $\langle 110 \rangle$ and $\langle 100 \rangle$ channels, and only along the $\langle 111 \rangle$ direction was a relatively undisturbed lattice seen, in agreement with the Willis model. Figure 28 of Chapter 1 (Sørensen) shows clusters in UO_{2+x} in more detail.

In the substoichiometric region of fluorite-type oxides MO_{2-x}, cluster formation has been reported for the most extensively studied case of $(U, Pu)O_{2-x}$ (Manes and Manes-Pozzi, 1976; De Franco and Gatesoupe, 1976), for PuO_{2-y} (Chereau, 1974), and for $(Ce, Y)O_{1.95}$ (Fender, 1973). The simplest cluster is shown in Fig. 5b. It consists of one oxygen vacancy and two trivalent (reduced) cations, thus forming a nucleus of M_2O_3. These nuclei can grow stepwise by doubling, trebling, etc., their size. This model can describe specific features of thermodynamic functions and explains the phase diagrams of fluorite-type substoichiometric oxides (Manes *et al.*, 1979). More details are given in Chapters 1 and 3 (Sørensen and Manes) of this book.

A third example is that of the metal-deficient oxide $Fe_{1-x}O$ [see also Chapter 2 (Catlow) and Fig. 26 in Chapter 1 (Sørensen)]. The structure of cation-deficient wüstite and its further oxidation to higher oxides such as Fe_3O_4 are classic problems of short-range order. Hobbs (1976) has summarized the previous literature. Fe^{2+} ions oxidized to Fe^{3+} move into interstitial positions. The simplest stable cluster is the 4 : 1 cluster consisting of trivalent Fe_i^{3+} surrounded by four cation vacancies. With this cluster as a basis, the association of defect units can again occur. For instance, three edge-sharing defect clusters form a 8 : 3 cluster.

These clusters unavoidably imply a degree of covalent bonding. Their thermal stability is largely unknown. Many observations on clusters (or shear planes; see Section III.D) have been made on samples quenched under a specific $p(O_2)$. Some indications at relatively low temperatures are available (e.g., Hayakawa *et al.*, 1974, for $Fe_{0.9}O$ at 840°C; Willis, 1964a, for $UO_{2.13}$ at 800°C), but at the high temperatures of diffusion studies information is still lacking. First indications of stable clusters or ordering at temperature have recently been obtained, at 1050°C for TiO_{2-x} and at 1500°C for $(U, Pu)O_{2-x}$, in careful diffusion work (Sections V and VI), but more information is definitely needed. Clusters might also dissociate during irradiation (Onozuma *et al.*, 1977). Their stability toward irradiation should be investigated further.

C. Long-Range Order

An example of long-range ordering can again be found in the U–O system. Oxidation of UO_2 to UO_{2+x} by incorporation of O_i leads to superlattice reflections as soon as $UO_{2.25}$ ($= U_4O_9$) is reached (see Blank and Ronchi, 1968, for a review). At room temperature, 64 excess O_i are distributed in a large unit cell, with $a = 4a_0$, of space group $I\bar{4}$ 3d. Between 550 and 700°C, an order–partial disorder transformation occurs, the high-temperature phase becoming partly disordered.

D. Shear Phases

A particularly interesting form of point defect ordering is the formation of crystallographic shear (CS) structures (for critical reviews, see Magnéli, 1970; Anderson, 1972; Anderson and Tilley, 1974). Electron microscopy has shown that the reduction of some oxides can result in the formation of planar faults or shear planes. Basically, the collapse of a nonstoichiometric crystal containing many point defects to produce a CS structure eliminates the point defects. Thus most CS phases do not appear to contain significant concentrations of point defects.

A possible sequence for forming shear planes could be: Oxygen atoms diffuse out of the crystal thus leaving oxygen vacancies. These become ordered into "walls" or collapse into planar disks (Wadsley defects) across which the crystal then shears and annihilates the vacant sites. The original structure is thus divided into equidistant lamellae separated by shear planes. The different models of CS formation are discussed in detail by Gado (1965), Anderson and Hyde (1965, 1967), Andersson and Wadsley (1966), Hyde and Bursill (1970), and Matzke and Ronchi (1972). O'Keeffe (1970, 1971) has further discussed kinetic questions related to extended defects. The interesting questions involving shear-plane formation and lateral migration cannot be discussed here. Though there is evidence that shear planes can move by a concerted motion of ions bordering the CS planes (Bursill *et al.*, 1971), it is still unclear how this contributes to bulk tracer diffusion. Also, the implications of the important observation of fast solid-state reactions between substances prone to CS formation (e.g., $6Nb_2O_5 + MoO_3 \rightarrow MoNb_{12}O_{33}$) cannot be discussed here either. Rather, the discussion will be focused on the problem of what a tracer atom does within a *preformed equilibrated* CS structure or how it reacts to a flux of Wadsley defects, which can be thought to sweep the tracer through the crystal somewhat analogously to the vacancy wind of Manning (1961).

The substances most prone to CS are reduced TiO_2, Nb_2O_5 (or NbO_2F), WO_3 (also doped with Nb or Ta), MoO_3, and V_2O_5. It can also occur three-dimensionally as a block structure, as in Nb_2O_{5-x}, and has been reported for UO_{2-x} (Matzke and Ronchi, 1972). In the present context, TiO_{2-x} (and to a minor degree UO_{2-x} and Nb_2O_{5-x}) is most interesting, since diffusion data for this material are available. The important questions are

Can point defects and shear structures coexist?

How does a tracer atom move within a CS structure? Can it penetrate from one lamella into another one or is it repelled by the shear plane?

In shear planes, metal–metal distances are inevitably shorter, possibly connected with a change in bonding from largely ionic (e.g., in UO_2) to largely covalent (e.g., in UO). Note also that the spacing of CS planes can be quite short, with distances of 20–100 Å only.

At low degrees of reduction, as well as at high enough temperatures, isolated point defects certainly exist or shear planes dissociate into point defects. For instance, in the case of TiO_{2-x}, Baumard *et al.* (1977) observed pronounced changes in the shape of plots of log conductivity versus x at $x \sim 0.01$, indicating that shear planes start to form at $TiO_{1.99}$ (see also Section VI). No information seems to be available on thermal dissociation of CS planes. In this context, however, the problem of quenching a bulk piece of a nonmetal should be mentioned. Many studies on CS structures (or other extended defects) are performed at room temperature using samples more or less effectively quenched from high temperatures. Obviously, switching off the furnace and cooling the specimen in the equilibrium atmosphere of the high temperature is unsuitable because of the long time involved and because not only the defect ordering but also the O/M ratio of the sample will change since the oxygen potential is a function of temperature. Typically, the sample will increase its O/M ratio during such a treatment. But quenching attempts should also be regarded with care: O'Keeffe (1970) has estimated that a sphere of a typical ceramic of radius 1 mm, the surface of which is instantancously cooled by $\Delta T - 1000°C$, still has a cooling time of ~ 1 sec. This allows a typical defect to perform about 10^7 jumps yielding a mean displacement on the order of 1 μm, a large distance indeed. More strikingly, if there is a gradient or driving force directing the jumps in one direction, the total distance moved can be greater than or approximately equal to the sample dimensions. Thus, if there is a tendency to cluster, it may be impossible to freeze in high-temperature randomness. Since most of the above jumps occur in the first cooling period, the structure observed (or frozen in) will be characteristic of some fairly high temperature that, however, neither corresponds to nor necessarily depends on the initial annealing temperature. Further aspects of shear structures are discussed in Section VI.

E. Lambda Transitions

A lambda (order–disorder) transition, i.e., the occurrence of a large degree of positional or configurational disorder above about $0.8\,T_m$ (the melting point in K) as observed in nonoxides (CaF_2, SrF_2, $SrCl_2$, BaF_2, PbF_2, K_2S) of the fluorite structure (Dworkin and Bredig, 1968; Lidiard, 1974), has been postulated to exist in the oxides UO_2 and ThO_2 as well (Bredig, 1970), since the diffusional properties in both types of substance are very similar. Temperatures T_c of 2400 (UO_2) and 2600°C (ThO_2) were suggested. A lambda transition was previously thought to be a cooperative process somewhat resembling fusion. It is usually evidenced by pronounced anomalies in heat content and heat capacity and was originally interpreted as being due to a transition from the normal fluorite structure MB_2 to a disordered structure where the B atoms are randomly distributed over both the tetrahedrally and the octahedrally coordinated sites. The observed increasing anharmonicity of oxygen displacements with increasing temperature (Willis, 1963) may be a preamble to these transitions. Recent work by Schoonman (1979, 1980) on ionic conductivity in fluorites presents another picture. Schoonman suggests that at T_c only small concentrations of *mobile* defects are present, typically 0.5 to 2 mole %, which, however, show an unusually large defect mobility at T_c. Both anion vacancies and interstitials are said to contribute to the conductivity.

Experimental work on this question is still lacking. An attempt (Matzke, 1979b) to perform self-diffusion measurements with UO_2 above the lambda transition temperature has failed so far because of pronounced evaporation effects masking diffusion. Measurements of fluorine self-diffusion in CaF_2 and BaF_2 were unfortunately not extended to the lambda transition temperature, but Ca diffusion in CaF_2 (Matzke and Lindner, 1964) and conductivity measurements on BaF_2 (Carr *et al.*, 1976) showed no anomaly, though neutron-scattering work (Dickens *et al.*, 1976) revealed a dynamic disorder in BaF_2 at ~960°C. Obviously, more studies will be needed to confirm the existence and understand the mechanism and features of the transition. Also, the influence of deviations from stoichiometry on the possible transition should be elaborated. Short-range phenomena in microdomains might possibly occur in nonstoichiometric oxides. It would be most interesting to learn more about the possible consequences of the dynamic disorder in the *anion* sublattice above about $0.8\,T_m$ for the disorder in the *metal* sublattice. If one is dealing with isolated randomly distributed anion defects, then, as shown in Section III.A, via the Frenkel defect equilibrium between anion defects and the Schottky equilibrium between anion vacancies and cation vacancies, there should be a corresponding change in disorder in the metal sublattice. This question also has technological relevance. In advanced nuclear reactors, some of the UO_2 or $(U, Pu)O_2$ fuel may be above the postulated transition temperature. None of the fuel performance codes, so far, allows for this disorder.

IV. Tracer Diffusion and Related Processes

Many structural changes in solids occur by diffusion. Therefore, to understand fully and possibly predict processes such as phase changes, homogenization, sintering, grain growth, and creep, a thorough knowledge of diffusion is essential. For compounds like oxides, usually the slower of the two species is rate determining, since *all* and not just one atom species of a compound must be moved to achieve matter transport. The slower-moving species can be the metal (e.g., U in UO_2) or the oxygen (as in TiO_2). Exceptions to this rule are possible. The above processes can follow the kinetics of volume diffusion but can be rate-controlled by the diffusion of the faster species if the other species has another means of transport, e.g., via surface diffusion, grain boundary diffusion or evaporation–condensation (an example is UC where U is transported in the gas phase and C by

TABLE III

Common Symbols and Terms for Diffusion Coefficients[a]

D^T, or simply D	The tracer or self-diffusion coefficient represents only the random-walk diffusion process in the absence of chemical potential or temperature gradients.
D_l or D_v	The lattice diffusion coefficient describes any diffusion process within the bulk, lattice, or volume of the crystal.
D_s	The surface diffusion coefficient represents diffusion along a surface or in an open interconnected porosity.
D_b	Describes diffusion along an interface, a grain boundary, or dislocations (dislocation pipe diffusion).
$\tilde{D}$	The chemical or interdiffusion coefficient refers to diffusion in a chemical potential gradient.
	Intrinsic diffusion refers to a process when only native point defects (thermally created) are the vehicle for transport.
	Extrinsic diffusion refers to diffusion via defects not created from thermal energy, e.g., impurities.
D_a	The *apparent diffusion coefficient* includes the contributions from several diffusion paths in one net diffusion coefficient.
D^O, D^M, D^{Mf}	A superscript can describe the diffusing species, such as oxygen, the metal atom of the oxide MO, or an impurity Mf.

[a] All diffusion coefficients are given in cm^2/sec. To obtain the unit m^2/sec multiply by 10^4.

solid-state diffusion and C is rate controlling through $D^{C}/D^{U} \sim 10^{6}$). The relation between the above processes and the diffusion coefficients are briefly explained. Different diffusion coefficients can be involved, as indicated in Table III. Furthermore, some basic mathematics of diffusion are given and some important points regarding experimental techniques are made in this section.

A. Theoretical Background

The basic process of diffusion can be treated in two ways. The substance in which diffusion is to be studied can be regarded as a continuum (ignoring the atomic structure), or else the elementary random walk with individual jumps over atomic distances can be considered. The latter approach yields information on the atomic mechanism, but the former approach is more generally applicable. The former approach dates back to Fick and consists essentially of solving appropriate differential equations usually based on Fick's second law, which in one dimension is

$$\frac{\partial c}{\partial t} = \frac{\partial}{\partial x}\left(D \frac{\partial c}{\partial x}\right) \tag{15}$$

where c is the concentration of the diffusing atoms, x the diffusion distance, and D the diffusion coefficient (here always in cm^2/sec; see Table III). Some typical and useful solutions are given in Table IV.

Essentially the same kinetics are obtained by applying the random walk theory. There, jumps are assumed to be made in all directions, with high

TABLE IV

Frequently Used Solutions of the Diffusion Equation

Condition	Solution[a]	
Thin film of tracer	$c(x,t) = \frac{c_0}{2\sqrt{\pi Dt}} \exp(-\frac{x^2}{4Dt})$	(16)
Constant surface concentration c_s	$\frac{c(x,t)}{c_s} = 1 - \text{erf}(\frac{x}{2\sqrt{Dt}})$	(17)
Thick tracer source (thickness h)	$c(x,t) = \frac{1}{2}c_0(\text{erf}\frac{h-x}{2\sqrt{Dt}} + \text{erf}\frac{h+x}{2\sqrt{Dt}})$	(18)

[a] c_0 = Concentration (tracer activity) at $x = 0$ and $t = 0$; c_s = constant surface concentration; x = penetration distance; t = annealing time; $\text{erf}(x) = (2/\sqrt{\pi}) \int_0^x \exp(-\eta^2)\,d\eta$.

jump frequencies of about 10^8 jumps/sec at the melting point. The problem of calculating how far an individual atom migrates from its initial site after a given time (given number of random jumps) was solved by Einstein, yielding a relation between the continuum diffusion coefficient D and the average sum of the squares of the individual jumps $\overline{x^2}$

$$\overline{x^2} = 2Dt \tag{19}$$

Here $(\overline{x^2})^{1/2}$ is also called the mean square displacement and gives the average diffusion distance in time t. It is thus a handy rule of thumb for estimating distances involved in diffusion studies.

Diffusion in the solid itself occurs by jumps on an atomic scale via the movement of point defects. The mechanisms of practical importance are:

(i) Vacancy mechanism: An atom on a normal site jumps into an adjacent vacancy. The atom thus moves in a direction opposite that of the vacancy movement.

(ii) Interstitial mechanism: An atom of an interstitial site jumps into another interstitial site via one of (usually) several possible saddle points, thus distorting the lattice considerably.

(iii) Interstitialcy mechanism: An atom at an interstitial site pushes a neighboring lattice atom into another interstitial site and itself jumps into the vacancy thus created. The net result is the same as in (ii), but excessive lattice distortions are avoided.

In all practical cases, a given type of atom (M or O) diffuses only within its own sublattice. Atoms can also diffuse along surfaces, grain boundaries, and dislocation lines, involving more complicated mechanisms. Space does not allow these cases to be discussed in more detail.

B. Experimental Techniques

The conventional technique of measuring tracer self-diffusion involves the thin-layer method combined with serial sectioning. A thin layer of radioactive tracer is deposited on a well-prepared face of a diffusion sample. Evaporation, electrodeposition, sputter deposition, or application of a drop of a tracer-containing solution can be used. The thickness d of this layer should be $\leqslant 0.1\sqrt{Dt}$, the mean diffusion distance, in order to be able to apply Eq. (16) in Table IV. Otherwise, Eq. (18) is applicable. The surface of the diffusion specimen must be well defined. Surface roughness leads to errors in D, and chemically different surface layers (e.g., air-grown Al_2O_3 layers on Al) can make all measurements meaningless. Usually, the surfaces are well polished or lapped at different stages with down to $\leqslant$ 1-μm diamond paste to achieve a mirrorlike surface. For recovery of mechanical polishing damage in near-surface layers and to restore the original perfect lattice, the samples should be preannealed before applying the tracer. This is particularly important for nonstoichiometric oxides, since meaningful results require a constant O/M ratio during the diffusion anneal. This anneal has to be performed in the appropriate atmosphere to maintain the constant O/M ratio, e.g., in a CO–CO_2 or a H_2–H_2O mixture. Care must be taken to maintain the O/M ratio while cooling the diffusion specimens to room temperature following the diffusion anneal, since otherwise chemical analysis yields incorrect results. Either the oxygen potential of the annealing atmosphere should be continuously changed with the cooling temperature or, more conveniently, the sample should be pulled out of the furnace under high vacuum, since then slow surface reactions and fast cooling rates will guarantee freezing in of the high-temperature equilibrium O/M ratio.

Usually two specimens, each containing a tracer layer, are placed on top of each other to form a *diffusion couple* in order to minimize evaporative losses of tracer and to avoid chemical reaction of the diffusion zone with the container or furnace material.

Following the diffusion anneal, the tracer concentration profile is determined in order to calculate D with the aid of Eqs. (16)–(18) in Table IV. Usually the sample is destroyed by serial sectioning, i.e., by removing parallel layers from the surface and determining their tracer concentration by measuring the radioactivity. In the most straightforward case of an original thin tracer layer and in the absence of any short-circuiting paths for the tracer (such as grain boundaries, dislocation lines, phase boundaries, porosity), a straight line is obtained in a plot of log(tracer concentration) versus (penetration depth $x)^2$ [see Eq. (16) in Table IV and Fig. 6b]. If grain boundary diffusion contributes to volume diffusion, such a plot shows curvature at larger penetration depths until finally, if the tracer penetrates only along grain boundaries, a straight line is obtained in a plot of $\log c$ versus $x^{6/5}$. For a complete analysis, see the work of Fisher (1951), Whipple (1954), and Suzuoka (1961, 1964, 1965). With sinters, the D values obtained are often high because of pore and grain boundary contributions. With a small enough grain size (or a high enough dislocation density), linear plots of $\log c$ versus x^2 can easily be obtained, which can be misinterpreted as proof that volume diffusion dominates. Such a value should be called the apparent diffusion coefficient D_a rather than D or D_1 (Table III). Finally, Table IV and Fig. 6 also include the solution for the case of constant surface tracer concentration, e.g., for heating the sample in a large reservoir of tracer.

Recently, a new and nondestructive method was introduced to determine concentration profiles of tracers decaying by α-particle emission (Schmitz and Lindner, 1965; Höh and Matzke, 1973). High-resolution α spectroscopy can be used to determine the depth of the emitting tracer

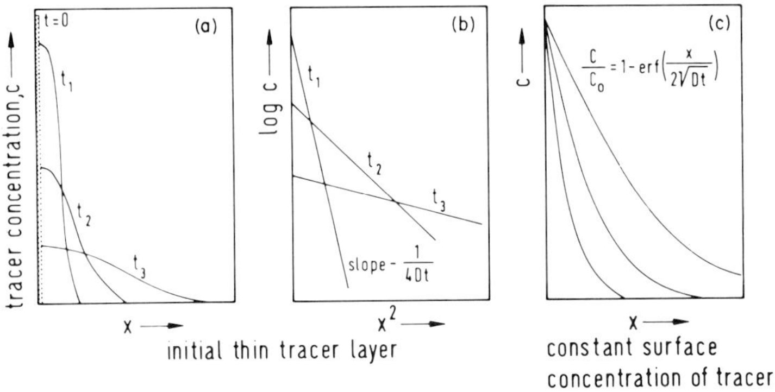

Fig. 6. Schematic tracer penetration profiles. (a) Gaussian profiles for the starting condition of a thin tracer layer; (b) same, but in a plot of $\log c$ versus x^2; (c) error function-type penetration profiles for the case of constant surface concentration of tracer.

atom, hence the diffusion profile. The method has predominantly been used for actinide dioxides having the fluorite structure (Section V.C). The diffusion specimens can be used for different annealing steps at a given temperature. This allows the different contributions to D_a to be separated. For D_l, the tracer penetration is proportional to $t^{1/2}$, for D_b, to $t^{1/4}$, etc. For most materials studied with this technique, a further complicating phenomenon was observed (see, e.g., Matzke, 1976a, b): A plot of the variance of the penetration profile $4Dt$ versus t gave a first fast tracer penetration followed by much slower diffusion. The initial penetration was explained to be due to "surface smoothing" or "accommodation of the tracer layer to the surface," i.e., to a type of surface relaxation, possibly similar to the observed step formation on surfaces of UO_2 annealed in H_2 (see replicas in Fig. 7). These surface steps are formed as a result of local changes in oxygen content and in the orientation-dependent evaporation rates and the anisotropy in surface energy. Figure 8 shows the errors that can easily be introduced in measuring D if no time dependence is measured and the initial "pseudo-$4Dt$" is not subtracted from the subsequent penetration due to volume diffusion. At short annealing times, the error can be such that the data become completely meaningless, as is true of much of the published data on metal diffusion in UO_2 and ThO_2 (Matzke, 1976a, b). Even at long annealing times, the error can still be significant.

In summary, modern standards call for exact control of annealing atmospheres, for extremely well-polished surfaces [of preferably single-crystal specimens, the quality and surface conditions of which should be verified with channeling techniques (Matzke, 1969b)], and extremely thin (and carrier-free) tracer layers, together with verification of the fact that the calculated D value does not vary with time.

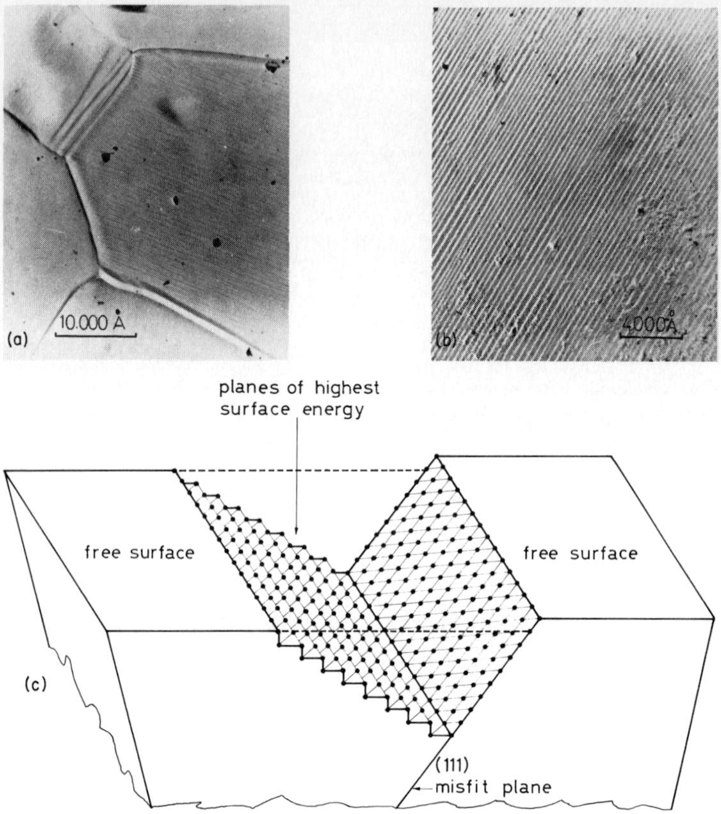

Fig. 7. Formation of relaxed surfaces, hence step formation, on surfaces of UO_2 and possible explanation of step formation. (From Matzke and Ronchi, 1972.) (a) Five minutes at 1500°C; (b) 30 min at 1950°C; (c) schematic representation of the model for formation of surface steps.

The above statements all hold for metal atom diffusion. They also apply to a large extent to oxygen diffusion which, however, causes additional difficulties. Since there is no suitable radioactive isotope of oxygen, the above conventional methods for measuring diffusion coefficients cannot be used to obtain results for oxygen diffusion. Therefore a number of modified or new techniques had to be developed (Table V).

The most frequently used method is gas–solid isotope exchange. The oxide is heated in an atmosphere enriched in ^{18}O or, alternatively, the oxide is enriched in ^{18}O and heated in natural oxygen, mainly ^{16}O. The change in ^{18}O concentration in the gas phase is measured by mass spectrometry (see, e.g., Haul and Dümbgen, 1965). This (normal) method is indirect, and a major disadvantage is that the surface exchange must be rapid compared with the bulk diffusion rate. This is often not the case, in particular at low

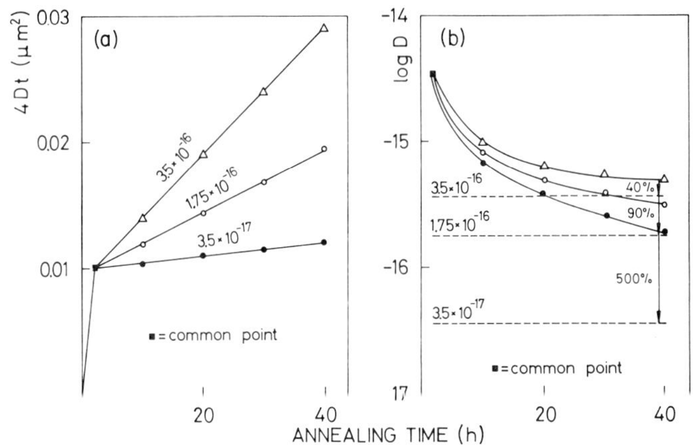

Fig. 8. (a) Calculated cases for three diffusion coefficients and initial surface relaxation leading to 0.1-μm tracer penetration. (b) The simulated decrease in D with time if no correction is made for the initial surface effect. The errors involved are large (see arrows) even for a relatively long anneal of 40 h. (From Matzke, 1976b. Reproduced by permission of North-Holland Publ. Co., Amsterdam.)

temperatures, thus yielding D values that are too low and activation enthalpies that are incorrect. Also, powdered samples are often used for which the assumption of an idealized spherical shape of constant and uniform radius is frequently questionable. Combined with sectioning of the sample (Simpson and Carter, 1966), the depth profile can also be obtained.

Thin-layer methods, which avoid problems involving surface exchange rates, have so far been applied extensively only to UO_2 (Marin and Contamin, 1969). Fairly thick layers of oxide enriched in ^{18}O must be used, which are badly defined in terms of stoichiometry and purity.

Nuclear reactions are frequently used. A number of suitable reactions exist and have been used (Table V). Depth profiles can be obtained if the reaction product has a high stopping power, e.g., α particles in the reaction $^{18}O(p,\alpha)^{15}N$. Alternatively, with the same reaction, the proton capture resonance of 1.765 MeV can be used (Derry *et al.*, 1971). If the incident beam has the resonance energy, the ^{18}O concentration of the surface is measured. At higher energies, the beam penetrates into the solid and loses the additional energy. In this way, the depth profile can be obtained. Since the method is nondestructive, the time dependence of diffusion at constant temperature can be measured with one sample (Calvert *et al.*, 1974). Of the remaining techniques mentioned in Table V, the $^{18}O(p,n)^{18}F$ reaction is worthwhile mentioning since ^{18}F is produced *in situ*. It has a half-life of 112 min and decays by β^+ emission. The 0.51-MeV γ rays from positron annihilation are easy to measure. When beveled samples and autoradiography are used, preferential penetration of oxygen along grain boundaries can convincingly be demonstrated (Condit and Holt, 1964).

TABLE V

Techniques for Measurement of Oxygen Diffusion Coefficients

Method	Remarks	System studied	Diffusion	Oxygen analysis	Reference
1 *Gas–solid isotope exchange*	Mass spectrometer needed; most	Al_2O_3	+		Oishi and Kingery (1960)
	widely used technique. Good for	UO_{2+x}	+		Auskern and Belle (1961)
(in ^{18}O-containing	low diffusion rates ($D \sim 10^{-16} cm^2/sec$).	TiO_2	+		Haul and Dümbgen (1965)
O_2 or Co–Co_2)	Measurement at diffusion temperature.	Cr_2O_3	+		Hagel (1965)
	Concentration profile not directly determined. Surface exchange reaction may be rate controlling	ThO_2	+		Ando and Oishi (1979)
With sectioning and					
Mass spectrometer	Depth profile measured	$(Zr, Ca)O_{2-x}$	+		Simpson and Carter (1966)
Ion beam mass spectrometry	High depth resolution, possibly selective sputtering	ZrO_2 layers	+		Cox and Pemsler (1968)
2 *Thin tracer layer of ^{18}O-containing oxide*	Fairly thick (10–20 μm) layer				
with		UO_2	+		Marin and Contamin (1969)
Spark mass spectrometry	Not very accurate, overestimates D	UO_2	+		Contamin and Stefani (1967)
Laser source mass spectrometry	Accurate localization possible	UO_2	+		Eloy *et al.* (1968)
Ion mass analyzer	Versatile, but sputter rate depends on grain orientation	UO_2	+		Contamin and Slodzian (1968)
Nuclear reaction	See below	UO_2	+		

3	*Electrical conductivity*	Indirect and application limited to ionic conductors	$(Zr, Ca)O_{2-x}$	(+)		Simpson and Carter (1966)
			CeO_{2-x}	(+)		Steele and Floyd (1971)
	Electro transport	Indirect and of limited application	UO_2	(+)		Dornelas and Lacombe (1967a, b)
4	*Nuclear reactions*	Need good spectroscopy				
	$^{16}O(d, n)^{17}F$	Uses common isotope; not applicable to self-diffusion in oxides	Layers on Ta, etc.		+	Möller *et al.* (1967)
	$^{16}O(d, p)^{17}O$	Uses common isotope; not applicable to self-diffusion in oxides	O in U–C–O		+	Matzke (1969a)
			O position in U_4O_9		+	Matzke *et al.* (1971)
	$^{17}O(^{3}He, \alpha)^{16}O$	Needs rare isotope	ZrO_2 films	+		Cox and Roy (1966)
			ZrO_2 films	+		Ollerhead *et al.* (1966)
	$^{18}O(p, n)^{18}F$	^{18}F is radioactive and can be measured by autoradiography. Depth distribution on beveled samples	ZrO_2 films	+		Holt and Himmel (1969)
	$^{18}O(p, \gamma)^{19}F$	Surface concentration measured ($\sim 10\ \mu m$ depth)		+		Zelinski *et al.* (1970)
		Distribution by repeated grinding and remeasuring	UO_2	+		Hadari *et al.* (1971)
	$^{18}O(p, \alpha)^{15}N$		ion implantation		+	Whitton *et al.* (1971)
		For small diffusion distances follows profile from shape of α spectrum	UO_2	+		Ligeon *et al.* (1970)
			Si	+		Gass *et al.* (1973)
			Zr–O, Ti–O	+		Price and Bird (1969)
	$^{18}O(\alpha, n)^{12}N$	Replaces mass spectrometry in gas–solid exchange; gives depth profile if combined with grinding and used as (p, α) reaction	PuO_2	+		Deaton and Wiedenheft (1973)
			Oxidation of Fe–Cr	+		Bruckman *et al.* (1972)

V. Diffusion in the Fluorite Structure: UO_2, $(U,Pu)O_2$, CeO_2, ZrO_2, ThO_2

The predominant defects in oxides of the fluorite structure, hence in nonstoichiometric $MO_{2\pm x}$, are oxygen defects. Substoichiometric MO_{2-x} contains predominantly oxygen vacancies rather than metal interstitials. This was shown for PuO_{2-x} by density measurements (Atlas *et al.*, 1966) and for CeO_{2-x} by x-ray and neutron diffraction (Faber *et al.*, 1967a, b). Similarly, MO_{2+x} contains predominantly oxygen interstitials rather than metal vacancies, which was again demonstrated by measurements of density and lattice parameters (as reported, e.g., for UO_{2+x}, in Belle, 1961). Note that for high defect concentrations (e.g., $\geqslant 0.5$ at. %), clustering of these defects occurs as discussed in Section III.

In addition to direct tracer diffusion measurements, information on point defects and diffusion can also be obtained from quenching work and from recovery (i.e., thermal annealing) of irradiation damage, as well as from studies on diffusion-related processes, such as ionic conductivity, grain growth, creep, and sintering. These processes can be used to obtain information on diffusion mechanisms, activation enthalpies, and even diffusion coefficients. Typical relations are shown in Table VI.

A. Oxygen Self-Diffusion

Oxygen diffusion is much faster than metal diffusion in the fluorite structure. It is therefore of interest in oxidation and reduction kinetics but not in other mass transport processes involving transport of *all* components of the oxide, such as sintering. If these processes are rate-controlled by diffusion, the diffusion of the slower metal atom determines the kinetics. Oxygen diffusion data exist for ZrO_2, CeO_2, ThO_2, UO_2, $(U,Pu)O_2$, and PuO_2, both for stoichiometric and nonstoichiometric compositions. Because of the high concentration of oxygen defects in nonstoichiometric fluorite-type oxides, diffusion in both MO_{2+x} and MO_{2-x} is faster than diffusion in the stoichiometric oxides, MO_2. This behavior is illustrated in Fig. 9 for the case of UO_2 and UO_{2+x}. Here UO_2 was chosen as an example since more data exist for this oxide than for any other fluorite-type oxide because of its importance as fuel for nuclear reactors. Three recent reviews and models have adequately treated the dependence of oxygen self-diffusion (D) and chemical diffusion ($\tilde{D}$) on deviations from stoichiometry (Matzke, 1976b; Breitung, 1978; Murch and Thorn, 1978). Figure 9 shows experimental results for both D and $\tilde{D}$ as scatter bands based on data published by different authors (Auskern and Belle, 1961; Dornelas and Lacombe, 1967a, b; Contamin and Stefani, 1967; Contamin and Slodzian, 1968; Roberts *et al.*, 1963; Marin and Contamin, 1969; Ligeon *et al.*, 1970; Hadari *et al.*,

TABLE VI

Relations for Deducing Diffusion Coefficients from Diffusion-Controlled Processes: Ionic Conductivity, Sintering, and Creep[a]

1. *Ionic Conductivity* (Nernst–Einstein relation): $\sigma = Ne^2 D/nkT$ (20)

N = number of ions/cm^3; n = transference number; e = electronic charge (valid if mechanisms of conductivity and diffusion are the same, not necessarily valid if clustering occurs and is affected by electric fields).

2. *Initial Sintering; Shrinkage:* $y = (CKD)^P t^P$ (21)

Mechanism	C	P	Reference
Bulk, D_l	$31/\pi^2$, 3.0	0.47	Johnson and Cutler (1963)
	$20/\sqrt{2}$	0.40	Kingery and Berg (1955)
	2	0.50	Coble (1958)
Grain boundary, D_b	$7.1b/r$	0.31	Johnson and Cutler (1963)
	$1.2b/r$	0.316	Berrin and Johnson (1965)
	$1.5b/r$	0.33	Coble (1958)

y = fractional shrinkage; b = apparent grain boundary width; $K = \gamma\Omega/kr^3T$ with γ = surface energy; Ω = vacancy volume; r = radius of (assumed) sphere.

3. *Second-Stage Sintering:* $dP/dt = -AK'D$ (22)

Mechanism	A	Reference
Bulk, D_l	33	Coble (1961, 1965)

P = porosity; $K' = \gamma\Omega/kl^3T$ with γ = surface energy; Ω = vacancy volume; l = edge length of assumed tetrakaidecahedron ($l = 0.36d$, grain diameter).

4. *Diffusional Creep:* $\epsilon = A\Omega/kTd^2(D_v + A'bD/d)\sigma$ (23)

Mechanism	A	A'	Reference
Bulk, D_l	14	—	Nabarro (1948); Herring (1950)
Grain boundary, D_b	—	π	Coble (1963, 1977)

ϵ = creep rate; d = grain size; σ = applied stress.

[a] t = time; T = temperature; D = diffusion coefficient (see Table III); k = Boltzmann constant.

1971; Contamin *et al.*, 1972; Murch *et al.*, 1975, for D in UO_{2+x}; Lay, 1970; Marin, 1968; Bittel *et al.*, 1969; Carter and Lay, 1970; Taskinen and Kullberg, 1979, for $\tilde{D}$ in $UO_{2\pm x}$). Included are also $\tilde{D}$ values for (U, Pu) $O_{2\pm x}$ labeled as $MO_{2\pm x}$ (D'Annucci and Sari, 1977; Sari, 1978; Bayoglu and Lorenzelli, 1979).

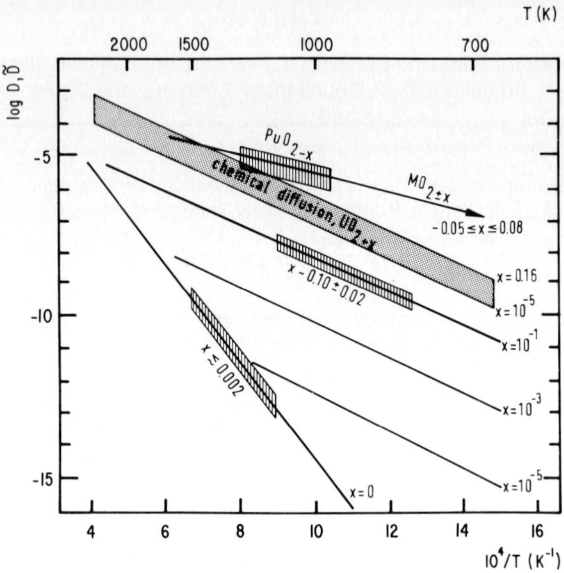

Fig. 9. Oxygen diffusion in $MO_{2\pm x}$. Measured results are given as bands, and the lines represent calculated values (Breitung, 1978). Both self-diffusion coefficients D and chemical diffusion coefficients $\tilde{D}$ are given.

The value of D is seen to depend strongly on x; empirically, and at a first approximation, $D/x = \text{const.}$ at a given temperature (Matzke, 1976b). More precisely, D can be described by the relation

$$D = D_0^{\mathrm{v}}[\mathrm{V_O}]\exp(-\Delta H_{\mathrm{m}}^{\mathrm{v}}/RT) + 2D_0^{\mathrm{i}}[\mathrm{O_i}]\exp(-\Delta H_{\mathrm{m}}^{\mathrm{i}}/RT) \qquad (24)$$

where D^{v} and D^{i} are the diffusion coefficients, $[\mathrm{V_O}]$ and $[\mathrm{O_i}]$ the concentration of oxygen vacancies (v) and interstitials (i), and ΔH_{m} the migration enthalpy. Based on the defect model for $UO_{2\pm x}$ (Section III, Table II, and below), Breitung (1978) has calculated both D and $\tilde{D}$. The results are included in Fig. 9 as lines. The quantities in Eq. (24), as used by Breitung and as determined from experiments, are $D_0^{\mathrm{v}} = 1.44$, $D_0^{\mathrm{i}} = 0.013\ \mathrm{cm^2/sec}$, $\Delta H_{\mathrm{m}}^{\mathrm{v}} = 63$, and $\Delta H_{\mathrm{m}}^{\mathrm{i}} = 23.8$ kcal/mole ($\Delta H_{\mathrm{m}}^{\mathrm{v}}$ may be high, as discussed below). In MO_{2+x}, $[\mathrm{O_i}]$ is determined by x and $[\mathrm{V_O}]$ is negligible (interstitial diffusion mechanism), whereas in MO_{2-x} (M = U + Pu, usually $U_{0.8}Pu_{0.2}$) $[\mathrm{V_O}]$ is given by x and $[\mathrm{O_i}]$ becomes very small (vacancy diffusion mechanism).

For stoichiometric oxides, defects are created thermally, and their concentration increases with temperature proportional to $\exp(-\Delta G_{\mathrm{FO}}/RT)$. The crucial quantity, the free energy ΔG required to create oxygen Frenkel defects is usually taken as ~70 kcal/mole. The chemical diffusion coefficients $\tilde{D}$ are less dependent on x but are very high indeed. As shown by

Breitung, they are essentially obtained by multiplying D by a thermodynamic factor containing the change in oxygen potential $\Delta G(O_2)$ with stoichiometry:

$$\tilde{D} = D \frac{(2+x)}{2RT} \frac{d[\Delta G(O_2)]}{dx} \tag{25}$$

Since the oxygen potential $\Delta G(O_2)$ varies strongly with x as well, particularly when approaching the stoichiometric composition (e.g., Woodley and Adamson, 1979), the product $D \times d[\Delta G(O_2)]/dx$ is nearly constant. Basically, similar results are obtained when statistical thermodynamic relations are used (Murch and Thorn, 1978).

Figure 10 gives a summary of available oxygen diffusion data on fluorite-type materials. Included are data for fluorine diffusion in CaF_2 and BaF_2 which also have predominant anti-Frenkel disorder. For ease of comparison, a normalized temperature scale was chosen, i.e., T_m/T, with T_m = melting point in K. The right ordinate plots $\log(D/x)$, which allows the data for nonstoichiometric oxides to be shown in one band, averaged for UO_{2+x} by the dotted line labeled UO_{2+x}(12). This line adequately describes oxygen diffusion in UO_{2+x} for the range $0.006 \leqslant x \leqslant 0.16$. The

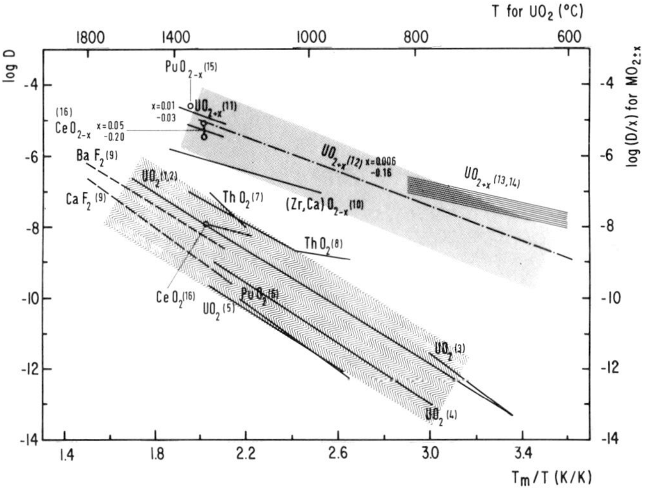

Fig. 10. Summary of oxygen and fluorine diffusion data in materials with the fluorite structure. For ease of comparison, a normalized temperature scale has been used (T_m = melting point in K). Stoichiometric materials: UO_2: (1) Belle (1969), (2) Roberts *et al.* (1963), (3) Auskern and Belle (1961), (4) Marin and Contamin (1969), (5) Hadari *et al.* (1971); PuO_2: (6) Deaton and Wiedenheft (1973); ThO_2: (7) Edwards *et al.* (1963), (8) Ando and Oishi (1979) and Ando *et al.* (1976); CeO_2: (16) Steele and Floyd (1971); CaF_2, BaF_2: (9) Matzke (1970). Nonstoichiometric oxides: $(Zr, Ca)O_{2-x}$ (10): Simpson and Carter (1966); UO_{2+x}: (11) Roberts *et al.* (1963), (12) Auskern and Belle (1961), (13, 14) Marin and Contamin (1969), Contamin *et al.* (1972), and Murch *et al.* (1975); PuO_{2-x}: (15) Chereau and Wadier (1973); CeO_{2-x}: (16) Steele and Floyd (1971).

two bands in Fig. 10 represent the data for stoichiometric and nonstoichiometric oxides, respectively. The fact that all data fall into these bands points to a similar behavior for all fluorite-type materials considered, hence to diffusion coefficients and activation enthalpies that are proportional to T_m.

The results for UO_2 [lines (1)–(5)] show some scatter which may be due to deviations from stoichiometry or to the inherent difficulties of the isotope exchange technique, which arise, for example, because of the finite rate of the gas–solid reaction. Such factors may have affected the results for ThO_2, as well. (Note that no radioactive tracer exists for oxygen; see also Table V.) The data of Marin and Contamin (1969) [line (4)] are considered the most representative for stoichiometric UO_2 (annealing in H_2, several different experimental methods), thus indicating a good melting point correlation with the fluorine tracer data for CaF_2 and BaF_2. The results for PuO_2 and ThO_2 (with the above reservations) also seem to follow this relation approximately. The results for CeO_2 probably represent extrinsic diffusion. Hence, for UO_2, the most probable relation seems to be

$$D^{O}_{(UO_2)} = 0.26 \exp(-59{,}300/RT) \quad \text{cm}^2/\text{sec}$$

For all nonstoichiometric oxides, diffusion in both hyper- and hypostoichiometric compositions is obviously faster than diffusion in stoichiometric compounds. For instance, the dashed-dotted line (12) in Fig. 10 can be described (Matthews, 1974) by the empirical relation

$$D^{O}_{(UO_2)} = 0.14 \exp(-31{,}000/RT) \quad \text{cm}^2/\text{sec}$$

In hyperstoichiometric oxides, oxygen interstitials control oxygen diffusion. This interstitial migration occurs by a concerted or interstitialcy mechanism: The migrating interstitial displaces a lattice oxygen ion into a neighboring interstitial site and itself occupies the site of the displaced lattice ion. In contrast, vacancies migrating by a direct jump of a lattice ion into the vacancy control diffusion in hypostoichiometric oxides and probably also in stoichiometric oxides.

Though more data exist on D^{O} in UO_{2+x} than on any other nonstoichiometric oxide, and though a number of different techniques (Table V) have been applied in order to avoid the drawbacks of the indirect isotope exchange method, there are still open questions. For UO_{2+x}, the treatment given in Section III.A and Table II and the above-cited one by Breitung (1978) describe most data adequately, yet they neglect the known cluster formation in UO_{2+x} (Section III.B). De Bruin and Murch (1973) have treated the influence of defect interactions and, in particular, of Willis 2:2:2 defect clusters on the dependence of D^{O} on x at constant temperature. A maximum value of D is always predicted, as would intuitively be expected, since defects forming clusters become immobile. The experimental results, however, tend rather to follow the relation D/x = const. used in Fig. 10. Probably, as suggested by Murch (1975), several defect clusters or higher-order clusters are responsible for the continued high diffusion rates. Vollath (1976) has obtained indications of the existence of a number of different clusters in his study on the recovery of quenched-in resistivity (see below). Alternatively, a dynamic exchange between mobile defects and defects temporarily immobilized in clusters or domains might be postulated. More knowledge is obviously needed on the thermal stability, the dissociation, and the high-temperature form of the clusters. These properties may depend on both T and x, as indicated, for example, by the high ΔH of ~41–44 kcal/mole in the two-phase region UO_{2+x} plus U_4O_9 below 550°C (Contamin *et al.*, 1972; see also below).

For the stoichiometric oxide UO_2, difficulty arises in defining the exact stoichiometry. The partial molar thermodynamic quantities, as well as the defect concentrations (here the ratio of O_i to V_O), change very rapidly with composition for $x \sim 0$ (Fig. 11). The presence of heterovalent cation impurities determines this ratio at low temperatures. As an example, Steel and Floyd (1971) report CeO_2 with ~1000 ppm total impurities to show extrinsic behavior over the entire investigated temperature range (800–1100°C). The same holds for UO_2 and explains part of the difficulty in determining ΔG_{FO}, the free energy of formation of oxygen Frenkel defects, which is the crucial quantity in the defect model of the fluorite structure (Table II).

Usually, ΔG_{FO} is determined by subtracting the activation enthalpy of extrinsic diffusion (here $\Delta H_m^{O_i}$) from ΔH(intrinsic) and multiplying by 2, since two defects are formed; hence $\Delta H_{diff}^{O} = \frac{1}{2}\Delta G_{FO} + \Delta H_m^{O_i}$, neglecting entropy effects. In this way, Auskern and Belle (1961) have deduced ΔG_{FO} from their oxygen diffusion measurements as amounting to 3.1 eV by taking the activation enthalpy of oxygen diffusion in UO_{2+x} as the migration energy of defects in UO_2. Catlow and Norgett (1973) have criticized this procedure, postulating that oxygen diffusion in UO_2 proceeds via oxygen vacancies. Using their *calculated* $\Delta H_m^{V_O}$ of only 0.25 eV, they revised the Auskern and Belle result to 5.1 eV. This in turn was criticized by Matzke (1973), who pointed out that $\Delta H_m^{V_O}$ would be expected to be essentially larger than 0.25 eV, based on a comparison with BaF_2, CaF_2, and ZrO_2 which show values of about 0.6–1.0 eV (CaF_2 and BaF_2) and 1.3 eV (ZrO_2); see Table VII. A melting point relation would thus confirm the value of 1.7 eV for UO_2 in Table I. Also, Vollath's quenching experiments (1972, 1973, 1976), as well as statistical and thermodynamic considerations (Aitken *et al.*, 1966), would lead to 2.6 eV $\leqslant \Delta G_{FO} \leqslant$ 3.2 eV, a range recently confirmed by results for PuO_2 (Table VII). Part of the supporting evidence, however, would be inadmissible, if the experimental $\Delta H_m^{V_O}$ values were to include an essential defect association energy (see below) or if

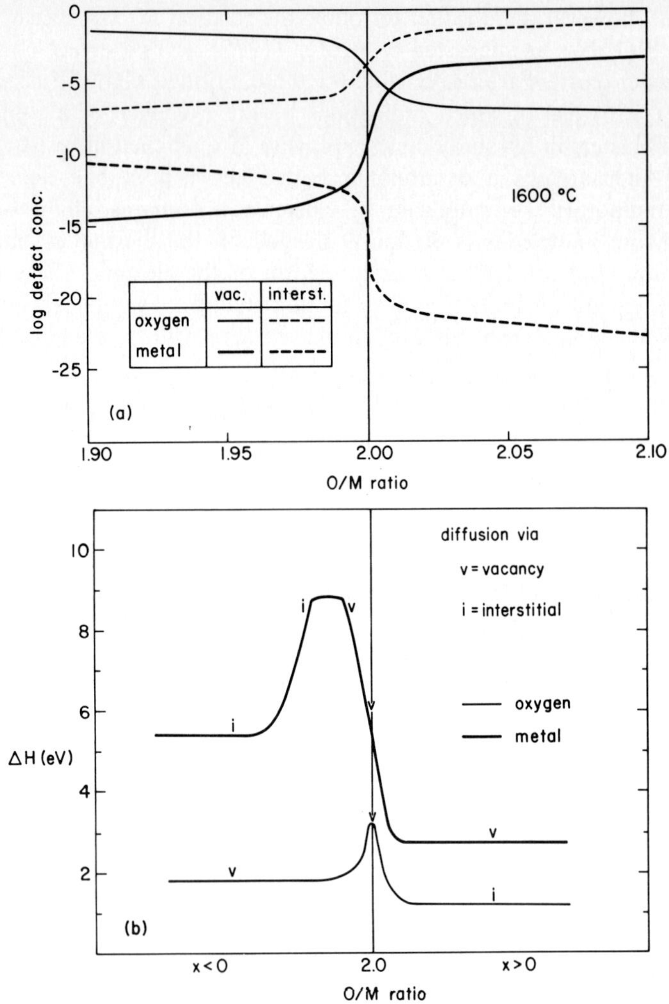

Fig. 11. Predicted defect concentrations (a) and activation enthalpies (b) for self-diffusion in $UO_{2\pm x}$. (From Matzke, 1976b. Reproduced by permission of North-Holland Publ. Co., Amsterdam.)

Vollath's assignment of recovery stages to defect mobilities were to need revision. In conclusion none of the existing evidence seems convincing enough to state an *exact* numerical value for ΔG_{FO}. This is a serious drawback for any defect model (Section III), but at present the *most probable* value for ΔG_{FO} is still ~70 kcal/mole or ~3 eV.

The related question concerning the enthalpy of migration $\Delta H_m^{V_O}$ of oxygen vacancies in the fluorite structure is also under debate. Vollath

TABLE VII

Defect Energies for Materials of the Fluorite Structure[a]

	BaF_2	CaF_2	$(Zr, Ca)O_{2-x}$	CeO_2, $(Ce^2, Y)O_{2-x}$	PuO_2	UO_2, $(U, Pu)O_2$	ThO_2, $(Th, Ca)O_{2-x}$, $(Th, Y)O_{2-x}$
T_m(K)	1830	1960	2980	~2870	2670	3150	3570
$T_m/T_m(UO_2)$	0.58	0.62	0.95	~0.91	0.85	1	1.13
Anions							
ΔH_{diff}	1.6	2.0	—	—	—	~2.6–3.0	2.1–2.8
ΔG_{FO}	1.6–1.9	2.2–3.1	—	—	2.7–3.0	3.1	2.8
ΔH_m							
interstitial	0.8	1.65	—	—	—	0.9–1.3	—
vacancy	0.6–0.8	0.4–1.0	1.2–1.4	1.0–0.4[b]	0.5	1.7	0.8–1.0
						0.25 2.7[c]	
Cations							
ΔH_{diff}	3.8	3.8–4.4	4.0–4.3	—	—	5.6	6.5
ΔH_m							
interstitial	—	—	—	—	—	(~0.2)[d]	—
vacancy	—	1.5–1.8	—	—	—	2.2–2.4	—
ΔG_{FC}	8.8	7.1	—	—	—	9.5	—
ΔG_S	5.2	5.8	—	—	—	6.4	—

[a]The data for BaF_2 and CaF_2 are due to Matzke (1970); for ZrO_2 to Simpson and Carter (1966), Rhodes and Carter (1966), and Kingery *et al.* (1959); for CeO_2 to Steele and Floyd (1971); for PuO_2 to Chereau and Wadier (1973) and Bayoglu and Lorenzelli (1979); for UO_2 and $(U, Pu)O_2$ to the review of Matzke (1976b); for ThO_2 to Ando *et al.* (1976), Ando and Oishi (1979), Oishi (1979), and Matzke (1976a).

[b]Decreases with x for CeO_{2-x}, but not for $(Ce, Y)O_{2-x}$ (Steele and Floyd, 1971).

[c]See discussion in text on oxygen diffusion in the fluorite structure. The behavior may be different in UO_2 and in UO_{2+x} (Vollath, 1972, 1973, 1976).

[d]From a study on loop growth, Soullard (1977) suggested mobility of U_i during low-temperature irradiation with $\Delta H_m^{U_i} \sim 0.1$–$0.4$ eV. See also Table II.

(1972, 1973, 1976) suggested high values (1.7 eV for UO_2; 2.7 eV for UO_{2-x}) from his quenching work (see below), whereas Catlow and Norgett (1973) calculated a very low value (0.25 eV). However, the calculated energies of these authors for metal defects using the same (ionic) model are unrealistic, thus casting some doubt on the ionic model they used for UO_2. There are no direct diffusion data for UO_{2-x}, since the UO_{2-x} phase exists only at very high temperatures. There is evidence for other fluorite-type oxides that oxygen anion vacancy migration can be related to a concomitant process of electron hopping, such as $Ce^{3+} \rightarrow Ce^{4+}$ in CeO_{2-x} (Floyd, 1973). Since $D^{O} \gg D^{M}$ in the fluorite structure, the clusters mentioned in Section III.B for MO_{2-x} would be thought to affect D^{O} much more than D^{M} (see below). Some low values of $\Delta H_{m}^{V_O}$ were reported in the literature, such as 4–6 kcal/mole for CeO_{2-x} (Steele and Floyd, 1971) and PuO_{2-x} (Chereau and Wadier, 1973), but these represented compositions far from stoichiometry. For such compositions, vacancy ordering, cluster formation, formation of ordered phases (e.g., such as Ce_nO_{2n-2}, postulated by Baenzinger *et al.*, 1961, or impurity vacancy interactions [such as Ca–V_O pairs in $(Ce, Ca)O_{2-x}$, according to Nowick *et al.* (1979)] affect the effective $\Delta H_{m}^{V_O}$ value. Recent results for PuO_{2-x} and $(U, Pu)O_{2-x}$ favor a value between 0.5 and 0.75 eV (D'Annucci and Sari, 1977; Sari, 1978; Bayoglu and Lorenzelli, 1979). Note also, in this respect, that the alternative to V_O, i.e., M_i (e.g., Ce^{3+} and Ce^{4+} interstitials in CeO_{2-x}) was suggested for substoichiometric CeO_2 and PuO_2 (Kofstad and Hed, 1967; Kofstad, 1972). Though it seems unlikely that M_i is the dominating defect in the total defect structure in fluorite-type substances (see also Chapter 1), M_i can be the dominating defect in the metal sublattice (see below). Metal interstitials have been suggested by Soullard (1977) to have a low migration enthalpy (Table VII). Evidently, more work is needed, in particular for small x values in MO_{2-x}, to settle the question of a reliable enthalpy for oxygen vacancy migration.

Most of the above discussion was based on UO_2. The few relevant results for CeO_{2-x} and $(Zr, Ca)O_{2-x}$ mentioned in the text are included in Table VII.

B. Quenching and Recovery of Irradiation Damage

Further insight into oxygen diffusion rates can be obtained by irradiating or by quenching samples and measuring the recovery of a suitable property (e.g., electrical conductivity) during controlled annealing. This latter type of experiment offers special advantages because of its high resolving power. Results for UO_2 have been reported by Nagels *et al.* (1966) and by Vollath (1972, 1973, 1976). The latter author has performed very careful experiments with UO_2, UO_{2+x}, and UO_{2-x}. He succeeded, for instance, in resolving three recovery steps with activation energies of 20, 22, and 23 kcal/mole between 156 and 223°C, which he attributed to the precipitation of excess oxygen to form U_4O_9. In contrast to the situation in tracer experiments, it is not obvious how recovery stages can be attributed to defects of a specific kind of atom (e.g., O or U atoms); the assignment must rather be deduced from other available knowledge. Accepting Vollath' s interpretation implies that in reality the complexes in the oxygen sublattice of UO_{2+x} and U_4O_9 are even more complex than suggested by Willis in his model (see above). The remaining defect energies deduced from other recovery stages are included in Table VII and confirm the deductions from diffusion experiments (see also Matzke, 1979a).

For irradiated specimens, the most obvious change in physical properties is an increase in lattice parameters caused by the accumulation of defects. The electric resistivity and the thermal conductivity are two further physical properties that are affected. Isochronal recovery curves show anneal stages which, however, are less well defined than those in quenched samples, especially when saturation damage occurs. Usually, a stage at about 300–400°C and another at (very roughly) 600°C are found in UO_2 and attributed to oxygen and uranium defect mobility, respectively. As an example, Vollath (1976) has suggested a uranium vacancy migration enthalpy of 55 kcal/mole for the recovery step at 600°C in irradiated UO_2.

Irradiation itself enhances self-diffusion at low temperatures. Most effective are fission events of U or Pu atoms. Data are available for U and Pu diffusion in UO_2 and $(U,Pu)O_2$ (Höh and Matzke, 1973; Matzke, 1979a). An athermal temperature-independent diffusion is observed between 130 and 1000°C, with a certain thermal contribution between 1000 and 1300°C. The radiation-enhanced diffusion coefficient D^* increases linearly with the fission rate. The effect could be shown to be due to fission spikes. A typical value for D^* is 10^{-16} cm^2 sec.

C. *Metal Self-Diffusion*†

Metal self-diffusion data are available for $(Zr,Ca)O_{2-x}$, $UO_{2\pm x}$, $(U,Pu)\cdot O_{2\pm x}$, and ThO_2 [and $(U,Th)O_2$]. As for oxygen diffusion, these results will not be discussed separately for each material, but rather together in terms of a common behavior within the fluorite structure. Metal diffusion is

†Diffusion of U and other actinides in UO_2: Lindner and Schmitz (1961); Auskern and Belle (1961); Schmitz and Lindner (1963, 1965); Marin and Michaud (1965); Yajima *et al.* (1966); Hawkins and Alcock (1968); Belle (1969); Marin and Contamin (1969); Matzke (1969a); Reimann and Lundy (1969); Lee and Alcock (1971); Matzke (1973, 1976b); Lambert (1979); in UO_{2+x}: Hawkins and Alcock (1968); Marin *et al.* (1967); Marin and Contamin (1969); Matzke (1969a, 1970); Lee and Alcock (1971); Lambert (1979); UO_2 containing

(*Footnote continued on page 188*)

generally much slower than oxygen diffusion and therefore difficult to measure because of the low D values involved.

Unfortunately, most of the existing metal diffusion data on fluorite oxides must be rejected either because of the use of outdated techniques, because of poor specimen quality, or because they represent artifacts. Some of the early data on UO_2 and ThO_2 were obtained with the surface decrease method, where the decrease in α activity at the surface is used to calculate diffusion coefficients. A correction for short-circuit paths is not possible and, with actinide oxides, the data *must* be affected by evaporative loss of tracer. The reported diffusion coefficients are therefore upper limits and can easily be too high by several orders of magnitude. In other work, ^{235}U was used as tracer. This can be acceptable if mass spectrometry is used to determine the tracer profile, but if α counting is used, the thin-layer condition must be violated because of the long half-life (low specific activity) and the natural abundance of ^{235}U. Hence incorrect mathematics were applied, and the results were affected in an unknown and irreproducible way by all the impurities deposited with the *thick* tracer layer and by its deviation from the stoichiometric composition. Other experiments must have been affected by short-circuit paths. Recent experiments over long annealing times with a careful analysis of the shape of the tracer profiles and investigation of the time dependence of the various contributions (Lambert and Matzke, 1980) have shown that, in normal sinters of actinide oxides (grain sizes of 5–20 μm) and at the most frequently used experimental temperatures (1300–1800°C), 50–80% of the tracer penetrates the sintered specimens along short-circuit paths.

Therefore reliable work should be done with single crystals. Single crystals of varying quality are available for CeO_2, UO_2, $(U, Pu)O_2$, PuO_2, and ThO_2, usually obtained from arc-fused material with the floating zone technique or grown from the vapor. They usually contain 50–500 ppm impurities (mainly Fe), and even with these single crystals great care must be taken to analyze the diffusion profile properly because, here also, short-circuit phenomena cause deeply penetrating tails at $T \lesssim 1700°C$ (King, 1971; Matzke, 1976a, b). Other difficulties are introduced by initial non-volume-diffusion phenomena at or near the surface (Section IV.B and Figs. 7 and 8). If the published data are critically reviewed considering the

---

impurities: Matzke (1966a, b). ThO_2: Matzke (1967b); Hawkins and Alcock (1968); Furuya and Yajima (1968); Morgan and Poteat (1968); King (1970, 1971); Matzke (1976a); $(U, Pu)O_{2+x}$: Lindner *et al.* (1967); Riemer and Scherff (1971); Matzke and Lambert (1973, 1974, 1976, 1980); Matzke (1973, 1976b); Lambert (1979). $\tilde{D}$ in $(U, Pu)O_{2-x}$: Theisen and Vollath (1967); Butler and Meyer (1973); Chilton and Edwards (1978); Matzke (1979a). $(Zr, Ca)O_{2-x}$: Rhodes and Carter (1966). Related material, CaF_2: Matzke and Lindner (1964); Matzke (1970); Berard (1971); King and Moermann (1974).

above sources of error, most of them must unfortunately be rejected. Some of the few remaining data are also of questionable quality. This is due to another phenomenon: peak or tracer layer mobility. As a result of evaporation–condensation processes, part of the tracer becomes embedded (Matzke, 1973). This phenomenon has recently been used to explain and predict pore mobility in actinide carbides (Matzke, 1975). In the oxides, which are known to show very large pore mobilities by evaporation–condensation (see below), this effect causes uneven surfaces and distorted tracer profiles because of the pronounced anisotropy of the surface energy. Without correcting for this phenomenon, too high D values are again obtained. Therefore, to measure reliable diffusion coefficients, single crystals should be used, the time dependence of the tracer penetration should be carefully followed, and the consistency of the results *across the specimen surface* should be verified. This has only been done very recently.

Figure 12 summarizes the state of knowledge on metal diffusion in stoichiometric substances of the fluorite structure, again, as for oxygen diffusion, on a normalized temperature scale. The shaded areas show the commonly accepted range of data for U diffusion in UO_2 and Th diffusion in ThO_2, data that should be rejected. Most of these data are for temperatures below $0.6T_m$. It is therefore not surprising that they are not typical for volume diffusion, especially since in addition to grain boundary contributions the above-mentioned artifacts cause the data to be misleading. In addition to experimental difficulties, unnoted deviations from stoichiometry

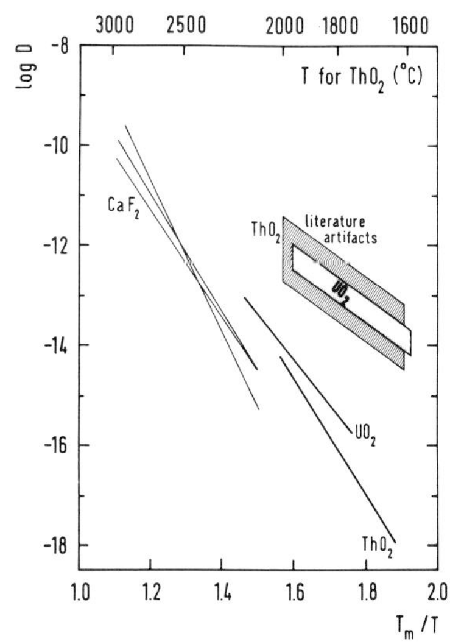

Fig. 12. Arrhenius diagram of metal atom diffusion in the fluorite structure. As in Fig. 10, a normalized temperature scale has been used.

have a pronounced effect on diffusion coefficients. The reliable diffusion coefficients for UO_2 (annealed under nonoxidizing or reducing conditions) (Reimann and Lundy, 1969; Lambert and Matzke, 1980; Lambert, 1979), ThO_2 (King, 1971; Matzke, 1976b), and $(U, Pu)O_2$ (again under neutral or reducing conditions) (Schmitz and Marajofsky, 1975; Lambert and Matzke, 1980; Matzke, 1979b) are all lower by two to four orders of magnitude than the previously accepted data. Also, the data follow the approximate melting point rule already observed for oxygen diffusion (Fig. 10), as indicated in Fig. 12 by the approximate agreement of the reliable data for CaF_2 and those for UO_2, ThO_2, and $(U, Pu)O_2$. Simultaneously, the activation enthalpies are higher than previously reported—about 130 kcal/mole (UO_2) and 150 kcal/mole (ThO_2) rather than ~100 kcal/mole as claimed in studies which have now been rejected.

The dependence of metal diffusion on deviation from stoichiometry (the O/M ratio or x in $MO_{2\pm x}$) will be discussed together with the defect structure of fluorite-type substances in the following section.

D. Atomistic Model of the Defect Structure

The first model of the defect structure of fluorite-type dioxides was due to Lidiard (1966) and Matzke (1966b) and is based on the assumption of randomly distributed point defects and ionic bonding. The anion Frenkel, Schottky, and cation Frenkel product relations were solved to yield the temperature dependencies of the concentrations of vacancies and interstitials in both the oxygen and the metal sublattices (Table II). An important feature of the model is the crucial role of ΔG_{FO}, the free energy of oxygen Frenkel defect formation. The model predicts the dependencies of the activation enthalpies of diffusion-controlled processes in stoichiometric and nonstoichiometric materials of the fluorite structure shown in the accompanying tabulation, where ΔG is the free energy of formation for Schottky defects (S), oxygen Frenkel defects (FO), and metal Frenkel defects (FM), and ΔH^m is the migration enthalpy of metal (M) and oxygen (O). (See also Table VII.)

	Oxygen	Metal	
MO_{2+x}	$\Delta H^m_{O_i}$	$\Delta G_S - 2\Delta G_{FO} + \Delta H^m_{V_M}$	~2.6 eV
MO_2	$\frac{1}{2}\Delta G_{FO} + \Delta H^m_{V_O}$	$\Delta G_S - \Delta G_{FO} + \Delta H^m_{V_M}$	~5.7 eV
MO_{2-x}	$\Delta H^m_{V_O}$	$\Delta G_S + \Delta H^m_{V_M}$	~7.8 eV ($x \lesssim 0.02$)
		$\Delta G_{FM} - \Delta G_S + \Delta H^m_{M_i}$	(~5 eV) ($x \gtrsim 0.02$)
		or a cluster mechanism	

Hence the activation enthalpy for metal diffusion should increase or decrease, respectively, by ΔG_{FO} when passing from MO_2 to MO_{2+x} or MO_{2-x}, respectively, if the metal diffuses via a vacancy mechanism (the reverse would be true for diffusion via metal interstitials). Simultaneously, at a constant temperature, the metal defect concentrations (hence the rates of metal diffusion-controlled processes) should depend on x^2. These two features are shown in Fig. 11 (Matzke, 1973). The predicted dependence of metal self-diffusion in MO_{2+x} on x^2 has been convincingly confirmed for U in UO_{2+x} (see Matzke, 1976b). An approximate x^2 dependence has also been reported for sintering, grain growth, and creep, as indicated in Fig. 13b for rates of grain growth (MacEwan and Hayashi, 1967; Sari, 1975) and creep (Burton, 1975; Routbort *et al.*, 1972), for metal atom diffusion, and for evaporation (Rand and Markin, 1968).

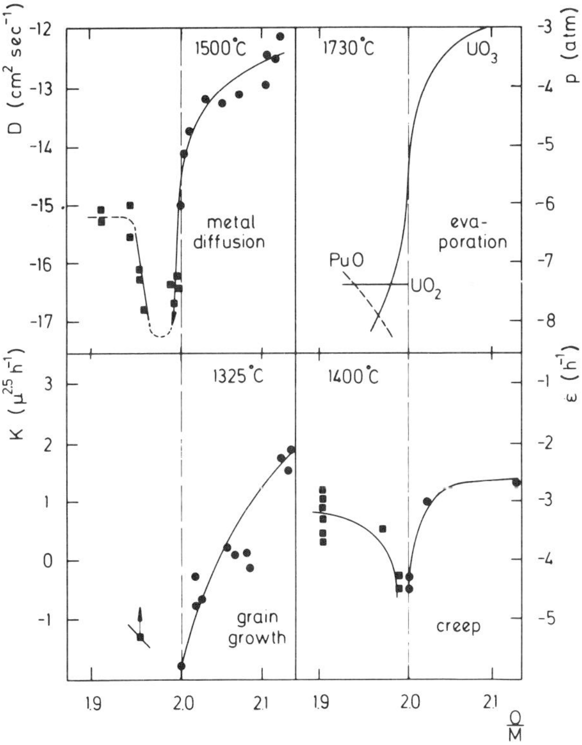

Fig. 13b. Kinetic processes in $MO_{2\pm x}$. Same data as in Fig. 13a for self-diffusion (upper left), but as function of the O/M ratio, together with dependence of grain growth, creep, and evaporation on the O/M ratio (the experimental temperatures are not identical).

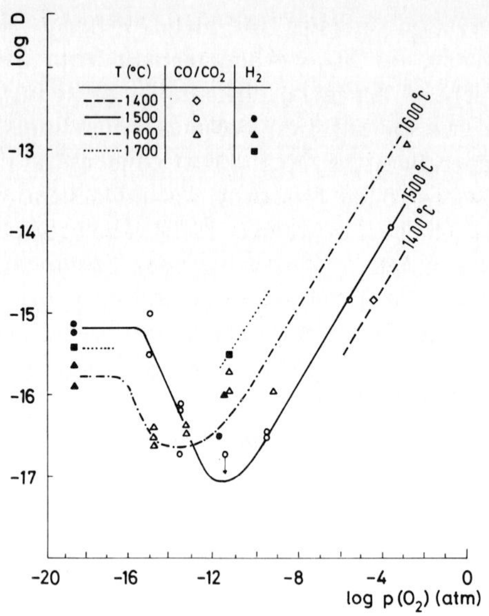

Fig. 13a. Plutonium self-diffusion in $(U_{0.8}Pu_{0.2})O_{2\pm x}$ as a function of oxygen partial pressure, for $T = 1500$ and 1600°C with three additional results for 1400 and 1700°C.

The expected strong dependence of D^{M} on x in $MO_{2\pm x}$ was also confirmed recently in experiments on the diffusion of Pu in $(U, Pu)O_{2\pm x}$ (Matzke and Lambert, 1980). Since the vacancy concentration increases with x, whereas the metal interstitial concentration increases with decreasing x, a minimum is expected and observed. The location of the minimum in Fig. 13a (at an O/M ratio of ~1.98 for 1500°C) is probably temperature-dependent. Also, temperature-dependent defect clustering, such as the Willis-type clusters of oxygen vacancies and interstitials in MO_{2+x}, shear planes in UO_{2-x}, and Pu_2O_3-type clusters in MO_{2-x} (Section III), affect the shape of the troughlike function. A similar dependence on the O/M ratio has also been reported for U–Pu interdiffusion (Chilton and Edwards, 1978).

Because of the minimum, three possibilities are given for MO_{2-x}—one for a vacancy, one for an interstitial, and the third for a cluster mechanism. Numerical values for ΔH can be found in Fig. 11 and Table VII. Figure 11 shows also the predicted diffusion mechanism (vacancies for O/M ratios of $\gtrsim 1.98$, interstitials for lower O/M values) and the calculated metal atom defect concentrations. Cluster mechanisms are thought to dominate at lower O/M values (< 1.95), as indicated by the absence of any further dependence of D^{Pu} on $\log p(O_2)$ or x.

The above model assumes that the migration enthalpy ΔH_{M}^{m} is independent of x in MO_{2+x}. It also assumes that changes in D are only due to changes in defect concentrations. This implies that the migration enthalpies for differently charged ions, e.g., U^{4+}, U^{5+}, and U^{6+}, should be very similar. This was indeed recently confirmed by studying U diffusion in ThO_2 as a function of $p(O_2)$. No obvious influence on D^U was observed (Matzke and Cole, 1980). The model is thus assumed to hold for stoichiometric oxides and for small deviations from stoichiometry. Figure 11 then indicates that truly intrinsic behavior of UO_2 has never been measured at normal experimental temperatures. For instance, at 1500°C an excess oxygen content of $x \lesssim 10^{-5}$ or an impurity level of $\lesssim 10$ ppm would have been needed.

Recent experimental evidence still supports the defect formation energies ΔG_{FO}, ΔG_S, and ΔG_{FC} in Table VII. The alternative of accepting the recently *calculated* values (Sugisaki, 1979) of $\Delta G_{FO} = 5.2$ eV, $\Delta G_S = 10.3$ eV, and $\Delta G_{FC} = 18.2$ eV seems less attractive: There would be no means of explaining mass transport in MO_{2-x} with a $\Delta H < 10$ eV. Obviously, the fully ionic model exaggerates the ionic charges in UO_2, which is particularly important for cation interstitials. The theoretical value $\Delta G_{FC} = 18.5$ eV would preclude any possibility of metal interstitials contributing to or controlling mass transport, since the predicted concentrations would be negligibly small (e.g., $\sim 10^{-26}$ for $MO_{1.98}$ at 1600°C, compared to $\sim 10^{-12}$ for the much lower value of 9.5 eV in Table VII).

Schmitz and Marajofsky (1975) have provided an interesting alternative explanation of the observed increase in matter transport with x in MO_{2-x}: a complex of the type described in Section III; hence two Pu^{3+} atoms (or a Pu_2O_3 molecule) connected to an oxygen vacancy would provide enhanced metal mobility via a type of ring mechanism.

In conclusion, the important points to note for metal diffusion in nonstoichiometric fluorite-type oxides are

1. The true D values for stoichiometric oxides are lower than the ones reported in the literature before about 1975. This is partly due to (i) outdated techniques in the earlier work, (ii) lack of recognition of and correction for surface effects, and (iii) deviation from stoichiometry and/or poor quality of the specimens available at the time. The true volume diffusion coefficients in UO_2 can be described by the Arrhenius law:

$$D = 0.65 \exp(-129.000/RT) \quad \text{cm}^2/\text{sec}$$

2. Therefore the activation enthalpy ΔH for volume diffusion is higher than previously reported (~ 130 rather than 90–110 kcal/mole).
3. There is a strong dependence on x, the deviation from stoichiometry.
4. Similar arguments hold for ThO_2 and $(U, Pu)O_2$.

E. Grain Boundary and Surface Diffusion

As typical examples of fast diffusion phenomena in nonstoichiometric oxides, and to indicate the difficulties encountered when attempting to obtain quantitative data, the cases of UO_2 and ThO_2 will be discussed. Also, hardly any data exist on other nonstoichiometric oxides. A number of studies have been published on both surface and grain boundary diffusion in UO_2 and ThO_2. Knowledge of these two phenomena is important in understanding the high-temperature behavior of the oxides: For instance, grain boundary diffusion offers an alternative to volume diffusion as a rate-controlling mechanism for creep (Coble creep, see Table VI), and surface diffusion can be rate controlling for the mobility of fission gas bubbles in nuclear fuels.

To *deduce grain boundary diffusion coefficients* D_b from non-Gaussian portions of diffusion profiles, the Fisher–Whipple (Fisher, 1951; Whipple, 1954) or the Suzuoka analysis (Suzuoka, 1961, 1964, 1965) are usually applied. These methods, however, do not yield D_b directly, but rather $\Delta = D_b b / D_v$. To obtain D_b, therefore, an assumption must be made about the effective grain boundary width b (frequently, $b = 5$ Å), and the volume diffusion coefficient must be known. Values for D_b have been published for UO_2, UO_{2+x}, ThO_2, and (U, ThO_2) (Yajima *et al.*, 1966; Hawkins and Alcock, 1968; Marin *et al.*, 1967; Furuya and Yajima, 1968; Marin, 1970). If the D_v values used for calculating D_b are rejected, it is logical to reject the D_b and their ΔH values as well. If the D_v values for UO_{2+x} of Marin *et al.* (1967) are accepted (despite the fact that thick tracer layers were used), at least one set of grain boundary diffusion data can also be accepted (Marin, 1970). This implies that, though grain boundary diffusion certainly occurs to a large extent in sinters of actinide dioxides, consistent quantitative data are not available. Rather, a reevaluation of published data and further investigations are necessary.

Surface diffusion coefficients D_s are often deduced from *surface relaxation* studies based on measurements of surface shape changes induced by the tendency of the specimen to minimize the total surface free energy of the system. The most widely used methods involve studying the changes in surface profiles occurring in (a) the formation and growth of grain boundary grooves, (b) the decay of isolated scratches, or (c) the relaxation of a sinusoidal profile. The mathematics for evaluating these changes have been given by Mullins (1959, 1961). Basically, relaxation can proceed via vapor transport, volume diffusion in the solid, or surface diffusion. For UO_2, evaporation–condensation can frequently be rate determining, thus masking surface diffusion. To obtain reliable data, the time dependence of the process should be carefully studied, preferably by the decay of sinusoidal profiles of different wavelengths. The results obtained are always the product of diffusion coefficient and surface free energy $D\gamma$. Because of the anisotropy of γ for UO_2, very irregular time dependencies can originate.

Figure 14 contains published surface diffusion data for UO_2 (Amato *et al.*, 1966; Robertson, 1969; Henney and Jones, 1968; Reynolds, 1967; Marlowe and Kaznoff, 1968; Berman, 1969; Hodkin and Nicholas, 1973; Furuya and Koizumi, 1975; Maiya, 1971a, b) corresponding to lines 1–9, corrected where necessary for evaporation or volume diffusion. The arrows indicate the growing effect of vapor-phase transport with increasing temperature. Line 9, labeled "average" corresponds to the relation (Maiya, 1971a) (valid for $\gamma = 1000$ ergs/cm^2)

$$D_s = 5 \times 10^5 \exp(-108{,}000/RT) \quad \text{cm}^2/\text{sec}$$

and can be considered to represent best the current knowledge of surface diffusion of UO_2. The total of lines 1 to 9 may be considered to represent a scatter band. Data for other nonstoichiometric fluorite-type oxides are still lacking.

Surface diffusion is probably also influenced by the reported surface step formations on UO_2 observed with LEED techniques and electron microscopy (Matzke and Ronchi, 1972; Ellis and Schwoebel, 1968; Taylor and Ellis, 1978; see also Fig. 7), and both phenomena probably affect the surface smoothing reported above to have invalidated many of the metal self-diffusion results.

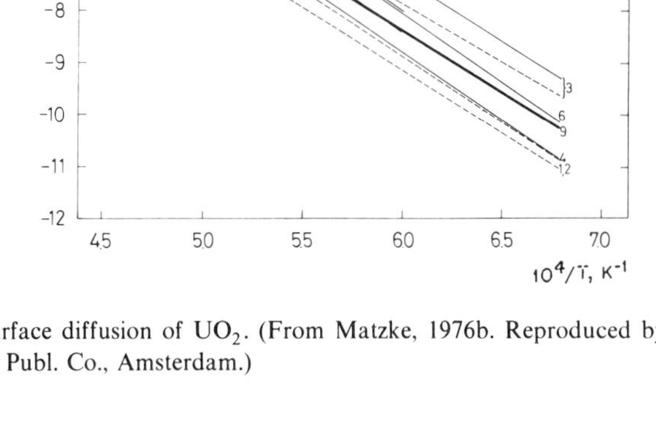

Fig. 14. Surface diffusion of UO_2. (From Matzke, 1976b. Reproduced by permission of North-Holland Publ. Co., Amsterdam.)

F. *Diffusion-Related Processes: Grain Growth, Creep, Sintering*

Many papers have been published in which measurements of diffusion-related properties have been used to deduce diffusion coefficients (see also Table VI). Two examples are careful studies on thermally activated densification of sintered UO_2 (Assmann *et al.*, 1975) and final-stage densification of $(U, Pu)O_2$ (Routbort *et al.*, 1979). Frequently, such high-temperature kinetic processes are reported to be rate-controlled by the slower-moving lattice species, hence by metal atom diffusion. If both the previous D and ΔH values for metal diffusion are rejected, many of the deductions of the above studies must also be reconsidered. Before doing so, one should realize that many of the studies of the above type report results for three to five temperatures only, often in a very narrow temperature range, thus that the reported ΔH values are subject to considerable error.

Creep studies (e.g., Vollath, 1971; Seltzer *et al.*, 1972; Routbort *et al.*, 1972; Burton, 1975) give the best support to the atomistic model and the ΔH values reported in Section VI.D for the fluorite-type oxides UO_{2+x} and $(U, Pu)O_{2\pm x}$: Both an x^2 dependence of the creep rate for $x > 0$, a minimum for $x < 0$, and approximate numerical agreement in ΔH values for creep and diffusion are observed. The details have been discussed by Matzke (1976b).

Grain growth results are instructive in demonstrating the difficulties encountered. Because of the x^2 dependence of the rate constant k for grain growth in UO_{2+x} (Fig. 13b), volume diffusion was thought to control grain growth. Usually, grain growth data are fitted to an equation of the type

$$d^n - d_0^n = kt^m \exp(-\Delta H/RT)$$

where d_0 is the initial grain size and d the grain size after annealing at T for time t. (Ideally, and for metals, $m = 1$ and $n = 2$.)

Unfortunately, the value of ΔH calculated from grain growth experiments depends on the explanation offered for a nonideal value of m (hence $m \neq 1$) or the value chosen for n. There are always reasons for selecting specific values; for instance, $m \neq 1$ if the driving force decreases with time, or a cubic law ($n = 3$) will be applicable if grain growth is controlled by the migration of pores at equilibrium pressure, or one might choose $n = 4$ for pores with constant pressure or even $m = 5$ for rate control by surface diffusion (Nichols, 1966, 1968).

Figure 15 illustrates the difficulties encountered. It shows a plot of the results for UO_2 at 1832°C of MacEwan and Hyashi (1967) for different fits. These data can be safely considered to be some of the most carefully and accurately measured results. Yet, a unique decision on whether a fit to $n = 2$, 2.5, 3, or even 4 is best is not possible on the basis of how well a straight line fits the data. The time exponent m could be chosen as a

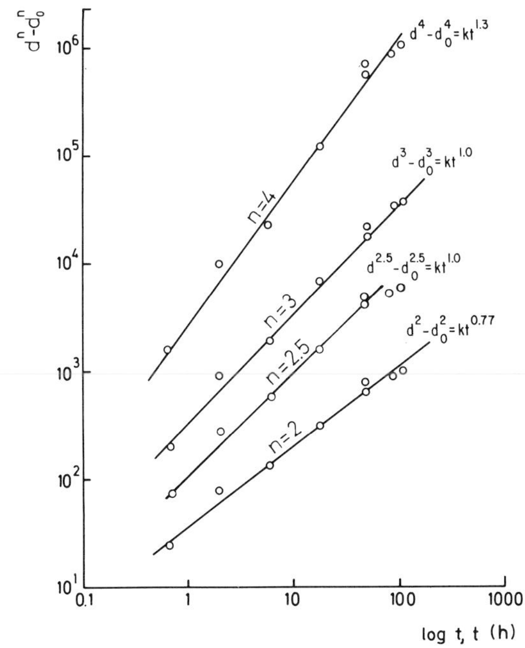

Fig. 15. Grain growth data for UO_2 plotted in different ways. The data are from MacEwan and Hyashi (1967). (From Matzke, 1976b. Reproduced by permission of North-Holland Publ. Co., Amsterdam.)

criterion, but good fits for $m = 1$ and both $n = 2.5$ and $n = 3$ are possible. Finally, there are reasons for accepting $m = 0.8$ as well, hence to present a $d^2 \sim t$ plot.

The data of MacEwan *et al.* (MacEwan, 1962; MacEwan and Hyashi, 1967) have been previously fitted to $n = 2$ and $n = 2.5$ by the authors. Values for ΔH of 87 and 109 kcal/mole, respectively, were obtained, which might be taken to indicate control by uranium self-diffusion when two sets of data were available, those of Auskern and Belle (1961) with 88 kcal/mole and those of Lindner and Schmitz (1961) with 104 kcal/mole. Therefore, by deciding that one type of presentation is preferable, one could use the data as a criterion for selecting one of two self-diffusion studies as more representative. However, we have argued above that neither is representative for volume diffusion. (Note also that, strictly speaking, impurity diffusion should control grain growth.) Alternatively, by selecting a $d^3 \sim t$ fit one obtains $\Delta H = 124$ kcal/mole for the same data, and one might argue in favor of an evaporation–condensation mechanism of pore mobility controlling grain growth, since the reported value for the heat of vaporization is 138 kcal/mole for $d^4 \sim t$. Or, if in the work of both

MacEwan and Hyashi (1967) and of Stehle (1963), ΔH is reported to be 87 kcal/mole, this should not be taken as agreement but rather as important disagreement, since either a $d^2 \sim t$ or a $d^3 \sim t$ law was assumed.

Consequently, studies on diffusion-related processes can be very helpful in yielding information on self-diffusion, but care should be taken in deciding whether

1. the data are experimentally reliable (enough results in a not too narrow temperature range);
2. the mathematics applied are adequate; and
3. the attribution of the rate-controlling process has been properly made.

An important alternative to metal self-diffusion in actinide dioxides is mass transport via evaporation–condensation. Total vapor pressures show a minimum in their dependence upon x, which resembles that of metal diffusion (Fig. 13). In fact, evaporation–condensation phenomena are rate controlling for bulk mass transport in oxide fuel elements for nuclear reactors. As an example, within the temperature gradient existing in these fuels, much of the sintering porosity is moved up the temperature gradient to produce a central hole in the fuel within the first hours of operation in the nuclear reactor (for review, see Nichols, 1979).

VI. Diffusion in the Rutile Structure: TiO_2

A. *Ordered Phases*

The TiO system is very complex and has a large number of compounds (Fig. 16) such as Ti_2O, TiO, Ti_2O_3, Ti_3O_5, TiO_2, and the Magnéli phases Ti_nO_{2n-1}. Andersson *et al.* (1957) originally reported phases with $n = 4, \ldots, 10$ ($TiO_{1.9}$), whereas Bursill *et al.* (1969) more recently described phases with n up to 38 ($TiO_{1.974}$) and Hyde and Bursill (1970) with $n = 4$–53 ($TiO_{1.981}$). Figure 17 shows some of these phases schematically together with relevant data for Ti and O diffusion in reduced TiO_{2-x}. Data for $TiO_{1\pm x}$ are included in Section VII.

Though the defect properties of rutile have been extensively studied, there is still disagreement on its defect structure. TiO_3 is oxygen-deficient when nonstoichiometric, but there is still disagreement as to the extent of nonstoichiometry, possibly because the formation of phase transitions between the oxides of the homologous series Ti_nO_{2n-1} is slow and sluggish. Hyde and Bursill (1970), for instance, obtained a three-phase solid with $n = 18$, 19, and 20 after equilibrating TiO_2 with H_2 plus H_2O at 1100°C for 3.5 days, a time much longer than usually applied in thermodynamic or

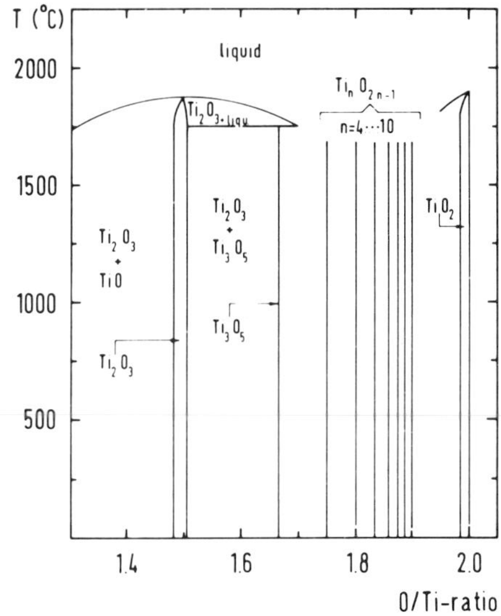

Fig. 16. The oxygen-rich part of the Ti–O phase diagram.

diffusion studies. The question whether or not equilibrium is achieved is thus primarily a question of accuracy in the determination of x in TiO_{2-x}. For instance, phases of Ti_nO_{2n-x} for $n \sim 30$ differ by a Δx of ~ 0.001 and a change in oxygen partial molar free energy $\Delta\overline{G}(O_2)$ of only ~ 4 kcal/mole. To establish stationary compositions, an accuracy in Δx of 0.0001 and in $\Delta[\Delta\overline{G}(O_2)]$ of 1 kcal/mole seems necessary, thus straining measurement techniques to the utmost. It is recommended that electron microscope examinations be performed on all specimens used for diffusion (and thermodynamic) work. One should note that, in this respect, it is much

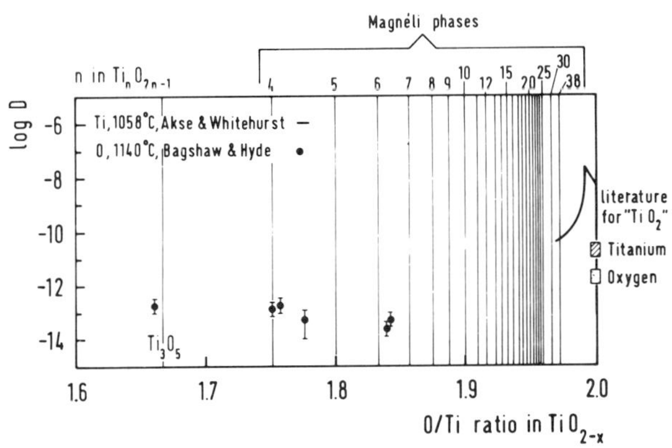

Fig. 17. Schematic presentation of Magnéli phases Ti_nO_{2n-1} with relevant data for oxygen and titanium diffusion.

more difficult to obtain equilibrium in a macroscopic specimen of, say 0.5 cm^3, for diffusion work than in a sample for electron diffraction (say 10^{-10} mm^3). In this sense, much of the reported diffusion work on TiO_2 should be regarded with care.

In the composition range $TiO_{1.75}$–$TiO_{1.90}$ ($n = 4$–10) the homologous series is based on (121) CS planes, whereas between $TiO_{1.938}$ and $TiO_{2-\epsilon}$ ($n = 16$ to ~ 38), (132) CS planes were found ordered at the lower end of the range and disordered at the upper end. In the range $TiO_{1.90}$–$TiO_{1.93}$, the CS plane orientation swings from that of the (121) ordered $Ti_{10}O_{19}$ phase to the (132) CS orientation of $Ti_{16}O_{31}$ (Bursill *et al.*, 1971).

B. Defect Types

The dominant defects in rutile are still under discussion, though much work has been devoted to obtaining definite conclusions and a wide variety of techniques including diffusion studies, conductivity or resistivity measurements, Hall effect, electron paramagnetic resonance, dielectric relaxation, internal friction or channeling studies were applied. The two alternatives are oxygen vacancies, V_O, and titanium interstitials, Ti_i. One or the other has been suggested during the past three decades, and at the moment titanium interstitials are favored. Literature on this point includes Iguchi and Yajima (1972a, b), resistivity work concluding that V_O is the dominant defect; Baumard *et al.* (1977), conductivity work favoring Ti_i; Yagi *et al.* (1977), channeling work confirming Ti_i as dominating; Akse and Whitehurst (1978), diffusion studies also favoring Ti_i; and Catlow (1979), theoretical predictions that V_O dominates.

In particular, the channeling study gives a direct proof of Ti interstitials, at least for the type of specimen used: 99.99% pure TiO_2, vacuum-reduced (10^{-6} torr) at 1100°C for 240 h and cooled under unreported conditions leading to an estimated composition of $TiO_{1.994}$.

Most probably, the suggestion of Kofstad's (1967) analysis is still to be recommended. Kofstad proposed that the defect structure simultaneously is comprised of doubly charged oxygen vacancies and interstitial titanium ions with three and four effective charges. It is suggested that V_O predominates only at high oxygen partial pressures and low temperatures, whereas Ti_i^{3+} and Ti_i^{4+} are suggested to be the dominant defects if the temperature is raised or the oxygen pressure lowered.

At first, the electron microscope investigations of Eikum and Smallmann (1965) and Hyde and Bursill (1970) seem to suggest that both the anion vacancy and the interstitial cation protagonists may be incorrect. The latter authors observed "a maze of disordered planar faults in at least 5 different orientations" in H_2-reduced (1123 K) $TiO_{1.995}$. In later work, Bursill and

Hyde (1971) prepared a $TiO_{1.9986}$ specimen coulometrically, which they equilibrated for 3 months at 1000°C with practically the same results as before. However, at small enough deviations from stoichiometry and at high enough temperatures, single point defects must exist. It is fortunate that diffusion studies have provided the first answer to this problem (see below).

C. Impurity Cation Diffusion in Rutile

As discussed in Section II, diffusion is expected to be anisotropic in the tetragonal rutile structure. Because of the small ratio of the edges of the unit cell, $c/a = 0.64$, open channels exist parallel to the c axis, which allow fast diffusion of small (interstitial) ions. Thus Johnson (1964) has reported that lithium diffuses along the c axis at least 10^8 times faster than perpendicular to the c axis. Parallel to the c axis, good Arrhenius-type behavior with $D_0 \sim 0.3$ cm^2/sec and a low ΔH of 0.33 eV was observed. Though only an indirect method (optical absorption) was used and Li was present in bulk quantities ($\sim 10^{19}$ Li atoms/cm^3, hence diffusion in a chemical gradient was measured), the results clearly show the strongly anisotropic mobility of small (Li^+, radius 0.62 Å) interstitial ions. Bogomolov (1963), in a qualitative study, reported a similar behavior of boron, and Wittke (1966), in a qualitative observation based on coloration of the originally straw-yellow rutile, concluded that transition metal ions such as Cr, Fe, Co, and Ni also showed high mobility along the c axis. The observation of high mobility of Fe along the c axis was confirmed by Steele and McCartney (1969). It is also known that surface damage (e.g., produced by polishing) inhibits the rapid anisotropic diffusion (Johnson, 1964). If the Ti–O bonds are assumed to be 60% ionic and, using a linear interpolation of Pauling's (1960) ionic and covalent radii for Ti and O [hence $R(\text{Ti}) = 0.90$ Å and $R(\text{O}) - 1.06$ Å] together with the known c and a values of 2.959 and 4.593 Å, the channels parallel to the c axis (Fig. 4) have a radius of ~ 0.77 Å. Thus both Li and B as well as Cr^{3+} (0.63 Å), Fe^{3+} (0.64 Å), Co^{2+} (0.74 Å), and Ni^{2+} (0.72 Å), would be expected to diffuse through these channels quickly, in contrast to oxygen or larger impurities. No numerical diffusion coefficient for the direction perpendicular to the c axis has been reported, but optical effects were below detection level. Tracer diffusion measurements are still to be performed to confirm these observations and to provide quantitative numerical data.

Theoretical considerations confirming the high mobility of Li (Kingsbury *et al.*, 1968) and other small interstitial ions (Ajayi *et al.*, 1976) have also been published. In the latter case an increase in ΔH was calculated for interstitial mobility along the c axis with the radius of the mobile cation R

R (Å)	0.6	0.9	1.1	1.3
ΔH(kcal/mole)	17	34	51	79

This conflicts somewhat with the above simple picture, since impurities with $R \sim 1$ Å could also diffuse interstitially. In fact, additional qualitative work on the mobility of Na [$R(Na^{+}) = 0.97$ Å] and even Ba [$R(Ba^{2+}) =$ 1.34 Å] has indicated higher mobility along the c axis (Kingsbury *et al.*, 1968; Nakayama and Sasaki, 1963) though neither D values nor the magnitude of the difference between D_{zz} and D_{xx} was given. The only tracer data are due to Lutze-Birk *et al.* (1967), yielding a high D value of $\sim 10^{-9}$ cm^2/sec at 800°C for ^{110}Ag, but in polycrystalline TiO_2 [$R(Ag^{2+})$ $= 0.98$ Å]. This value is four or five orders of magnitude, respectively, above those for Ti and O.

In conclusion, small metal impurities show an extreme anisotropic mobility in rutile. Quantitative data for this interesting phenomenon are still lacking. Also, the question of whether and how shear planes or Ti interstitials block this fast "chimney diffusion" is still open. Yagi *et al.* (1976) conclude from their channeling work that Ti_i are located within ~ 0.4 Å of the [001] midchannel axis. These questions are further treated in Section VI.F.

D. Oxygen Self-Diffusion

In contrast to the situation in the fluorite structure, oxygen diffusion in rutile is slower than metal (Ti) diffusion, as shown in Fig. 18. The first and most extensive data are due to Haul and Dümbgen (1962, 1965) and Haul *et al.* (1962). Their data are for TiO_2 powders heated in O_2 and are thus expected to give an average D for the c and a axes. These data follow the equation

$$D_{O}^{TiO_2} = 2.0 \times 10^{-3} \exp[(-60{,}000 \pm 1500)/RT] \quad \mathrm{cm^2/sec}$$

Experiments with single crystals at 1347°C showed a somewhat faster diffusion along the a axis ($D_{xx}/D_{zz} = 1.6$), in contrast to the high interstitial mobility along the c axis of metal impurities (Section VI.C) and also in contrast to the observed larger electric conductivity along the c axis (Blumenthal *et al.*, 1967).

This different behavior is probably due to the shorter O–O distances (2.52 versus 2.96 Å) along the a axis, since because of the large size of the oxygen ion fast diffusion through the channels of the c axis would not be expected. Haul and Dümbgen (1965) observed no change in D^{O} while varying the oxygen pressure at 897°C between 760 and 10^{-3} torr. This indicates extrinsic behavior due to a relatively high Al_2O_3 content (100–200 ppm) of their diffusion specimens and suggests an impurity-controlled

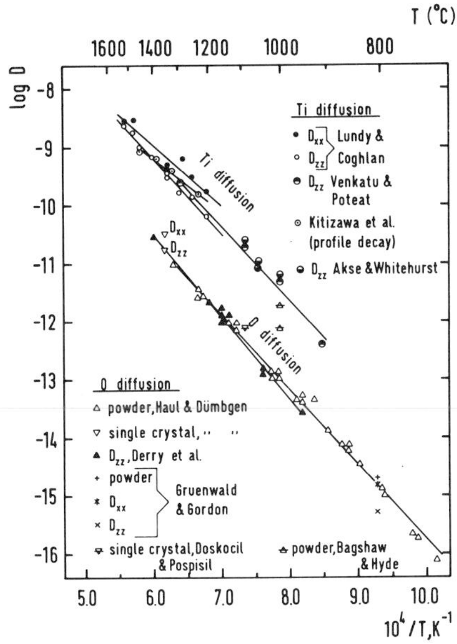

Fig. 18. Arrhenius diagram of oxygen and titanium self-diffusion in nominally stoichiometric rutile, TiO_2.

vacancy diffusion mechanism for their measurements. Oxygen vacancies, $V_O^{\cdot\cdot}$, are introduced by the reaction

$$Al_2O_3 \rightleftharpoons 2Al_{Ti} + V_O + 3O_O$$

The rutile defect model of Kofstad (1967) allows a value of $\sim 10^{-5}$ to be extrapolated for the native defect (V_O) concentration in pure rutile at 1100°C. Since substitutionally dissolved Al introduces one oxygen vacancy per Al_2O_3 molecule for charge equilibrium, the specimens of Haul and Dümbgen (1965) would be expected to show extrinsic behavior up to ~1400°C, which is indeed above the upper limit (1300°C) of the temperature range used by Haul and Dümbgen. In fact, TiO_2 of normal purity (99.99%) should show intrinsic behavior only above 1500°C. The accessible temperature range for intrinsic diffusion would thus be quite small (~350°C), and no data have been reported for these high temperatures. Later work by Doskocil and Pospisil (1972), as well as Derry *et al.* (1971) and Calvert *et al.* (1974), for D_{zz} confirmed the results of Haul and Dümbgen. The results of Calvert *et al.* (1974) are described by $D = 0.017 \exp(-66{,}000/RT)$ cm^2/sec. Since widely different techniques were used (mass spectrometry, nuclear reactions), the data on D^O can be accepted as being representative. Further results reported by Gruenwald and Gordon (1971) confirmed the powder data of Haul and Dümbgen but gave a slightly lower value for D_{zz} and the D_{xx}/D_{zz} ratio was 5.3, again confirming somewhat faster oxygen diffusion along the *a* axis. These authors also reported measurements on "air- or vacuum-reduced" TiO_{2-x}. It is impossible to attribute accurate x values to these treatments, but a rough guess of the order of magnitude could be $x \sim 10^{-5}$ and $x \sim 10^{-3}$, respectively. The D values for the vacuum-reduced samples were very similar (possibly ~6% smaller) to those of air-reduced samples, and both sets of data were higher by about a factor of 3.5 compared to those for stoichiometric TiO_2. It is difficult to explain these results quantitatively. Qualitatively, an increase in D with x would be expected, given the fact that the pressure–temperature conditions were those resulting in the predominance of V_O defects.

The mobility of oxygen in reduced TiO_{2-x} (or chemical diffusion of oxygen $\tilde{D}$) was studied by Barnabel and Bogomolov (1969), Iguchi and Yajima (1972a, b), and Baumard (1976). All three papers report data deduced from measurements of electrical conductivity σ. The oxygen pressure dependence, σ versus $p(O_2)^{-1/n}$, generally yields values of n between 4 and 6 (see also Fig. 19), leading different authors to suggest V_O or Ti_i as defects (see above). In two of these studies the mobile defects were assumed to be V_O, and in the third one Ti_i was assumed. The resultant activation enthalpies were 19 kcal/mole between 300 and 800°C (Barnabel and Bogomolov, 1969); 55 kcal/mole between 800 and 1100°C (Iguchi and Yajima, 1972a, b); and 9 kcal/mole between 900 and 1200°C (Baumard, 1976). This time D_{xx} is smaller than D_{zz} by about a factor of 2 (Iguchi and Yajima, 1972a, b). Clearly, these results indicate disagreement. However, one should note that different temperature ranges and deviations from stoichiometry (oxygen pressures) were used. The high-temperature data of Baumard (1976), when converted from the chemical diffusion coefficient $\tilde{D}$

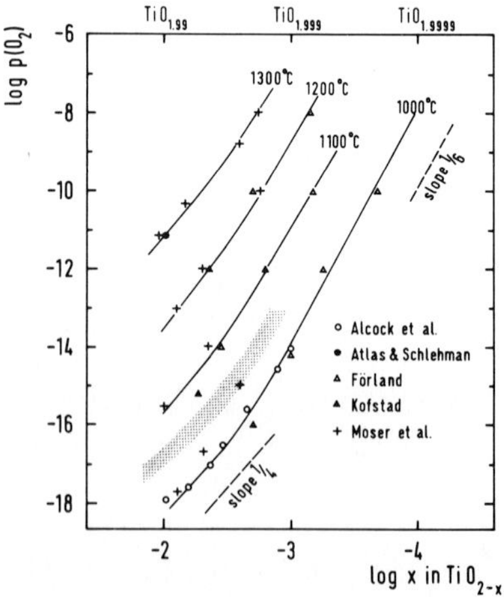

Fig. 19. Deviations from stoichiometry for different temperatures and oxygen partial pressures in rutile.

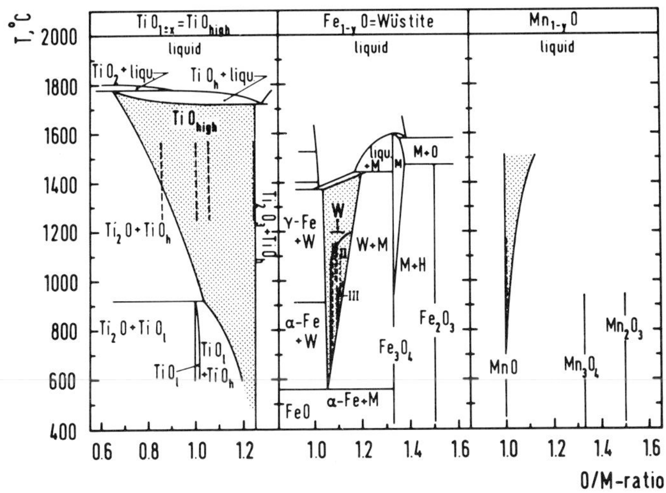

Fig. 20. Partial phase diagrams of the systems Ti–O, Fe–O, and Mn–O showing the phase fields of the NaCl-type monoxides together with the temperature and O/M ratios of available diffusion data.

to the tracer diffusion coefficient D by using the defect concentration data deduced from the conductivity measurements of Baumard *et al.* (1975), were compatible with the diffusion data for Ti (Fig. 18), whereas Barnabel and Bogomolov (1969), at their lower temperatures could have dealt with *oxygen* vacancies. The D values reported by Iguchi and Yajima are about four orders of magnitude lower (at 1000°C) than those of Baumard, indicating that they dealt with specimens containing extended defects, i.e., (132) CS planes. Their atmosphere was 4×10^{-9} torr obtained with an ion pump,† which, if identical to $p(O_2)$ would indicate a small reduction only (Fig. 19). Deviations from stoichiometry were not reported.

A direct study on oxygen tracer diffusion in reduced TiO_{2-x} is due to Bagshaw and Hyde (1976). Their results are presented in Fig. 17. They used oxygen for their measurements on TiO_2, whereas they used hydrogen–water vapor gas buffers for studies on the Magnéli phases Ti_nO_{2n-1}. Thus the data are not necessarily completely comparable. Their data for stoichiometric TiO_2 appear in Fig. 19 and are obviously much higher than any other reported data.

†Note that much of the published literature cannot be critically evaluated because O/Ti ratios were not determined or reduction conditions were not properly described. For instance, hydrogen reduction of TiO_2 at 1050°C can lead to any composition between $TiO_{1.75}$ and $TiO_{1.97}$ depending on the water content of the H_2 (Anderson and Khan, 1970). Also, at constant $p(O_2)$, reduction conditions of different atmospheres may still not be comparable: Reduction in H_2 leads to uptake of hydrogen, reduction in a diffusion pump vacuum causes surface contamination with deposits of diffusion pump oil, etc.

The results for Magnéli phases and for Ti_3O_5 ($TiO_{1.66}$) are lower than those for stoichiometric TiO_2 by factors between 5 and 30 (or an additional order of magnitude if the data for Magnéli phases are as high as the above authors' data for TiO_2). This tends to show that oxygen mobility in shear structures is retarded, Unfortunately, CS phases far from stoichiometry have been used [and one result is irrelevant to the problem since it is for another oxide, Ti_2O_5, which has a different crystal structure (monoclinic)].

In summary, oxygen diffusion data for stoichiometric rutile are only available for the extrinsic range. Within this range, the data of different authors using different techniques show remarkable agreement. In low-*n* Magnéli phases, Ti_nO_{2-1}, the shear structure retards diffusion. It would be very helpful to obtain data for TiO_2 at higher temperatures (intrinsic diffusion), as well as to study oxygen diffusion at the lower limit of the TiO_{2-x} homogeneity range, i.e., at about $TiO_{1.99}$ (Blumenthal and Whitmore, 1963; Zador, 1969; recently confirmed by Kozlowska-Rog *et al.*, 1976), where the nonstoichiometry causes a defect concentration of $\sim$0.5%. The effects of impurities would thus become negligible. Also, data for high-*n* Magnéli phases are needed to achieve a full understanding of oxygen diffusion (see also Section VIII: Nb_2O_5).

E. Titanium Self-Diffusion

As mentioned before, Ti interstitials are presently favored as dominating defects in TiO_{2-x}, at least for high temperatures and low oxygen partial pressures. Thus one should expect Ti diffusion to be faster than oxygen diffusion unless the migration energies overcompensate for the effect of defect concentration on diffusion. Figures 17 and 18 summarize the available data. For an unstated atmosphere and composition, Venkatu and Poteat (1970) reported diffusion of Ti^{44} along the *c* axis to follow the relation

$$D_{zz} = 0.064 \exp(-61{,}400/RT) \quad \text{cm}^2/\text{sec}$$

Later work by Lundy and Coghlan (1973) performed in air confirmed these results and showed in addition that Ti diffusion along the *a* axis was somewhat faster:

$$D_{zz} = 0.046 \exp(-59{,}900/RT) \quad \text{cm}^2/\text{sec}$$

$$D_{xx} = 0.002 \exp(-48{,}500/RT) \quad \text{cm}^2/\text{sec}$$

The similar diffusion rates (or the slight increase along the *a* axis as compared to the *c* axis) for Ti diffusion are compatible with the results of O diffusion and point to a substitutional character of the diffusing defect. Thus Ti ions in stoichiometric rutile appear to have no significant interstitial site residence. Lundy and Coghlan also consider the jump process of the Ti ions. They conclude that diagonal jumps dominate up to about 1300°C. At higher temperatures, significant numbers of jumps along the x axis occur in addition.

Venkatu and Poteat (1970) regard their data as representing extrinsic diffusion due to the presence of $\sim$150 ppm Al_2O_3. As discussed in Section VI.D, this introduces a number of oxygen vacancies in excess of those created thermally, which in turn (Section III.A) must affect the Ti defect concentration. As a matter of fact, it should decrease the concentration of Ti vacancies and increase that of Ti interstitials. Thus the slope in the Arrhenius diagram should be $\Delta H_m/R$. A fourth value of $\Delta H_m = 57$ kcal/mole for D_{zz}, in good agreement with the above two values of 59.9 and 61.4 kcal/mole, was recently reported by Akse (1977) and Akse and Whitehurst (1978) for a material containing $\sim$50 ppm Al.

A further increase in the concentration of Ti_i should occur at lower oxygen partial pressures, where the native defect concentration exceeds that introduced by impurities. For instance, at the phase boundary of O/Ti$\sim$ 1.99, a native defect concentration on the order of $\sim$0.5% will exist. Akse and Whitehurst (1978) have extended their measurements to a range of low $p(O_2)$ between 10^{-13} and 10^{-18} atm. Their results for 1058°C are shown on the right-hand side of Fig. 17. These data are very helpful in understanding the oxygen-rich part of the Ti–O system. Experimental data on self-diffusion experiments relate directly to one type of defect only; e.g., the above measurements relate exclusively to the titanium interstitial. In contrast, thermogravimetric or conductivity data can originate from any one of several types of defects. Attributing a specified defect type to an observed process is often arbitrary [e.g., based on the slope of a plot versus log $p(O_2)$ only], and usually this specific defect, though hopefully the major contributor to the observed process, seldom constitutes 100% of the measured value. Also, while reaching its limit of rate determination [e.g., by lowering temperatures and increasing $p(O_2)$, oxygen vacancies will be formed at increasing rates], small shifts in a few experimental points in conductivity and/or thermogravimetric studies can change the conclusions regarding the different contributions of Ti_i and V_O.

As shown in Fig. 17, Ti diffusion is faster in TiO_{2-x} than in TiO_2, and it increases further when $p(O_2)$ decreases. This clearly excludes Ti vacancies as a transport vehicle, since in this case D should be directly proportional to $p(O_2)$. At an oxygen pressure of 10^{-16} atm, D^{Ti} decreases sharply. The corresponding composition is $TiO_{1.994}$. This indicates the lower limit of the TiO_{2-x}-phase field. Note that this result was obtained at elevated temperature and not following cooling or quenching of the specimen.

Also, from the measured slope (1/4.8) of the plot of $\log D^{Ti}$ versus $\log p(O_2)$, the conclusion emerges that the titanium interstitial at 1058°C is tetravalent, Ti_i^{4+}. A decrease in slope between 1100 and 1000°C indicates that, at lower temperatures and in the same pressure range, Ti_i^{3+} contributes more at lower temperatures. Thus Ti_i becomes increasingly more ionized as the temperature is raised.

Akse and Whitehurst (1978) measured the temperature dependence of Ti diffusion at a constant (low) $p(O_2)$ of 1.7×10^{-16} atm and obtained $\Delta H = 112$ kcal/mole, which they assumed to be $\Delta H_f/5 + \Delta H_m$, the factor 5 being due to the dependence on $p(O_2)^{-1/5}$. This yields a formation energy for Ti_i^{4+} of 276 ± 15 kcal/mole. Blumenthal *et al.* (1966) and Kofstad (1967) had previously suggested a value of 245 kcal/mole. Note, however, that at a constant pressure the composition varies with the temperature (Fig. 19) and that, with varying temperature, the relative contribution of Ti_i^{4+} and Ti_i^{3+} changes.

The sharp decrease in D^{Ti} at $p(O_2) < 10^{-16}$ atm (O/Ti < 1.992) indicated in Fig. 17 for 1058°C, and the fact that the curve is smooth though it crosses several phases of the homologous series, show that these transitions must be second order. Also, a pretransition drop in D^{Ti} associated with short-range order is indicated. The case of TiO_{2-x} is more complicated than the normal order to short-range order transition, because two ordering processes occur simultaneously: the ordering of interstitials into clusters or isolated CS planes (Wadsley defects) and the ordering of these clusters and point defects into CS phases. Evidently, with increasing order, D^{Ti} decreases, as would be expected because of a decrease in free defect concentration and because the correlation factor f decreases as the defect ordering begins. Since certain configurations of defect clusters are more stable than others, jumps that tend to destroy these complexes are highly correlated, resulting in a decrease in f and thus a decrease in D.

Though this extensive diffusion study of Akse and Whitehurst (1978) has made an important contribution to the understanding of the high-oxygen part of the Ti–O system, there remain some questions: The authors observed no indication of a decrease in D in the same pressure range (10^{-13}–10^{-18} atm) at the only slightly different temperatures of 1000 and 1100°C. Also, when a different boule of rutile was used, D^{Ti} at 1100°C changed by a factor of 3, which is unexpected for intrinsic behavior. For the extrinsic range, on the other hand, where differences for different crystals would be expected, the agreement of all the data of different authors is surprisingly good (Fig. 18). Also, the penetration profiles (Venkatu and Poteat, 1970) were reported to be not completely linear, and they resembled those typical of sinters rather than those usually obtained for single crystals. A very high dislocation density would have to be assumed to account for the non-Gaussian contributions. Furthermore, Venkatu and Poteat (1970) reported a slow "near-surface process" in one experiment. As shown for UO_2 (Sections IV.B and V), such processes can represent true volume diffusion and should not be discarded a priori. Obviously, more work is needed to understand fully the complicated behavior of TiO_2 and TiO_{2-x}.

F. Diffusion-Related Processes

A number of diffusion-related kinetic processes have been investigated. A Japanese study (Kitazawa *et al.*, 1977) measuring the decay of sinusoidal surface profiles on the *c* surface of TiO_2 in air shows volume diffusion to be the rate-determining process. The deduced D values (Fig. 18) coincide with those of Ti diffusion. This leads to the conclusion that oxygen is transported faster than Ti, either by surface diffusion or by an evaporation–condensation process.

The creep investigations of Hirthe and Brittain (1963) and Bell *et al.* (1971), the deformation studies of Ashbee and Smallman (1963) and of Farb *et al.* (1964), and the sintering work of Whitmore and Kawai (1962) and of Anderson (1967) are less easily attributed to a specific mass transport process. Also, these studies were performed before the Ti diffusion data became available, necessitating some reinterpretation (see also Venkatu, 1972). Finally, as is often the case in this type of work, only limited temperature ranges were covered, often with only three or four experimental temperatures being used. An agreement of deduced activation energies, e.g., those from the creep studies yielding 66 and 67 kcal/mole, respectively, in the work of Hirthe and Brittain (1963) and of Bell *et al.* (1971), with that for oxygen diffusion is certainly not reason enough to state that creep is controlled by oxygen diffusion. First, single crystals were used rather than sinters, the first data of Haul *et al.* (1962) of 72 kcal/mole for oxygen diffusion were later corrected to 60 kcal/mole, and the more recent data for Ti diffusion (57, 59.9, 61.4 kcal/mole; see above) are so similar that no unambiguous statement can be made on the basis of a comparison of ΔH values. The creep work of Hirthe and Brittain (1963) also indicated differences between vacuum- and H_2-reduced TiO_2 affecting the apparent ΔH. Farb *et al.* (1964) reported a temperature dependence of the apparent ΔH, which also varied with atmosphere (O_2 or N_2), doping (Fe and Al), and applied stress or strain rate. This indicates a very complex creep behavior and rules out a single, simple, rate-limiting process, as would also be expected from present knowledge of the complex behavior of the TiO_2–Ti_nO_{2n-1} system.

Studies on initial sintering of TiO_2 indicate a similarly complicated behavior. Volume diffusion, surface diffusion, grain boundary diffusion, and viscous flow were all suggested as being rate controlling. Anderson

(1967) covered a comparatively large temperature range (700–1130°C) and observed the same shrinkage for annealing in oxygen, air, vacuum (10^{-3} torr), or N_2. This surprising phenomenon was concluded to be simultaneously due to (i) impurity-generated vacancies and (ii) grain boundary diffusion. Obviously, reinterpretations are necessary.

Last, the technique of using inert gases as a probe for defects (Matzke, 1968; Kelly and Jech, 1969) has been applied to TiO_2 and H_2-reduced TiO_{2-x} to study the differences between rutile and Magnéli structures, hence to obtain information on diffusion processes in shear structures (Matzke and Whitton, 1966; Matzke, 1967). The release of Xe bombarded into rutile single crystals followed volume diffusion kinetics. The ΔH was 78 kcal/mole and thus similar to that of O or Ti self-diffusion, indicating vacancy-assisted release. At higher ion doses (4×10^{13} ions/cm^2), enough radiation damage was produced during bombardment to retard release by gas–defect interactions (Kelly and Matzke, 1965, 1966). At still higher doses, bombardment led to an amorphous surface layer, and recrystallization at temperatures between about 400 and 600°C released much of the gas. In contrast to the behavior of stoichiometric TiO_2, annealing prebombarded (stoichiometric) TiO_2 in H_2 led to reduction and the formation of shear structures during release, which swept out all of the gas at low temperatures (e.g., fractional release, $F = 1$, at 800°C for the high doses; compared to $F \leqslant 0.2$ for TiO_2 and 4×10^{13} ions/cm^2). In contrast, bombarding *prereduced* TiO_{2-x}, hence a material with a preexisting shear structure led to very low release rates (e.g., $F = 0.4$ at 1300°C) despite the annealing in air. Diffusion rates are thus drastically increased during *formation* of a shear structure but reduced in a *preexisting* shear structure. This type of work looks promising in obtaining further information on the interesting question of diffusional processes in shear structures.

In conclusion, the defect structure of rutile is very complex. Magnéli phases Ti_nO_{2n-1} and Wadsley defects dominate in TiO_{2-x} for $x \gtrsim 0.01$. For $x < 0.01$, Ti_i^{4+} seems to be the dominant defect at higher temperatures (in contrast to the calculations of Catlow *et al.*, 1976), with Ti_i^{3+} contributing at lower temperatures ($\lesssim 1000$°C). For very small values of x, at high values of $p(O_2)$ and low temperatures, oxygen vacancies occur. Much of the diffusion work represents extrinsic behavior (e.g., behavior due to Al as an impurity), but some of the diffusion studies belong to the investigations most helpful for understanding the complicated defect behavior and defect interactions in nonstoichiometric rutile. Future work should accurately control nonstoichiometry and impurity content and, if possible, be supported by electron microscope characterization of the specimens. Care should also be taken to maintain a given composition during anneals at different temperatures.

VII. Diffusion in the Rock-Salt Structure: FeO, CoO, NiO

The NaCl-type oxides of the group VII and VIII elements, NiO, FeO, CoO (and MnO), are metal-deficient, $M_{1-x}O$, nonstoichiometric compounds. The nonstoichiometry is most pronounced with $Mn_{1-x}O$ ($x \sim 0.18$; see Table I), whereas in the series $Ni_{1-x}O$, $Co_{1-x}O$, and $Fe_{1-x}O$, x is about 0.001, 0.01, and 0.1–0.15, respectively. An oxide of a group IVA element, $TiO_{1\pm x}$, will also be mentioned. Most of these M–O systems show different stable (or also nonstoichiometric) phases. Rutile, TiO_{2-x} (Section VI) and Fe_3O_4 (Section VIII) are two examples that will be discussed here. Diffusion properties have most extensively been studied in NiO which, however, shows the smallest degree of nonstoichiometry. Therefore more emphasis will be put on FeO, CoO, and MnO (and some mixed oxides). Figure 20 shows partial phase diagrams of three of the M–O systems considered. Vertical or horizontal dashed lines indicate temperature–composition combinations investigated in diffusion work. Some of the higher or lower oxides are also indicated in Fig. 20. Diffusion data exist only for higher iron oxides (M = magnetite, Fe_3O_4, with a spinel structure; H = hematite, Fe_2O_3, with a corundum structure).

A. FeO

Wüstite, $Fe_{1-x}O$, is stable only above ~560°C (Fig. 20). The deviation from stoichiometry increases with decreasing temperature at constant oxygen partial pressure, in contrast to the situation in many other oxides. Early work has indicated that wüstite can accommodate up to ~4% doubly charged Fe vacancies (vacant octahedral sites in the oxygen sublattice). At higher x values, the vacancies become singly charged (Kofstad and Hed, 1968), and for $x \gtrsim 0.08$ they are neutralized (Burgmann, 1975). Structural studies have indicated that tetrahedral sites may also be occupied by interstitial Fe ions, nucleating more complex defects or microdomains of magnetite in the wüstite structure (see also Section III.B, and Roth, 1960; Fender and Riley, 1969; Burgmann, 1975).

Diffusion data for wüstite are available for Fe tracer diffusion (Himmel *et al.*, 1953; Carter and Richardson, 1954; Desmarescaux *et al.*, 1965; Hembree and Wagner, 1969; Leroy *et al.*, 1970, 1973, 1976; Chen and Peterson, 1973b, 1975), from Mössbauer studies with Fe tracers (see, e.g., Greenwood and Howe, 1972a, b), and from chemical diffusion work (Fujii and Meussner, 1967; Campbell, 1969; Childs *et al.*, 1971; Laub and Wagner, 1973; Rickert and Weppner, 1974). Figure 21 includes some of the results obtained in the form of an Arrhenius diagram; the dependence of the activation enthalpy on deviation from stoichiometry x is shown in the

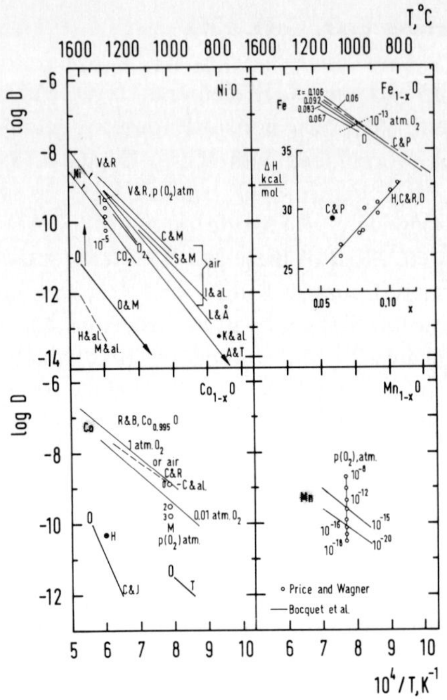

Fig. 21. Arrhenius diagrams of self-diffusion in NiO, $Fe_{1-x}O$, $Co_{1-x}O$ and $Mn_{1-x}O$.

inset. The letters refer to the authors (H = Himmel *et al.*, 1953; C&R = Carter and Richardson, 1954; D = Desmarescaux *et al.*, 1965; C&P = Chen and Petersen, 1973b, 1975). Figure 22 shows the dependence of D^{Fe} on x at constant temperatures covering most of the wüstite phase field (Fig. 20).

Iron diffusion is seen to be very fast. Early work indicated an increase in D^{Fe} with x in $Fe_{1-x}O$. A relation of the type $D^{Fe} \propto x^2$ was suggested. The activation enthalpy ΔH was reported to increase with increasing deviation from stoichiometry. These observations indicated that the iron ions did not move independently of each other. Kofstad (1972) combined the thermodynamic data of x as a function of $p(O_2)$ with the Fe diffusion data to produce the dotted line in Fig. 21 labeled "10^{-13} atm O_2," hence showing diffusion at constant oxygen partial pressure rather than constant composition. At constant $p(O_2)$, D^{Fe} *decreases* with *increasing* temperature. This a priori unusual behavior was said to be due to the fact that the defect concentration decreases with increasing temperature to an even larger extent than the mobility of the defects increases.

The work of Chen and Peterson (1973b, 1975), however, is at variance to some degree with the above results. While the order of magnitude of their D values is very similar to previous results, the dependence of D^{Fe} on x is irregular: D^{Fe} decreases with increasing x at 802°C, it is rather insensitive to any change in x at 1003°C, and it increases very slightly with increasing

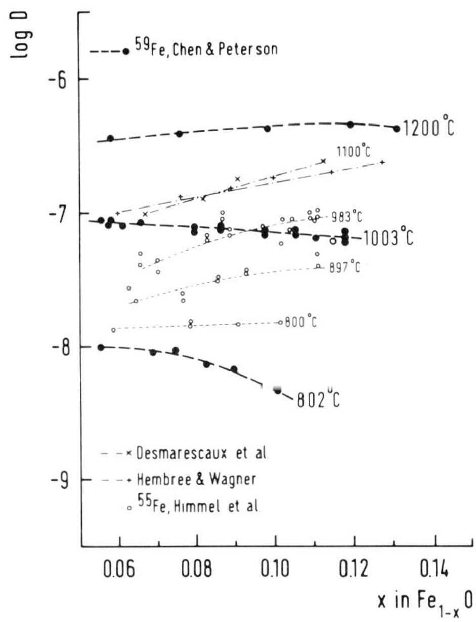

Fig. 22. Diffusion coefficients of Fe in $Fe_{1-x}O$ as function of deviation from stoichiometry according to different authors.

x at 1200°C (Fig. 22). The experiments of Chen and Peterson were performed very carefully, using high-purity single crystals, and the penetration profiles were followed and observed to be linear in a plot of $\log c$ versus x^2 over more than two orders of magnitude in concentration, whereas most previous work was done with the less reliable surface decrease technique.

Also, the isotopic mass effect was measured by simultaneously applying ^{52}Fe and ^{59}Fe as tracers. The isotopic mass effect for two isotopes 1 and 2 with masses m_1 and m_2 is

$$f\Delta K = \frac{1-(D_2/D_1)}{1-(m_1/m_2)^{1/2}}$$

where $f = [1-(D_2/D_1)]/(1-\Gamma_2/\Gamma_1)$ is the correlation factor as expressed by Tharmalingam and Lidiard (1959) (Γ_ν = jump frequencies) and ΔK the fraction of the total translational kinetic energy at the saddle point, associated with motion in the direction of the diffusional jump, that belongs to the jumping atom. If D_2/D_1 is measured, the diffusion mechanism can be deduced by comparison with calculated values of f and ΔK. Similar measurements have shown that, in $Co_{1-x}O$ and $Ni_{1-x}O$, where x is much smaller than in $Fe_{1-x}O$, $f\Delta K$ is independent of temperature and compatible with a single-vacancy mechanism for cation diffusion (Chen *et al.*, 1969a; Volpe *et al.*, 1971; see also Sections VIIB and C). In contrast, $f\Delta K$ was smaller for $Fe_{1-x}O$ than for $Ni_{1-x}O$ or $Co_{1-x}O$, and it decreased with increasing defect concentration at 1000°C.

The combined evidence is thus in favor of a diffusion mechanism involving complexes. For $x = 0.05$ and an fcc lattice, the random probabilities that a vacancy will have no, one, or two other vacancies as nearest neighbors are 0.54, 0.24, and 0.11, respectively. The irregular dependence of D^{Fe} on x (Fig. 22; data in Chen and Peterson, 1973b) is then qualitatively understandable. At low temperatures (e.g., 802°C), D^{Fe} decreases with x because an increasing number of vacancies form immobile clusters, whereas at higher temperatures (e.g., 1200°C), D^{Fe} tends to increase with x since the binding energy of the clusters becomes less important relative to kT with increasing temperature T. The decrease in $f\Delta K$ with increasing defect concentration might point to a decrease in the correlation factor f, implying a certain contribution of divacancies to Fe diffusion. The fact that linear Arrhenius plots were measured for constant compositions (Fig. 21) indicates a balance between binding energies and migration energies of mono- and divacancies.

Leroy *et al.* (1973, 1976) have performed calculations on f for $Fe_{1-x}O$ assuming a Roth-type defect (Roth, 1960) consisting of two octahedral V_{Fe} and one tetrahedral $I_{Fe^{3+}}$. Their f increased with x, which may be an indication that larger clusters of the Koch–Cohen (1969) type or those mentioned in Section III.B are operative. On the other hand, De Bruin and Murch (1973) have shown that, in a system where short-range repulsive defect–defect interaction exists, a vacancy may prevent a certain number of neighboring sites from being occupied by other vacancies, the so-called site-blocking effect. In this way, both tracer and vacancy jumps become correlated, thus causing f to decrease with x.

Early investigations of chemical diffusion (see, e.g., Fujii and Meussner, 1967; Campbell, 1969, as reported in O'Keeffe, 1970; Childs *et al.*, 1971; Laub and Wagner, 1973), deduced from, for example, weight changes of a crystal with time following a change in $p(O_2)$, yielded $\tilde{D}$ values that decreased with increasing x, which was attributed to a decrease in vacancy mobility. The deduced tracer diffusion coefficients also decreased with x (see, e.g., Landler and Komarek, 1966). The $\tilde{D}$ values of different authors varied by a factor of ~ 4. In contrast, a more recent study using emf techniques, which are not affected by surface exchange problems (Rickert and Weppner, 1974), reported higher $\tilde{D}$ values which increased with increasing x.

In conclusion, the data of Chen and Peterson are thought to be the most reliable because of the advanced techniques used and because of the high quality of their specimens. In consequence, iron diffusion in $Fe_{1-x}O$ most probably proceeds via free mobile vacancies coexisting with defect clusters. Thus only a fraction of the total vacancies contributes to cation diffusion in $Fe_{1-x}O$. The Fe ions diffuse by exchanging sites with free mobile vacancies. Because of the site-blocking effect, the mobility of these free mobile vacancies and the apparent correlation factor for Fe tracer diffusion decrease with an increase in x.

Because of the probable coexistence of small Roth (1960)-type clusters and larger Koch–Cohen (1969)-type clusters, as well as the postulated (Fender and Riley, 1969) existence of three separate wüstite regions separated by order–disorder transitions (see I to III in Fig. 20), a detailed quantitative fit between a defect model and the experimental diffusion data has not been attempted yet.

There appear to be no quantitative data on oxygen diffusion in wüstite, but indirect evidence and the analogy with CoO and NiO show that oxygen diffusion is much slower than iron diffusion.

B. CoO

The largest deviation from stoichiometry in $Co_{1-x}O$ is much smaller than that in $Fe_{1-x}O$, x being $\lesssim 0.01$. As a consequence, defect clustering is much less likely. Self-diffusion data exist for both Co and O and yield a quite consistent picture though, as with most other physical property data, most diffusion results apply to the high-oxygen-activity part of the stability range only. Cobalt diffusion is much faster than oxygen diffusion. Figure 21 includes typical results in the form of an Arrhenius diagram.

The tracer data for Co labeled C&R, C&*al*, M, D&S, R&B, and D in Fig. 21 are due to Carter and Richardson (1955), Chen *et al.* (1969b), Morin (1975), Dieckmann and Schmalzried (1975a), Rahman and Berard (1977) and Dieckmann (1977). Similar results were deduced from a study on the growth kinetics of grain boundary grooves by Maiya *et al.* (1970). The activation enthalpies ΔH are similar to those for $Fe_{1-x}O$; $\Delta H = 34.5$ kcal/mole in 1 atm O_2 (C&R), 38.4 kcal/mole in air (C&*al*), or 31.6 kcal/mole for $Co_{0.995}O$ (R&B). These values represent the enthalpy of motion for Co ions moving via a vacancy jump mechanism. The D^{Co} value increases with increasing $p(O_2)$ (increasing x in $Co_{1-x}O$) as indicated in Fig. 21 by the different Arrhenius lines as well as by the isothermal data of Morin (1975) at $T = 1000°C$ and Dieckmann (1977) at $T = 1400°C$. The arrow labeled D shows the range of D^{Co} values for the entire phase field of $Co_{1-x}O$.

The value of D^{Co} increases approximately with $p(O_2)^{1/4}$ [Rahman and Berard (1977) give an exponent $1/n$ with $n = 0.279$ for 1150°C], which might suggest that the predominant mobile defects are singly ionized Co vacancies. A complete analysis of all diffusion data (see, e.g., Dieckmann, 1977) indicates, however, that differently charged vacancies in the cation sublattice are responsible for cobalt diffusion. The isotope effect measurements of Chen *et al.* (1969b), which yield a temperature-independent $f\Delta K = 0.582$ ($1081°C \leqslant T \leqslant 1411°C$), confirm the assumption that cation diffusion occurs by a vacancy mechanism only. Thus, in contrast to $Fe_{1-x}O$ with its larger deviation from stoichiometry ($0.05 \leqslant x \leqslant 0.15$), no defect clustering is indicated. Both $V_O^{\cdot\cdot}$ and $Co_i^{\cdot\cdot}$ exist only as minority defects in $Co_{1-x}O$.

Results for chemical diffusion reported by Price and Wagner (1966), Morin (1975), Wimmer *et al.* (1975), and Mrowec and Przybylski (1977a, b, c, d) confirm the above result. The $\tilde{D}$ values were obtained from the oxidation of Co metal. Experiments with Pt markers prove that oxidation is due to outward diffusion of Co through the $Co_{1-x}O$ layers. Values of D^{Co} calculated from the $\tilde{D}$ values agree very well with tracer diffusion data (Mrowec and Przybylski, 1977a, b, c, d). The latter results and those of Fryt (1976) were said to indicate that, at $T \geqslant 1100°C$, cation vacancies due to nonstoichiometry start to overlap with intrinsic Frenkel disorder. However, care must be taken in analyzing data involving chemical gradients (and additional electrical driving forces). Dieckmann (1977) has discussed this point. The tracer diffusion data are still compatible with differently charged metal vacancies without any significant contribution of Co_i.

Oxygen diffusion is slower than Co diffusion by up to four orders of magnitude. Experimental data on D^O are included in Fig. 21 labeled with C&J, H, and T (Chen and Jackson, 1969; Holt, 1967; Thompson, 1962). The high data of Thompson were obtained with polycrystals and may have been influenced by grain boundaries or other drawbacks of the gaseous isotope-exchange technique (Section IV.B), as also indicated by his low preexponential D_0 of 6×10^{-6} cm^2/sec. The remaining data are for single crystals and follow the relation $D^O = 50 \exp(-95{,}000/RT)$ cm^2/sec (annealing in air). The mechanism of oxygen diffusion is still unclear, though some indications support an interstitial mechanism. Diffusion-related properties, such as creep, are rate-controlled by the slower moving ion, hence oxygen in $Co_{1-x}O$. The creep work of Clauer *et al.* (1969) yields a dependence of creep rate $\epsilon \propto (O_2)^{0.48}$ that corresponds to the $p(O_2)^{1/2}$ dependence expected for neutral (zero effective charge) oxygen interstitials as rate-controlling species. More work on oxygen diffusion as function of $p(O_2)$ is needed to decide conclusively on the diffusion mechanism.

C. *NiO*

The maximum deviation from stoichiometry in nickel oxide is even smaller than that in $Co_{1-x}O$ [typically $x \sim 10^{-4}$ at 1000°C and $\sim 10^{-3}$ at 1500°C (Sockel and Schmalzried, 1968)]. This necessitates low concentrations of heterovalent impurities in order to obtain meaningful data on the intrinsic diffusion behavior, and it helps to explain the large scatter of experimental diffusion coefficients (by one order of magnitude for D^{Ni} and by up to two orders of magnitude for D^{O}). For NiO, more diffusion data exist than for any other nonstoichiometric NaCl-type oxide. Again, as with FeO and CoO, cation (Ni) diffusion is much faster than oxygen diffusion. Figure 21 shows typical self-diffusion data for D^{Ni} due to Shim and Moore (1957) (S&M in Fig. 21), Lindner and Åkerström (1957) (L&Å), Choi and Moore (1962) (C&M), Volpe and Reddy (1970) (V&R), Klotsman *et al.* (1972) (K&*al*), Ikeda *et al.* (1974, 1976) (I&*al*), and Atkinson and Taylor (1978, 1979) (A&T). Also included are data for D^{O} due to O'Keeffe and Moore (1961) (O&M), Hallwig *et al.* (1977) (H&*al*) and Monty *et al.* (1977) (M&*al*). The early data of O&M were corrected by O'Keeffe and Moore (1961) to higher values as indicated by the arrow pointing upward from the line O&M. The more recent data (M&*al*, H&*al*) are closer to the original results of O&M. More work is needed to determine reliable values of D^{O} and the mechanism of oxygen diffusion. Yet, the relation $D^{O} \ll D^{Ni}$ certainly holds, showing that chemical diffusion, oxidation, etc., are mainly due to D^{Ni}.

The values of D^{Ni} at constant T and $p(O_2)$ scatter by about one order of magnitude, and the reported activation enthalpies range between 44 and 63 kcal/mole. The latter value (62.7 kcal/mole for CO_2, 60.8 kcal/mole for atmospheric O_2) is due to Volpe and Reddy (1970). who extended their measurements to very high temperatures (1180–1760°C corresponding to $0.9T_m$, the melting point of the oxide), used single crystals, and also reported the lowest D values (which usually are the most representative ones). The older data were extended only to ~1400°C and were therefore more subject to the effects of impurities, grain boundaries, dislocations, etc. Also, the soft radiation of the commonly used tracer, Ni^{63} (β rays of only 63 keV maximum energy) causes difficulties. The most recent data of Atkinson and Taylor (1978, 1979) extend the data of Volpe and Reddy (1970) down to 522°C by using sputtering to section the samples. Atkinson and Taylor report somewhat smaller ΔH and D_0 values than Volpe and Reddy [58–59 kcal/mole for air and O_2, $D_0 = 0.015$ cm²/sec (air) and 0.022 cm²/sec (O_2) for A&T, $D_0 = 0.048$ cm²/sec (O_2) and 0.021 cm²/sec (CO_2) for V&R], but the D^{Ni} values in the overlapping temperature range show very good agreement with one another. This gives NiO a somewhat unique position within the nonstoichiometric oxides, since agreeing diffusion data from different authors are available for a very broad temperature range, 522–1760°C, without an indication of knee or pronounced curvature in the Arrhenius diagram. A surprising fact is the absence of a noticeable impurity effect in the work of Atkinson and Taylor, at least at atmospheric oxygen pressure. Only at $p(O_2) < 10^{-10}$ atm did the lack of a dependence of D^{Ni} on $p(O_2)$ indicate extrinsic behavior and suggest a concentration of heterovalent impurities (Al^{3+}, Cr^{3+}) of ~1 ppm in their samples. Co impurities did not appreciably affect D^{Ni}, at least at concentrations of ~200 ppm (Volpe and Reddy, 1970; see also Section VII.D).

The dependence of D^{Ni} on $p(O_2)$ [$D^{Ni} \propto p(O_2)^{1/n}$] in the intrinsic region was neither $n = 4$ nor $n = 6$ but varied between 5 and 6, decreasing with temperature. The total range of variation in D^{Ni} is shown in Fig. 21 for $T = 1380°C$ (data of V&R). Measurements of the isotope effect and thus of the correlation factor for Ni diffusion (Volpe *et al.*, 1971) showed that $f \Delta K = 0.613$ was temperature-independent ($1201°C \leqslant T \leqslant 1678°C$). This proves that Ni diffusion in NiO, like Co diffusion in $Co_{1-x}O$, proceeds via a vacancy mechanism and any contribution of Ni_i is negligible. The predominant defect in NiO is thus the nickel vacancy, electrically compensated for by electron holes, which causes $Ni_{1-x}O$ to be a *p*-type semiconductor. Both singly and doubly ionized vacancies exist, depending on $p(O_2)$ and *T*, and heterovalent impurities may interfere. In contrast to $Fe_{1-x}O$, there is no need to invoke defect clustering, though some clusters may exist as postulated by Roth (1970), who reported antiferromagnetic domains in NiO and clusters of the type $Ni^{3+}-V_{Ni}-Ni^{3+}$. These affect magnetic properties but have a negligible effect on diffusion.

The above conclusions are confirmed by data on impurity and chemical diffusion.† These results also show the effect of heterovalent impurities, such as Cr, and of grain boundaries and short circuits, particularly in the low-temperature oxidation of nickel. Recent channeling work shows that, though NiO layers growing on Ni are partly epitaxial, the NiO is distorted and contains excess oxygen, possibly in voids, at least during the initial growth stages (Roulet *et al.*, 1977). This helps to explain why values deduced from oxidation kinetics using Wagner's theory are high, though a quantitative explanation of the high *D* values at low temperatures requires further microstructural examination of oxide scales.

D. *Mixed Oxides (Co, Ni)O*

A few of the large number of diffusion experiments on mixed oxides (A, B)O of the rock-salt structure will be treated in this section, since they are often relevant to or confirm the above discussions.

The only system considered here is $(Ni_{1-c}Co_c)O$. Chen and Peterson (1973b) have shown that Co additions up to $c = 0.02$ enhance both Ni and

†Impurity diffusion in NiO: e.g., Fe in NiO (Stewart and Wagner, 1975); Cr in NiO (Seltzer, 1971; Perkins and Rapp, 1973; Chen *et al.*, 1973); Co in NiO (Chen and Peterson, 1972). Chemical diffusion in NiO: e.g., Price and Wagner (1966); Fueki and Wagner (1966); Deren *et al.* (1971); Herchl *et al.* (1972); Koel and Gellings (1973); Christian and Gilbreath (1975); Farhi and Petot-Ervas (1976, 1978). Grain boundary effects: e.g., Graham *et al.* (1972); Atkinson and Taylor (1978, 1979).

Co diffusion by increasing the concentration of cation vacancies with the addition of cobalt [a larger x in $(Ni, Co)_{1-x}O$ at constant $p(O_2)$ and T]. Vacancy–solute interactions were negligible at these concentrations and at the experimental temperatures (1179–1649°C), and the vacancy mechanism for metal diffusion was confirmed. Chen and Peterson (1973a) investigated also the effect of higher Co concentrations, as well as interdiffusion. The apparent number of effective negative charges on a cation vacancy varied almost linearly from 2 in NiO to ~1 in CoO for annealing in air. The interdiffusion experiments also showed a net flow of cations from CoO (the material showing faster diffusion) to NiO, producing a gradient in oxygen activity that can alter the diffusion behavior of the samples (see also Yurek and Schmalzried, 1975, for marker experiments). Dieckmann and Schmalzried (1975a) have extensively discussed the appreciable correlation effects and the coupling of the self-diffusion coefficients of the cations in the solid solution (Ni, Co)O. Space does not allow these interesting questions to be considered in detail.

E. Other Materials (MnO, TiO, NbO)

Other NaCl-type monoxides are of interest because of their extended nonstoichiometry, such as VO, TiO, and NbO (see Table I) which are unusual on account of the large number of vacant sites present in *both* the metal and oxygen sublattices. Unfortunately, no relevant diffusion data are available for VO, and the data for TiO and NiO are limited.

TiO exists in a high-temperature form with a defective NaCl structure and in a low temperature form, the structure of which is still under discussion and is probably monoclinic or of the NaCl type with an ordered arrangement of the vacancies. The high-temperature form, TiO_{high} or TiO_h in Fig. 20, has a very large homogeneity range (shaded area in Fig. 20) at high temperatures, which extends from about $TiO_{0.89}$ to $TiO_{1.25}$ at 1300°C. The self-diffusion of Ti^{44} was measured for four compositions (vertical dashed lines in Fig. 20). The results (Lundy *et al.*, 1973) can be summarized as shown in the accompanying tabulation and indicate that Ti diffusion depends very little on x. The concentration of V_{Ti} increases with x, and so does D, but only slightly. More data are needed to quantitatively understand diffusion processes in $TiO_{1\pm x}$.

x in $TiO_{1\pm x}$	−0.193	0.006	0.06	0.259
D_O (cm²/sec)	0.3 ± 0.1	1.0 ± 0.2	0.6 ± 0.1	0.4 ± 0.02
ΔH (kcal/mole)	65 ± 14	65 ± 7	63 ± 6	57 ± 2
$D \times 10^{-9}$ at 1400°C (cm²/sec)	1.18	3.63	3.62	13.6

For NbO, which has also a defective NaCl structure in which one-fourth of all cation and anion sites are vacant, some results for Nb^{95} exist (Gel'd and Lyubimov, 1963) yielding a much lower $\Delta H = 29.6$ kcal/mole. Again, more data are needed.

More information exists for $Mn_{1-x}O$; its defect structure is complicated. Complex defects and clustering of defects may occur, and three different regions may exist within the $Mn_{1-x}O$-phase region as reported for $Fe_{1-x}O$ (Fender and Riley, 1970). Note that, in contrast to NiO and CoO, $Mn_{1-x}O$ can have x values $\geqslant 0.1$. Data on Mn tracer diffusion have been reported by Bocquet *et al.* (1967), Price (1968), and Price and Wagner (1970). These data show that the predominant defects in $Mn_{1-x}O$ are, again, metal vacancies, and that their concentration increases with $p(O_2)$. Figure 21 contains some of the diffusion results, both as a function of $p(O_2)$ at constant T and as a function of T at constant $p(O_2)$ (for the difficulties of obtaining ΔH values from Arrhenius plots for constant $p(O_2)$ rather than constant composition, see Section VIII). Price and Wagner (1970) found no change in $p(O_2)$ dependence in the pressure region, where the electric conductivity had a minimum, and explained their data by Mn diffusion via doubly charged vacancies over the entire phase field. Bocquet *et al.* (1967), in contrast, found a change in oxygen dependence and considered the possibility of manganese interstitials, which, however, seems less likely as pointed out by Kofstad (1972). Clearly, more experiments are needed to answer this open question.

There appear to be no data on oxygen diffusion in these materials.

VIII. Results for More Complex Nonstoichiometric Oxides

Some scattered diffusion results exist for more complex nonstoichiometric oxides of structures different from those treated above. Two higher iron oxides, hematite (Fe_2O_3, corundum structure) and magnetite (Fe_3O_4, spinel structure), as well as Nb_2O_5, will be discussed here.

Hematite, Fe_2O_3, is slightly oxygen-deficient, Fe_2O_{3-x} ($x \lesssim 0.01$). This substance is interesting for diffusion, since it represents the rare case of approximately equal diffusion coefficients for metal ions and oxygen at high temperatures (D^{Fe}: Lindner, 1952; Himmel *et al.*, 1953; Izvekov *et al.*, 1962; Chang and Wagner, 1972; D^{O}: Kingery *et al.*, 1960; Hagel, 1966). Also, D values deduced from sintering agree with the tracer values (Coble, 1958). With decreasing $p(O_2)$, the value of D^{Fe} increases with a pressure dependence of -0.75, which is compatible with interstitial migration of iron ions.

For Nb_2O_5, only oxygen diffusion data are available. Nb_2O_5 is oxygen-deficient. The NbO_2–Nb_2O_5 region has a homologous series of oxides of the formula $Nb_{3n+1}O_{8n-2}$, with Nb_2O_5 corresponding to $n = 9$ and the next lower oxide, with $n = 8$, being very close: $Nb_{25}O_{62} = NbO_{2.48}$. The Nb–O system thus resembles in some respects the Ti–O system (Section VI), and shear structures have also been suggested to occur. At near atmospheric pressure, the oxygen vacancy concentration becomes small ($\sim 10^{-6}$ at $\sim 1000°C$), and extrinsic behavior can dominate unless the samples are pure. The oxygen diffusion data (Chen and Swalin, 1966; Chen and Jackson, 1967; Sheasby *et al.*, 1968; Massiani *et al.*, 1978) give a good example of the difficulties of diffusion work with nonstoichiometric oxides. For greatly varying and, in particular, low values of $p(O_2)$, and at low temperatures ($\lesssim 900°C$), results were obtained that were difficult to reconcile, probably for one or more of the following reasons: extrinsic behavior, presence of Wadsley defects or even shear structures, or oxygen gradients. As expected, therefore, only data at $T \geqslant 1000°C$ and for near-atmospheric pressure ($\sim 10^{-2}$–1 atm O_2) yielded consistent results (Chen and Jackson, 1967; Massiani *et al.*, 1978). The value of D^O increased with decreasing oxygen pressure and the pressure dependence, $D^O \propto p(O_2)^{-1/4}$, indicated that singly charged oxygen vacancies predominate. More such data are needed before definite conclusions can be drawn for data at lower pressures and about the possible effect of shear structures on diffusion (see also Sections III.D and VI).

The last example to be discussed is magnetite, Fe_3O_4. Magnetite can have a metal deficit, $Fe_{3-x}O_4$, which increases with increasing $p(O_2)$. There are two sites for iron ions in the spinel structure of magnetite; Fe^{2+} ions occupy octahedral sites, whereas Fe^{3+} ions occupy octahedral and tetrahedral sites in equal numbers at room temperature. At higher temperatures, randomization might occur as a result of electron exchange. The early Fe diffusion work (Himmel *et al.*, 1953; Izvekov, 1958; Klotsman *et al.*, 1960) was evaluated in terms of conventional Arrhenius diagrams (see dashed area in Fig. 23, right), with only the results of Himmel *et al.* (1953) being made at constant composition. The oxygen diffusion results (isotope exchange with ^{18}O-enriched water vapor, Castle and Surman, 1967, 1969) showed a certain dependence of D^O on $p(O_2)$; i.e., the authors reported a relation $D^O \propto [p(H_2)/p(H_2O)]^{0.27}$, hence D increased with decreasing $p(O_2)$, indicating that results at constant $p(O_2)$ would not necessarily give the right ΔH.

Most interesting in the present context, however, are recent metal diffusion data. Following the results of Schmalzried (1962) indicating that D^{Fe} increases with $p(O_2)$, i.e., $D^{Fe} \propto p(O_2)^{0.4}$, obtained with the contact method (radioactive pellets are pressed against inactive pellets and the transfer of radioactivity is followed), an important breakthrough was

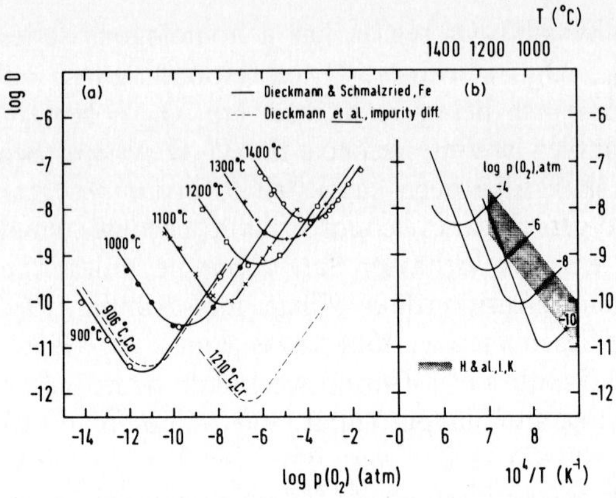

Fig. 23. Diffusion coefficients of Fe, Co, and Cr in magnetite $Fe_{3-x}O_4$ as a function of oxygen partial pressure (a) according to Dieckmann *et al.* 1981; Dieckmann, 1977). The same data together with previous results are plotted as a function of inverse temperature in (b).

achieved by Dieckmann and Schmalzried *et al.* (Dieckmann and Schmalzried, 1975b, 1977; Dieckmann *et al.*, 1978; Hodge, 1978). Careful tracer diffusion experiments for Fe (and also Cr and Co) showed pronounced minima in plots of $\log D$ versus $\log a(O_2)$ or $\log p(O_2)$. The slopes were $\frac{2}{3}$ at high oxygen activity $a(O_2)$ and $-2/3$ at low-oxygen activity (Fig. 23, left). As with the data for $(U,Pu)O_{2-x}$ in Fig. 13, a point defect analysis has shown that a vacancy mechanism dominates at high oxygen activities and that an interstitial (or interstitialcy) mechanism dominates at low oxygen activities with a change in mechanism in the middle of the magnetite-phase field. This has two important consequences: (i) It shows that a single point defect model explains the extensive experimental results very well; i.e., no complex formation or clustering has to be invoked in spite of the relatively high defect concentrations, possibly because of the high buffer and screening capacity of the Fe^{2+}–Fe^{3+} ion mixture. (ii) The data as plotted versus $1/T$ in an Arrhenius-type diagram (Fig. 23, right) clearly demonstrate that the determination of activation enthalpies ΔH from a limited number of measurements with nonstoichiometric oxides can lead to completely meaningless results—an error that certainly sometimes was made in the past. Finally, the conclusion can be made that stoichiometric Fe_3O_4 has a predominant Frenkel disorder and that the divalent ions (Fe^{2+}) are mainly responsible for metal diffusion ($D^{Cr^{3+}} \ll D^{Fe^{2+}} \sim D^{Co^{2+}}$ in Fig. 23; also, for the apparent motion enthalpy ΔH_m, $\Delta H_m^{Cr^{3+}} \gg \Delta H_m^{Fe^{2+}} \sim \Delta H_m^{Co^{2+}} \sim 33$ kcal/mole for the vacancy regime).

The minimum is temperature-dependent [again as in the case of $(U,Pu)O_{2-x}$], and it is interesting and possibly surprising to note that D decreases with increasing T at constant oxygen pressure.

IX. Concluding Remarks

The above discussion has focused on some aspects of diffusion in nonstoichiometric oxides. Some emphasis was put on experimental difficulties, which are very serious indeed since these substances easily change composition and structure with small changes in oxygen activity. Often compositions at a specific temperature cannot be maintained during cooling. Evaporation is usually much more important than with, for example, metals, and it is often incongruent, causing changes in bulk and surface compositions. In future work, therefore, oxygen potentials should be carefully controlled and measured, experiments should be done at both constant pressure and, if possible, constant composition, and modern techniques should be used to follow the time and depth dependence of tracer penetration. Supporting data are also needed. These can be obtained from electron microscopy for defect interactions but also from other physical measurements, notably electrical conductivity (including recovery work on quenched samples). Furthermore, more effort should be devoted to obtaining reliable supporting evidence for diffusion-controlled measurements such as creep, sintering, and possibly grain growth, as well as from theoretical studies and calculations of defect formation and interaction energies. Needless to say, the quality of samples should be improved. Preferably, high-purity single crystals should be used.

We have attempted to cover most materials of interest and to give extensive and typical references. Obviously, not all aspects and materials could be treated. The selection was based on the technological importance of the oxides as well as on a (partly subjective) estimation of the scientific interest. Doubtless, some interesting references were not or could not be given. Also, since a compromise had to be made to meet space requirements, diffusion theory was not as adequately treated as the author would have liked. Nevertheless, it is hoped that this chapter will serve as a useful basis for determining the reliability of given data and for pointing out the gaps in available data and in current knowledge. Certainly, there is ample space for more good, careful investigations in the field of diffusion in nonstoichiometric oxides.

References

Aitken, E. A., Brassfield, H. C., and Fryxell, R. E. (1966). *In* "Thermodynamics of Nuclear Materials," Vol. II, p. 453. IAEA, Vienna.

Ajayi, O. B., Nagel, L. E., Raistrick, I. D., and Huggins, R. A. (1976). *J. Phys. Chem. Solids* **37**, 167–172.

Akse, J. R. (1977). Ph.D. Thesis, Arizona State Univ., Tempe.

Akse, J. R., and Whitehurst, H. B. (1978). *J. Phys. Chem. Solids* **39**, 457–465.

Amato, K., Colombo, R. L., and Grappiolo, G. C. (1966). *Solid State Commun.* **4**, 237–239.
Anderson, H. V. (1967). *J. Am. Ceram. Soc.* **50**, 225–238.
Anderson, J. S. (1972). *In* "Surface and Defect Properties of Solids" (M. W. Roberts and G. M. Thomas, eds.), Vol. 1, p. 1. Chem. Soc., London.
Anderson, J. S., and Hyde, B. G. (1965). *Bull. Soc. Chim. Fr.* pp. 1215–1216.
Anderson, J. S., and Hyde, B. G. (1967). *J. Phys. Chem. Solids* **28**, 1393–1408.
Anderson, J. S., and Khan, A. S. (1970). *J. Less-Common Met.* **22**, 219–223.
Anderson, J. S., and Tilley, R. J. D. (1974). *In* "Surface and Defect Properties of Solids" (M. W. Roberts and G. M. Thomas, eds.), Vol. 3, pp. 1–50. Chem. Soc., London.
Andersson, S., and Wadsley, A. D. (1966). *Nature (London)* **211**, 581–583.
Andersson, S., Collén, B., Kuylenstierna, V., and Magnéli, H. (1957). *Acta Chem. Scand.* **11**, 1641–1652.
Ando, K., and Oishi, Y. (1979). *J. Nucl. Sci. Technol.* **16**, 214–220.
Ando, K., Oishi, Y., and Hidaki, Y. (1976). *J. Chem. Phys.* **65**, 2751–2755.
Ashbee, K. H. G., and Smallman, R. E. (1963). *J. Am. Ceram. Soc.* **46**, 211–214.
Assmann, H., Maier, G., and Stehle, H. (1975). *Proc. Reaktortagung Nürnberg, Ger. Atomforum* pp. 391–394.
Atkinson, A., and Taylor, R. I. (1978). *J. Mater. Sci.* **13**, 427–432.
Atkinson, A., and Taylor, R. I. (1979). *Philos. Mag. A* **39**, 581–595.
Atlas, L. M., Schlehman, G. J., and Readey, D. W. (1966). *J. Am. Ceram. Soc.* **49**, 624–625.
Auskern, A. B., and Belle, J. (1961). *J. Nucl. Mater.* **3**, 267–276, 311–319.
Baenzinger, N. C., Eick, H. A., Schuldt, H. S., and Eyring, L. (1961). *J. Am. Chem. Soc.* **83**, 2219–2223.
Bagshaw, A. N., and Hyde, B. G. (1976). *J. Phys. Chem. Solids* **37**, 835–838.
Barnabel, V. I., and Bogomolov, V. N. (1969). *Fiz. Tverd. Tela* **11**, 2671–2672.
Baumard, J. F. (1976). *Solid State Commun.* **20**, 859–862.
Baumard, J. F., Panis, D., and Ruffier, D. (1975). *Rev. Int. Hautes Temp. Refract.* **12**, 321–327.
Baumard, J. F., Panis, D., and Anthony, A. M. (1977). *J. Solid State Chem.* **20**, 43–51.
Baur, W. H. (1956). *Acta Crystallogr.* **9**, 515–520.
Bayoglu, A. S., and Lorenzelli, R. (1979). *J. Nucl. Mater.* **79**, 437–438; **82**, 403–410.
Bell, H., Krishnamachari, V., and Jones, J. T. (1971). *J. Am. Ceram. Soc.* **54**, 359–360.
Belle, J., ed. (1961). "Uranium Dioxide: Properties and Nuclear Applications." U.S. Gov. Print. Off., Washington, D.C.
Belle, J. (1969). *J. Nucl. Mater.* **30**, 3–15.
Berard, M. F. (1971). *J. Am. Ceram. Soc.* **54**, 144–146.
Berman, R. M. (1969). Rep. WAPD-TM 843. Westinghouse Atomic Power Department.
Berrin, L., and Johnson, D. L. (1965). *Proc. Conf. Sintering Relat. Phenom.* Gordon & Breach, New York.
Bevan, D. J. M., and Kordis, J. (1964). *J. Inorg. Nucl. Chem.* **26**, 1509–1523.
Bittel, J. H., Sjodahl, L. H., and White, J. F. (1969). *J. Am. Ceram. Soc.* **52**, 446–451.
Blank, H., and Ronchi, C. (1968). *Acta Crystallogr., Sect. A* **24**, 657–666.
Blumenthal, R. N., and Whitmore, D. H. (1963). *J. Electrochem. Soc.* **110**, 92–93.
Blumenthal, R. N., Coburn, J., Baukus, J., and Hirthe, W. M. (1966). *J. Phys. Chem. Solids* **27**, 643–654.
Blumenthal, R. N., Baukus, J., and Hirthe, W. M. (1967). *J. Electrochem. Soc.* **114**, 172–176.
Bocquet, J. P., Kawahara, M., and Lacombe, P. (1967). *C. R. Acad. Sci.* **265**, 1318–1321.
Bogomolov, V. N. (1963). *Fiz. Tverd. Tela* **5**, 2011 [*Sov. Phys.—Solid State* **5**, 1468–1469 (1964)].
Bredig, M. A. (1970). *Colloq. Int. CNRS* **205**, 184.
Breitung, W. (1978). *J. Nucl. Mater.* **74**, 19–26.
Bruckman, A., Emmerich, R., and Mrowec, S. (1972). *Oxid. Met.* **5**, 137–147.

Burgmann, W. (1975). *Met. Sci. J.* **9**, 169–175.

Bursill, L. A., and Hyde, B. G. (1971). *Philos. Mag.* **23**, 3–15.

Bursill, L. A., Hyde, B. G., Terasaki, O., and Watanabe, D. (1969). *Philos. Mag.* **20**, 347–359.

Bursill, L. A., Hyde, B. G., and Philp, D. K. (1971). *Philos. Mag.* **23**, 1501–1513.

Burton, B. (1975). *In* "Thermodynamics of Nuclear Materials 1974," Vol. I, pp. 415–435. IAEA, Vienna.

Butler, E. M., and Meyer, R. O. (1973). *J. Nucl. Mater.* **47**, 229–240.

Calvert, J. M., Derry, D. J., and Lees, D. G. (1974). *J. Phys. D* **7**, 940–953.

Campbell, R. H. (1969). Thesis, Arizona State Univ., Tempe.

Carr, V. M., Chadwick, A. V., and Figueroa, D. R. (1976). *J. Phys. (Paris), Colloq.* **C7**, 337–341.

Carter, R. E., and Lay, K. W. (1970). *J. Nucl. Mater.* **36**, 77–86.

Carter, R. E., and Richardson, F. D. (1954). *Trans. AIME* **200**, 1244–1257.

Carter, R. E., and Richardson, F. D. (1955). *Trans. AIME* **203**, 336–343.

Castle, J. E., and Surman, P. L. (1967). *J. Phys. Chem.* **71**, 4255–4259.

Castle, J. E., and Surman, P. L. (1969). *J. Phys. Chem.* **73**, 632–634.

Catlow, C. R. A. (1981). *In* "Proc. Nato Advanced Study Institute on Physics and Chemistry of Refractory Oxides." Aleria, Sijthoff & Nordhoff, Netherlands.

Catlow, C. R. A., and Norgett, M. J. (1973). *J. Phys. (Paris), Colloq.* **C9**, 45–48.

Catlow, C. R. A., James, R., and Norgett, M. J. (1976). *J. Phys. (Paris), Colloq.* **C7**, 443–448.

Chang, R. H., and Wagner, J. B., Jr. (1972). *J. Am. Ceram. Soc.* **55**, 211–213.

Cheetham, A. K. (1973). *In* "Chemical Applications of Thermal Neutron Scattering" (B. T. M. Willis, ed.), pp. 225–249. Oxford Univ. Press, London and New York.

Cheetham, A. K., Fender, B. E. F., Steele, D., Taylor, R. I., and Willis, B. T. M. (1970). *Solid State Commun.* **8**, 171–173.

Chen, W. K., and Jackson, R. A. (1969). *J. Phys. Chem. Solids* **30**, 1309–1314.

Chen, W. K., and Jackson, R. A. (1967). *J. Chem. Phys.* **47**, 1144–1148.

Chen, W. K., and Peterson, N. L. (1972). *J. Phys. Chem. Solids* **33**, 881–892.

Chen, W. K., and Peterson, N. L. (1973a). *J. Phys. Chem. Solids* **34**, 1093–1108.

Chen, W. K., and Peterson, N. L. (1973b). *J. Phys. (Paris), Colloq.* **C9**, 303–307.

Chen, W. K., and Peterson, N. L. (1975). *J. Phys. Chem. Solids* **36**, 1097–1103.

Chen, W. K., and Swalin, R. A. (1966). *J. Phys. Chem. Solids* **27**, 57–64.

Chen, W. K., Peterson, N. L., and Reeves, W. T. (1969a). *Phys. Rev.* **186**, 887–891.

Chen, W. K., Peterson, N. L., and Reeves, W. T. (1969b). *J. Met.* **21**, 50A.

Chen, W. K., Peterson, N. L., and Robinson, L. C. (1973). *J. Phys. Chem. Solids* **34**, 705–709.

Chereau, P. (1973). Rep. CEA-R-4402.

Chereau, P., and Wadier, J. P. (1973). *J. Nucl. Mater.* **46**, 1–8.

Childs, P. E., Laub, L. W., and Wagner, J. B., Jr. (1971). *Proc. Br. Ceram. Soc.* **19**, 29–53.

Chilton, G. R., and Edwards, J. (1978). *J. Nucl. Mater.* **78**, 182–191.

Choi, J. S., and Moore, W. J. (1962). *J. Phys. Chem.* **66**, 1308–1311.

Christian, J. D., and Gilbreath, W. P. (1975). *Oxid. Met.* **9**, 1–25.

Clauer, A. H., Seltzer, M. S., and Wilcox, B. H. (1969). *NASA Contract. Rep.* **NASA CR-72590**.

Coble, R. L. (1958). *J. Am. Ceram. Soc.* **41**, 55–63.

Coble, R. L. (1961). *J. Appl. Phys.* **32**, 787–792.

Coble, R. L. (1963). *J. Appl. Phys.* **34**, 1679–1682.

Coble, R. L. (1965). *J. Appl. Phys.* **36**, 7327.

Coble, R. L. (1977). *In* "Reactivity of Solids" (J. Wood, L. Lindqvist, C. Helgesson, and N. G. Vannerberg, eds.), pp. 669–688. Plenum, New York.

Condit, R. H., and Holt, J. B. (1964). *J. Electrochem. Soc.* **111**, 1192–1194.

Contamin, P., and Slodzian, G. (1968). *C. R. Acad. Sci., Ser. C* **267**, 805–808.

Contamin, P., and Stefani, R. (1967). Rep. CEA-R 3179. Grenoble.
Contamin, P., Bacmann, J. J., and Marin, J. F. (1972). *J. Nucl. Mater.* **42**, 54–64.
Cox, B., and Pemsler, J. P. (1968). *J. Nucl. Mater.* **28**, 73–78.
Cox, B., and Roy, C. (1966). *Electrochem. Technol.* **4**, 121–127.
D'Annucci, F., and Sari, C. (1977). *J. Nucl. Mater.* **68**, 357–359.
Deaton, R. L., and Wiedenheft, C. J. (1973). *J. Inorg. Nucl. Chem.* **35**, 649–650.
De Bruin, H. J., and Murch, G. E. (1973). *Philos. Mag.* **27**, 1475–1488.
De Franco, M., and Gatesoupe, J. P. (1976). *In* "Plutonium 1975 and Other Actinides" (H. Blank and R. Lindner, eds.), pp. 133–143. North-Holland Publ., Amsterdam.
Deren, J., Jarzebski, Z. M., and Mrowec, S. (1971). *Bull. Acad. Pol. Sci. Chim.* **19**, 153–157.
Derry, D. J., Lees, D. G., and Calvert, J. M. (1971). *Proc. Br. Ceram. Soc.* **19**, 77–83.
Desmarescaux, P., Bocquet, J. P., and Lacombe, P. (1965). *Bull. Soc. Chim. Fr.* No. 15, 1106–1111.
Dickens, M. H., Hayes, W., and Hutchings, M. T. (1976). *J. Phys. (Paris), Colloq.* **C7**, 353–358.
Dieckmann, R. (1977). *Z. Phys. Chem. (Frankfurt am Main)* **107**, 189–210.
Dieckmann, R., and Schmalzried, H. (1975a). *Ber. Bunsenges. Phys. Chem.* **79**, 1108–1115.
Dieckmann, R., and Schmalzried, H. (1975b). *Z. Phys. Chem. (Frankfurt am Main)* **96**, 331–333.
Dieckmann, R., and Schmalzried, H. (1977). *Ber. Bunsenges. Phys. Chem.* **81**, 344–347, 414–419.
Dieckmann, R., Mason, T. O., Hodge, J. D., and Schmalzried, H. (1978). *Ber. Bunsenges. Phys. Chem.* **82**, 778–783.
Dornelas, W., and Lacombe, P. (1967a). *C. R. Acad. Sci., Ser. C* **265**, 359–362.
Dornelas, W., and Lacombe, P. (1967b). *J. Nucl. Mater.* **21**, 100–104.
Doskocil, J., and Pospisil, Z. (1972). *Silikaty* **16**, 113–123.
Dworkin, A. S., and Bredig, M. A. (1968). *J. Phys. Chem.* **72**, 1277–1281.
Edwards, H. S., Rosenberg, A. F., and Bittel, J. T. (1963). Rep. **ASD-TDR-63-635**.
Eikum, A., and Smallman, R. E. (1965). *Philos. Mag.* **11**, 627–632.
Ellis, W. P., and Schwoebel, R. L. (1968). *Surf. Sci.* **11**, 82–98.
Eloy, J. F., Contamin, P., Stefani, R., and Cornu, A. (1968). *Ann. Conf. Mass Spectrom. Appl. Top., 16th, Pittsburgh, Pa.* 000–000.
Faber, J., Jr., Seitz, M. A., and Müller, M. H. (1976a). *J. Phys. Chem. Solids* **37**, 903–907.
Faber, J., Jr., Seitz, M. A., and Müller, M. H. (1976b). *J. Phys. Chem. Solids* **37**, 909–915.
Farb, N. E., Johnson, O. W., and Gibbs, P. (1964). *J. Appl. Phys.* **36**, 1746–1751.
Farhi, R., and Petot-Ervas, G. (1976). *J. Phys. (Paris), Colloq.* **C7**, 438–442.
Farhi, R., and Petot-Ervas, G. (1978). *J. Phys. Chem. Solids* **39**, 1169–1173.
Fender, B. E. (1973). *In* "Chemical Applications of Thermal Neutron Scattering" (B. T. M. Willis, ed.), pp. 250–269. Oxford Univ. Press, London and New York.
Fender, B. E., and Riley, F. D. (1969). *J. Phys. Chem. Solids* **30**, 793–798.
Fender, B. E., and Riley, F. D. (1970). *In* "Chemistry of Extended Defects in Non-Metallic Solids" (L. Eyring and M. O'Keeffe, eds.), pp. 54–61. North-Holland Publ., Amsterdam.
Fisher, J. C. (1951). *J. Appl. Phys.* **22**, 74–77.
Floyd, J. M. (1973). *Indian J. Technol.* **11**, 589–594.
Fryt, E. (1976). *Oxid. Met.* **10**, 311–327.
Fueki, K., and Wagner, J. B., Jr., (1966). *J. Electrochem. Soc.* **112**, 384–388.
Fujii, C. T., and Meussner, R. A. (1967). Prog. Rep. MO 1-12, pp. 27–28. U.S. Nav. Res. Lab.
Furuya, H., and Koizumi, M. (1975). *In* "Thermodynamics of Nuclear Materials 1974," Vol. I, pp. 447–456. IAEA, Vienna.
Furuya, H., and Yajima, S. (1968). *J. Nucl. Mater.* **25**, 38–44.

Gado, P. (1965). *Acta Phys. Acad. Sci. Hung.* **18**, 111–117.
Gass, J. E., Müller, H. H., Schmied, H., Jörissen, L., and Ziffermayer, G. (1973). *Nucl. Instrum. Methods* **106**, 109–113.
Gel'd, P. V., and Lyubimov, V. D. (1963). *Poroshk. Metall., Akad. Nauk. Ukr. SSSR, Ser. Khim. Nauk.* **3**, 76–78.
Graham, N. J., Caplan, H., and Cohen, M. (1972). *J. Electrochem. Soc.* **119**, 1265–1267.
Greenwood, N. N., and Howe, A. T. (1972a). *In* "Reactivity of Solids" (J. S. Anderson, M. W. Roberts, and F. S. Stone, eds.), pp. 240–251. Chapman & Hall, London.
Greenwood, N. N., and Howe, A. T. (1972b). *J.C.S. Dalton Trans.* 122–126.
Gruenwald, T. B., and Gordon, G. (1971). *J. Inorg. Nucl. Chem.* **33**, 1151–1155.
Hadari, Z., Kroupp, M., and Wolfson, Y. (1971). *J. Appl. Phys.* **42**, 534–535.
Hagel, W. C. (1965). *J. Am. Ceram. Soc.* **48**, 70–75.
Hagel, W. C. (1966). *Trans. AIME* **236**, 179–184.
Hallwig, D., Sockel, H. G., and Monty, C. (1977). *In* "Reactivity of Solids" (J. Wood *et al.*, eds.), pp. 631–634. Plenum, New York.
Haul, R., and Dümbgen, G. (1962). *Ber. Bunsenges. Phys. Chem.* **66**, 636–641.
Haul, R., and Dümbgen, G. (1965). *J. Phys. Chem. Solids* **26**, 1–10.
Haul, R., Just, D., and Dümbgen, G. (1962). *Z. Phys. Chem. (Frankfurt am Main)* **31**, 309–320.
Hawkins, R. J., and Alcock, C. B. (1968). *J. Nucl. Mater.* **26**, 112–122.
Hayakawa, M., Morinaga, M., and Cohen, J. B. (1974). *In* "Defects and Transport in Oxides" (A. S. Seltzer and R. I. Jaffe, eds.), pp. 177–203. Plenum, New York.
Hembree, P., and Wagner, J. B., Jr. (1969). *Trans. Metall. Soc. AIME* **245**, 1547–1552.
Henney, J., and Jones, J. W. S. (1968). *J. Mater. Sci.* **3**, 158–164.
Herchl, R., Khoi, N. N., Homma, T., and Smeltzer, W. W. (1972). *Oxid. Met.* **4**, 35–49.
Herring, C. (1950). *J. Appl. Phys.* **21**, 437–445.
Himmel, L., Mehl, R. F., and Birchenall, C. E. (1953). *Trans. AIME* **197**, 827–843.
Hirthe, W. M., and Brittain, J. O. (1963). *J. Am. Ceram. Soc.* **46**, 411–417.
Hobbs, L. W. (1976). *J. Phys. (Paris), Colloq.* **C7**, 3–26.
Hodge, J. D. (1978). *J. Electrochem. Soc.* **125**, 55C–57C.
Hodkin, E. N., and Nicholas, M. G. (1973). *J. Nucl. Mater.* **47**, 23–30.
Höh, A., and Matzke, Hj. (1973). *J. Nucl. Mater.* **48**, 157–164.
Holt, J. B. (1967). *Proc. Br. Ceram. Soc.* **11**, 157–170.
Holt, J. B., and Himmel, L. (1969). *J. Electrochem. Soc.* **116**, 1569–1580.
Hyde, B. G., and Bursill, L. A. (1970). *In* "The Chemistry of Extended Defects in Non-Metallic Solids" (L. Eyring and M. O'Keeffe, eds.), pp. 347–373. North-Holland Publ., Amsterdam.
Hyde, B. G., Bevan, D. J. M., and Eyring, L. (1966). *Proc. R. Soc. London* **259**, 583–614.
Iguchi, E., and Yajima, K. (1972a), *J. Phys. Soc. Jpn.* **32**, 1415–1421.
Iguchi, E., and Yayima, K. (1972b). *Trans. Jpn. Inst. Met.* **13**, 45–49.
Ikeda, Y., and Nii, K. (1976). *Trans. Jpn. Inst. Met.* **17**, 419–425.
Ikeda, Y., Nii, K., Béranger, G., and Lacombe, P. (1974). *Trans. Jpn. Inst. Met.* **15**, 441–446.
International Atomic Energy Agency (1967). Technical Reports Series, No. 79. IAEA, Vienna.
Izvekov, V. I. (1958). *Dokl. Akad. Nauk B. SSR* **1**, 64.
Izvekov, V. I., Gorbunov, N. S., and Babad-Zakhrapin, A. A. (1962). *Fiz. Met. Metalloved.* **14**, 195–199.
Johnson, D. L., and Cutler, I. B. (1963). *J. Am. Ceram. Soc.* **46**, 541–545, 545–556.
Johnson, O. W. (1964). *Phys. Rev.* **136**, A284–A290.
Kelly, R., and Jech, C. (1969). *J. Nucl. Mater.* **30**, 122–133.
Kelly, R., and Matzke, Hj. (1965). *J. Nucl. Mater.* **17**, 179–191.
Kelly, R., and Matzke, Hj. (1966). *J. Nucl. Mater.* **20**, 171–183.

King, A. D. (1970). Rep. AECL-3655. Chalk River, Ontario.
King, A. D. (1971). *J. Nucl. Mater.* **38**, 347–349.
King, A. D., and Moermann, J. (1974). *Phys. Status Solidi A* **22**, 455–463.
Kingery, W. D., and Berg, M. (1955). *J. Appl. Phys.* **26**, 1205–1212.
Kingery, W. D., Pappis, J., Daty, M. E., and Hill, D. C. (1959). *J. Am. Ceram. Soc.* **42**, 393–398.
Kingery, W. D., Hill, D. C., and Nelson, R. P. (1960). *J. Am. Ceram. Soc.* **43**, 473–476.
Kingsbury, P. I., Jr., Ohlensen, W. D., and Johnson, O. W. (1968). *Phys. Rev.* **175**, 1091–1101.
Kitazawa, K., Kuriyama, T., Fueki, K., and Mukaibo, T. (1977). *J. Am. Ceram. Soc.* **60**, 363–366.
Klotsman, S. M., Timofeev, A. N., and Trakhtenberg, I. S. (1960). *Phys. Met. Metallogr. (USSR)* **10**(5), 93–96.
Klotsman, S. M., Timofeev, A. N., and Traklitenberg, I. S. (1972). *Sov. Phys.—Solid State* **14**, 763–765.
Koch, F., and Cohen, J. B. (1969). *Acta Crystallogr., Sect. B* **25**, 275–287.
Koel, G. J., and Gallings, D. J. (1972). *Oxid. Met.* **5**, 185–203.
Kofstad, P. (1967). *J. Less-Common Met.* **13**, 635–638.
Kofstad, P. (1972). "Nonstoichiometry, Diffusion and Electrical Conductivity in Binary Metal Oxides." Wiley (Interscience), New York.
Kofstad, P., and Hed, A. Z. (1967). *J. Am. Ceram. Soc.* **50**, 681–682.
Kofstad, P., and Hed, A. Z. (1968). *J. Electrochem. Soc.* **115**, 102–104.
Kozlowska-Rog, A., Kozinski, S., and Rog, G. (1976). *Rocz. Chem.* **50**, 2153–2159.
Kröger, F. A. (1964). "The Chemistry of Imperfect Crystals," North-Holland, Amsterdam.
Kröger, F. A., and Vink, H. J. (1956). *Solid State Phys.* **3**, 310–435.
Lambert, R. A. (1979). Ph.D. Thesis, Univ. of Surrey. U.K.
Lambert, R. A., and Matzke, Hj. (1980). To be published.
Landler, P. F. J., and Komarek, K. L. (1966). *Trans. Metall. Soc. AIME* **236**, 138–149.
Laub, L. W., and Wagner, J. B., Jr. (1973). *Oxid. Met.* **7**, 1–22.
Lay, K. W. (1970). *J. Am. Ceram. Soc.* **53**, 369–373.
Lee, H. M., and Alcock, C. B. (1971). *Proc. Br. Ceram. Soc.* **19**, 85–93.
Leroy, B., Beranger, G., and Lacombe, P. (1970). *Ann. Chim. (Paris)* **5**, 275–282.
Leroy, B., Beranger, G., and Lacombe, P. (1973). *J. Phys. (Paris), Colloq.* **C9**, 309–316.
Leroy, B., Beranger, G., and Lacombe, P. (1976). *Colloq. Metall.* **2**, 843–864.
Lidiard, A. B. (1966). *J. Nucl. Mater.* **19**, 106–108.
Lidiard, A. B. (1974). *In* "Crystals with the Fluorite Structure" (W. Hayes, ed.), pp. 101–184. Oxford Univ. Press (Clarendon), London and New York.
Ligeon, E., Fontenille, G., Bontemps, A., and Contamin, P. (1970). Rep. CEA-R-4095. Grenoble.
Lindner, R. (1952). *Ark. Kemi* **4**, 381–384.
Lindner, R., and Åkerström, Å. (1957). *Discuss. Faraday Soc.* **23**, 133–136.
Lindner, R., and Schmitz, F. (1961). *Z. Naturforsch., Teil A* **16**, 1373–1378.
Lindner, R., Reimann, D., and Schmitz, F. (1967). *In* "Plutonium as a Reactor Fuel," pp. 265–272. IAEA, Vienna.
Lundy, T. S., and Coghlan, W. A. (1973). *J. Phys. (Paris), Colloq.* **C9**, 299–302.
Lundy, T. S., Padget, R. A., and Banus, M. D. (1973). *Metall. Trans.* **4**, 1179–1180.
Lutze-Birk, A., Wezranowski, E., and Radwan, M. (1967). *Arch. Elektrotech.* **16**, 807–828.
MacEwan, J. R. (1962). *J. Am. Ceram. Soc.* **45**, 37–41.
MacEwan, J. R., and Hayashi, J. (1967). *Proc. Br. Ceram. Soc.* **7**, 245–272.
Magnéli, A. (1970). *In* "The Chemistry of Extended Defects in Non-Metallic Solids" (L. Eyring and M. O'Keeffe, eds.), pp. 148–162. North-Holland Publ., Amsterdam.

Maiya, P. S. (1971a). *J. Nucl. Mater.* **40**, 57–65.
Maiya, P. S., (1971b). *Acta Met.* **19**, 255–262.
Maiya, P. S., Chen, W. K., and Peterson, N. L. (1970). *Metall. Trans.* **1**, 1801–1809.
Manes, L., and Manes-Pozzi, B. (1976). *In* "Plutonium 1975 and Other Actinides" (H. Blank and R. Lindner, eds.), pp. 145–163. North-Holland Publ., Amsterdam.
Manes, L., Sørensen, O. T., Mari, C., and Ray, I. (1979). *Int. Symp. Thermodyn. Nucl. Mater.*, IAEA, Jülich Pap. IAEA-Sm-236/6.
Manning, J. R. (1961). *Phys. Rev.* **124**, 470–482.
Marin, J. F. (1968). Rep. CEA-N-883.
Marin, J. F. (1970). *J. Nucl. Mater.* **34**, 348–350.
Marin, J. F., and Contamin, P. (1969). *J. Nucl. Mater.* **30**, 16–25.
Marin, J. F., and Michard, H. (1965). *C. R. Acad. Sci.* **261**, 693–696.
Marin, J. F., Michaud, H., and Contamin, P. (1967). *C. R. Acad. Sci., Ser. C* **264**, 1633–1636.
Marlowe, M. O., and Kaznoff, A. I. (1968). *J. Nucl. Mater.* **25**, 328–333.
Massiani, Y., Crousier, J. P., and Streiff, R. (1978). *J. Solid State Chem.* **24**, 85–95.
Matthews, J. R. (1974). *U.K. At. Energy Res. Establ., Memo.* **AERE-M 2643**. Her Majesty's Stationary Office.
Matzke, Hj. (1966a). *Nucl. Appl.* **2**, 131–137.
Matzke, Hj. (1966b). Rep. AECL-2585. Chalk River, Ontario.
Matzke, Hj. (1967a). *J. Mater. Sci.* **2**, 444–456.
Matzke, Hj. (1967b). *J. Nucl. Mater.* **21**, 190–198.
Matzke, Hj. (1968). *Can. J. Phys.* **46**, 621–634.
Matzke, Hj. (1969a). *J. Nucl. Mater.* **30**, 110–121.
Matzke, Hj. (1969b). *J. Appl. Phys.* **40**, 3819–3824.
Matzke, Hj. (1970). *J. Mater. Sci.* **5**, 831–836.
Matzke, Hj. (1973). *J. Phys. (Paris), Colloq.* **C9**, 317–325.
Matzke, Hj. (1975). *J. Nucl. Mater.* **57**, 180–186.
Matzke, Hj. (1976a). *J. Phys. (Paris), Colloq.* **C7**, 452–457.
Matzke, Hj. (1976b). *In* "Plutonium 1975 and Other Actinides" (H. Blank and R. Lindner, eds.), pp. 801–831. North-Holland Publ., Amsterdam.
Matzke, Hj. (1979a). *In* "Fission Gas Behavior in Nuclear Fuels" (C. Ronchi, Hj. Matzke, J. van der Laar, and H. Blank, eds.), European Applied Research Reports, Vol. 1, Euratom Report EUR 6600 EN, pp. 289–306, 337–349. Luxembourg.
Matzke, Hj. (1979b). Unpublished results.
Matzke, Hj., and Cole, S. (1980). To be published.
Matzke, Hj., and Lambert, R. A. (1973–1974). *J. Nucl. Mater.* **49**, 325–328.
Matzke, Hj., and Lambert, R. A. (1976). *J. Nucl. Mater.* **64**, 211–219.
Matzke, Hj., and Lambert, R. A. (1980). To be published.
Matzke, Hj., and Lindner, R. (1964). *Z. Naturforsch., Teil A* **19**, 1178–1180.
Matzke, Hj., and Ronchi, C. (1972). *Philos. Mag.* **26**, 1395–1407.
Matzke, Hj., and Whitton, J. L. (1966). *Can. J. Phys.* **44**, 995–1010.
Matzke, Hj., Davies, J. A., and Johansson, G. E. (1971). *Can J. Phys.* **49**, 2215–2226.
Möller, E., Nilsson, L., and Starfelt, N. (1967). *Nucl. Instrum. Methods* **50**, 270–276.
Monty, C., Dubois, C., Talon, R., and Berbezat, S. (1977). *Colloq. Metall. Diffus. Condensed Mediums, 19th* Rep. ISBN2-7272-0016-1, pp. 813–838.
Morgan, C. S., and Poteat, C. E. (1968). Rep. ORNL-4370. Oak Ridge Natl. Lab., Oak Ridge, Tennessee.
Morin, F. (1975). *Can. Met.* **14**, 105–109.
Mrowec, S., and Przybylski, K. (1977a). *Oxid. Met.* **11**, 383–403.
Mrowec, S., and Przybylski, K. (1977b). *Rev. Int. Hautes Temp. Refract.* **14**, 285–288.

Mrowec, S., and Przybylski, K. (1977c). *Bull. Acad. Pol. Sci., Ser. Sci. Chim.* **25**, 869–875.

Mrowec, S., and Przybylski, K. (1977d). *In* "Reactivity of Solids" (J. Wood *et al.*, eds.), pp. 177–182. Plenum, New York.

Mullins, W. W. (1959). *J. Appl. Phys.* **30**, 77–83.

Mullins, W. W. (1961). *Trans. Metall. Soc. AIME* **221**, 1039–1046.

Murch, G. E., (1975). *Philos. Mag.* **32**, 1129–1140.

Murch, G. E., and Thorn, R. J. (1978). *J. Nucl. Mater.* **71**, 219–226.

Murch, G. E., Bradhurst, D. H., and de Bruin, H. J. (1975). *Philos. Mag.* **32**, 1141–1150.

Nabarro, F. R. N. (1948). *Rep. Conf. Strength Solids, Univ. Bristol* pp. 75–90. Phys. Soc., London.

Nagels, P., van Lierde, W., De Batist, R., Denayer, M., De Jonghe, L., and Gevers, R. (1966). *In* "Thermodynamics of Nuclear Materials," Vol. II, pp. 311–326. IAEA, Vienna.

Nakayama, T., and Sasaki, T. (1963). *Bull. Chem. Soc. Jpn.* **36**, 569–574.

Nichols, F. A. (1966). *J. Appl. Phys.* **37**, 4599–4602.

Nichols, F. A. (1968). *J. Am. Ceram. Soc.* **51**, 468–469.

Nichols, F. A. (1979). *J. Nucl. Mater.* **84**, 1–25.

Nowick, A. S., Wang, D. Y., Park, D. S., and Griffith, J. (1979). *J. Electrochem. Soc.*

Oishi, Y. (1979). Personal communication.

Oishi, Y., and Kingery, W. D. (1960). *J. Chem. Phys.* **33**, 480–486.

O'Keeffe, M. (1970). *In* "The Chemistry of Extended Defects in Non-Metallic Solids" (L. Eyring and M. O'Keeffe, eds.), pp. 609–627. North-Holland Publ., Amsterdam.

O'Keeffe, M. (1971). *Proc. Br. Ceram. Soc.* **19**, 1–13.

O'Keeffe, M., and Moore, W. J. (1961). *J. Phys. Chem.* **65**, 1438–1439; corrections p. 2277.

Ollerhead, R. W., Almquist, E., and Kuehner, J. A. (1966). *J. Appl. Phys.* **37**, 2440–2444.

Onozuma, T., Koiwa, M., Ishikawa, J., Yamaguchi, S., and Hibarayashi, M. (1977). *Rad. Effects* **31**, 117–123.

Pauling, L. (1960). "The Nature of the Chemical Bond," 3rd ed. Cornell Univ. Press, Ithaca, New York.

Perkins, R. A., and Rapp, R. A. (1973). *Metall. Trans.* **4**, 193–205.

Price, J. B. (1968). Ph.D. Thesis, Northwestern Univ., Evanston, Illinois (Abstr. 69-6979, p. 141).

Price, J. B., and Wagner, J. B., Jr. (1966). *Z. Phys. Chem. (Frankfurt am Main)* **49**, 257–270.

Price, J. B., and Wagner, J. B., Jr. (1970). *J. Electrochem. Soc.* **117**, 242–247.

Price, P. B., and Bird, J. R. (1969). *Nucl. Instrum. Methods* **69**, 277–281.

Rahman, S. R., and Berard, M. F. (1977). *J. Am. Ceram. Soc.* **60**, 67–71.

Rand, M. H., and Markin, T. L. (1968). *In* "Thermodynamics of Nuclear Materials 1967," pp. 637–650. IAEA, Vienna.

Reed, T. B. (1970). *In* "The Chemistry of Extended Defects in Non-Metallic Solids" (L. Eyring and M. O'Keeffe, eds.), pp. 21–36. North-Holland Publ., Amsterdam.

Reimann, D. K., and Lundy, T. S. (1969). *J. Am. Ceram. Soc.* **52**, 511–512.

Reynolds, G. L. (1967). *J. Nucl. Mater.* **24**, 69–73.

Rhodes, W. H., and Carter, R. E. (1966). *J. Am. Ceram. Soc.* **49**, 244–249.

Rickert, H., and Weppner, W. (1974). *Z. Naturforsch., Teil A* **29**, 1849–1859.

Riemer, G., and Scherff, H. L. (1971). *J. Nucl. Mater.* **39**, 183–188.

Roberts, L. E. J., Wheeler, V. J., and Perrin, A. (1963). Unpublished results. Cited in Belle (1969).

Robertson, W. M. (1969). *J. Nucl. Mater.* **30**, 36–49.

Roth, W. L. (1960). *Acta Crystallogr.* **13**, 140–149.

Roth, W. L. (1970). *In* "The Chemistry of Extended Defects in Non-Metallic Solids" (L. Eyring and M. O'Keeffe, eds.), pp. 455–486. North-Holland Publ., Amsterdam.

Roulet, M., Jaccard, C., and Rudolf, F. (1977). *Thin Solid Films* **44**, 163–168.
Routbort, J. L., Javed, N. A., and Voglewede, J. C. (1972). *J. Nucl. Mater.* **44**, 247–259.
Routbort, J. L., Voglewede, J. C., and Wilkinson, D. S. (1979). *J. Nucl. Mater.* **80**, 348–355.
Ruh, R., and Garrett, H. J. (1967). *J. Am. Ceram. Soc.* **50**, 257–261.
Sari, C. (1975). Personal communication.
Sari, C. (1978). *J. Nucl. Mater.* **78**, 425–426.
Schmalzried, H. (1962). *Z. Phys. Chem. (Frankfurt am Main)* **31**, 184–197.
Schmitz, F., and Lindner, R. (1963). *Radiochim. Acta* **1**, 218–220.
Schmitz, F., and Lindner, R. (1965). *J. Nucl. Mater.* **17**, 259–269.
Schmitz, F., and Marajofsky, A. (1975). *In* "Thermodynamics of Nuclear Materials 1974," Vol. I, pp. 457–467. IAEA Vienna.
Schoonman, J. (1979). *In* "Fast Ion Transport in Solids" (Vashista *et al.*, eds.), pp. 631–636. Elsevier, Amsterdam.
Schoonman, J. (1980). *J. Solid State Ionics* **1**, 121–131.
Seltzer, M. S. (1971). *J. Am. Ceram. Soc.* **118**, 802–803.
Seltzer, M. S., Clauer, A. H., and Wilcox, B. A. (1972). *J. Nucl. Mater.* **44**, 43–56.
Sheasby, J. S., Smeltzer, W. W., and Jenkins, A. E. (1968). *J. Electrochem. Soc.* **115**, 338–343.
Shim, M. T., and Moore, W. J. (1957). *J. Chem. Phys.* **26**, 802–804.
Simpson, L. A., and Carter, R. E. (1966). *J. Am. Ceram. Soc.* **49**, 139–144.
Sockel, H. G., and Schmalzried, H. (1968). *Ber. Bunsenges. Phys. Chem.* **72**, 745–754.
Soullard, J. (1977). Rep. CEA-R-4882.
Steele, B. C. H., and Floyd, J. M. (1971). *Proc. Br. Ceram. Soc.* **19**, 55–76.
Steele, J. L., and McCartney, E. R. (1969). *Nature (London)* **222**, 79.
Stehle, H. (1963). *Ber. Dtsch. Keram. Ges.* **40**, 129–134.
Stewart, W. D., and Wagner, J. B., Jr. (1975). *J. Electrochem. Soc.* **122**, 570–572.
Sugisaki, M. (1979). *J. Nucl. Mater.* **79**, 338–344.
Suzuoka, T. (1961). *Trans. Jpn. Inst. Met.* **2**, 25, 176.
Suzuoka, T. (1964). *J. Phys. Soc. Jpn.* **19**, 839–851.
Suzuoka, T. (1965). *J. Phys. Soc. Jpn.* **20**, 1259–1270.
Taskinen, A., and Kullberg, H. (1979). *J. Nucl. Mater.* **83**, 333–334.
Taylor, T. N., and Ellis, W. P. (1978). *Surf. Sci.* **77**, 321–326.
Tharmalingam, K., and Lidiard, A. B. (1959). *Phil. Mag.* **4**, 899–906.
Theisen, R., and Vollath, D. (1967). *In* "Plutonium as a Reactor Fuel," pp. 253–264. IAEA Vienna.
Thompson, B. A. (1962). Ph.D. Thesis, Rensselaer Polytech. Inst., Troy, New York.
Venkatu, D. A., and Poteat, L. E. (1970). *Mater. Sci. Eng.* **5**, 258–262.
Venkatu, H. A. (1972). *J. Am. Ceram. Soc.* **55**, 60.
Vollath, D. (1971). *Proc. Reaktortagung, Ger. Atomforum* pp. 558–561.
Vollath, D. (1972). Rep. KfK-1621. Karlsruhe, Germany.
Vollath, D. (1973). Rep. KfK-1884. Karlsruhe, Germany.
Vollath, D. (1976). *In* "Plutonium 1975 and Other Actinides" (H. Blank and R. Lindner, eds.), pp. 843–850. North-Holland Publ., Amsterdam.
Volpe, M. L., and Reddy, J. (1970). *J. Chem. Phys.* **53**, 1117–1125.
Volpe, M. L., Peterson, N. L., and Reddy, J. (1971). *Phys. Rev. B* **3**, 1417–1421.
Whipple, R. T. P. (1954). *Philos. Mag.* **45**, 1225–1236.
Whitmore, D. H., and Kawai, T. (1962). *J. Am. Ceram. Soc.* **45**, 375–379.
Whitton, J. L., Mitchell, I. V., and Winterbon, K. B. (1971). *Can. J. Phys.* **49**, 1225–1232.
Willis, B. T. M. (1963). *Proc. Roy. Soc. (London)* **A274**, 133–144.
Willis, B. T. M. (1964a). *Proc. Br. Ceram. Soc.* **1**, 9–19.
Willis, B. T. M. (1964b). *J. Phys. (Paris)* **25**, 431–439.

Willis, B. T. M. (1965). *Acta Crystallogr.* **18**, 75–76.

Wimmer, J. M., Blumenthal, R. N., and Branski, L. (1975). *J. Phys. Chem. Solids* **36**, 269–275.

Wittke, J. P. (1966). *J. Electrochem. Soc.* **113**, 193–194.

Woodley, R. E., and Adamson, M. G. (1979). *J. Nucl. Mater.* **82**, 65–75.

Yagi, E., Koyama, A., Sakairi, H., and Hasiguti, R. R. (1976). *Inst. Phys. Conf. Ser.* No. 31, 485–490.

Yagi, E., Koyama, A., Sakairi, H., and Hasiguti, R. R. (1977). *J. Phys. Soc. Jpn.* **42**, 939–953.

Yajima, S., Furuya, H., and Hirai, T. (1966). *J. Nucl. Mater.* **20**, 162–170.

Yurek, G. J., and Schmalzried, H. (1975). *Ber. Bunsenges Phys. Chem.* **79**, 255–262.

Zador, S. (1969). Ph.D. Thesis, Univ. of London, London.

Zelinski, V. F., Kharkov, O. N., Kukalov, U. S., and Shakun, N. A. (1970). *Prot. Met. (USSR)* **6**, 235–238.

5

Mass Transport in Anion-Deficient Fluorite Oxides

J. A. Kilner and **B. C. H. Steele**
Wolfson Unit for Solid State Ionics
Department of Metallurgy and Materials Science
Imperial College
London, United Kingdom

I. Introduction

The technological applications of ceramic electrolytes based on zirconia solid solutions are well known, but confusion remains as to the fundamental aspects of the defect and transport properties of these highly defective fluorite oxides. An understanding of these properties is essential if

NONSTOICHIOMETRIC OXIDES

ISBN 0-12-655280-0

progress is to be made in the synthesis of alternative oxide materials with higher oxygen ion conductivities than those presently available. Fortunately experimental techniques have improved over the last decade, so that conductivity measurements on polycrystalline samples can be interpreted to provide reliable data on bulk behavior without the influence of grain boundary and electrode phenomena. We have analyzed these data and provided a consistent model for the values associated with oxygen ion migration energies that emphasizes the role of the energy of association between an oxygen ion vacancy and an aliovalent dopant cation in these electrolyte materials.

It is relevant to note that extensive experimental data are also available for the thermodynamic, structural, and transport properties of solid solutions of the fluorite oxide systems UO_2 and UO_2–PuO_2, arising from the technological interest in these materials shown by the nuclear power industry. Moreover a theoretical survey (Catlow, 1977) has suggested new interpretations of the transport data on UO_2 with revised values for defect formation and migration energies. This analysis suggests that the activation energy for anion vacancy migration in stoichiometric UO_2 has a value of about 0.25 eV. Although it is difficult to obtain oxygen diffusion data for hypostoichiometric anion-deficient UO_{2-x}, it is relevant to note that an experimental study (Chereau and Wadier, 1973) on anion vacancy migration in PuO_{2-x} also obtained low values, 0.2–0.3 eV, for the migration activation energies. A value of 0.25 eV was also recently selected (Murch and Thorn, 1978) for a theoretical study on oxygen diffusion in near-stoichiometric UO_2. Available evidence to date for UO_2 systems thus indicates a relatively low value (~0.25 eV) for the anion vacancy migration energy. It is unlikely that this low value also applies to oxygen vacancy migration in other anion-deficient fluorite solid solutions that do not exhibit the special features of UO_2 crystals, namely, highly polarizable U^{4+} cations and high electronic disorder (Chereau and Wadier, 1973). Indeed, some preliminary studies (Badwal and Bevan, 1979) on $U_{0.3}Y_{0.7}O_{2-x}$ suggest activation energies for oxygen migration of about 1 eV. These data could be accounted for by a much higher value for the oxygen vacancy migration energy (~0.6 eV) together with a major contribution from a defect association energy. Such a situation would be in accordance with the analysis of ionic conductivity measurements presented in Section VI.

II. Cation Mobility in Anion-Deficient Fluorite Oxides

Although this chapter is principally concerned with an analysis of oxygen ion transport data, a few comments about cation mobility are also relevant. It is well known that values of cation diffusion coefficients are much lower than those associated with anion diffusion coefficients, and at 1000°C, for example, the difference is at least six orders of magnitude (Etsell and Flengas, 1970). For many processes therefore it is the cation transport that is the rate-determining step. These processes include the rate of homogenization and structural rearrangement (Bevan and Summerville, 1978; Dell and Hooper, 1978); kinetics of sintering, grain growth, and creep (Matthews, 1974); and rate mechanisms involved in the formation of toughened partially stabilized zirconia (Garvie *et al.*, 1977).

Detailed structural investigations of ZrO_2-based solid solutions often reveal the presence of small, coherent microdomains of ordered phases (e.g., $CaZr_4O_9$), and it now appears that a broad spectrum of results can be obtained from a phase analysis experiment depending upon the degree of heterogeneity of the sample. Because cation mobility in the system is so low, this degree of heterogeneity is not changed significantly except by annealing at very high temperatures; complete equilibrium is attainable only under these conditions. Bevan and Sommerville (1978) have pointed out that, where the "characteristic temperatures" (to use a spectroscopic term) of the two sublattices are about 1000°C apart, low-temperature equilibrium is unattainable by conventional techniques. They introduce the phrase operational equilibrium to describe the state achieved after a low-temperature anneal when the anion sublattice adjusts to a very high-temperature cation distribution. This situation should be reproducible, otherwise, the defect structure is a function of the relevant heat treatment. It is not surprising therefore that differences in anionic conductivity behavior are recorded for solid solutions having the same nominal composition. It would be interesting to know how the activation energy for cation migration varies with composition in anion-deficient ($A_{1-y}B_yO_{2-x}$) solid oxide electrolytes, but relevant data are not available. Some information, however, is available on cation mobility in anion-deficient hypostoichiometric UO_2 solid solutions. Cation diffusion coefficients and the associated activation energies as a function of the oxygen-to-metal (O/M) ratio in UO_2–PuO_2 solid solutions (Matthews, 1974) are depicted in Figs. 1 and 2. It has been suggested by Catlow (1977) that the minima in Fig. 1 and the maxima in Fig. 2 do not correspond to a slightly anion-deficient crystal but to the stoichiometric point. This is quite possible, as it is always difficult to measure the exact stoichiometry of UO_2. If this proposal is accepted, then the cation diffusion mechanism changes from a vacancy to an interstitial process on passing from the stoichiometric to the hypostoichiometric region. With this new interpretation theoretical calculations (Catlow, 1977) are now generally compatible with available cation diffusion data on anion-excess and stoichiometric UO_2. However, only poor numerical agreement is obtained for the calculations for anion-deficient oxides. No theoretical model gives a calculated activation energy of less than 10 eV

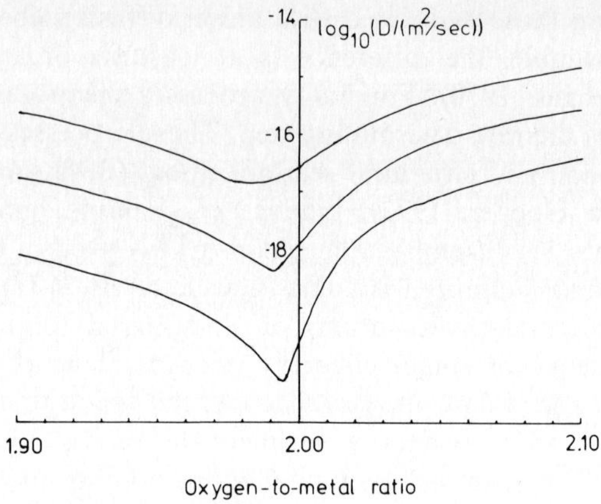

Fig. 1. Cation self-diffusion coefficients as a function of the O/M ratio in mixed UO_2–PuO_2 solid solutions (From Matthews, 1974.)

(cf. experimental values of about 4 eV). It should be noted that creep activation energies for anion-deficient $U_{1-y}Pu_yO_{2-x}$ samples are also indicative (Matthews, 1974) of cation migration energies in the range 3–5 eV. This discrepancy between theory and experiment could be due to a failure of the ionic model for cation interstitials or, alternatively, it could reflect the complex nature of the defect structure in anion-deficient crystals.

There appear to be no relevant data for grain growth and sintering kinetics in anion-deficient hypostoichiometric crystals, but it has been observed (Killeen, 1975) that small additions of La_2O_3 to UO_2 can significantly increase the rate of grain growth.

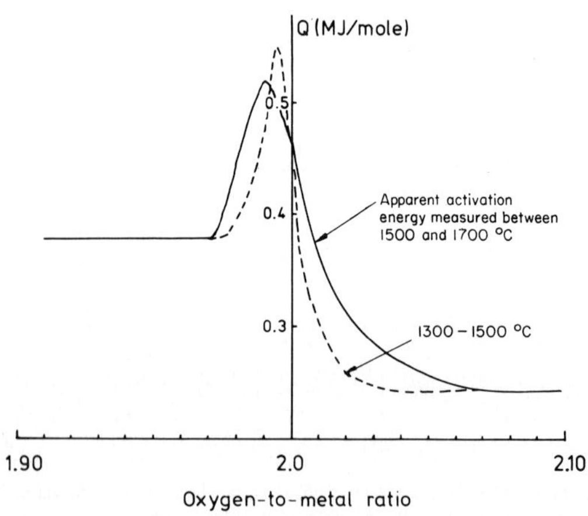

Fig. 2. Apparent activation energy for cation self-diffusion in mixed UO_2–PuO_2 solid solutions (From Matthews, 1974.)

A final comment concerns the relative mobilities of the constituent cations of zirconia-based solid oxide electrolytes. Many years ago Bray and Merten (1964) speculated that different cation mobilities could produce an unmixing of the solid solution electrolyte when it was subjected to high current densities for extended periods of time. The fact that such a process has not been observed in the intervening years during the many investigations on high-temperature fuel cells (Rohr, 1978) suggests that the likelihood of this phenomenon occurring is remote.

III. Anion Mobility in Anion-Deficient Oxides

In this section we shall describe the ionic conductivity of oxides in which the host lattice has been made oxygen-deficient by the introduction of lower-valent cations. For such oxides two charge-compensating defects can occur: oxygen vacancies or cation interstitials. However, for the oxides we shall describe, oxygen vacancies are the only charge-compensating defects. We shall also restrict our discussion to systems where oxygen vacancies are the majority charge carriers; mixed conductors are described elsewhere in this book.

Many publications can be found that deal with the large amount of data available on anionic conductivity in these oxides, for example, Dell and Hooper (1978) have reviewed the field. Takahaski (1972), Kvist (1972), and Steele (1972) have all contributed to the discussion of the many problems associated with the study of these oxides. The largest compilation of electrical data is the review by Etsell and Flengas (1970) who have collected an impressive amount of information from publications up to mid-1969. We do not intend to repeat the analyses undertaken in these works but more to emphasize recent data that have become available by the use of complex plane impedance techniques to measure the ionic conductivity of oxides.

In the following sections we shall describe the methods by which conductivity can be measured, with particular reference to ac methods used on ceramic samples. We shall discuss the effects of microstructure, temperature, ambient atmosphere, and dopant type upon the conductivity of these oxides and apply this analysis to recent experimental data.

A. Electrical Conductivity

The electrical conductivity of any solid can be expressed as

$$\sigma_{tot} = \sum_j \sigma_j \tag{1}$$

where σ_j are the partial conductivities of each of the charge carriers (in siemens per meter) j present in the solid. These charge carriers can be either electronic in nature (i.e., either electrons or electron holes) or atomic in nature and usually are defects of either the anionic or cationic sublattice. We can expand each of the partial conductivities as

$$\sigma_j = C_j Z_j e \mu_j \tag{2}$$

where C_j is the number of carriers of type j (per meter^3), Z_j the number of electronic charges e of these carriers (coulombs), and μ_j their mobility in the solid (meter^2 per siemen per volt). Substitution of Eq. (2) into Eq. (1) gives

$$\sigma_{\text{tot}} = \sum_j C_j Z_j e \mu_j \tag{3}$$

We can see from Eq. (2) that each partial conductivity is a product of the number of carriers multiplied by a mobility; consequently, if two carriers are present in equal concentrations, it will be their mobilities that determine which mode of conductivity will predominate. For oxides the mobilities of electronic defects are approximately 10^2–10^3 times greater than the mobilities of ionic defects, and therefore a relatively small concentration of electrons can have a large effect on the total conductivity.

When describing a solid in which there is more than one charge carrier, a useful concept is the transport number for each carrier. The transport number t_k for any carrier k is defined by

$$t_k = \sigma_k \Big/ \sum_j \sigma_j \tag{4}$$

This number clearly describes the proportion of the total conductivity due to this particular charge carrier. In cases where mixed electronic and ionic conductivities occur an ionic transport number is often defined as

$$t_{\text{ion}} = \sigma_{\text{ion}} / (\sigma_{\text{ion}} + \sigma_{\text{elec}}) \tag{5}$$

The effects of dopant and ambient atmosphere on t_{ion} and σ_{tot} will be given in Section V.A.

B. Experimental Methods for the Determination of Conductivity

Historically measurements of the conductivity of oxides have been made using dc methods on sintered ceramic samples. This can lead to errors because of effects due to the behavior of the grain boundaries and electrodes, which may mask the true behavior of the bulk. This uncertainty can be removed by use of the complex plane impedance analysis first applied to solid electrolytes by Bauerle (1969). In this method the complex impedance

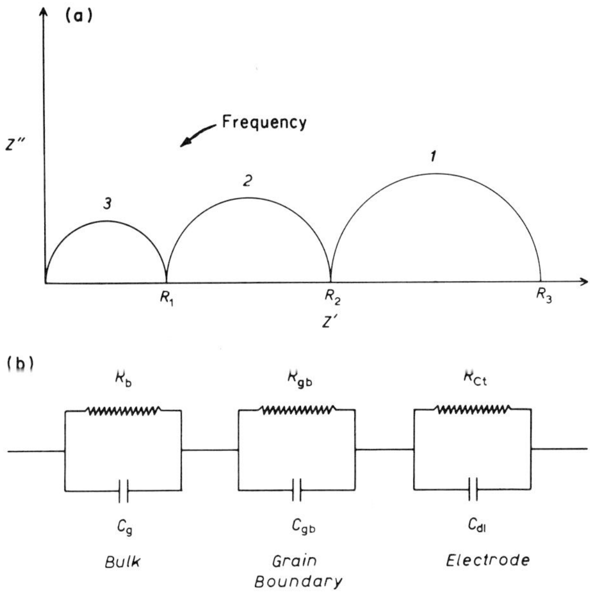

Fig. 3. (a) An idealized complex impedance plot for a ceramic oxide with partially blocking electrodes. (b) An equivalent circuit representing the ceramic specimen corresponding to 3(a).

Z or, equivalently, the complex admittance Y ($Z = 1/Y$) of a specimen is plotted on an Argand diagram for a large range of frequencies, usually from ~1 Hz to 10^5 Hz. An idealized plot for the spectrum of a ceramic oxide specimen with partially blocking electrodes is shown in Fig. 3a. The spectrum consists of three semicircles which may be tilted below the real axis by differing amounts. The number of semicircles seen depends upon the temperature range and frequency range used.

The analysis of this type of plot usually follows the original example of Bauerle (1969) who assigned an equivalent circuit to the specimen consisting of lumped RC elements. Each RC element gives rise to a semicircle with a particular time constant or relaxation time. The frequency corresponding to the crest of each of the semicircles n is given by

$$\omega_{\max}^{n} = (R_n C_n)^{-1} \tag{6}$$

The semicircles will be well-defined provided their relaxation times are well separated. The tilting of the relaxation semicircles is usually taken as an indication of the spread of the values of the elements contributing to a lumped circuit unit.

An equivalent circuit for a ceramic oxide corresponding to the complex impedance plot in Fig. 3b is shown in Fig. 3b. Schouler *et al.* (1973) have pointed out that these equivalent circuits are not unique, however, they all show many of the same general characteristics. In Fig. 3b the electrodes are taken to be partially blocking and are represented in the equivalent circuit by C_{dl}, the double-layer capacitance, and R_{ct}, the charge transfer resistance. The behavior of the grains of the ceramic is represented by a bulk resistance R_b in parallel with a geometric capacitance C_g. The grain boundaries can be a little more complicated because they represent a component both in parallel and in series with the grains. Normally the boundary resistance is much higher than that of the grains and only the series contribution need be considered. This resistance is shunted by a capacitance C_{gb}. We can therefore represent the electrical behavior of a ceramic sample in a very simple manner by the series of *RC* circuits shown in Fig. 3b.

The relative values of the *RC* units for each component determine the frequency range of the semicircle. For ceramic oxides the low-frequency semicircle is due to electrode dispersion, the intermediate frequency circle is due to the grain boundaries, and the high-frequency semicircle is due to the bulk behavior. The intercepts on the real axis of Fig. 3a, R_1, R_2, R_3, can be related to the equivalent circuit as

$$R_1 = R_b, \qquad R_2 = R_b + R_{gb}, \qquad R_3 = R_2 + R_{ct}$$

It is therefore theoretically possible, by using complex plane impedance analysis, to separate bulk and grain boundary resistances in a simple and elegant manner.

With reference to Fig. 3b we can see that using four-probe dc techniques, which eliminate electrode effects, the measured resistance is given by $R_{dc} = R_b + R_{gb}$. This can be quite different from R_b if the grain boundary resistance is high. Figure 4 shows this error quite strikingly for a sample of $Ce(Y)O_{2-x}$ where the dc four-point resistance is compared with resistances obtained by complex plane impedance measurements (Nowick *et al.*, 1979). The bulk behavior is quite different from that measured by a four-point dc method which is clearly dominated by a grain boundary contribution. Therefore any attempt to correlate this four-point dc behavior with the stoichiometry of the bulk can lead to error, especially in the lower temperature ranges.

As well as four-point dc methods, it is also quite common to find examples in the literature where ac conductivity measurements have been made at a single fixed frequency, usually on the order of 1–20 kHz, again to overcome the effects of electrodes. These measurements can also be in error, as has been stressed by Schouler *et al.* (1973, 1976) and Hohnke (1980). This error is caused by the fixed-frequency point moving around the semicircle as the resistances change with temperature; indeed, this point

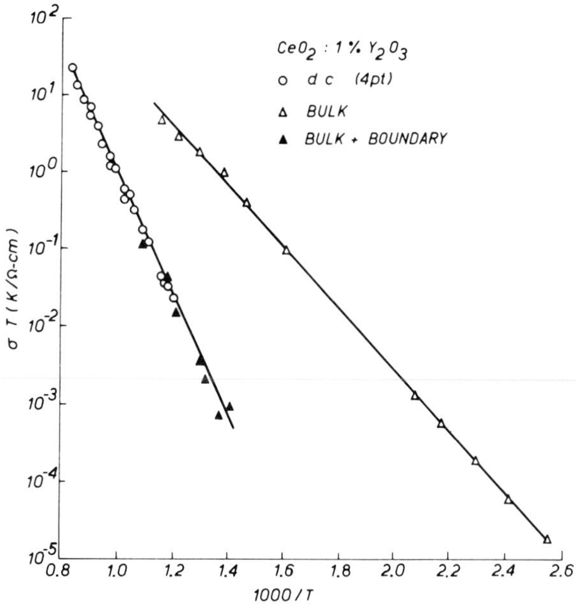

Fig. 4. The ionic conductivity of a CeO_2–1% Y_2O_3 sample showing the bulk, total, and four-point dc conductivities. (From Nowick *et al.*, 1979. Reproduced by permission of North-Holland Publ. Co., Amsterdam.)

may well change from one semicircle to another. Schouler *et al.* (1973) have plotted the difference between the 10-kHz point and the true bulk resistance of a $Zr(Y)O_{2-x}$ sample, and these data are shown in Fig. 5. The 10-kHz data show a break in slope that is not part of the true bulk behavior; furthermore these data can be considered only to approximate the true bulk behavior at high temperatures. Clearly great care must be

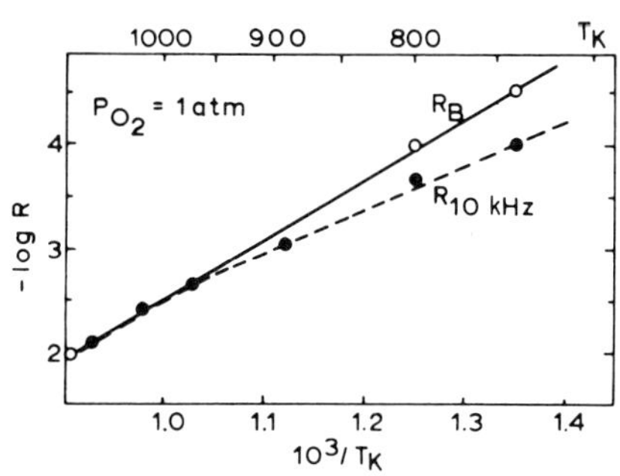

Fig. 5. The resistance of a $Zr(Y)O_{2-x}$ sample obtained using complex plane analysis compared with that obtained using a fixed frequency of 10 kHz. (From Schouler *et al.*, 1973.)

taken when interpreting the results of dc and fixed-frequency ac measurements made on ceramic samples.

IV. The Effects of Microstructure

The advent of complex plane analysis has allowed the study of the effects of microstructure on the total (four-point dc) and bulk conductivities. It is not within the scope of this chapter to review these phenomena fully; however, we shall mention the main points emerging from these analyses that have a bearing on previously published dc conductivities.

The conductivity of the boundaries is a complex function of many parameters, and there are differences between different oxides; however, several general points seem clear. The purity of the samples has a large effect on the boundary conductivity in zirconia-based electrolytes (Bauerle, 1969; Inozemtsev and Perfil'ev, 1975). This phenomenon has been attributed to the segregation of impurities, particularly Si and Ca, to the grain boundaries, leading to the formation of boundary phases (Schouler *et al.*, 1973; Bauerle, 1969; Beekmans and Heyne, 1976). High-purity samples show very low resistance, and in these cases the four-point dc resistance compares well with the bulk behavior (Bauerle, 1969).

The effects of porosity have been studied by Inozemtsev *et al.* (1974) who observed the grain boundary resistance for various sintered compacts. Figure 6 shows the effect of porosity on the grain boundary resistance for fine-grained $Zr(Sc)O_{2-x}$ samples, and clearly the porosity has a marked effect upon the magnitude of the boundary resistance in this case. The bulk resistance of these samples (where it could be measured) was found to vary very little, therefore for samples with any appreciable porosity it is clear that this boundary resistance can dominate the total dc resistance of the sample.

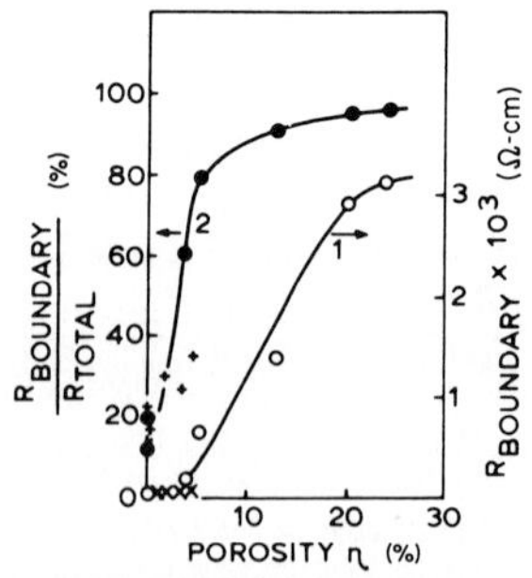

Fig. 6. The variation in the boundary contribution to the total conductivity as a function of porosity. (From Inozemtsev *et al.*, 1974. Translation copyrighted by Plenum Publishing Corp.; reprinted with permission.)

Grain size also affects the boundary resistance, as shown by Inozemtsev *et al.* (1974), Chu and Seitz (1978), and Ioffe *et al.* (1975) for $Zr(M)O_{2-x}$ samples. The dependence on grain size is small compared with the effect of porosity and is noticeable only for small-grained samples. The boundary resistance increases when the grain size decreases, as would be expected.

Two investigations (Schouler *et al.*, 1973; Inozemtsev *et al.*, 1976) have shown the presence of aging phenomena involving grain boundaries. Schouler *et al.* (1973) attribute this aging to the dissolution of impurities at the grain boundaries. Both investigations report the presence of two aging phenomema of different types, one concerned with grain boundaries and the other with the bulk material. The magnitude of the changes and the contribution to the total resistance is much higher in the case of grain boundary aging. The presence of these two phenomena may account for the confusion over aging phenomena found in the literature, especially since the two aging phenomena follow the different time dependences

$$\sigma_{gb} = f(\sqrt{t}) \qquad \text{and} \qquad \sigma_b = f[\exp(-kt)]$$

where σ_b and σ_{gb} represent the bulk and grain boundary conductivities, k is a constant, and t is time (Schouler *et al.*, 1973).

Chu and Seitz (1978) and El Adham (1978) have reported that grain boundary conductivity is also a function of dopant concentration. El Adham (1978) has correlated this fact with the distribution of dopant between the grain boundaries and the bulk found by using microprobe analysis on $Ce(Ca)O_{2-x}$ samples.

Clearly the boundary resistance is a complex function of many microstructural parameters, and consequently this contribution to the dc conductivity is very difficult to assess. This in turn makes the interpretation of dc and fixed-frequency ac data somewhat uncertain. However, there appear to be some conditions under which the boundary resistance is negligible:

1. As the activation energy for the grain boundary conductivity is either equal to or higher than that for the bulk conductivity, measurements at high temperatures ($T \gtrsim 1000 \pm 200$ K) will tend to reflect the bulk behavior.
2. At low temperatures only measurements on single crystals or large-grained, high-purity, fully dense ceramics are unambiguous.

V. Defect Chemistry and Conductivity

In this section we shall derive the equations that describe the effect of the atmosphere and the dopant on the electronic and ionic charge carrier concentrations. The notation used in the following equations is due to Kröger and Vink (1956). For a full explanation of the type of analysis employed here see, for example, Brook (1974).

A. The Effect of the Ambient Atmosphere on the Conductivity

Let us consider the oxide of the cation M^{m+}, i.e., M_2O_m. We shall assume that the introduction of the lower-valent cation $Mf^{(m-n)+}$ (i.e., the effective charge of the dopant is n') into the oxide will cause the formation of oxygen vacancies as charge-compensating defects. In the absence of any defect reactions or electronic defects the relationship between the total concentration of vacancies and the dopant is simply given by the condition that the crystal as a whole be electrically neutral, i.e.,

$$n[Mf_M^{n'}] = 2[V_O^{\cdot\cdot}](m/2) \tag{7}$$

This is slightly modified when electronic defects are introduced, caused by equilibration of the lattice with the ambient oxygen. This equilibrium can be expressed in the following equations.

For high partial pressures of oxygen, where oxygen is being incorporated into the lattice, we can write

$$\tfrac{1}{2}O_2(g) + V_O^{\cdot\cdot} \rightarrow O_O^x + 2h^{\cdot} \tag{8}$$

where $h^{\cdot}$ represents an electron hole. By applying the law of mass action to Eq. (8) we obtain

$$\frac{[O_O^x][h^{\cdot}]^2}{[V_O^{\cdot\cdot}]P_{O_2}^{1/2}} = K_1(T) \tag{9}$$

Rewriting $[h^{\cdot}]$ as p and noting that approximately $[O_O^x] \simeq 1$,

$$p^2 = [V_O^{\cdot\cdot}]P_{O_2}^{1/2}K_1(T) \tag{10}$$

The condition for electroneutrality gives

$$2[V_O^{\cdot\cdot}](m/2) + p = n[Mf_M^{n'}] \tag{11}$$

however, $p \ll [V_O^{\cdot\cdot}]$, therefore $[V_O^{\cdot\cdot}] = (n/m)[Mf_M^{n'}]$, and Eq. (10) becomes

$$p = \{K_1(T)(n/m)[Mf_M^{n'}]\}^{1/2}P_{O_2}^{1/4} \tag{12}$$

Thus at high oxygen pressures the concentration of electron holes is proportional to $P_{O_2}^{1/4}$.

At low oxygen pressures we can write a similar equation describing the loss of oxygen from the lattice:

$$O_O^x \rightarrow V_O^{\cdot\cdot} + 2e' + \tfrac{1}{2}O_2(g) \tag{13}$$

where e′ represents an electron. The above analysis can be repeated to yield an equation that describes the electron concentration (n_e) as a function of oxygen partial pressure:

$$n_e = \left\{ \frac{K_2(T)}{(n/m)[\mathrm{Mf}_M^{n'}]} \right\}^{1/2} P_{O_2}^{-1/4} \tag{14}$$

$K_1(T)$ and $K_2(T)$ are related through the requirement that

$$n_e p = K_i(T) \tag{15}$$

where K_i is the equilibrium constant for intrinsic electron–hole pair formation.

We can now plot the concentration of the various defects in the oxide by the use of a Brouwer diagram (Brouwer, 1954). In this type of diagram the logarithmic defect concentrations are plotted as a function of oxygen partial pressure. This variation is shown in Fig. 7a for the case under consideration; note that because the concentrations of the electronic de-

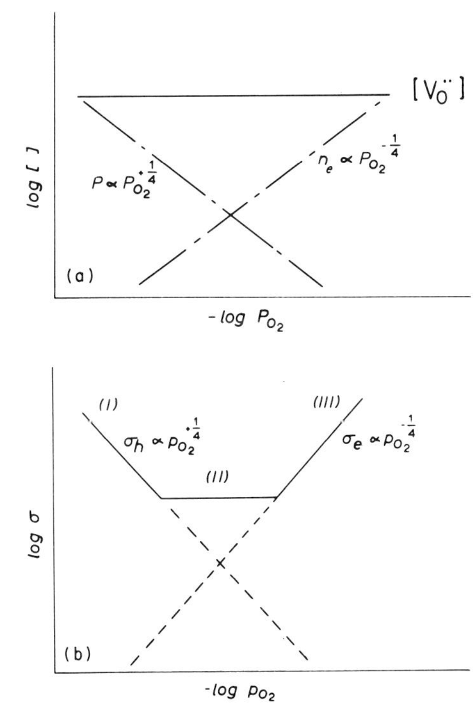

Fig. 7. (a) The variation in the electronic defect concentrations with oxygen partial pressure. (b) The variation in the electric conductivity of the sample in (a) with oxygen partial pressure.

fects are very small the oxygen vacancy concentration can be considered a constant fixed by the dopant level.

The electrical conductivity of this oxide can be simply obtained by multiplying the defect concentrations by the appropriate mobilities (remembering that $\mu_e, \mu_h > \mu_{V_O^{\cdot\cdot}}$) and summing the relative contributions. Figure 7b shows the result of this manipulation for the electronic ionic and total conductivities. In the total conductivity three distinct ranges are visible:

I. at high P_{O_2} where the conductivity is due to electron holes and varies as $P_{O_2}^{1/4}$;
II. where the conductivity is due to the oxygen vacancies generated by doping;
III. at low partial pressures of oxygen where the conductivity is due to electrons and varies as $P_{O_2}^{-1/4}$.

The central range is often described in the literature as the ionic plateau or ionic domain, and its extent is dependent upon temperature and dopant concentration. Because n_e and p both increase with temperature, the ionic domain becomes smaller or may even vanish at high temperatures.

B. The Effect of Temperature on Ionic Conductivity

Empirically it has been found that the temperature dependence of the conductivity can be expressed by an equation of the form

$$\sigma T = A \exp(-E_a/kT) \tag{16}$$

where A is a constant, E_a the activation energy, and k the Boltzmann constant.

When $\log(\sigma T)$ is plotted against $1/T$, a single straight line is usually observed with slope $-E_a/k$ and an intercept on the $\log(\sigma T)$ axis of A, although it is quite common to find two straight-line portions of different slope reported in the literature.

In this subsection we shall derive expressions for the temperature dependence of the ionic conductivity and examine the significance of both A and E_a for the different possible defect situations.

We recall Eq. (2):

$$\sigma_j = C_j e Z_j \mu_j$$

For the migration of oxygen vacancies we can expand the mobility

$$\mu = \gamma(Ze/kT)a_0^2\nu_0\exp(-\Delta G_m/kT) \tag{17}$$

where γ is a geometric factor, ν_0 the jump attempt frequency, ΔG_m the

Gibbs free energy for the jump, e, k, and T all have their normal meanings, a_0 is the jump distance, and $Z = 2$.

Now $\Delta G_m = \Delta H_m - T\Delta S_m$, thus Eq. (17) can be expanded to give

$$\mu = (B/T)\exp\left(-\frac{\Delta H_m}{kT}\right)\exp\left(\frac{\Delta S_m}{k}\right) \quad \text{or} \quad \mu = (B'/T)\exp\left(-\frac{\Delta H_m}{kT}\right) \tag{18}$$

where the constant B' contains the temperature-independent terms, assuming of course ΔS_m to be temperature-independent. Substituting Eq. (18) into Eq. (2) we obtain

$$\sigma = (B'/T)CeZ\exp(-\Delta H_m/kT) \tag{19}$$

Now the number of charge carriers is simply given by

$$C = [V_O^{\cdot\cdot}]\{1 - [V_O^{\cdot\cdot}]\}N_0$$

which for small values of $[V_O^{\cdot\cdot}]$ approximates

$$C = [V_O^{\cdot\cdot}]N_0 \tag{20}$$

where N_0 is the number of anion sites per unit volume

$$\sigma = (B'/T)[V_O^{\cdot\cdot}]N_0eZ\exp(-\Delta H_m/kT) \tag{21}$$

or

$$\sigma T = B''[V_O^{\cdot\cdot}]\exp(-\Delta H_m/kT) \tag{22}$$

where

$$B'' = N_0eZB' \tag{23}$$

C. *The Effect of Dopants*

In order to further expand Eq. (22) we must now define the free oxygen vacancy concentration for each of the possible defect situations. We shall not consider the case of the generation of charge carriers by intrinsic atomic disorder because, in refractory oxides, defect formation energies are high and consequently significant carrier concentrations can occur only at very high temperatures.

We shall again consider the oxide M_2O_m doped with the lower-valent cation $Mf^{(m-n)+}$.

1. *Associates*

We shall first assume that the vacancies are not free but are bound to dopant cations to form defect associates. This binding energy is mainly due to the coulombic attraction of the defects caused by their effective charges in the lattice; however, it also contains terms due to relaxation of the lattice around the defect, which depend upon the effective charge, the size of the dopant, and the cation polarizability.

Experimental evidence for the existence of these defect associates for oxides that adopt the fluorite structure has been compiled by Nowick and Park (1976) and Nowick *et al.* (1979). They have shown, by examining the evidence from dielectric and anelastic relaxation phenomena that in doped fluorites, at low temperatures ($T < 1200$ K) and low dopant concentrations, all the oxygen vacancies form complexes with dopant cations. Nowick *et al.* (1979) differentiate between the associates formed for the most widely used dopants in fluorite oxides, i.e., Mf^{3+} (e.g., Y^{3+}, Sc^{3+}) and Mf^{2+} (e.g., Ca^{2+}, Sr^{2+}). As the host cation in fluorite oxides is tetravalent, this corresponds, in our notation, to $n = 2$ for divalent dopants and $n = 1$ for trivalent dopants.

For the case $n = 2$, only one simple defect associate is possible, the neutral associate

$$\{Mf_M'' V_O^{\cdot\cdot}\}^x \rightarrow Mf_M'' + V_O^{\cdot\cdot} \tag{24}$$

This means that when no dissociation occurs all the dopant cations are bound to oxygen vacancies. The evidence supporting this assumption has been provided by Nowick and Park (1976) by an analysis of the dielectric relaxation in the oxide $Ce(Ca)O_{2-x}$. They found that the range of this simple behavior extended from $x = 0$ to $x \sim 0.04$ (i.e., 2% anion sites vacant). In materials with a higher dopant concentration extra interactions occur.

For the case $n = 1$ two defect associates are possible: a charged associate

$$\{Mf_M' V_O^{\cdot\cdot}\}^{\cdot} \rightarrow Mf_M' + V_O^{\cdot\cdot} \tag{25}$$

or a neutral associate

$$\{Mf_M' V_O^{\cdot\cdot} Mf_M'\}^x \rightarrow 2Mf_M' + V_O^{\cdot\cdot} \tag{26}$$

However, of the two possibilities it seems that the charged defect would be much more favorable for the following reasons.

At the high temperatures required to fabricate these oxides, the cations are mobile in the lattice, and consequently the distribution of the dopant among the host lattice sites can be regarded as being essentially random. This randomness is frozen in as the sample is cooled to room temperature. Clearly this does not favor formation of the neutral defect, as this requires the two dopant cations to be situated in adjacent anion coordination polyhedra. At low concentrations this must be discounted, at least below temperatures of 1200 K, where cation mobilities are low. Thus at low temperatures and low concentrations only the charged associate forms. This means that only half the dopant cations will be complexed with anion vacancies and the other half will be free.

Evidence supporting this assumption comes from ESR measurements of the trivalent dopants in fluorite oxides (Abraham *et al.*, 1966), which have shown a large fraction of dopant ions to have cubic environments, i.e., no anion vacancies in the first nearest-neighbor shell.

Using these two distinct defects we can now determine the expected variation in oxygen vacancy concentration with temperature.

a. Case 1: n = 2, e.g., Ca^{2+} *in* CeO_2 *or* ZrO_2. We can write the equation for the dissociation of these defects (Eq. (24):

$$\{Mf_M''V_O^{\cdot\cdot}\}^x \rightarrow Mf_M'' + V_O^{\cdot\cdot}$$

Applying the law of mass action to this equation gives

$$[Mf_M''][V_O^{\cdot\cdot}]/[\{Mf_M''V_O^{\cdot\cdot}\}^x] = K_{A2}(T) \tag{27}$$

Electroneutrality in the crystal dictates that

$$2[V_O^{\cdot\cdot}] = 2[Mf_M''](2/m) \tag{28}$$

Substituting Eq. (28) in Eq. (27) gives

$$[V_O^{\cdot\cdot}]^2 = (2/m)[\{Mf_M''V_O^{\cdot\cdot}\}^x]K_{A2}(T) \tag{29}$$

For full association of defects,

$$[\{Mf_M''V_O^{\cdot\cdot}\}^x] \gg [V_O^{\cdot\cdot}] \qquad \therefore [\{Mf_M''V_O^{\cdot\cdot}\}^x] \cong C_M \tag{30}$$

where C_M is the total dopant concentration expressed as a site fraction of the cation lattice.

The equilibrium constant $K_{A2}(T)$ can be expanded as

$$K_{A2}(T) = (1/W)\exp(-\Delta H_{A2}/kT) \tag{31}$$

where W is the number of orientations of the associate and ΔH_{A2} the enthalpy of association. Substituting Eqs. (31) and (30) in Eq. (29) we obtain

$$[V_O^{\cdot\cdot}]^2 = C_M(2/m)(1/W)\exp(-\Delta H_{A2}/kT) \tag{32}$$

$$\therefore [V_O^{\cdot\cdot}] = (2C_M/mW)^{1/2}\exp(-\Delta H_{A2}/2kT) \tag{33}$$

Substitution of Eq. (33) into Eq. (22) gives

$$\sigma T = B''\left(\frac{2C_M}{mW}\right)^{1/2}\exp\left[-\frac{1}{kT}\left(\Delta H_m + \frac{\Delta H_{A2}}{2}\right)\right] \tag{34}$$

By comparison with Eq. (16)

$$A \equiv B''\{2C_M/mW\}^{1/2}, \qquad E_a \equiv \Delta H_m + \Delta H_{A2}/2 \tag{35}$$

b. Case 2, n = 1, e.g., Y^{3+} *in* CeO_2, ThO_2, ZrO_2. We recall Eq. (25):

$$\{Mf_M'V_O^{\cdot\cdot}\}^{\cdot} \rightarrow Mf_M' + V_O^{\cdot\cdot} \tag{36}$$

By applying the law of mass action we obtain

$$[\mathrm{Mf}'_{\mathrm{M}}][\mathrm{V}_{\mathrm{O}}^{\cdot\cdot}]/[\{\mathrm{Mf}'_{\mathrm{M}}\mathrm{V}_{\mathrm{O}}^{\cdot\cdot}\}^{\cdot}] = K_{\mathrm{A1}}(T) \tag{37}$$

Here the condition for electroneutrality gives

$$[\mathrm{Mf}'_{\mathrm{M}}] = 2(m/2)[\mathrm{V}_{\mathrm{O}}^{\cdot\cdot}] + [\{\mathrm{Mf}'_{\mathrm{M}}\mathrm{V}_{\mathrm{O}}^{\cdot\cdot}\}^{\cdot}] \tag{38}$$

Again we can assume full defect association:

$$[\{\mathrm{Mf}'_{\mathrm{M}}\mathrm{V}_{\mathrm{O}}^{\cdot\cdot}\}^{\cdot}] \gg [\mathrm{V}_{\mathrm{O}}^{\cdot\cdot}], \qquad \therefore [\mathrm{Mf}'_{\mathrm{M}}] = [\{\mathrm{Mf}'_{\mathrm{M}}\mathrm{V}_{\mathrm{O}}^{\cdot\cdot}\}^{\cdot}] \tag{39}$$

Substituting Eq. (39) into Eq. (37)

$$[\mathrm{V}_{\mathrm{O}}^{\cdot\cdot}] = K_{\mathrm{A1}}(T) \tag{40}$$

Again K_{A1} can be expanded to give

$$[\mathrm{V}_{\mathrm{O}}^{\cdot\cdot}] = (1/W)\exp(-\Delta H_{\mathrm{A1}}/kT) \tag{41}$$

Substitution into Eq. (16) leads to

$$\sigma T = (B''/W)\exp[-(1/kT)(\Delta H_{\mathrm{m}} + \Delta H_{\mathrm{A1}})] \tag{42}$$

By comparison with Eq. (16) we obtain

$$A \equiv B''/W, \qquad E_{\mathrm{a}} \equiv \Delta H_{\mathrm{m}} + \Delta H_{\mathrm{A1}}$$

c. Case 3: Free Vacancies. Here we shall consider all the oxygen vacancies to be free, i.e., not associated with dopant cations, a situation most likely to occur at high temperatures. In this case the condition of electroneutrality determines that

$$2(m/2)[\mathrm{V}_{\mathrm{O}}^{\cdot\cdot}] = n[\mathrm{Mf}^{n'}_{\mathrm{M}}] \tag{43}$$

Substituting Eq. (43) into Eq. (22) we obtain

$$\sigma T = B''(n/m)[\mathrm{Mf}^{n'}_{\mathrm{M}}]\exp(-\Delta H_{\mathrm{m}}/kT) \tag{44}$$

Comparing this with the empirical relation for the conductivity Eq. (16),

$$A \equiv B''(n/m)C_{\mathrm{M}}, \qquad E_{\mathrm{a}} = \Delta H_{\mathrm{m}} \tag{45}$$

2. *Summary and Discussion*

The results of the three discussed cases are summarized in Table I. We can see that the preexponential factors show a distinct concentration dependence. This concentration dependence will be reflected in the isothermal conductivity, provided E_{a} remains constant. In this case we can write

$$(\sigma T)_{T=\mathrm{const}} \propto A \propto [\mathrm{Mf}^{n'}_{\mathrm{M}}]^{r} \tag{46}$$

where $r = 0$ for $n = 1$ at low temperatures and concentrations, $r = \frac{1}{2}$ for $n = 2$ at low temperatures and concentrations, and $r = 1$ for any n at high temperatures.

TABLE I

The Form of the Preexponential Factor and Activation Energy for Defect Cases 1–3

Case	Preexponential factor, A	Activation energy, E_a
1	$B''(2C_M/mW)^{1/2}$	$\Delta H_m + \Delta H_{A2}/2$
2	$B''(1/W)$	$\Delta H_m + \Delta H_{A1}$
3	$B''(n/m)C_M$	ΔH_m

One further interesting point to emerge from this analysis concerns the value of the activation energy for the two associate ranges. The contribution from the associate binding energy is different in the two cases, being only half the binding energy for the case $n = 2$. It is interesting to develop this a little further by assuming that this binding energy is totally due to the coulombic forces between the charged defects and neglecting relaxation effects.

We can estimate the binding energy by using the relationship

$$\Delta H_A \cong Q_1 Q_2/\epsilon r \tag{47}$$

where Q_1 and Q_2 are the two effective charges of the defects, ϵ the dielectric constant of the material, and r the equilibrium separation of the two components of the associate. For $n = 2$ we have

$$\Delta H_{A2} \sim 4e^2/\epsilon r \tag{48}$$

where e represents the electronic charge. For $n = 1$,

$$\Delta H_{A1} \sim 2e^2/\epsilon r \tag{49}$$

thus to a first approximation $\Delta H_{A2} \sim 2\Delta H_{A1}$.

This means for both $n = 1$ and $n = 2$ that the value of E_a is approximately $\Delta H_m + \Delta H_{A1}$, a rather surprising result.

Nowick and Park (1976) have calculated the temperature at which the vacancy population can essentially be regarded as being free, i.e., the limit of the associate behavior. They did this for a fluorite oxide doped with CaO, using as a break point between the two ranges the temperature at which half the total vacancy population is free. The results of this calculation are shown in Fig. 8. The break point temperature is strongly dependent upon the value of ΔH_{A2} (denoted E_A in Fig. 8) and the concentration of defects. As an estimate of this temperature we can say that ΔH_{A2} will be on the order of 0.3–0.5 eV, which with reference to Fig. 8 means that the break point is expected to occur at a temperature of $\sim$800 K or greater.

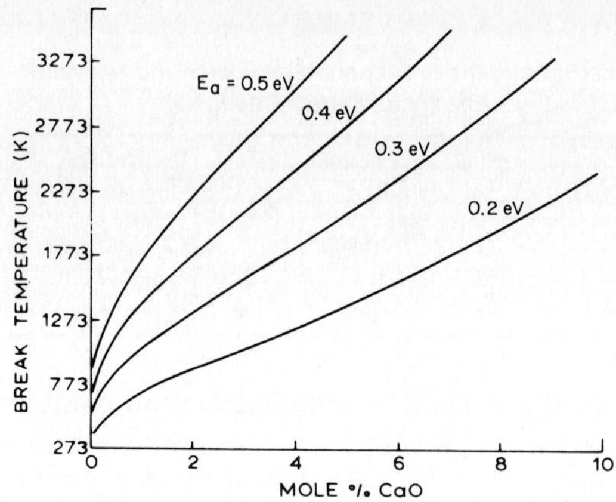

Fig. 8. The calculated temperature for the break between associated and free vacancy behavior for a fluorite oxide as a function of dopant concentration and association energy. (From Nowick and Park, 1976.)

The analysis presented in this section is valid only for dilute concentrations of dopant. In solids where the deviation from stoichiometry is large this analysis will break down. In the following section we shall review the published data for both the dilute and concentrated ranges, with particular reference to fluorite oxides.

VI. Fluorite Oxides

Fluorite oxides can show extensive nonstoichiometry of both of the types mentioned in Chapter 1. The crystal structure is shown in Fig. 9 for a half unit cell showing the position of a dopant cation and an anion vacancy in the cubic lattice. This structure is adopted by the binary oxides CeO_2 and ThO_2 at room temperature, but ZrO_2 exists in this structure only at high temperatures. ZrO_2 and HfO_2 can be stabilized in this cubic phase by the addition of divalent or trivalent dopants. The phase diagram for these pseudobinaries can be rather complex, especially at low dopant contents. This can lead to difficulties in the interpretation applied to conductivity data for these regions. On the other hand both CeO_2 and ThO_2 form solid solutions with many divalent and trivalent cations which have the fluorite structure from zero to quite large dopant concentrations. Thus these oxides can be used to study both dilute and concentrated regions.

The conductivity of these fluorite structure oxides is typified by a peak in the conductivity-versus-dopant concentration plot as shown in Fig. 10. In order to analyze this type of data we shall divide the discussion into two

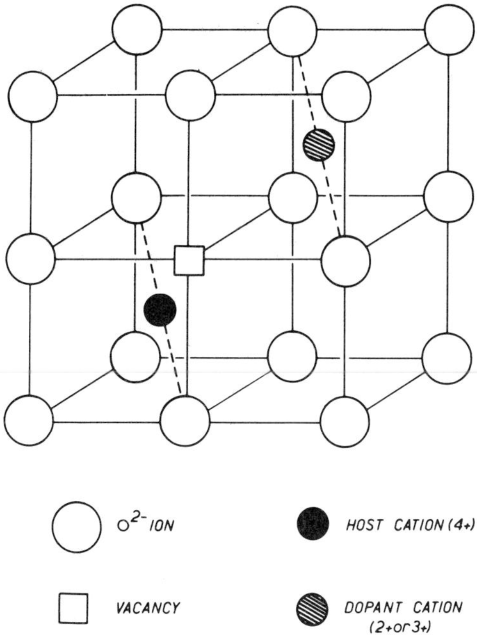

Fig. 9. A half unit cell of the fluorite structure showing the position of a dopant cation–oxygen vacancy associate.

parts: a discussion of data for the dilute region below the maximum and a discussion of data for the concentrated range at the maximum and above. We shall define this dilute solution to be in the range $0 < x < 0.08$ and the concentrated range as $x > 0.08$ following an analysis by Hohnke (1979).

A. The Dilute Solution Range

Any discussion of the conductivity of these oxides can be classified in one of two ways: by dopant type (i.e., the value of n) or by host lattice type. As we expect the behavior of these oxides to be different for each type of dopant, we have divided the discussion into two parts corresponding to dopant type.

1. Dopants for Which $n = 2$

(a) In the case of doped CeO_2 two investigations have been performed to determine the composition dependence of the conductivity. Blumenthal *et al.* (1973) have studied the high-temperature dc electric conductivity of $Ce(Ca)O_{2-x}$ with x varying from 0.001 to 0.16 in the temperature range

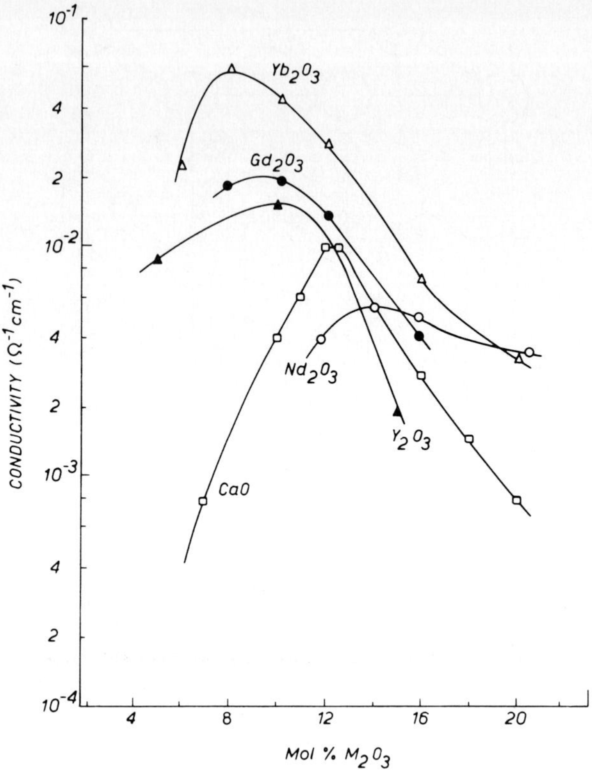

Fig. 10. The variation in the conductivity with concentration for various zirconia solid solutions showing the maximum in the conductivity.

973–1773 K. X-ray and pycnometric density measurements confirmed that oxygen vacancies were the charge-compensating defects. Two domains of conductivity were clearly established: At high oxygen pressures and low temperatures ionic conductivity predominated, whereas at low oxygen pressures and high temperatures electronic conductivity predominated.

Two points emerge from an analysis of the ionic conductivity: (1) For the composition range 1.0–14 mole% CaO ($x = 0.01$–0.14) the activation energy for the samples is constant at 0.62 ± 0.03 eV. (2) The isothermal conductivity in the range $x = 0.01$–0.08 shows a linear variation with composition. At higher compositions a maximum is seen in the isothermal conductivity (Fig. 11).

A very similar type of behavior has been found for $Ce(Sr)O_{2-x}$ solid solutions by Blumenthal and Garnier (1976). Figure 12 shows the variation in the isothermal ionic conductivity found with high-temperature dc measurements. Again the value of the activation energy was found to be

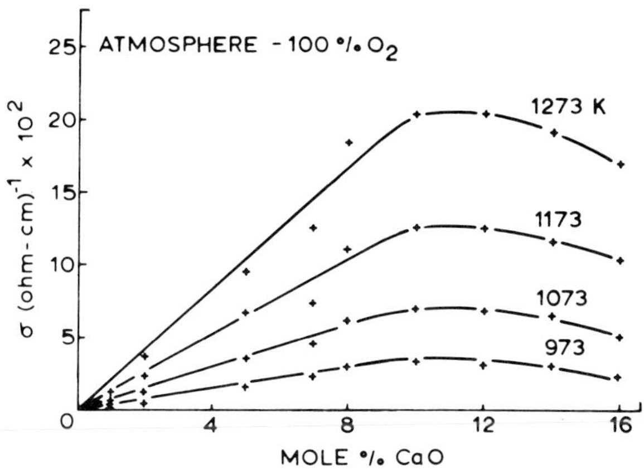

Fig. 11. The isothermal conductivity of $Ce(Ca)O_{2-x}$ solid solutions. (From Blumenthal *et al.*, 1973. Reprinted by permission of the publisher, The Electrochemical Society, Inc.)

independent of x, giving a value of 0.58 ± 0.03 eV, essentially the same as that found for the CaO-doped samples. The isothermal conductivity shows a linear relationship with x, however, the data are not as convincing as those for $Ce(Ca)O_{2-x}$.

Low-temperature studies on $Ce(Ca)O_{2-x}$ using complex plane analysis have been performed by Nowick *et al.* (1979) and El Adham (1978). Nowick *et al.* (1979) have measured samples with x values from 0.003 to 0.15. The activation energy for the bulk conductivity is essentially a constant in this range with an average value of 0.93 eV. The isothermal bulk conductivity for these samples measured at 500 K is proportional to

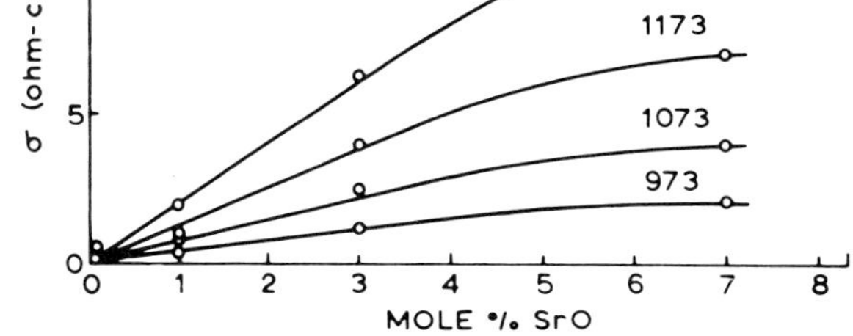

Fig. 12. The isothermal conductivity of $Ce(Sr)O_{2-x}$ solid solutions. (From Blumenthal and Garnier, 1976.)

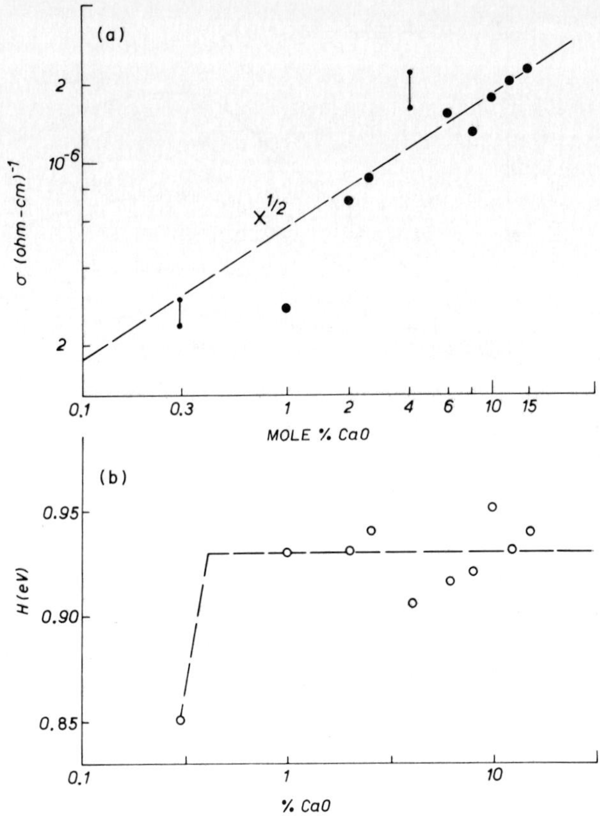

Fig. 13. The composition dependence of (a) the isothermal conductivity (500 K) and (b) the activation energy for $Ce(Ca)O_{2-x}$ solid solutions. (From Nowick *et al.*, 1979. Reproduced by permission of North-Holland Publ. Co., Amsterdam.)

$x^{1/2}$, not to x (Fig. 13). El Adham (1978) has found very similar behavior for the bulk conductivity of his samples. The activation energy is essentially constant over the composition range $x = 0.04$–0.11 with a deviation at low concentration to higher values of E_a. The value of E_a over the constant region is ~0.8 eV. The variation in the isothermal conductivity at 553 K is rather complex, with linear portions and a clear maximum at $x \sim 0.11$, however, this variation is not as unambiguous as the high-temperature variation seen by Blumenthal *et al.* (1973). From these data we can draw several conclusions.

1. The high-temperature dc data on $Ce(Mf)O_{2-x}$ appear to correspond well to the behavior expected for free vacancies given in Section IV.C.1.a.

2. The low-temperature complex plane data are unfortunately not as clear, however, the $x^{1/2}$ variation in the isothermal conductivity found by Nowick *et al.* (1979) indicates that the vacancies are not free but bound to the dopant cations to form associates. This argument is further strengthened by the relative values of the experimental activation energies seen in the two temperature ranges.

3. We can thus equate ΔH_m with the high-temperature value of $E_a \sim 0.6$ eV and $\Delta H_m + \Delta H_{A2}/2$ with the low-temperature value of $E_a \sim 0.8$–0.9 eV. This leads to a value of ΔH_{A2} in the range 0.4–0.6 eV.

(b). The situation for ThO_2 samples doped with divalent dopants is rather confused because most recent studies, for example, Maiti and Subbarao (1976) and Hartmanova *et al.* (1978), used fixed-frequency methods which are open to errors in interpretation. One point that emerges from these data is the high value of all the measured activation energies, even at high temperatures. The most reliable values of the activation energy must be taken to be the high-temperature values found by Maiti and Subbarao (1976). They found that for solid solutions of $Th(Ca)O_{2-x}$ with $x = 0.01$–0.07 the values of E_a fell within the range 1.16–1.42 eV and were much higher than the values found for $Ce(Mf)O_{2-x}$ solid solutions at high temperatures.

2. *Dopants for Which $n = 1$*

The amount of reliable conductivity data available for these solid solutions is rather more than in the previous case. We shall begin this subsection by examining data presented by Hammou (1975a, b) on the high-temperature conductivity of $Th(Y)O_{2-x}$ solid solutions in the composition range $x = 0.025$–0.11. He notes that the isothermal conductivity does not show a linear dependence except at very low concentrations (Fig. 14) and shows a peak at $x \sim 0.07$. He also notes the following rather interesting conclusions shown in Fig. 15.

For $x < 0.01$ the activation energy is constant at ~ 1.1 eV and the preexponential factor increases approximately linearly with x. For $0.075 > x > 0.01$ the preexponential factor remains constant with x and the activation energy drops to a minimum of ~ 0.95 eV at $x = 0.075$. For more concentrated solutions, i.e., $x > 0.075$, both the activation energy and the preexponential factor increase with x. This flatness of the preexponential factor and composition dependence of E_a has been noted before for other $n = 1$ systems such as $Ce(Y)O_{2-x}$ and $Zr_{1-p}Ce(Y)_pO_{2-x}$. The data for these systems, given by Hammou (1975b), are shown in Fig. 16. All show minima for the E_a values in the range $0.075 > x > 0.01$.

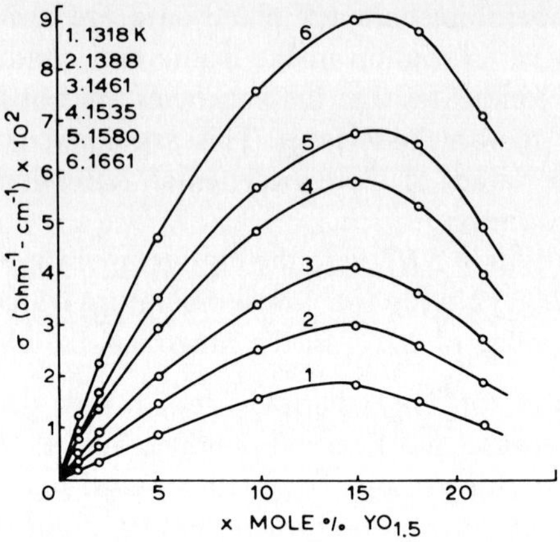

Fig. 14. The isothermal conductivity of $Th(Y)O_{2-x}$ solid solutions. (1) 1318 K; (2) 1388 K; (3) 1461 K; (4) 1535 K; (5) 1580 K; (6) 1661 K. (From Hammou, 1975a.)

Complex plane analysis of the conductivity of the $Th(Y)O_{2-x}$ system has been carried out by Schouler *et al.* (1976) in the temperature range 673–1473 K and the composition range $0.0005 < x < 0.075$. They found essentially the same behavior as that observed by Hammou (1975a) except that the activation energies were about 0.1 eV higher than in the dc studies. The minimum in activation energy was confirmed by these studies.

Nowick *et al.* (1979) have studied the low-temperature conductivity of $Ce(Y)O_{2-x}$ samples in the composition range $0 < x < 0.29$ using complex

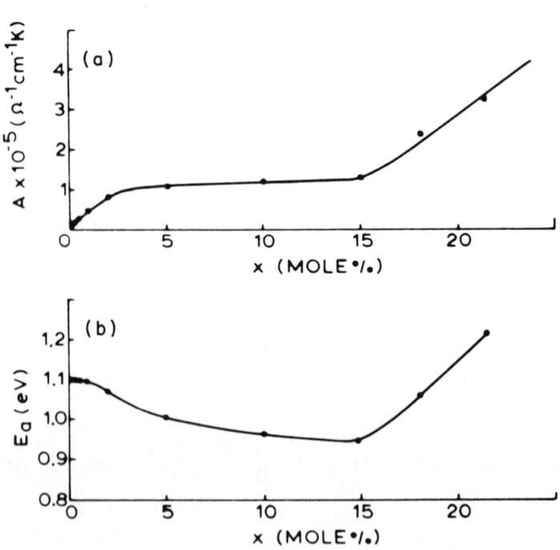

Fig. 15. The composition dependence of (a) the preexponential factor and (b) activation energy for $Th(Y)O_{2-x}$ solid solutions. *X* refers to the concentration in mole percent $YO_{1.5}$. (From Hammou, 1975b.)

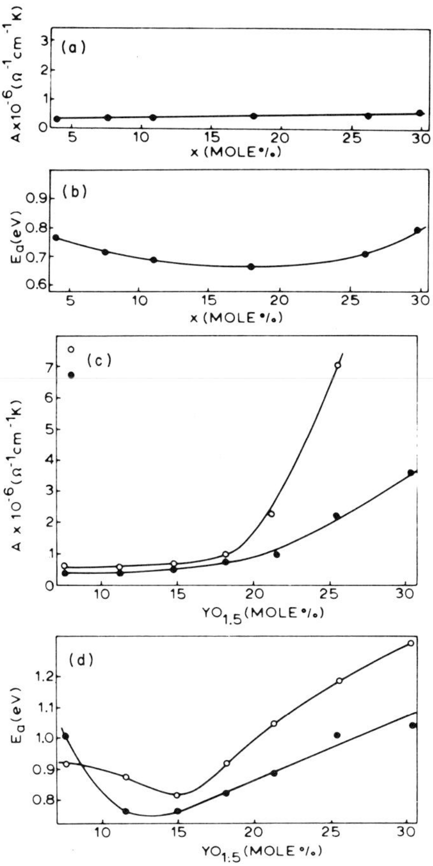

Fig. 16. The composition dependence of the preexponential factor and activation energy, respectively, for (a), (b) $(CeO_2)_{1-x}$–$(YO_{1.5})_x$ solid solutions and (c), (d) $(ZrO_2)_{1-P}$–$(CeO_2)_{P-x}$–$(Y_2O_3)_x$ solid solutions. Open circles, $P = 0.3$; solid circles, $P = 0.75$. (From Hammou, 1975b.)

plane analysis. The activation energy and the isothermal conductivity at 454 K for these samples are shown in Fig. 17. Again there is a maximum in the conductivity coupled to a minimum in the activation energy as seen by Hammou (1975a) and Schouler *et al.* (1976) for $Th(Y)O_{2-x}$. The data of El Adham (1978) for $Ce(Y)O_{2-x}$ solutions are essentially the same as those of Nowick *et al.* (1979), but there are some discrepancies, especially at low dopant concentrations.

The interpretation of these results is somewhat more complicated than for the case $n = 2$. The high-temperature data for thoria solid solutions presented by Hammou (1975b) is in good agreement with the composition-

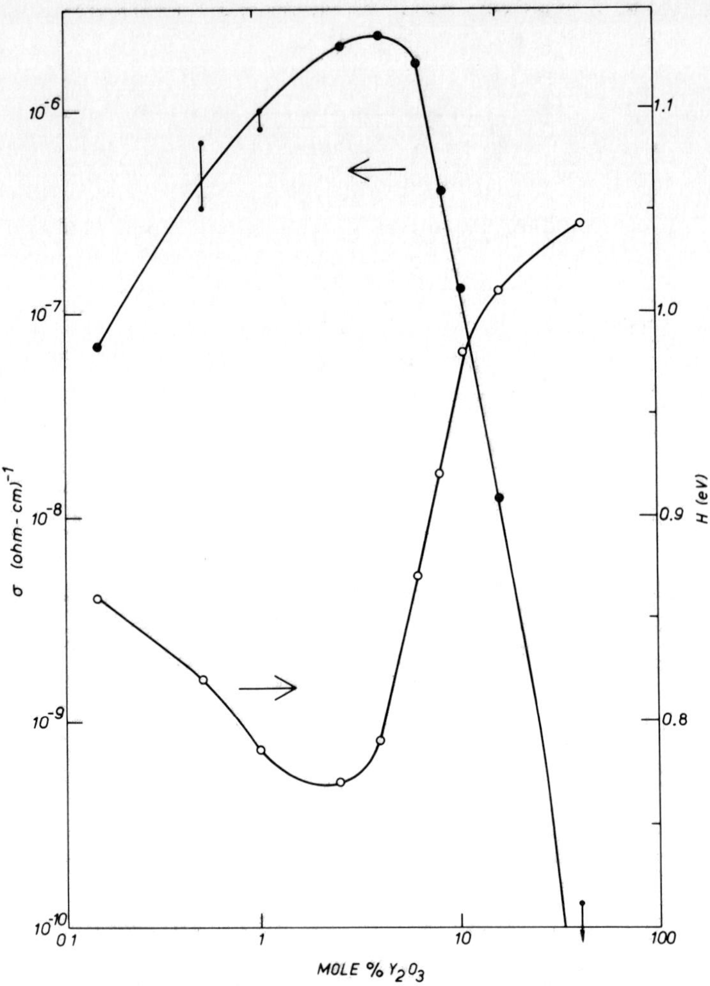

Fig. 17. The isothermal conductivity and activation energy for $Ce(Y)O_{2-x}$ solid solutions. (From Nowick *et al.*, 1979. Reproduced by permission of North-Holland Publ. Co., Amsterdam.)

independent preexponential factor expected from associate formation. Nowick *et al.* (1979) also conclude that the behavior of their $Ce(Y)O_{2-x}$ samples corresponds to the association of defects. They suggest that the changing activation energy is caused by a composition dependence of ΔH_{A1} due to the coulombic interactions of the free Y'_{Ce} dopant ions and the charged associates $\{Y'_{Ce}V_{O}^{\cdot\cdot}\}^{\cdot}$. Using the previously determined value of $\Delta H_m = 0.6$ eV for $Ce(Mf)O_{2-x}$ we can calculate ΔH_{A1} from the dilute limit of the Nowick *et al.* data. This gives a value of ~0.25 eV. We can compare this with the value of $\Delta H_{A2} = 0.4$–0.6 eV calculated in Section V.C.2, and we can see that the expected relationship between ΔH_{A1} and ΔH_{A2} may be justified.

3. *Conclusions for the Dilute Range, $x < 0.08$*

The behavior presented in these studies corresponds well with that predicted by the theory presented in Section V.C. One unexpected result is the composition dependence of the activation energy for dopants with $n = 1$, which leads to a composition dependence of the isothermal conductivity. It is also surprising that the high-temperature ionic conductivity behavior of $Ce(Mf)O_{2-x}$ corresponds to free vacancies, while that of $Th(Mf)O_{2-x}$ seems to correspond to associated defects. This may reflect the fact that ΔH_A for ThO_2 is greater than that found for CeO_2.

Several authors have noticed that at very dilute solutions, i.e., $x < 0.001$, the conductivity and activation energy are different from those seen in the rest of the dilute range. The explanation for this is not clear; it may be that there is a critical value of dopant concentration at which oxygen vacancies form, or it may simply be due to differences in sample preparation.

B. The Concentrated Range, $x > 0.08$

Most data on the concentrated range involve $Zr(Mf)O_{2-x}$ solid solutions. This is mainly because the cubic fluorite solid solution is stabilized only by relatively large additions of dopant, which place the conductivity near the maximum. The physical properties of these solid solutions at compositions near or beyond the maximum are quite complex. Aging phenomena have been reported by Schouler *et al.* (1973) and Inozemtsev *et al.* (1976) for these concentrated solutions, which appear to be correlated with the ordering of dopant cations and vacancies observed in diffraction studies (Carter and Roth, 1968; Steele and Fender, 1974; Hudson and Moseley, 1976). The kinetics of these ordering phenomena appear to be very slow indeed, reflecting the low cation mobilities in the fluorite structure at moderate temperatures. This means that it is very difficult to obtain an equilibrium structure for these materials, which may account for the many discrepancies noted in the literature.

Some recent studies on the conductivity data obtained by complex plane impedance analysis are of interest. Ioffe *et al.* (1978) have shown that the activation energy for conductivity increases approximately linearly with x for $Zr(Y)O_{2-x}$ materials once $x > 0.08$. Hohnke (1980) has recently used

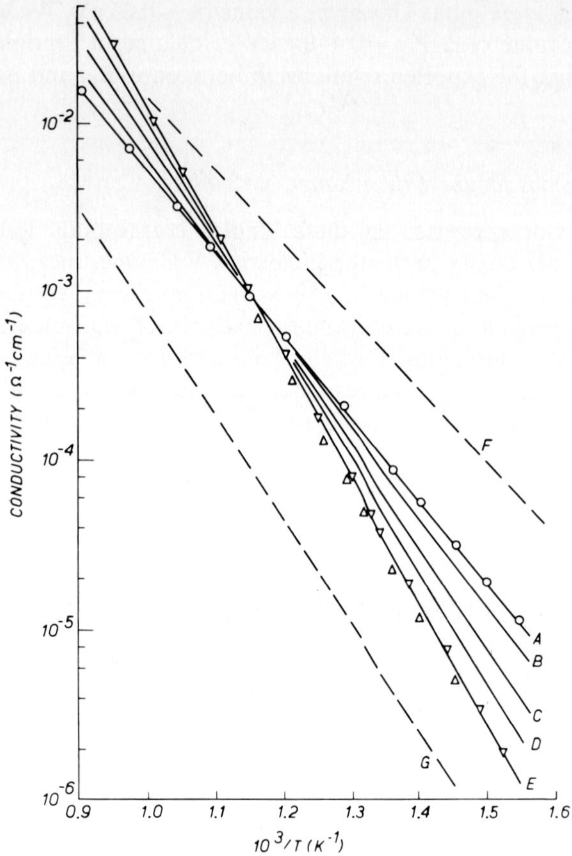

Fig. 18. The temperature dependence of the conductivity for $Zr(In)O_{2-x}$ solid solutions. A: $x = 0.083$; B: $x = 0.095$; C: $x = 0.105$; D: $x = 0.135$; E: $x = 0.160$ compared with the conductivity of $Zr(Y)O_{2-x}$; F: $x = 0.091$; G: $x = 0.167$. (From Hohnke, 1980.)

complex plane analysis to investigate the behavior of $Zr(In)O_{2-x}$ solid solutions in the concentration range $x = 0.08$–0.16. He has found the surprising result that these solid solutions show a conductivity focus at $T = 863$ K. This behavior is shown in Fig. 18. The high-temperature data of Vishnevskii *et al.* (1974) for $Zr(Ca)O_{2-x}$ and $Zr(Y)O_{2-x}$ samples show a similar focusing behavior, only at much higher temperatures, ~2600 and 2200 K, respectively. When the activation energy is plotted against the preexponential factor for $Zr(In)O_{2-x}$, a straight line is observed, suggesting some form of compensation law. Hohnke (1979) has extended this analysis to other fluorite solid solutions, using previously published data, and has shown two main points:

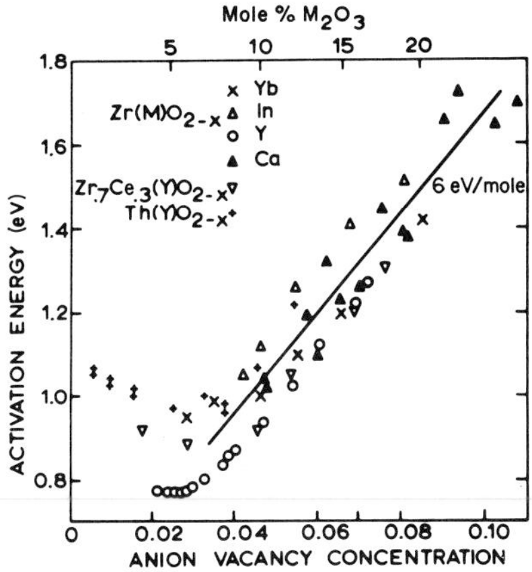

Fig. 19. The relationship between the preexponential factor and the activation energy for various $Zr(M)O_{2-x}$ solid solutions. (From Hohnke, 1979.)

1. For anion vacancy concentrations greater than $x = 0.08$ the activation energy becomes concentration-dependent and shows the same slope of 6 eV/mole (Fig. 19).

2. The activation energy-versus-preexponential factor curve for this region is a straight line; both increase with concentration. The preexponential factors become anomalously large compared to those expected from the type of analysis given earlier (Fig. 20).

Hohnke has shown that this variation in preexponential factor and activation energy is consistent with the introduction of an extra Gibbs free energy-like term which is dependent on composition. He calls this function

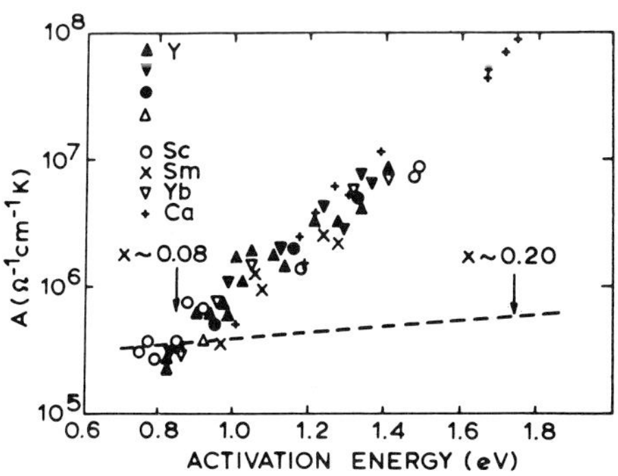

Fig. 20. The composition dependence of the activation energy for various fluorite solid solutions. (From Hohnke, 1979.)

$\Delta G(x)$, and the conductivity becomes, for $x > 0.08$,

$$\sigma = \sigma' \exp\{-\Delta G(x)/kT\} \tag{50}$$

where σ' is the conductivity for $x < 0.08$.

Hohnke associates this free energy term with the free energy for long-range ordering. Previously attempts were made to reproduce the conductivity of doped zirconia using models based on defect clusters or zones (Carter and Roth, 1968) or simple statistical models (Barker and Knop, 1971). All these models reproduce the main features of the variation in σ with concentration and temperature, however, some discrepancies remain.

Hohnke has calculated values of the isothermal conductivity of $Zr(Y)O_{2-x}$ solid solutions at high temperatures using an equation of the form given by Eq. (50) and obtained excellent agreement with the experimental data. This very close agreement had not been previously achieved by any other model.

Nakamura and Wagner (1980) have analyzed the conductivity of $Zr(Ca)O_{2-x}$ solid solutions using the idea of effective coordination numbers for the anion sites to describe the effects of association. Three such anion sublattices are distinguished, those which are 0-, 1-, or 2-fold coordinated by the dopant Ca_{Zr}'' cations, and a mobility is described for oxygen vacancies in each sublattice. It should also be noted that these coordination numbers are described as "effective," i.e., they take into account second nearest as well as nearest neighbors.

The conductivity of the system is described in terms of the mobilities of the vacancies within these sublattices, the greatest mobility being in the 1-fold coordinated sublattice. The maximum conductivity, here given at $x = 0.125$, occurs when all the anion sites are 1-fold coordinated. One problem associated with this type of analysis is that it does not take into account the observed microstructure, e.g., microdomains, seen at these high concentrations.

C. Conclusions for Fluorite Oxides

The use of complex plane analysis has enabled the bulk properties of these materials to be measured with a high degree of certainty. The compilation and analysis of these data have revealed several important points.

a. For $x < 0.08$

1. Association of defects takes place at low temperatures.
2. The type of associate formed depends upon the effective charge of the dopant $Mf_{M}^{n'}$ within the lattice.

3. The composition dependence of the isothermal conductivity arises for different reasons for the two cases $n = 1$ and $n = 2$.

(a) For $n = 2$ the dependence upon $x^{1/2}$ seems to be valid and is caused solely by the changing preexponential factor, the activation energy remaining constant.
(b) For $n = 1$ the reverse seems to be true; i.e., the preexponential factor remains constant and the activation energy decreases with composition. This leads to an increasing isothermal conductivity with x.

4. The values of the activation energies found in the associate regime are similar for any particular host lattice. This similiarity arises because the contribution of the association enthalpy ΔH_{A} to the measured activation energy E_{a} is of the form

$$E_{\mathrm{a}} = \Delta H_{\mathrm{m}} + \Delta H_{\mathrm{A}n}/n \tag{51}$$

where $n = 1, 2$. And, as shown earlier, $\Delta H_{\mathrm{A1}} \sim \frac{1}{2}\Delta H_{\mathrm{A2}}$.
5. The lowering of E_{a} found for solid solutions with $n = 1$ is quite unexpected and is thought to arise from an interaction of the free dopant cations with the associates. The nature of this interaction is as yet unknown and presents an interesting problem for both theoretical and experimental investigation.
6. It is also interesting to speculate on the relative magnitude of the activation energies seen for different host lattices. Both $\mathrm{Ce(Mf)O_{2-x}}$ and $\mathrm{Zr(Mf)O_{2-x}}$ show lower values of E_{a} than a $\mathrm{Th(Mf)O_{2-x}}$ solid solution. One possible explanation lies in the magnitude of ΔH_{A} for these solid solutions.

The high-temperature conductivity data for $\mathrm{Ce(Mf)O_{2-x}}$ correspond well to the case for free vacancies. In contrast the high-temperature conductivity data for $\mathrm{Th(Mf)O_{2-x}}$, obtained within the same range of x, seem to correspond well to association of vacancies. This discrepancy indicates that the association enthalpies (ΔH_{A}) for thoria solid solutions are higher than those for ceria solid solutions simply because, for the latter, association is important at higher temperatures. Thus any explanation of the variation in the experimentally determined activation energies should take into account the difference between the association enthalpies of dopant cations and vacancies in addition to the differences in migration enthalpies.

b. For $x > 0.08$. In this range of concentration the dilute solution behavior of noninteracting defects seems to break down, and the following is observed for both cases $n = 1, 2$.

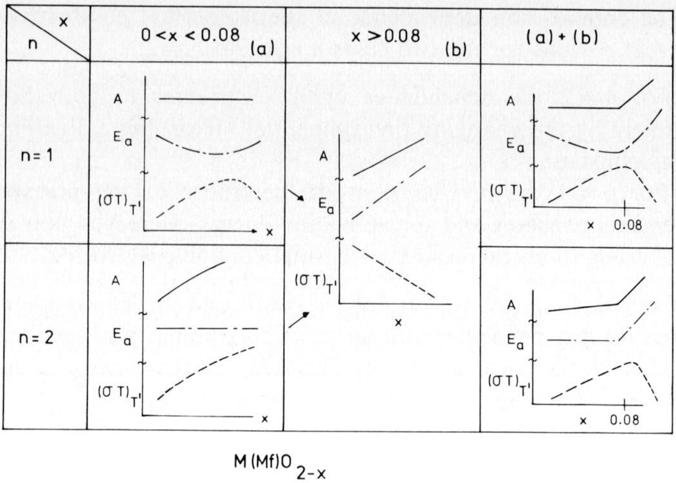

Fig. 21. A schematic representation of the variation in preexponential factor, activation energy, and isothermal conductivity in doped fluorite oxides.

1. E_a becomes strongly composition-dependent, increasing linearly with x.
2. E_a is also proportional to the preexponential factor A.
3. The composition dependence of these parameters can be explained in terms of an extra, composition-dependent, free energy-like term in the conductivity. This term may be related to the free energy for ordering in this type of oxide.

The total composition dependence of the preexponential, activation energy, and isothermal conductivity found for this type of oxide when associate behavior takes place is shown in schematic form in Fig. 21.

VII. Other Oxygen Ion Conductors

Several other oxygen ion conductors are known, but as yet the electrical properties of these oxides are not as well-characterized as fluorite structure oxides. Much interest has been shown in perovskite structure oxides (Takahashi and Iwahara, 1971; Kilner *et al.*, 1978; Browall *et al.*, 1976) in the hope that these materials may provide alternative electrolytes for high-temperature fuel cells. However, these oxides may suffer from the same limitations as fluorite structure oxides because, like fluorites, perovskites have to be doped to promote oxygen ion vacancies, and this introduces the possibility of associate formation.

An interesting group of compounds are the highly defective oxides based on the δ Bi_2O_3 structure (see, e.g., Takahashi and Iwahara, 1978). Essentially this structure is the fluorite structure with 25% of the anion sites vacant and, because no dopant is necessary to promote vacancies, it might be expected that low activation energies would be found for these materials. The complex impedance of pure δ Bi_2O_3 has recently been measured by Harwig and Gerards (1978). The activation energy they found for the δ phase was 0.4 eV, which is very low indeed for an oxygen ion conductor, and clearly this class of material merits further investigation. Finally it is also interesting to note that recent measurements on materials with the pyrochlore structure $A_2B_2O_7$, which can again be thought of as a type of defective fluorite, show low activation energies for oxygen conductivity (Shinozaki *et al.*, 1979; Uematsu *et al.*, 1979; De Vries *et al.*, 1979; Van Dijk *et al.*, 1980).

VIII. General Conclusions

Study of the conductivity of these nonstoichiometric oxides reveals complex and interesting behavior. The increasing use of complex plane analysis has revealed a wide spectrum of behavior for both the bulk and grain boundary contributions to the total conductivity. The continuing study of both these components is important for both scientific and technological reasons.

References

Abraham, M. M., Weeks, R. A., Clark, G. W., and Finch, C. B. (1966). *Phys. Rev.* **148**, 350–352.

Badwal, S. P. S., and Bevan, D. J. M. (1979). *J. Mater. Sci.* **14**, 2353–2365.

Barker, W. W., and Knop, O. (1971). *Proc. Br. Ceram. Soc.* **19**, 15–27.

Bauerle, J. E. (1969). *J. Phys. Chem. Solids* **30**, 2657–2670.

Beekmans, N. M., and Heyne, L. (1976). *Electrochim. Acta* **21**, 303–310.

Bevan, D. J. M., and Summerville, E. (1978). *In* "Handbook of Physics and Chemistry of Rare Earths" (K. A. Gschneider and L. Eyring, eds.), Vol. 4, North-Holland, Amsterdam.

Blumenthal, R. N., and Garnier, J. E. (1976). *J. Solid State Chem.* **16**, 21–34.

Blumenthal, R. N., Brugner, F. S., and Garnier, J. E. (1973). *J. Electrochem. Soc.* **120**, 1230–1237.

Bray, D. T., and Merten, U. (1964). *J. Electrochem. Soc.* **111**, 447–452.

Brook, R. J. (1974). *In* "Electrical Conductivity in Ceramics and Glass" (N. M. Tallan, ed.), Part A, pp. 179–267. Dekker, New York.

Brouwer, G. (1954). *Philips Res. Rep.* **9**, 366–376.

Browall, K. W., Muller, O., and Doremus, R. H. (1976). *Mater. Res. Bull.* **11**, 1475–1482.

Carter, R. E., and Roth, W. L. (1968). *In* "Electromotive Force Measurements in High Temperature Systems" (C. B. Alcock, ed.), pp. 125–144. Elsevier, Amsterdam.

Catlow, C. R. A. (1977). *Proc. R. Soc. London, Ser. A* **353**, 533–561.
Chereau, P., and Wadier, J. P. (1973). *J. Nucl. Mater.* **46**, 1–8.
Chu, S. H., and Seitz, M. A. (1978). *J. Solid State Chem.* **23**, 297–314.
Dell, R. M., and Hooper, A. (1978). *In* "Solid Electrolytes" (P. Hagenmuller and W. Van Gool, eds.), pp. 291–311. Academic Press, New York.
De Vries, K. J., Van Dijk, T., and Burggraaf, A. J. (1979). *In* "Fast Ion Transport in Solids" (P. Vashista, J. N. Mundy, and G. K. Shenoy, eds.), pp. 679–683. Elsevier/North-Holland, Amsterdam.
El Adham, K. (1978). Thesis, Univ. of Grenoble, Grenoble.
Etsell, T. H., and Flengas, S. N. (1970). Chem. Rev. **70**, 339–376.
Garvie, R. C., Hannink, R. H. J., Hughan, R. R., McKinnon, N. A., Pascoe, R. T., and Stringer, R. K. (1977). *J. Aust. Ceram. Soc.* **13**, 204.
Hammou, A. (1975a). *J. Chim. Phys.* **72**, 431–438.
Hammou, A. (1975b). *J. Chim. Phys.* **72**, 439–447.
Hartmanova, M., Hanic, F., Koller, A., and Janci, J. (1978). *Czech. J. Phys.* **28**, 414–433.
Harwig, H. A., and Gerards, A. G. (1978). *J. Solid State Chem.* **26**, 265–274.
Hohnke, D. K. (1980). *J. Phys. Chem. Solids* **41**, 777–784.
Hohnke, D. K. (1979). *In* "Fast Ion Transport in Solids" (P. Vashista, J. N. Mundy, and G. K. Shenoy, eds.), pp. 669–673. Elsevier/North-Holland, Amsterdam.
Hudson, B., and Moseley, P. T. (1976). *J. Solid State Chem.* **19**, 383–389.
Inozemtsev, M. V., and Perfil'ev, M. V. (1975). *Elektrokhimiya* **11**, 1031–1036.
Inozemtsev, M. V., Perfil'ev, M. V., and Lipilin, A. S. (1974). *Elektrokhimiya* **10**, 1471–1476.
Inozemtsev, M. V., Perfil'ev, M. V., and Gorelov, V. P. (1976). *Elektrokhimiya* **12**, 1231–1235.
Ioffe, A. I., Inozemtsev, M. V., Lipilin, A. S., Perfil'ev, M. V., and Karpachov, S. V. (1975). *Phys. Status Solidi A* **30**, 87–95.
Ioffe, A. I., Rutman, D. S., and Karpachov, S. V. (1978). *Electrochim. Acta* **23**, 141–142.
Killeen, J. C. (1975). *J. Nucl. Mater.* **58**, 39–46.
Kilner, J. A., Barrow, P., Brook, R. J., and Norgett, M. J. (1978). *J. Power Sources* **3**, 67–90.
Kröger, F. A., and Vink, H. J. (1956). *Solid State Phys.* **3**, 307–435.
Kvist, A. (1972). *In* "Physics of Electrolytes" (J. Hladik, ed.), Vol. 1, pp. 319–346. Academic Press, New York.
Maiti, H. S., and Subbarao, E. C. (1976). *J. Electrochem. Soc.* **123**, 1713–1718.
Matthews, J. R. (1974). *U.K. At. Energy Res. Estab., Memo.* **AERE-M 2643**.
Murch, G. E., and Thorn, R. J. (1978). *J. Nucl. Mater.* **71**, 219–226.
Nakamura, A., and Wagner, B. J. (1980). *J. Electrochem. Soc.* **127**, 2325–2333.
Nowick, A. S., and Park, D. S. (1976). *In* "Superionic Conductors" (G. Mahan and W. Roth, eds.), pp. 395–412. Plenum, New York.
Nowick, A. S., Wang, D. Y., Park, D. S., and Griffith, J. (1979). *In* "Fast Ion Transport in Solids" (P. Vashishta, J. N. Mundy, and G. K. Shenoy, eds.), pp. 673–679. Elsevier/North-Holland, Amsterdam.
Rohr, F. J. (1978). *In* "Solid Electrolytes" (P. Hagenmuller and W. Van Gool, eds.), pp. 431–449. Academic Press, New York.
Schouler, E., Giraud, G., and Kleitz, M. (1973). *J. Chim. Phys.* **70**, 1309–1316.
Schouler, E., Hammou, A., and Kleitz, M. (1976). *Mater. Res. Bull.* **11**, 1137–1146.
Shinozaki, K., Masahiro, M., Kuroda, K., Sakurai, O., Mizutani, N., and Kato, M. (1979). *J. Am. Ceram. Soc.* **62**, 538–539.
Steele, B. C. H. (1972). *In* "Solid State Chemistry" (H. F. Emeleus and L. E. J. Roberts, eds.), MTP International Review of Science, Inorganic Chemistry, Series One, Vol. 10, pp. 117–149. Butterworth, London.
Steele, D., and Fender, B. (1974). *J. Phys. C* **7**, 1–11.

Takahashi, T. (1972). *In* "Physics of Electrolytes" (J. Hladik, ed.), Vol. 2, pp. 989–1046. Academic Press, New York.
Takahashi, T., and Iwahara, H. (1971). *Energy Convers.* **11**, 105–111.
Takahashi, T., and Iwahara, H. (1978). *Mater. Res. Bull.* **13**, 1447–1453.
Uematsu, K., Shinozaki, K., Sakurai, K., Nobuyasu, M., and Kato, M. (1979). *J. Am. Ceram. Soc.* **62**, 219–221.
Van Dijk, T., De Vries, K. J., Burggraaf, A. J. (1980). *Phys. Status Solidi A* **58**, 115–125.
Vishnevskii, I. I., Sukharevskii, B. Y., and Shakhtin, D. M. (1974). *Sov. Phys.—Solid State* **16**, 951–953.

6

Mixed Conduction in Nonstoichiometric Oxides

H. L. Tuller
Department of Materials Science and Engineering
Massachusetts Institute of Technology
Cambridge, Massachusetts

I. Introduction

All solids may be induced to transport charge by placement within a potential field gradient. In predominantly ionic compounds, e.g., oxides, charge transport occurs by the simultaneous motion of ionic and

NONSTOICHIOMETRIC OXIDES

ISBN 0-12-655280-0

electronic charge carriers, making all such materials in principle mixed ionic and electronic conductors. Generally, only one type of carrier dominates charge transport, and so contributions from minority carriers are often ignored. Such compounds tend to be categorized either as semiconductors or as solid electrolytes, depending on whether electronic or ionic charge carriers predominate, and they have historically been studied by different groups of investigators often employing dissimilar experimental techniques.

For a considerably smaller subgroup of oxides, by circumstance of crystal structure, band gap, composition, and temperature, both ionic and electronic carriers contribute simultaneously in a major way to the overall electrical conductivity. Such mixed conductors are the subject of this chapter, with particular emphasis focused on the effects of variable stoichiometry on transport in these solids.

Simultaneous ionic and electronic conduction plays a critical role in a number of important processes occurring in oxides. For example, it accounts for the rate of formation of oxide layers on metals (Wagner, 1979): their oxidiation–reduction kinetics (Seltzer, 1976), and their gas permeation rates (Smith *et al.*, 1966). Efficient operation of certain catalytic oxide electrodes is believed to be a consequence of mixed conduction (Voorhoeve *et al.*, 1975), while at other times mixed conduction may be a detriment, as in solid electrolytes designed for fuel cells or chemical sensors (Tannhauser, 1978). In all cases the ability to predict and control the conditions under which oxides exhibit mixed conduction often determines their relative utility.

Although mixed conduction plays an important role in many processes, there are surprisingly few oxides about which more than a superficial knowledge of mixed transport exists. In the last decade, some progress has been made, particularly with oxides crystallizing in the cubic fluorite structure (e.g., ThO_2, CeO_2) and several of the more insulating materials such as Al_2O_3 and MgO. This has been a consequence of (a) the availability of improved materials, (b) the development of improved techniques for the characterization of ionic and electronic transport and means for their separation, and (c) the more extensive use of the principles of defect chemistry in the planning and subsequent analysis of experimental results.

I begin this chapter by reviewing the sources of ionic and electronic conduction in oxides. Attention is given to certain characteristics such as small polaron formation and fast ion conduction, which can dramatically alter estimates of the degree of mixed conduction. I then utilize defect relations of the type discussed in Chapter 1 to indicate how the concentrations of the various defects are interrelated, and derive relationships that predict how the relative mix of ionic and electronic defects varies with temperature, atmosphere, and impurity level. Special attention is given to the case where variable-valence impurities contribute to mixed conduction. These results are used to construct maps specifying the conditions under which an oxide may be expected to be an ionic, electronic, or mixed conductor. Relations describing ambipolar diffusion are developed and used to formulate expressions relevant to a number of electrochemical techniques useful in determining transference numbers of ions and electrons under varied circumstances including galvanic cells, polarization, and permeation techniques. For this subject I have drawn heavily on an excellent review by Heyne (1977). I also refer the reader to two additional reviews, one by Patterson (1974) in which a more general approach is taken, and one by Weppner and Huggins (1978) in which these and other electrochemical techniques are systematically reviewed. Other methods for evaluating transference numbers such as the impurity and atmosphere dependence of total conductivity, diffusion, and thermionic emission are also treated. In the final sections, we review the state of knowledge for a number of oxides systems in which mixed conduction is known to play an important role and conclude by discussing applications in which such materials are used.

II. Electrical Conduction in Oxides

The electrical conductivity of a solid is simply the sum of the partial conductivities associated with each type of charge carrier and is given by

$$\sigma = \sum_j \sigma_j \tag{1}$$

Each partial conductivity σ_j, which may represent transport of either ionic or electronic carriers, is defined by

$$\sigma_j = c_j Z_j q \mu_j \tag{2}$$

in which c_j is the carrier density (per meter^3), $Z_j q$ the effective charge (coulombs), and μ_j the mobility (meter^2 per volt second) of the jth species. The fraction of the total conductivity contributed by each charge carrier is

$$t_j = \sigma_j / \sigma \tag{3}$$

where t_j is known as the transference number. By definition the sum of the individual transference numbers must be unity.

When dealing with mixed ionic and electronic conductors one is generally concerned with the relative magnitude of the ionic transference number t_i, which includes cation and anion contributions, and the electronic transference number t_e which represents the sum of the electron and hole transference numbers $t_n + t_p$.

As noted earlier, since neither t_i nor t_e is ever truly zero, all oxides are in principle mixed conductors. In practice, it is useful to speak of mixed conduction only when both ions and electrons contribute in a significant manner to the total conductivity. How one determines what is significant depends on the subject of interest. Electronic conduction in excess of 1% is termed significant in the operation of, for example, solid electrolyte fuel cell elements (Obayashi and Kudo, 1977), while $\frac{1}{10}$ this value is significant when utilizing the same material as a sensor element (Worrell, 1977). Even smaller electronic transference numbers (e.g., $< 10^{-4}$) control, for example, the rate of gas permeation through solid electrolytes (Heyne and Beekmans, 1971). Often our choice of what is significant is limited only by the sensitivity of our instrumentation. In general, we shall consider situations in which, by reasonable choice of temperature, atmosphere, and/or composition, a material can be induced to exhibit ionic and electronic transference numbers on the order of 0.1%.

Mixed conduction implies near equality of t_i and t_e. If we assume for the moment that only one type of ionic and electronic carrier predominates, this condition requires that

$$c_i \mu_i \approx c_e \mu_e \tag{4}$$

Because electronic mobilities are typically 10^4–10^8 times greater than ionic mobilities at elevated temperatures, this has the important consequence that all mixed conductors must possess 10^4–10^8 times greater concentrations of ionic carriers than electronic carriers. This condition, as we shall see, is satisfied only in oxides in which high densities of mobile ions can be induced by doping or crystallographic disorder and which simultaneously possess relatively large energy band gaps and thereby low concentrations of intrinsic electronic carriers.

A schematic diagram which includes the most likely mechanisms controlling ionic and electronic carrier generation and their respective mobilities is given in Fig. 1. In general, it is necessary to discover the mechanisms that control charge formation and transport for each of the carriers and the parameters that characterize them (some of the most important are indicated) before one can hope to understand electrical conduction in specific oxides. In the following, we discuss those aspects that appear to be most relevant to an understanding of mixed conduction in nonstoichiometric oxides.

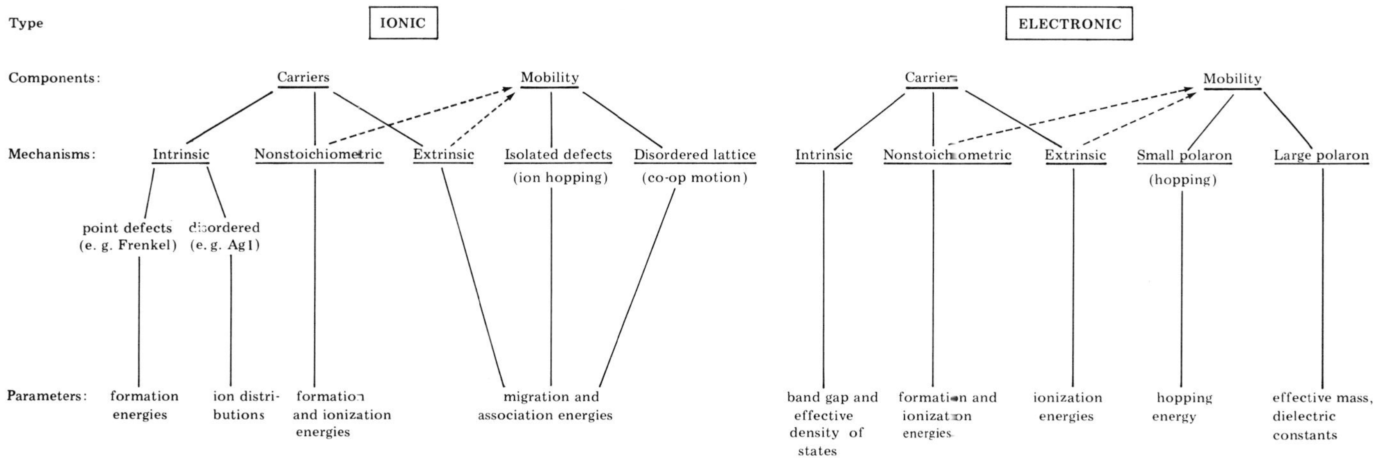

Fig. 1. The major sources of ionic and electronic carriers in oxides and their respective mobilities. Broken arrows indicate that large numbers of carriers generated via deviation from stoichiometry or doping may alter the carrier mobilities.

A. Ionic Conduction

Motion of ions is described by an activated jump process for which the diffusion coefficient is given by (Goodenough, 1978)

$$D = D_0 \exp(-\Delta G/kT) = \gamma(1-c)Za^2\nu_0 \exp(\Delta S/k)\exp(-E_m/kT) \quad (5)$$

where a is the jump distance, ν_0 the attempt frequency, and E_m the migration energy. The factor $(1-c)Z$ defines the number of neighboring sites that are unoccupied, while γ includes geometric and correlation effects. Since the ion mobility is defined by $\mu = qD/kT$ and the density of carriers of charge q is Nc, where N is the density of ion sites in the sublattice of interest, the ionic conductivity becomes

$$\begin{aligned}\sigma_{ion} &= \gamma(Nq^2/kT)c(1-c)Za^2\nu_0 \exp(\Delta S/k)\exp(-E_m/kT)\\ &= \sigma_0/T \exp(-E_m/kT) \quad (6)\end{aligned}$$

This expression shows that σ_{ion} is nonzero only when the product $c(1-c)$ is nonzero, i.e., for c not equal to one or zero. For classical ionic conductors c is generally either very close to one (e.g., vacancies in a normal sublattice) or to zero (e.g., ions at interstitial sites), this being the major reason for their relatively low ionic conductivity.

Since mixed conduction implies high ion carrier densities, thermal generation by instrinsic Schottky or Frenkel disorder (here c is temperature-dependent) characterized in oxides by large defect formation energies may be ignored. On the other hand, so-called fast ion conductors such as sodium β-alumina are known to possess extremely high intrinsic concentrations of mobile ions by virtue of incomplete occupation of certain crystallographic sublattices (Huggins, 1978). Such intrinsic crystallographic disorder results in values of $c \sim \frac{1}{2}$; this means that virtually all the ions in such sublattices are mobile, resulting in carrier densities of $\sim 10^{21}$–10^{22} cm^{-3}.

The formation of ionic defects that accompany excursions in composition away from stoichiometry may also be large. For example, CeO_2 may readily be reduced to $CeO_{1.8}$ at 1000°C, resulting in oxygen vacancy concentrations of 5×10^{21} cm^{-3} (i.e., $c = 0.9$) (Tuller and Nowick, 1979). These ionic defects are, however, accompanied by electronic defects of the same order of magnitude, as can be seen by examination of the reduction reaction

$$O_O \rightarrow V_O^{\cdot\cdot} + 2e' + \tfrac{1}{2}O_O \quad (7)$$

Since electron mobilities are usually orders of magnitude larger, such deviations from stoichiometry do not usually lead to mixed conduction.

Recent results show (Fujimoto and Tuller, 1979; Sempolinsky *et al.*, 1980; Dutt *et al.*, 1975), however, that an exception to the above rule can be made for oxides in which composition excursions are controlled by the addition of variable-valence ions. Such systems exhibit mixed ionic–electronic behavior different from those usually considered and are later treated in some detail.

Ionic defect densities may also be greatly enhanced above intrinsic levels by doping with aliovalent impurities. This may be at levels of hundred of ppm where the solute solubility is limited or as high as 10% or more when solid solution formation is extensive. By appropriate choice of dopant only the density of ionic defects is enhanced, leading to predominant ionic or mixed conduction. The popular solid electrolytes calcia-stabilized zirconia (CSZ) and yttria-doped thoria (YDT) belong to this category.

Ionic conduction in oxides tends to be controlled by the motion of ions in either the cation or oxygen sublattice. Loose, open structures, small ion size and charge, and high ion polarizability tend to lead to enhanced ion mobilities (Armstrong *et al.*, 1973). These factors favor higher cation mobilities, as can be seen by reference to Table I. A major exception is oxides that crystallize with the fluorite structure in which the 4-fold coordinated oxygen ions have roughly 10 orders of magnitude higher mobilities than their 8-fold coordinated cation counterparts (Etsell and Flengas, 1970). Clearly, oxides exhibiting high ion mobilities are most likely to possess the high ionic conductivities necessary for mixed conduction.

In summary, only oxides with high defect mobilities that are either sufficiently heavily doped or structurally disordered are likely to exhibit mixed conduction. In Table II the types of systems known to satisfy these requirements are listed along with examples of each.

B. Electronic Conduction

As discussed above, mixed conduction is most likely to occur for large ionic/electronic defect ratios. The minimum number of electronic defects is strongly dependent on the magnitude of the energy gap between the conduction and the valence band and is given by

$$n_\mathrm{i} = p_\mathrm{i} = \sqrt{N_\mathrm{c} N_\mathrm{v}}\,\exp(-E_\mathrm{g}/2kT) \tag{8}$$

in which n_i and p_i are the intrinsic density of electrons and holes and $N_{\mathrm{c,v}}$, the effective density of states in the conduction and valence bands, is (Kittel, 1976)

$$N_{\mathrm{c,v}} = 2(2\pi m^*_{\mathrm{e,h}} kT/h^2)^{3/2} \tag{9}$$

Although effective masses of electrons and holes $m^*_{\mathrm{e,h}}$ are rarely accurately known for oxides, substition of the free electron mass into the above

TABLE I

Ionic Mobilities[a]

Material and reference	Cation defect mobilities (cm^2/V sec)	Oxygen defect mobilities (cm^2/V sec)
BeO (Cline and Newkirk, 1968)	$\mu(V''_{Be}) = 2 \times 10^3 \exp(-2.83\ eV/kT)$ $= 7 \times 10^{-6}$ (1673 K)	
MgO (Sempolinsky and Kingery, 1980)	$\mu(V''_{Mg}) = 2 \exp(-2.13\ eV/kT)$ $= 7.8 \times 10^{-7}$ (1673 K)	
Al_2O_3 (Mohapatra and Kröger, 1977a)	$\mu(Al_i^{3\cdot}) = 7.13 \times 10^7 \exp(-4.6\ eV/kT)$[b] $= 1 \times 10^{-6}$ (1673 K)	
Naβ-Al_2O_3 (Kim *et al.*, 1979)	$\mu(Na_{Na}) = 6.3 \times 10^{-3} \exp(-0.14\ eV/kT)$ $= 3.6 \times 10^{-4}$ (573 K)	
CeO_2 (+5 mole % Y_2O_3) (Tuller *et al.*, 1980)		$\mu(V_O^{\cdot\cdot}) = 1.2 \exp(-0.87\ eV/kT)$ $= 4.3 \times 10^{-4}$ (1273 K)
ZrO_2 (+12 mole % Y_2O_3) (Stratton *et al.*, 1980)		$\mu(V_O^{\cdot\cdot}) = 4.2 \times 10^{-1} \exp(-0.99\ eV/kT)$ $= 5.1 \times 10^{-5}$ (1273 K)
ThO_2 (+$7\frac{1}{2}$ mole % Y_2O_3) (Iqbal and Baker, 1973)		$\mu(V_O^{\cdot\cdot}) = 1.5 \times 10^{-1} \exp(-1.0\ eV/kT)$ $= 1.7 \times 10^{-5}$ (1273 K)

[a] All values listed are "effective mobilities" obtained from ionic conductivity by assuming that all ionic defects generated by impurity compensation are in fact mobile.

[b] Indirect evidence.

TABLE II

Mixed Conductors[a]

Point defect type: Dilute	Point defect type: Concentrated	Disordered lattice type
$MgO(FeO)$ [$\sim 1 \times 10^{19}$ cm^{-3}] (Sempolinsky *et al.*, 1980)	$ZrO_2 \cdot CaO$ [$\sim 5 \times 10^{21}$ cm^{-3}] (Etsell and Flengas, 1972)	σ-Bi_2O_3 [1×10^{22} cm^{-3}] (Takashashi and Iwahara, 1978)
$Al_2O_3(CoO)$ [$\sim 1 \times 10^{17}$ cm^{-3}] (Dutt *et al.*, 1975)	$CeO_2 \cdot Y_2O_3$ [~ 1–5×10^{21} cm^{-3}] (Tuller and Nowick, 1975a)	$K_{1+x}Fe_{11}O_{17}$ [$\sim 5 \times 10^{21}$ cm^{-3}] (Dudley *et al.*, 1976)
$ThO_2(Y_2O_3)$ [$\sim 3 \times 10^{19}$ cm^{-3}] (Bauerle, 1966)	$ThO_2 \cdot Y_2O_3$ [~ 1–5×10^{21} cm^{-3}] (Lasker and Rapp, 1966)	
Ta_2O_5 [$\sim 3 \times 10^{18}$ cm^{-3}] (Stroud *et al.*, 1974)	$LaAlO_3 \cdot CaO$ [$\sim 1 \times 10^{21}$ cm^{-3}] (Kilner *et al.*, 1978)	

[a]Square brackets indicate estimated concentration of ionic defects.

TABLE III

Band Gaps[a] and Calculated Carrier Densities

	E_g (eV)	$n_i = p_i$ (m^{-3}) 773 K	1273 K	1773 K
Fe_2O_3	2.3	1×10^{18}	3×10^{21}	8×10^{22}
TiO_2	3.05	5×10^{15}	8×10^{19}	7×10^{21}
Ta_2O_5	4.6	4×10^{10}	7×10^{16}	4×10^{19}
ThO_2	5.75	8×10^{6}	4×10^{14}	1×10^{18}
MgO	6.8	3×10^{3}	3×10^{12}	3×10^{16}
Al_2O_3	9.5	5×10^{-6}	1×10^{7}	5×10^{12}
BeO	10.5	3×10^{-9}	2×10^{5}	2×10^{11}

[a] Strehlow and Cook, 1973.

relation is usually sufficient to obtain good approximations for n_i. Since n_i and p_i decrease exponentially with E_g at a given temperature, oxides with large gaps are most likely to exhibit mixed conduction even under conditions where the electronic conductivity is considerably above its minimum value. Table III lists the band gaps of a number of common oxides with values ranging from 2.3 to 10.5 eV, as well as intrinsic electronic carrier densities [Eq. (8)], for a number of temperatures. Note that a 4-fold increase in E_g results in a difference of nearly 30 orders of magnitude in n_i (773 K) for Fe_2O_3 and BeO.

The density of carriers may be significantly enhanced over intrinsic levels either by deviation from stoichiometry or by doping (Fig. 1). Ceria (CeO_2), for example, which has a band gap of ~5.5 eV (Strehlow and Cook, 1973), is calculated to have an intrinsic carrier concentration of $\sim 5 \times 10^{14}$ m^{-3} at 1273 K. CeO_2 may, however, as indicated earlier, be reduced to $CeO_{1.8}$ at this temperature to give $\sim 5 \times 10^{26}$ m^{-3} excess electrons or roughly 12 orders of magnitude more electrons than at stoichiometry. Similar levels of electron enhancement have been reported for stoichiometric CeO_2 doped with Nb_2O_5 (Naik and Tien, 1979).

In summary, although large band gaps by themselves do not guarantee mixed conduction, they favor it where care is taken to control impurities and limit excursions from stoichiometry.

For pure nonpolar broad-band semiconductors, electron drift mobilities are limited by acoustic phonon scattering and take the general form (Kittel, 1976)

$$\mu = \mu_0 T^{-3/2} \tag{10}$$

Values for μ in such systems vary from approximately 500 cm^2/V sec (ZnTe) to 65,000 (InSb) at room temperature.

In ionic compounds the electron mobility is determined in large part by its interaction with the polar modes of the crystal. In such solids the electron induces a deformation in the surrounding lattice, which tends to self-trap the electron in a potential well of radius r_p. The electron and its surrounding polarization cloud are commonly referred to as a polaron (Bransky and Tallan, 1971).

Two types of polarons are distinguished. Large polarons are formed for conditions of weak to intermediate coupling ($0 < \alpha < 6$) between electrons and optical phonons defined by Bosman and Van Daal, 1970)

$$\alpha = \frac{e^2}{2^{1/2}\hbar}\left(\frac{m^*}{\hbar\omega_0}\right)^{1/2}\left(\frac{1}{\epsilon_{\text{opt}}} - \frac{1}{\epsilon_{\text{stat}}}\right) \tag{11}$$

in which ω_0 is the mean optical phonon frequency and ϵ_{opt} and ϵ_{stat} are the high-frequency and static dielectric constants. For large polarons r_p is much greater than the interatomic spacing and the electron still moves in a band but with its mass enhanced above that of the effective density of states mass m^*. The large polaron mobility at temperatures above the Debye temperature θ is given by (Appel, 1968)

$$\mu = \mu_0' T^{-1/2} \tag{12}$$

and is expected to be $\sim$1–100 $\text{cm}^2/\text{V sec}$ at elevated temperatures.

For oxides with narrow bands and relatively strong electron–optical phonon coupling, small polarons are formed. In this case, the electron is self-trapped at a given lattice site and can move only to an adjacent site by an activated hopping process† similar to that exhibited by ionic diffusion. Consequently, the small polaron mobility takes the form (Tuller and Nowick, 1977)

$$\mu = [(1 - c)ea^2\nu_0/kT]\exp(-E_H/kT) \tag{13}$$

in which E_H is the hopping energy and $(1 - c)$ is the fraction of sites not already occupied by mobile electrons. As might be expected, small polaron mobilities are considerably reduced and are generally reported to be on the order of 10^{-4}–10^{-2} $\text{cm}^2/\text{V sec}$ at elevated temperatures—hundreds to thousands of times smaller than in large polaron conductors. For this reason a number of oxides [e.g., CeO_2 (Tuller and Nowick, 1975c)] exhibit mixed conduction under conditions where one would normally expect to find only semiconduction. Table IV lists electronic carrier mobilities for a number of oxides in which values for large or small polaron mobilities have been reported.

†At sufficiently low temperatures, usually $T < \theta/2$, tunneling behavior becomes possible.

TABLE IV

Electronic Mobilities

Material	μ_n (cm^2/V sec)	μ_p = (cm^2/V sec)
Fluorite structure		
CeO_{2-x} (Tuller and Nowick, 1977)	8.1×10^{-3} (1273 K, 0.4 eV, $x = 0.008$) 2.2×10^{-3} (1273 K, 0.47 eV, $x = 0.23$)	
HfO_2 (Tallan *et al.*, 1967)		3×10^{-4} (1273 K, 0.7 eV)
ThO_2 (Etsell, 1972; Bransky and Tallan, 1970; Tuller, 1977)	$\gtrsim 10^{-4}$ (1873 K)	> 6 (1273 K) ~ 100 (1273 K)
UO_2 (Dudney *et al.*, 1981)		7×10^{-2} (1273 K, 0.085 eV)
ZrO_2 + 10 mole % Y_2O_3 (Weppner, 1976)	1.5×10^{-2} (1073 K, 0.55 eV)	4.9×10^{-5} (1073 K, 1.4 eV)
NaCl structure		
CoO (Bransky and Tallan, 1971)		0.30–0.38 (1300–1700 K)
MgO (Sempolinsky *et al.*, 1980)	13 (1673 K)	4 (1673 K)
NiO (Bransky and Tallan, 1971)		0.33 (1300–1700 K)
Perovskite structure		
$BaTiO_3$ (Ihrig and Hennings, 1978)	0.6 (300 K) 0.1 (1370 K)	
$KTaO_3$ (Deputy and Vest, 1978)	0.5 (1273 K, 0.37 eV)	6.4×10^{-4} (1273 K)
$La_{1-x}Sr_xCrO_3$ (Karmin and Aldred, 1979)		4.6×10^{-3} (1273 K, 0.134 eV, $x = 0.02$)
ReO_3 (Pearsall and Lee, 1974)	2.3×10^3 (4.2 K) 33 (300 K)	
Others		
Al_2O_3 (Hughes, 1979)	3 ± 1 (100–350 K)	
Ta_2O_5 (Stroud *et al.*, 1974)	5×10^{-2} (1373 K)	
$Y_3Fe_5O_{12}$ (Metselaar and Huyberts, 1977)	1.4×10^{-2} (1273 K)	0.25 (1273 K)

III. Defect Equilibria

A. Low Defect Concentrations

In previous sections we emphasized the relative importance of the ionic/electronic carrier ratio for mixed conduction and remarked on the general dependence of both ionic and electronic carrier concentrations on variations in temperature, stoichiometry, and purity. In this section, we utilize defect relations of the type discussed in Chapter 1 to indicate how the concentrations of the various ionic and electronic defects are interrelated and derive quantitative expressions that predict how the relative mix of ionic and electronic defects depends on the above variables. Initially, a pure oxide with Frenkel disorder is considered. This is extended to the case in which fixed-valence impurities control defect levels in the vicinity of stoichiometry, as is commonly observed experimentally in solid electrolytes such as doped zirconia and thoria. Finally, this treatment is extended to include the effects of variable-valence impurities which have only recently been recognized to lead to mixed conduction in refractory oxides. In the foregoing analysis we begin by assuming that defect concentrations are sufficiently small so that activities may be replaced by concentrations in the mass-action relations. We later discuss the manner in which large defect levels are likely to modify these results.

Since many of the mixed oxide conductors of interest possess the fluorite structure, I have chosen an oxide with Frenkel disorder in the oxygen sublattice as our model. Ionic thermal disorder may in this case be written as

$$O_O \rightarrow O_i'' + V_O^{\cdot\cdot} \tag{14}$$

in which an oxygen ion leaves a normal lattice site vacant by moving into an interstitial position. Intrinsic thermal electron–hole generation is represented in similar manner by

$$O \rightarrow e' + h^{\cdot} \tag{15}$$

while equilibrium with the gas phase

$$O_O \rightarrow V_O^{\cdot\cdot} + 2e' + \tfrac{1}{2}O_2 \tag{16}$$

occurs by exchange of oxygen between the crystal lattice and the gas. To simplify analysis, all ionic defects are assumed to be in a maximum state of ionization.

If ideal behavior is assumed, Eqs. (14)–(16) may be replaced by the mass-action relations

$$[O_i''][V_O^{\cdot\cdot}] = K_F = K_F^0 \exp(-E_F/kT) \tag{17}$$

$$np = K_i = N_c N_v \exp(-E_g/kT) \tag{18}$$

$$[V_O^{\cdot\cdot}]nP_{O_2}^{1/2} = K_F = K_R^0 \exp(-E_R/kT) \tag{19}$$

Here [ ], n, and p indicate concentrations of ionic species, electrons, and holes, respectively, while K_F, K_i, and K_R are the appropriate temperature-dependent equilibrium constants. A fourth equation which follows from the requirement of electroneutrality

$$n + 2[O_i''] = p + 2[V_O^{\cdot\cdot}] \qquad (20)^\dagger$$

allows us in principle to solve for the four defect concentrations as a function of temperature and oxygen partial pressure.

In practice, it is considerably simpler to solve the above equations piecewise, following the approach of Brouwer (1954). Here one recognizes that only one term on either side of Eq. (20) is significant for a given set of experimental conditions. This allows one to solve readily the set of four equations for all the limiting forms of Eq. (20), each of which is appropriate for a range of temperature and P_{O_2}. Low oxygen pressures result in the formation of substantial concentrations of $V_O^{\cdot\cdot}$ and e′ [see Eq. (16)], leading Eq. (20) to be satisfied by $[V_O^{\cdot\cdot}] = n$. High oxygen pressures, on the other hand, produce O_i'' and h are thus characterized by $[O_i''] = p$. At intermediate pressures, i.e., near stoichiometry, intrinsic thermal defects predominate, giving $[O_i''] = [V_O^{\cdot\cdot}]$ (in mixed conductors $K_F \gg K_i$). Table V lists solutions for the major charge carriers in each of these three regimes [Eqs. (21), (22), and (25)]. These carrier concentrations are shown plotted in Fig. 2a as a function of P_{O_2} for arbitrary temperature and K values.

1. Fixed-Valence Impurities

More generally, mixed conduction in oxides is a result of impurity-enhanced ionic conductivity. The effect of impurities may readily be incorporated into the above treatment by modifying the electroneutrality relation, for example, to read

$$n + [N_M'] + 2[O_i''] = p + 2[V_O^{\cdot\cdot}] \tag{26}$$

in which N_M' is a substitutional impurity cation with valence one less than that of the host (e.g., Y^{3+} in ThO_2). Proceeding in a similar manner as above one obtains the limiting neutrality equations and defect relations given in Table V [Eqs. (21), (23)–(25)]. Rather than the single intermediate region obtained earlier for the pure oxide, there are now two intermediate

† Equations (21)–(25) appear in Table V.

TABLE V

Defect Concentrations for Pure and Doped Frenkel-Like Oxide[a]

	Limiting neutrality equation	P_{O_2} range	n	p	$[V_O^{\cdot\cdot}]$	$[O_i'']$
(21)	$n = 2[V_O^{\cdot\cdot}]$	Low	$(2K_R)^{1/3}P_{O_2}^{-1/6}$	$\dfrac{K_i P_{O_2}^{1/6}}{(2K_R)^{1/3}}$	$\dfrac{n}{2}$	$K_F\left(\dfrac{4}{K_R}\right)^{1/3} P_{O_2}^{1/6}$
(22)	$[O_i''] = [V_O^{\cdot\cdot}]$	Intermediate	$\left(\dfrac{K_R}{K_F^{1/2}}\right)^{1/2} P_{O_2}^{-1/4}$	$K_i\left(\dfrac{K_F^{1/2}}{K_R}\right)^{1/2} P_{O_2}^{1/4}$	$K_F^{1/2}$	$K_F^{1/2}$
(23)	$2[V_O^{\cdot\cdot}] = [N_M']$	Intermediate	$\left(\dfrac{2K_R}{[N_M']}\right)^{1/2} P_{O_2}^{-1/4}$	$\dfrac{K_i[N_M']^{1/2}}{(2K_R)^{1/2}} P_{O_2}^{1/4}$	$\dfrac{[N_M']}{2}$	$\dfrac{2K_F}{[N_M']}$
(24)	$p = [N_M']$	High	$\dfrac{K_i}{[N_M']}$	$[N_M']$	$\left(\dfrac{K_R[N_M']^2}{K_i^2}\right) P_{O_2}^{-1/2}$	$\left(\dfrac{K_F K_i^2}{K_R[N_M']^2}\right) P_{O_2}^{1/2}$
(25)	$p = 2[O_i'']$	High	$\left(\dfrac{K_R K_i}{2K_F}\right)^{1/3} P_{O_2}^{-1/6}$	$\left(\dfrac{2K_F K_i^2}{K_R}\right)^{1/3} P_{O_2}^{1/6}$	$\left(\dfrac{4K_R K_F^2}{K_i^2}\right)^{1/3} P_{O_2}^{-1/6}$	$\dfrac{p}{2}$

[a] The third and fourth rows apply only if $[N_M'] > K_F^{1/2}$, the intrinsic defect concentration. Otherwise row 2 is used.

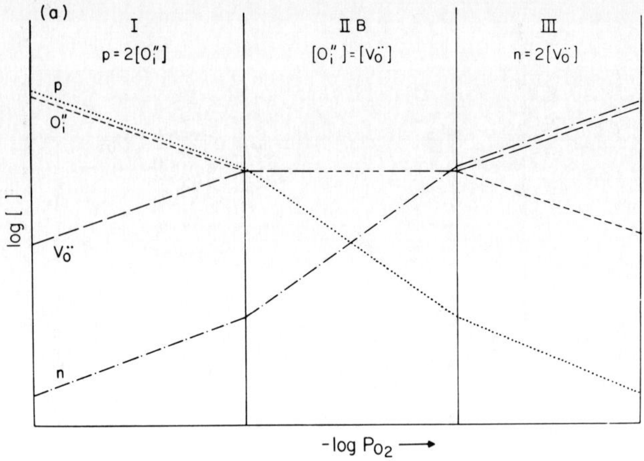

Fig. 2a. Defect concentration isotherm for pure oxide characterized by intrinsic Frenkel disorder.

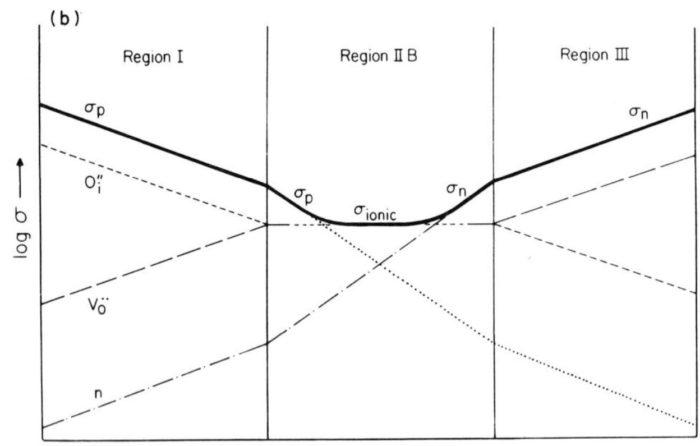

Fig. 2b. Corresponding conductivity isotherm. Ionic conductivity predominates only in the central portion of region IIB. The remainder of IIB is characterized by mixed conduction.

regions, both of which are impurity-controlled. Only the region characterized by the neutrality relation $2[V_O^{\cdot\cdot}] = [N'_M]$ is, however, likely to lead to mixed conduction. The manner in which the major defects depend on P_{O_2} is shown in Fig. 3a.

Increasing $[N'_M]$ results in a corresponding increase in $[V_O^{\cdot\cdot}]$, while maintaining $n = p = n_i$ constant at the center of region III. This serves to broaden the impurity-dominated regions as illustrated in Fig. 3a.

The partial conductivities in Figs. 2b and 3b are obtained by multiplying the carrier densities in Figs. 2a and 3a by their charges and mobilities,

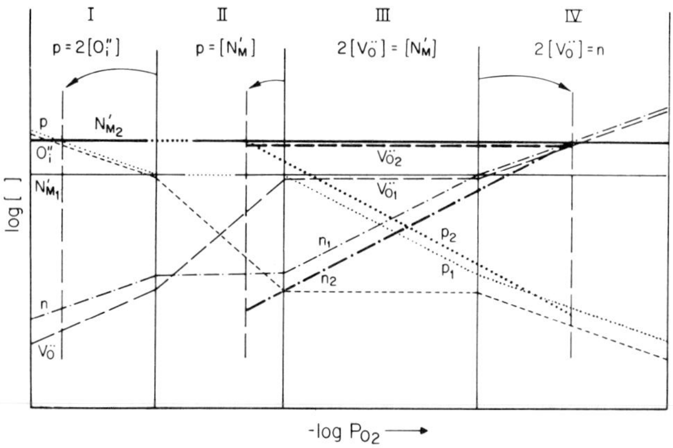

Fig. 3a. Defect concentration isotherm for acceptor-doped oxide with Frenkel disorder. Heavier curves illustrate the effects on defect concentrations upon increase of dopant concentration from $[N_M']_1$ to $[N_M']_2$, while arrows point to corresponding shifts in boundaries between defect regions.

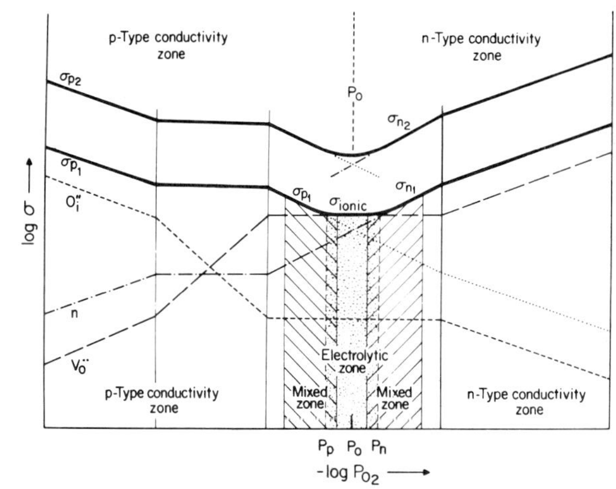

Fig. 3b. Corresponding conductivity isotherm. Solid curves correspond to the following cases: (1) $\mu_n, \mu_p > \mu(V_O^{\cdot\cdot})$, (2) $\mu_n, \mu_p \gg \mu(V_O^{\cdot\cdot})$. In case (1) five conductivity zones are obtained with the central zone (electrolytic) corresponding to predominantly ionic conduction. P_p and P_n correspond to the P_{O_2} values at which $\sigma_p = \sigma_{\text{ion}}$ and $\sigma_n = \sigma_{\text{ion}}$, respectively, while P_0 corresponds to the P_{O_2} at which $\sigma_n = \sigma_p$. Overlap of electrolytic and mixed zones is used to illustrate the imprecise nature of these zones.

respectively. Given the high electron/ion mobility ratios, this has the consequence of shifting the electronic curves upward relative to the ionic ones. In consequence, only the central portion of the region dominated by ionic disorder may exhibit predominantly ionic conductivity. Because oxides may be used as solid electrolytes within this more restricted region, it is commonly called the electrolytic domain. This description characterizes the lower solid conductivity curve found in Fig. 3b. Here we observe five possible zones as atmosphere varies from high to low P_{O_2}: (a) *p*-type electronic, (b) mixed *p*-type and ionic, (c) ionic (electrolytic zone), (d) mixed *n*-type and ionic, and (e) *n*-type electronic.

If the electronic/ionic carrier mobility ratio becomes sufficiently large, the upper solid curve in Fig. 3b is obtained. In this case, only *p*- and *n*-type conductivity zones are observed, with the transition between them characterized by a simple minimum rather than the P_{O_2}-independent region obtained previously.

The above results demonstrate a number of characteristic attributes of mixed conductors. First, given the high electon/ion mobility ratios, only in the regimes in which ionic species alone are included in the simplified neutrality equation, e.g., $[O_i''] = [V_O^{\cdot\cdot}]$ or $[N_M'] = 2[V_O^{\cdot\cdot}]$, can mixed conduction be expected. Second, in the regime where ions predominate, these majority defects do not usually exhibit a P_{O_2} dependence, while the minority electronic defects show a strong P_{O_2} dependence (e.g., $P_{O_2}^{\pm 1/4}$). Because mobilities are, in most cases, independent of P_{O_2}, this difference may be used to distinguish experimentally between ionic and electronic conductivity. Furthermore, these results emphasize the fact that the concentrations of minority electronic defects vary dramatically with experimental conditions even when ionic conductivity variations are imperceptibly small. Consequently, it is to be expected that the ionic/electronic defect ratio is itself a strong function of environment.

The minimum in electronic conductivity occurs at the center of the region dominated by ionic disorder (assuming $\mu_e = \mu_p$), as may be confirmed by reference to Figs. 2b and 3b. At this point, the electron and hole densities are equal and their magnitudes may be calculated from Eq. (8). This point also marks the position at which the ionic/electronic conductivity ratio is at its optimum. Deviation from this point in the direction of either increasing or decreasing P_{O_2} brings one closer to the condition of mixed conduction. As we shall see, the location and magnitude of this conductivity minimum play an important role in the relations describing mixed conduction in oxides..

Since the intrinsic electron–hole density is independent of doping, increases in $[N_M']$ and thereby $[V_O^{\cdot\cdot}]$ act to widen the electrolytic domain by shifting the boundaries separating regions II–III and III–IV to higher and lower P_{O_2} values, respectively (Fig. 3a). The location of these boundaries may be determined by solving for the conditions under which one majority defect is replaced by another in Eq. (26). Thus, for example, $p = 2[V_O^{\cdot\cdot}]$ at the transition from region II to III, while $n = [N_M']$ at the transition from

TABLE VI

Boundary Locations Defining Defect Transitions

Boundary location	Controlling defect relation	
$P_{O_2}(\text{I–II}) = \left(\dfrac{K_R[N'_M]^3}{2K_F K_i^2}\right)^2$	$[N'_M] = 2[O''_i]$	(27)
$P_{O_2}(\text{II–III}) = \left(\dfrac{2K_R[N'_M]}{K_i^2}\right)^2$	$p = 2[V_O^{\cdot\cdot}]$	(28)
$P_{O_2}(n\text{–}p) = \left(\dfrac{2K_R}{K_i[N'_M]}\right)^2$	$n = p$	(29)
$P_{O_2}(\text{III–IV}) = \left(\dfrac{2K_R}{[N'_M]^3}\right)^2$	$n = [N'_M]$	(30)

III to IV. The P_{O_2} values at which these occur, as well as the p–n transition within region III, are given in Table VI. The degree of ionic or mixed conduction relative to purely electronic conductivity is increased as region III is widened. With the use of Eqs. (28) and (30), the width of this region on a logarithmic scale is calculated to be

$$\log\left(\frac{P_{O_2}(\text{II–III})}{P_{O_2}(\text{III–IV})}\right) = 4\log\left(\frac{[N'_M]^2}{K_i}\right) \tag{31}$$

As expected, the electrolytic domain increases with increasing dopant and decreasing K_i $(= n_i^2)$. Since E_g is large for mixed conductors, K_i increases rapidly with increasing T, resulting in rapid shrinkage of the ionic domain with temperature.

When considering mixed conduction, only the values of $P_{O_2}(\text{II–III})$, $P_{O_2}(n\text{–}p)$, and $P_{O_2}(\text{III–IV})$ are of relevance. To evaluate these boundaries for a given material both K_i, K_r, and the dopant level must be known. The magnitude of K_i can usually be estimated from optically determined values of the high-frequency absorption edge (Table III). On the other hand, the equilibrium constant characterizing oxidation–reduction processes K_R has been determined experimentally for surprisingly few oxides. One of the recent exceptions has been the CeO_{2-x} system in which significant deviations from stoichiometry can be induced at readily accessible experimental conditions. Here $K_R(T)$ was found to be (Tuller and Nowick, 1979)

$$K_R(CeO_{2-x}) = 1.2 \times 10^{92}\exp(-4.67\ \text{eV}/kT) \quad \text{m}^{-9}\,\text{atm}^{1/2}$$

while in another study (Strehlow and Cook, 1973) a band gap of 5.5 eV was determined optically which, with Eqs. (9) and (18) and the free electron mass, leads to

$$K_i(CeO_2) = 2.3 \times 10^{43} T^3 \exp(-5.5 \text{ eV}/kT) \quad m^{-6}$$

In Fig. 4, the boundaries defining region III are plotted as $\log P_{O_2}$ versus $1000/T$ for $[N'_{Ce}]$ levels of 1 ppm and 1 mole%. One observes that the regime dominated by ionic disorder may cover an extensive range of atmospheres even for low impurity levels. In the case of CeO_2, this amounts to nearly 100 orders of magnitude of P_{O_2} at 600°C and 1 ppm impurity. It should be noted, however, that of this wide range of atmospheres only a small fraction are experimentally accessible at 600°C (i.e., $P_{O_2} \leqslant 1$ atm), while none are, e.g., at 1000°C. Therefore, except for termperatures below ~600°C and P_{O_2} values near 1 atm, one would expect CeO_2 to be a *n*-type semiconductor, as is in fact observed experimentally.

As predicted by Eq. (30) (i.e., $P_{O_2}(\text{III–IV}) \propto [N'_M]^{-6}$), the position of the III–IV boundary is extremely sensitive to impurity additions, so that a shift in $[N'_M]$ from 1 ppm to 1 mole% results in a 25 order of magnitude shift in $P_{O_2}(\text{III–IV})$ to more reducing values, as indicated in Fig. 4. Now, ionic disorder dominates the experimentally accessible regime below ~800°C and, depending on the μ_i/μ_e ratio, will be characterized by either ionic or mixed conduction. Since $P_{O_2}(\text{III–IV}) \propto \exp(-2E_R/kT) = \exp(-9.34 \text{ eV}/kT)$, the boundary shifts rapidly to a higher P_{O_2} with increasing *T*, so that by ~1200°C we expect that even this heavily doped material will become *n*-type semiconducting even at 1 atm. These predictions have been

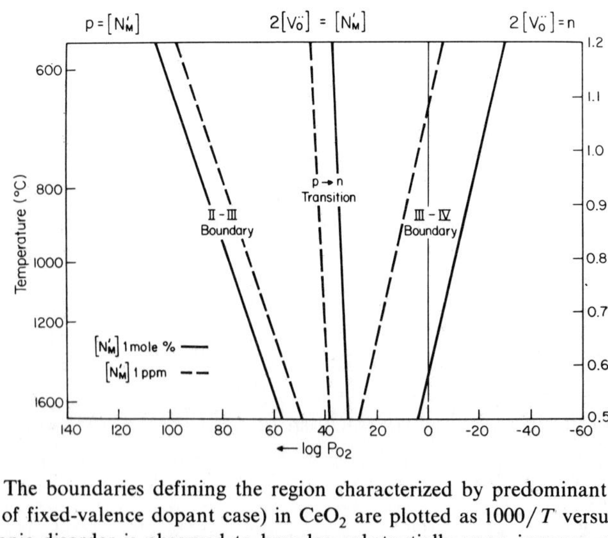

Fig. 4. The boundaries defining the region characterized by predominant ionic disorder (region III of fixed-valence dopant case) in CeO_2 are plotted as $1000/T$ versus $\log P_{O_2}$. The region of ionic disorder is observed to broaden substantially upon increase of $[N'_M]$ from 1 ppm to 1 mole %.

confirmed in a series of studies on CeO_2 doped with Y_2O_3, CaO, Nd_2O_3, La_2O_3, and Gd_2O_3 (Tuller *et al.*, 1981; Kudo and Obayashi, 1976).

For oxides without multivalent cations, for example, ThO_2 and Al_2O_3, E_R can be expected to be considerably larger in magnitude than the 4.67 eV obtained for multivalent CeO_2. This should result in an even more rapidly contracting ionic domain with increasing temperature. Similar considerations apply to the position of the II–III boundary.

2. *Mixed-Valence Impurities*

I next consider a somewhat more complex case in which the dopant exhibits variable valence. The importance of mixed-valence impurities in initiating mixed conduction, particularly in refractory oxides such as Al_2O_3, MgO, and ThO_2, is only now beginning to be appreciated. Consequently, little is available in the literature to guide one in predicting the properties of such systems. In the following I therefore extend the above treatment to such systems and emphasize the manner in which the resultant predictions differ from those for the fixed-valence case.

Let us choose an impurity N_M which may possess a charge equal to (N_M) or one less than (N'_M), that of the host. Under sufficiently reducing conditions, all dopant ions exist in the N'_M state, and so predictions should be indistinguishable from the fixed-valence case. At higher P_{O_2} values, however, $[N'_M]$ itself becomes a function of oxygen pressure and must thereby modify the defect relations as derived above.

The reduction of N^*_M

$$N^*_M + e' \rightarrow N'_M \tag{32}$$

may be described by the mass-action relation.

$$[N'_M]/[N^*_M]n = K_D(T) = K^0_D\exp(-E_D/kT) \tag{33}$$

in which $|E_D|$ is the (donor) ionization energy. Because the dopant is distributed between two valence states one requires a mass-balance relation given by

$$[N^*_M] + [N'_M] = [N_M]_{tot} \equiv N_t \tag{34}$$

in which the total dopant concentration $[N_M]_{tot}$ is designated by N_t. Equations (17)–(19) and (26) remain valid and, together with Eqs. (33) and (34), may be used to solve for the defect concentrations.

Following the Brouwer approach outlined earlier, we find five regions corresponding to unique combinations of limiting neutrality and mass-balance relations [i.e., Eqs. (26) and (34)], which are listed in Table VII. The solutions given in rows 1, 2, and 5 of this table are identical to the corresponding rows (1, 3, and 5) in Table IV. The difference shows up only

TABLE VII

Limiting neutrality and mass balance relations	P_{O_2} range	n	p
(35) $n = 2[V_O^{\cdot\cdot}]$; $[N_M'] = N_t$	Low	$(2K_R)^{1/3}P_{O_2}^{-1/6}$	$\dfrac{K_i}{(2K_R)^{1/3}}P_{O_2}^{1/6}$
(36) $[N_M'] = 2[V_O^{\cdot\cdot}]$; $[N_M'] = N_t$	Low	$\left(\dfrac{2K_R}{N_t}\right)^{1/2}P_{O_2}^{-1/4}$	$K_i\left(\dfrac{N_t}{2K_R}\right)^{1/2}P_{O_2}^{1/4}$
(37) $[N_M'] = 2[V_O^{\cdot\cdot}]$; $[N_M^*] = N_t$	Intermediate	$\left(\dfrac{2K_R}{K_D N_t}\right)^{1/3}P_{O_2}^{-1/6}$	$K_i\left(\dfrac{K_D N_t}{2K_R}\right)^{1/3}P_{O_2}^{1/6}$
(38) $[N_M'] = p$; $[N_M^*] = N_t$	High	$\left(\dfrac{K_i}{K_D N_t}\right)^{1/2}$	$(K_i K_D N_t)^{1/2}$
(39) $2[O_i''] = p$; $[N_M^*] = N_t$	High	$\left(\dfrac{K_R K_i}{2K_F}\right)^{1/3}P_{O_2}^{-1/6}$	$\left(\dfrac{2K_F K_i^2}{K_R}\right)^{1/3}P_{O_2}^{1/6}$

in the rows in which the majority of the dopant is isovalent with the host cation (i.e., rows 3 and 4). The results of Table VII are shown plotted in Fig. 5 as log[ ] versus $-\log P_{O_2}$ for arbitrary temperature and equilibrium constants.

There are a number of features that should be noted. First, in contrast to the cases treated previously, there are now two regions (III and IV) that are controlled by ionic disorder. Of special interest is the fact that in the first of these regions the majority ionic defect $V_O^{\cdot\cdot}$ is now a function of oxygen pressure. Because the ionic conductivity in oxides with fixed-valence impurities is independent of P_{O_2}, the tendency exists to identify any P_{O_2}-dependent conductivity as being electronic. Clearly such assumptions would lead to errors in the present case.

As before, we may obtain a representation of the P_{O_2} dependence of the partial conductivities by multiplying the carrier densities in Fig. 5 by their respective charges and mobilities. Again, because of the higher electronic mobilities, this has the effect of shifting the electronic components upward relative to the ionic ones. Depending on the relative mobilities of the various defects one of four possible $\log \sigma$ versus $\log P_{O_2}$ curves, as depicted in Fig. 6, can be imagined.

If $\mu(V_O^{\cdot\cdot})$ is not too much smaller than μ_p and μ_n, then curve 1 is obtained. Here the ionic conductivity predominates on both sides of the

Defect Concentrations for a Doped Frenkel-Like Oxide with a Variable-Valence Impurity

$[V_O^{\cdot\cdot}]$	$[O_i'']$	$[N_M]$
$\frac{n}{2}$	$K_F\left(\frac{4}{K_R}\right)^{1/3} P_{O_2}^{1/6}$	$\frac{N_t}{K_D(2K_R)^{1/3}} P_{O_2}^{1/6} = [N_M^*]$
$\frac{N_t}{2}$	$\frac{2K_F}{N_t}$	$\frac{N_t^{3/2}}{K_D(2K_R)^{1/2}} P_{O_2}^{1/4} = [N_M^*]$
$\left(\frac{K_R^{1/2}K_DN_t}{2}\right)^{2/3} P_{O_2}^{-1/6}$	$K_F\left(\frac{2}{K_R^{1/2}K_DN_t}\right)^{2/3} P_{O_2}^{1/6}$	$[N_M'] = 2[V_O^{\cdot\cdot}]$
$\frac{K_RK_DN_t}{K_i} P_{O_2}^{-1/2}$	$\frac{K_FK_i}{K_RK_DN_t} P_{O_2}^{1/2}$	$[N_M'] = p$
$\left(\frac{4K_RK_F^2}{K_i^2}\right)^{1/3} P_{O_2}^{-1/6}$	$\frac{p}{2}$	$[N_M'] = \left(\frac{K_RK_i}{2K_F}\right)^{1/3} K_DN_tP_{O_2}^{-1/6}$

III–IV boundary but changes its dependence on oxygen partial pressure from a $P_{O_2}^{-1/6}$ dependence in region III to a P_{O_2} independence in region IV.

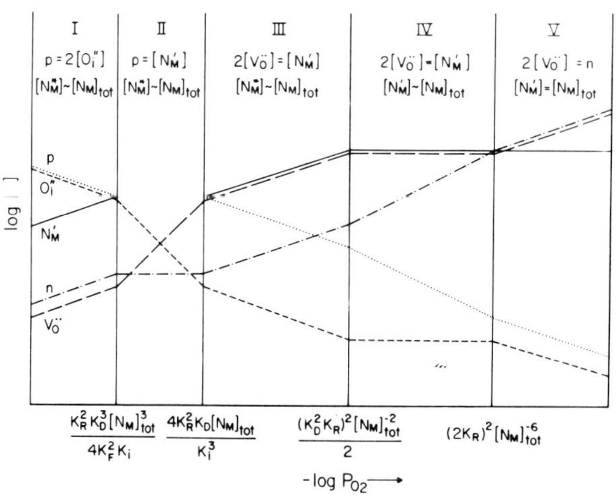

Fig. 5. Defect concentration isotherm for a variable-valence acceptor-doped oxide with Frenkel disorder. Two regions, III and IV, are dominated by ionic disorder, with $[V_O^{\cdot\cdot}]$ exhibiting a P_{O_2} dependence in the first of these regions. The expressions at the bottom of the figure define the P_{O_2} values at which transitions between the defect regions occur.

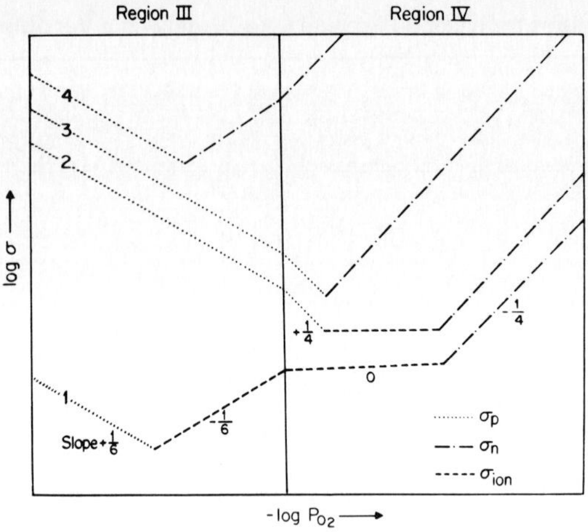

Fig. 6. Conductivity isotherms corresponding to the defect diagram in Fig. 5. Curves 1–4 correspond to the following conditions: (1) μ_n, $\mu_p > \mu(V_O^{\cdot\cdot})$; (2) $\mu_p > \mu_n > \mu(V_O^{\cdot\cdot})$; (3) μ_n, $\mu_p \gg \mu(V_O^{\cdot\cdot})$; (4) $\mu_n > \mu_p \gg \mu(V_O^{\cdot\cdot})$. For curves (1) and (2) an elecrolytic domain is obtained, while for (3) and (4) only semiconduction is obtained.

The conductivity minimum observed in III is now characteristic of a $p \rightarrow V_O^{\cdot\cdot}$ transition rather than the $p \rightarrow n$ transition to which such minima are normally attributed. This type of electronic-to-ionic transition, recently observed to occur in Ce-doped ThO_2 (Fujimoto and Tuller, 1979), is shown in Fig. 7. Ionic transference measurements, also shown, confirmed that the conductivity on the low-P_{O_2} side of the minimum was indeed ionic. Similar types of electron–ion transitions have recently been observed to occur in Co-doped Al_2O_3 (Dutt *et al.*, 1975) and Fe-doped MgO (Sempolinsky *et al.*, 1980). Since the sources of such conductivity minima have now been shown to result from either an electron–hole or electron–ion transition, care must be exercised when attempting to utilize such extrema to extract experimental parameters such as the band gap E_g. The P_{O_2} dependence of the conductivity alerts one to the defect regime one is operating in (i.e., III: $\sigma \propto P_{O_2}^{\pm 1/6}$; VI: $\sigma \propto P_{O_2}^{0}$ or $P_{O_2}^{\pm 1/4}$).

Increasing μ_p while maintaining $\mu(V_O^{\cdot\cdot})$ and μ_n approximately as before, one obtains the results in curve 2. Region III of Fig. 6 now becomes entirely p-type semiconducting, while substantial ionic conduction emerges only in region IV following a $\sigma_p \propto P_{O_2}^{1/4}$ segment. Under these circumstances the P_{O_2}-independent ionic conductivity plateau is bounded by P_{O_2}-dependent electronic conductivities in the same manner as found for electrolytes with fixed-valence impurities (Fig. 3b).

Curves 3 and 4 represent situations in which the electron mobility μ_n becomes successively larger in magnitude, resulting in a total masking of

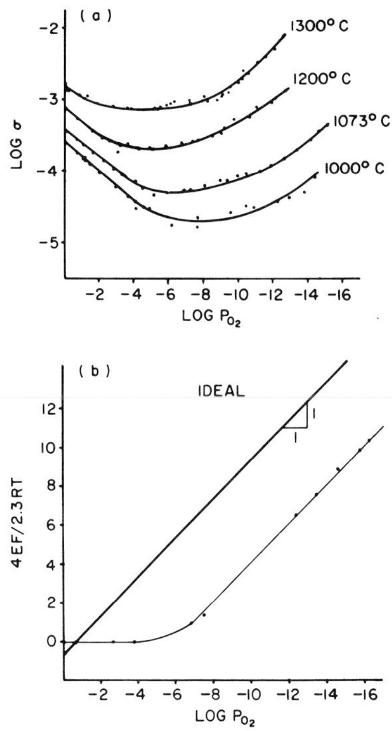

Fig. 7. (a) Total conductivity of ThO_2 plus 2 mole % Ce as a function of oxygen pressure. (From Fujimoto and Tuller, 1979.) (b) Oxygen concentration cell emf (E) measured across the same specimen at 1000°C. The slope at any pont determines t_{ion}. (From Fujimoto and Tuller, 1979.)

the ionic contribution. Depending on the relative electron/hole mobility ratio, the conductivity minimum may occur in either region III or region IV. This may be distinguished by noting whether the two branches of the V exhibit a $P_{O_2}^{\pm 1/6}$ or $P_{O_2}^{\pm 1/4}$ dependence. In either case, Eq. (48) may be used to derive the value of the band gap from the temperature dependence of the conductivity minimum after confirming experimentally that the minimum is indeed due to an electron–hole transition. Such an approach has most recently been used to obtain a value for the thermal band gap in MgO of 6.8 ± 0.5 eV (Sempolinsky *et al.*, 1980).

3. Defect Domain

a. Domain Boundaries. In the previous section we observed that transitions from one predominant defect to another could be induced by control of impurity level, temperature, and atmosphere. With the use of the defect relations listed in Tables V and VII, it was possible to predict the location of boundaries that define such transitions. Perhaps of greatest general utility are such mappings that specify the conditions under which an oxide may be expected to exhibit predominantly ionic conductivity and thus serve as a solid electrolyte in an electrochemical device. For this reason, such zones are usually referred to as the solid's ionic or electrolytic domain.

Three P_{O_2} values at each temperature are most useful in designating the domains of oxides. These are P_0, the atmosphere at which the electronic conductivity is at its minimum, and P_p and P_n, the atmospheres at both higher and lower P_{O_2} values at which the ionic conductivity equals the hole and electron conductivity, respectively (Fig. 3b). Thus, P_0 represents the point at which t_i is at its maximum, while P_p and P_n represent the conditions at high and low P_{O_2} values at which t_i falls to a value of $\frac{1}{2}$. Other P_{O_2} values at which t_i takes on other predetermined values (e.g., 0.99 or 0.9) are sometimes used, but these may easily be related to P_p and P_n.

Consider the form of the partial conductivities for the pure and fixed-valence dopant cases. The ionic conductivity is independent of the oxygen partial pressure and is given by

$$\sigma_i = \sigma_i^0 \exp(-E_i/kT) \tag{40}$$

while the electron and hole conductivities

$$\sigma_p = \sigma_p^0 P_{O_2}^{+1/4} \exp(-E_p/kT) \tag{41}$$

$$\sigma_n = \sigma_n^0 P_{O_2}^{-1/4} \exp(-E_n/kT) \tag{42}$$

can be distinguished by their opposite dependences on P_{O_2}. One obtains an expression for P_0 given by

$$\ln P_0 = -\frac{2(E_n - E_p)}{k}\frac{1}{T} + 2\ln\frac{\sigma_n^0}{\sigma_p^0} \tag{43}$$

by setting $\sigma_n = \sigma_p$ and solving for P_{O_2}. This equation defines the locus of points in $\ln P_{O_2}$–$1/T$ space that constitutes the boundary separating predominantly n-type from p-type conductivity. This boundary is observed to be a straight line characterized by slope $2(E_n - E_p)/k$ and intercept $2\ln(\sigma_n^0/\sigma_p^0)$.

In a similar manner, one obtains expressions for P_p and P_n:

$$\ln P_p = \frac{-4(E_i - E_p)}{k}\frac{1}{T} + 4\ln\left(\frac{\sigma_i^0}{\sigma_p^0}\right) \tag{44}$$

$$\ln P_n = \frac{-4(E_n - E_i)}{k}\frac{1}{T} + 4\ln\left(\frac{\sigma_n^0}{\sigma_i^0}\right) \tag{45}$$

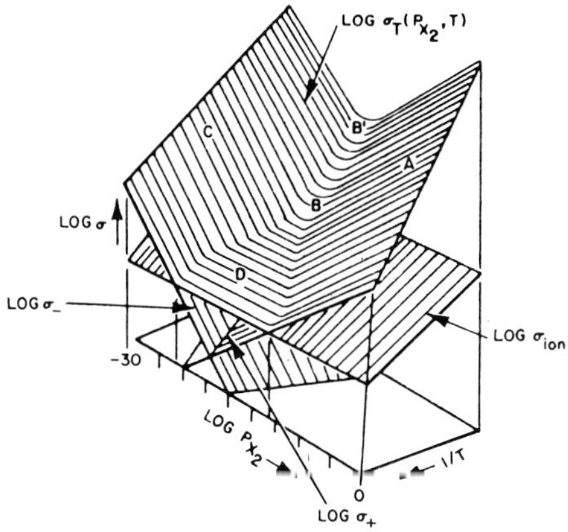

Fig. 8. Schematic representation of log σ surfaces over log P_{x_2}–$1/T$ space for $\sigma_{\rm ion}$, σ_p, σ_n, and $\sigma_{\rm tot}$. (Reprinted from Patterson, 1971, p. 135, by courtesy of Marcel Dekker, Inc.)

by setting $\sigma_p = \sigma_{\rm i}$ and $\sigma_n = \sigma_{\rm i}$, and solving for P_{O_2}. Again straight lines define the boundaries in ln P_{O_2}–$1/T$ space that separate regions of predominantly ionic from electronic conductivity. Consequently, the regime bounded by P_p at high pressures and P_n at low pressures comprises the ionic domain in which $t_{\rm i} \geqslant \frac{1}{2}$.

When Eqs. (40)–(42) are plotted as a function of P_{O_2} and $1/T$, one obtains a three-dimensional figure (Fig. 8) first introduced by Patterson (1971). The n-type, p-type, and ionic conductivities appear as planes whose intersections represent the boundaries between ionic and electronic conductivity domains defined by Eqs. (43)–(45). The intersection of planes A and C represents P_0, while the intersections of plane D with planes A and C represent P_p and P_n, respectively. A two-dimensional projection of Fig. 8 on the P_{O_2}, $1/T$ plane as shown in Fig. 9 is usually adequate, since it indicates the locations of the various domains.

Other three-dimensional figures that emphasize the effect of dopant on domain size have been introduced by Heyne (1977). The first of these is the isothermal log σ-versus-P_{O_2} and log$[N'_M]$ plot in Fig. 10. Here one observes that the ionic conductivity is represented by two intersecting planes, one of which is horizontal, showing $\sigma_{\rm i}$ to be independent of $[N'_M]$, and the other inclined at 45°, symbolizing a linear dependence of $\sigma_{\rm i}$ on $[N'_M]$. The transition occurs when impurity-generated defects become greater in concentration than those formed intrinsically (i.e., $2[N'_M] > K_F^{1/2}$), signaling the change in neutrality equation from Eq. (22) to Eq. (23). One also observes that the width of the electrolytic domain first remains constant and then widens once $\sigma_{\rm i}$ becomes dependent on $[N'_M]$. On the other hand, the

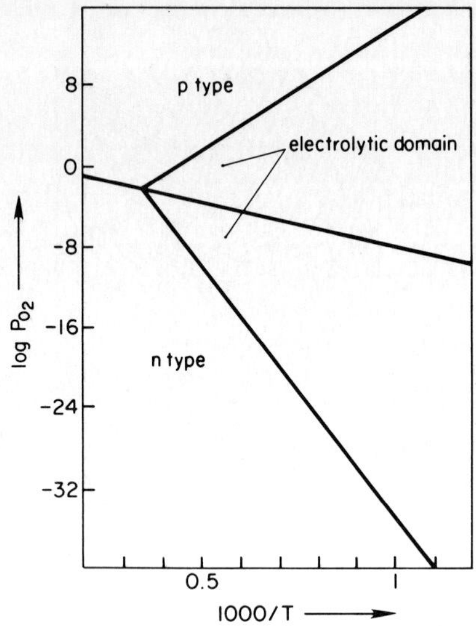

Fig. 9. Domain boundaries of stabilized zirconia as projected on the log P_{O_2}–$1000/T$ plane. (From Heyne, 1977.)

intrinsic electronic conductivity found at the intersection of the *n*- and *p*-type planes remains independent of $[N'_M]$, as expected from Eq. (8).

To see how the position of the domain boundaries depends on the three independent variables P_{O_2}, $1/T$, and $[N'_M]$ we construct the three-dimensional graph shown in Fig. 11 by superimposing the positions of the

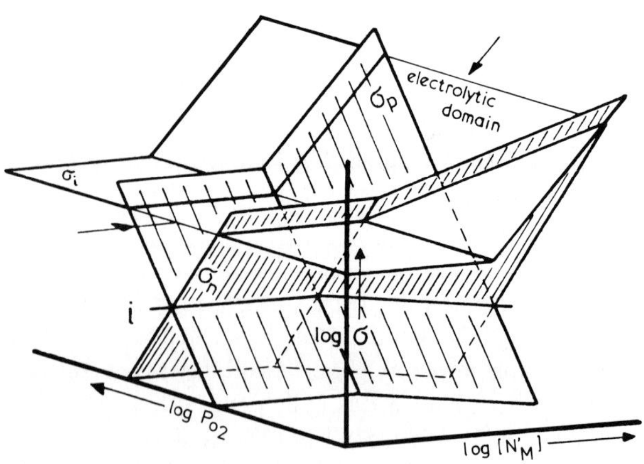

Fig. 10. Isothermal conductivity planes corresponding to electron, hole, and ion transport plotted as a function of log P_{O_2} and $\log[N'_M]$. The transition from horizontal to inclined σ_i plane occurs at $2[N'_M] = K_F^{1/2}$. (From Heyne, 1977.)

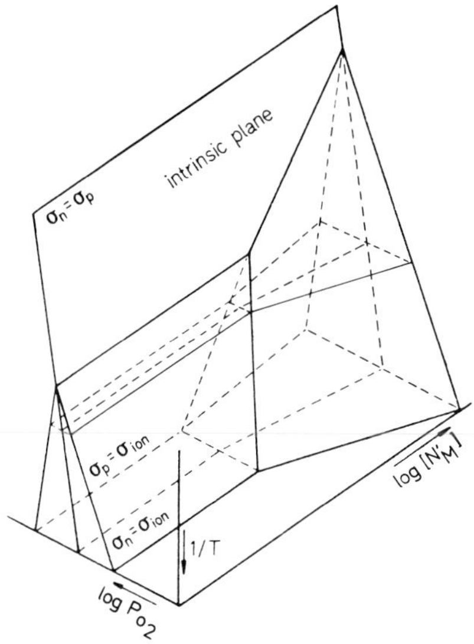

Fig. 11. Domain boundaries in $\log P_{O_2}$–$\log[N'_M]$–$1/T$ space. The interior of the tentlike structure corresponds to predominantly ionic conductivity. (From Heyne, 1977.)

domain boundaries of Fig. 10 vertically as a function of reciprocal temperature. A tentlike structure is generated, the interior of which corresponds to predominantly ionic conductivity. This figure clearly demonstrates that increasing levels of dopant increase not only the range of applicable P_{O_2} values but also the maximum operating temperatures within which the solid remains ionic.

The disappearance of the electrolytic domain at elevated temperatures is predicted on the assumption that the electronic conductivities possess higher activation energies than the ionic component; i.e., $E_n, E_p > E_i$. Although this is generally the case (E_i for solid electrolytes is usually low), there are exceptions, e.g., YDT, for which $E_p \cong E_i$ (Tuller, 1978).

Extending this type of analysis to oxides with variable-valence impurities introduces several complications. Two rather than one defect regime are now governed by ionic disorder. This manifests itself as a change in the P_{O_2} dependence of the ionic partial conductivity from $\sigma_i \propto P_{O_2}^{-1/6}$ to partial pressure independence as the boundary between the two regimes is crossed. In a three-dimensional representation such as Fig. 8, the electrolytic domain must now include an inclined plane at elevated P_{O_2} values in addition to the horizontal plane at lower P_{O_2} values. An appropriate $\log\sigma$–$\log P_{O_2}$–$1/T$ plot displaying these characteristics is presented in Fig. 12. It should be noted that σ_n and σ_p also change their dependence on T and P_{O_2} as the

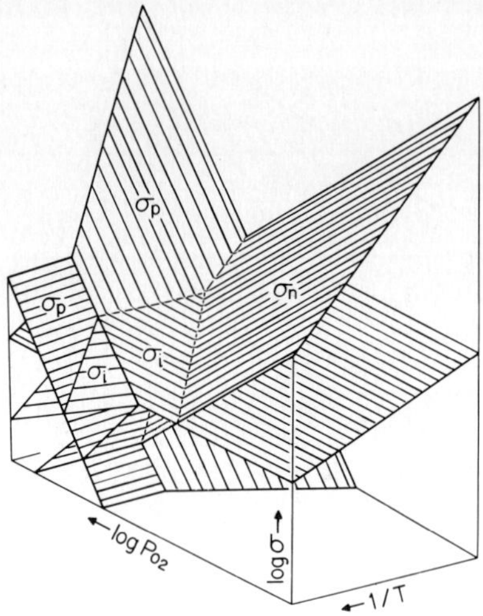

Fig. 12. Schematic representation of log σ surfaces over log P_{O_2}–$1/T$ space for the case of variable-valence impurities. Note that a component of the electrolytic domain at higher P_{O_2} values is inclined from the horizontal.

boundary between the two regimes of ionic disorder is crossed. For a reasonable choice of activation energies, the region of ionic disorder represented by $\sigma_i \propto P_{O_1}^{-1/6}$ disappears at elevated temperatures.

b. Minimum Requirements for the Existence of an Electrolytic Domain. The existence of an electrolytic domain implies that conditions exist under which the ionic conductivity is greater in magnitude than the combined n- and p-type conductivities. Clearly the optimum condition for this inequality exists at P_0, the P_{O_2} at which the electronic conductivity is at its minimum. For the most part, the intrinsic electronic conductivity is determined by the value of the band gap between the conduction and valence bands, as previously demonstrated in Table III. Heyne (1977) obtains an expression for the minimum allowable band gap energy consistent with the retention of predominantly ionic conduction in "good" electrolytes. He does this by requiring that the intrinsic electronic conductivity be less than 10^{-6} (Ω-cm)$^{-1}$ while assuming the electron and hole mobilities to be equal to 1 cm^2/V sec. In the following, we take a more general approach and derive a relation that can be applied to a much broader range of materials.

At the minimum in electronic conductivity, the n- and p-type conductivities are equal, so that

$$\tfrac{1}{2}\sigma_{\min} = nq\mu_n = pq\mu_p \tag{46}$$

Combining this with the fact that $np = K_i = N_c N_v \exp(-E_g/kT)$ we find

$$\sigma_{\min} = 2q(N_c N_v \mu_n \mu_p)^{1/2} \exp(-E_g/2kT) \tag{47}$$

Substituting for the constants in Eq. (9) one obtains

$$\begin{aligned}\sigma_{\min} &= 1540\, T^{3/2}(m_e^* m_h^*/m_0^2)^{3/4}(\mu_n \mu_p)^{1/2} \exp(-E_g/2kT) \quad \Omega^{-1}\,\mathrm{m}^{-1} \\ &= A(T)\exp(-E_g/2KT)\end{aligned} \tag{48}$$

The ionic conductivity, on the other hand, is given by

$$\sigma_i = q_i[\ \ ]_i(\mu_{i0}/T)\exp(-E_m/kT) = B(T)\exp(-E_m/kT) \tag{49}$$

where $[\ \ ]_i$ is the concentration of mobile ionic defects, q_i their effective charge, and E_m the migration energy. In the above equation one assumes the concentration of ionic defects is held constant by the concentration of fixed-valence impurities.

For a given ratio of ionic to electronic conductivity,

$$\frac{\sigma_i}{\sigma_e} = \frac{t_i}{(1-t_i)} = \frac{B}{A}\exp\left[\frac{-(E_m + E_g/2)}{kT}\right] \tag{50}$$

Solving for the band gap we obtain

$$E_g = 2kT\left[\ln\frac{t_i}{1-t_i} + \ln\left(\frac{A}{B}\right)\right] + 2E_m \tag{51}$$

which shows that the smallest possible energy gap consistent with the existence of an electrolytic domain varies linearly with temperature.

Taking appropriate values of μ_e, μ_h, $\mu(V''_{Mg})$, and E_g for MgO from Tables I, III, and IV and assuming a dopant level of 100 ppm we have generated such E_g-versus-T plots (Fig. 13) for three values of t_i, namely, 10^{-3}, 0.5, and 10^3. These values of t_i serve to divide the graph into regions that are heavily ionic ($t_i > 0.999$), mixed ionic and electronic ($0.001 < t_i < 0.999$), and heavily electronic ($t_i < 0.001$). We observe that the actual band gap of MgO, $E_g \approx 7.0$ eV (ignoring its temperature dependence), falls strictly within the ionic regime up to ~1300°C and remains above $t_i = 0.5$ up to the melting point. This picture of mixed conduction at elevated temperatures has recently been verified experimentally for MgO doped with Al and Sc (Sempolinsky *et al.*, 1980; Sempolinsky and Kingery, 1980).

Similar calculations have been made for CeO_2 plus 5 mole % Y_2O_3 which is known to be an excellent oxygen ion conductor at elevated temperatures (Tuller and Nowick, 1975c). Again, using Tables I, III, and IV to obtain values for the necessary variables in Eqs. (48), (49), and (51), we calculate E_g-versus-T curves for $t_i = 0.999$ and 0.5, which are plotted in Fig. 14. This time we observe that the actual band gap of ~5.5 eV never leads us into the mixed ionic–electronic regime at the electronic conductivity minimum.

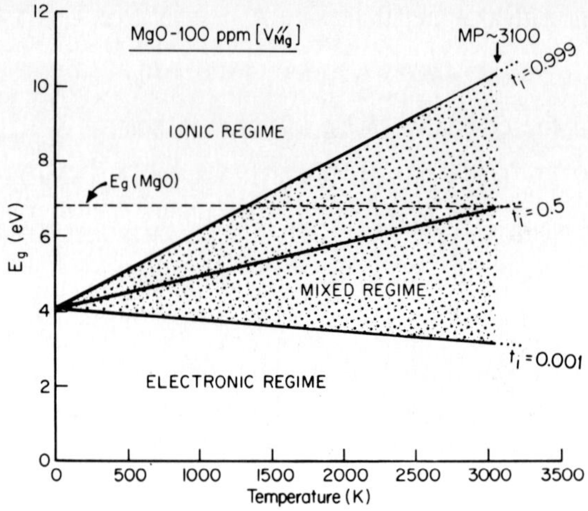

Fig. 13. The three curves correspond to the band gap that MgO would need to possess in order to be highly ionic ($t_i = 0.999$), mixed ($t_i = 0.5$), and highly electronic ($t_i = 0.001$) at the intrinsic condition $n_i = p_i$. These curves demonstrate that MgO with $[V''_{Mg}] = 100$ ppm will remain highly ionic ($t_i > 0.999$) up to ~1300 K and $t_i > 0.5$ up to the melting point at the electronic conductivity minimum.

It must be remembered, however, that composition excursions can shift the operating point far from this conductivity minimum, as previously illustrated in Fig. 4. As a case in point CeO_2 plus 5 mole % Y_2O_3 does in fact become a mixed conductor at $P_{O_2} < 10^{-4}$ atm at 1000°C (Tuller and Nowick, 1975c).

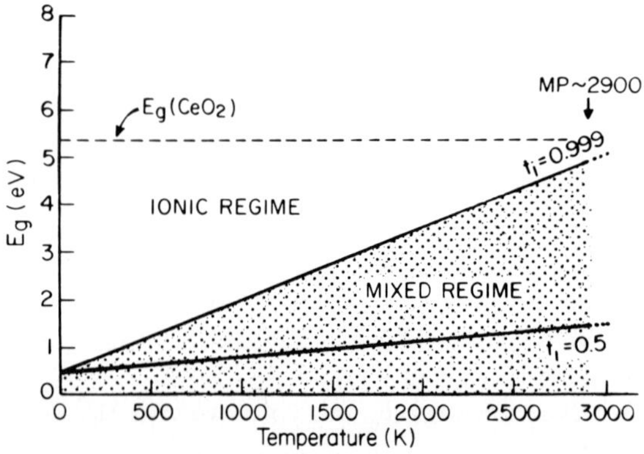

Fig. 14. Here the E_g-versus-T plot for CeO_2 plus 5 mole % Y_2O_3 demonstrates that at all temperatures, a material with high ionic conductivity remains an electrolyte in the vicinity of the electronic conductivity minimum. Only large excursions from stoichiometry or doping will change this picture.

B. High Defect Concentrations

We noted earlier that an essential criterion for the existence of predominantly ionic or mixed conductivity was for ion defect concentrations to be much greater in magnitude than the electronic defect levels. Thus in a large fraction of the cases of interest (Table II) this requires that ionic defect concentrations be at such high levels that the ideal gas laws on which the defect relationships derived earlier are based are no longer applicable. Under these circumstances, defect activities rather than their corresponding concentrations should properly be used in the appropriate mass-action relations.

Even for dopant levels on the order of $\lesssim 1$ mole % defect interactions can be expected. This may initially take the form of impurity–defect pair formation or the formation of extended defect complexes at even higher defect concentrations. The consequences that such complex formation has on ionic condition are discussed in detail in Chapter 5 by Kilner and Steele. Phenomenologically such complex formation serves to increase the activation energy associated with ionic conduction E_i [Eq. (40)] by including the energy of impurity–defect or complex disassociation. The general form of Eq. (40), however, remains unchanged.

Fortunately, the major features of our defect equilibria and domain analysis still remain intact. This follows from the fact that, in the vicinity of the electrolytic domain, defects formed due to excursions from stoichiometry remain negligible as compared to the level of ionic defects formed as a result of doping or structural disorder. Consequently, these defects may still be characterized by their concentrations within the mass-action relations. Furthermore, although ionic defects must now be treated in terms of their activities, because these usually remain constant over the composition range of interest, this leaves the general form of the relations describing defect formation and domain locations [Eqs. (43)–(45)] intact.

An additional consequence of the introduction of high levels of defect is to modify the values of the equilibrium constants characterizing the defect generation reactions, e.g., E_F, E_g, and E_R. Decreases in E_R have been reported, for instance, upon addition of > 1 mole % lower-valence dopants to CeO_2 (Tuller *et al.*, 1981; Kudo and Obayashi, 1976). This means that data characterizing mixed conduction in one composition cannot be extrapolated to other compositions, as recently pointed out for the cases of ceria- and thoria-based electrolytes (Tuller, 1978).

The consequences of high electronic defect levels in electrode materials have been recently treated by Heyne (1977) and are not discussed further in this chapter.

IV. Thermodynamic and Kinetic Processes

In the foregoing treatment we observed that the equilibrium concentration of ionic and electronic defects is a sensitive function of the activity of oxygen or equivalently the activity of the appropriate metal species in the vapor phase. Changes in stoichiometry and therefore in defect concentrations are induced by changing, for example, the P_{O_2} of the chamber in which the oxide is situated. These composition changes must proceed by mass transport via diffusion from the gas–solid interface into the bulk or from the bulk toward the gas–solid interface. The kinetics of these reduction–oxidation processes depend therefore on the rate of diffusion or mobility of the relevant defect species.

When considering reequilibration kinetics in semiconductors, a knowledge of the diffusion constants of the atomic species is adequate for predicting the rates of reaction. The situation is quite different when one is dealing with a solid electrolyte or mixed conductor. Here the diffusion rates of the electrons or holes may in fact be limiting. This may be appreciated by noting that, whatever the average composition of the oxide is at any point in time, it must remain electrically neutral. This requires that particles that diffuse inward toward or outward from the center of the solid do so effectively as neutral particles. Such ambipolar diffusion can therefore be expected to be limited by the minority species, which in the case of predominantly ionic conductors are electrons and holes.

In the following we consider the flow of charged particles under the combined influence of chemical and electric potential gradients. By considering various configurations that tend to suppress the flow of either electronic or ionic species within the oxide or the total external current to the oxide one arrives at relations that may be used to separate and analyze the partial ionic and electronic conductivities, as well as to predict other kinetic processes such as equilibration kinetics and gas permeation. These expressions also predict how the open-circuit voltages generated in galvanic cells, in which the oxide serves as the electrolyte, are influenced by nonnegligible electronic conductivity. These results are valuable not only in serving as a means for characterizing electronic conductivity but in predicting the relative utility of an oxide in such applications as fuel cells, batteries, and chemical sensors.

A. General Equations

Charge flows through solids under the combined influence of chemical and electrical potential gradients. If transport involves no volume changes, the effects of other forces and coupling between fluxes may be neglected

(Weppner and Huggins, 1978). The current density of the jth particle is related to its drift velocity v_j and the electrochemical potential gradient $\partial\eta_j/\partial x$ acting upon it by

$$\iota_j = c_j Z_j q v_j = -c_j Z_j q B_j(\partial\eta_j/\partial x) \tag{52}$$

where c_j is the carrier density, $Z_j q$ the charge, and $B_j = \mu_j/|Z_j|q$ the absolute mobility. The electrochemical potential

$$\eta_j = \bar{\mu}_j + Z_j q\phi \tag{53}$$

is composed of the electrostatic potential ϕ and the chemical potential† $\bar{\mu}$ which may be recast in terms of the activity a_j according to the relation

$$\bar{\mu}_j = \bar{\mu}_j^0 + kT\ln a_j \tag{54}$$

Application of Ohm's law to Eq. (52) after substitution of Eq. (53) leads to

$$\sigma_j = -(\iota_j/\Delta\phi)_{\partial\bar{\mu}_j/\partial x=0} = c_j Z_j^2 q^2 B_j = c_j Z_j q\mu_j \tag{55}$$

the same expression for the partial conductivity as assumed in Eq. (2). Combining Eqs. (55) and (52) we obtain a more compact relation for the current density given by

$$\iota_j = (-\sigma_j/Z_j q)(\partial\eta_j/\partial x) \tag{56}$$

The diffusion coefficient D_j may be introduced by way of the Nernst–Einstein relation

$$D_j = kTB_j = (kT/Z_j q)\mu_j \tag{57}$$

which when combined with Eqs. (53)–(56) results in an expression for the current density given by

$$\iota_j = -\sigma_j\frac{\partial\phi}{\partial x} - Z_j q D_j \frac{\partial\ln a_j}{\partial\ln c_j}\frac{\partial c_j}{\partial x} \tag{58}$$

The first term on the right, following Eq. (55), is a restatement of Ohm's law, while the second term is observed to satisfy Fick's first law for diffusion under a concentration gradient where D_{f}, the Fick's law diffusion coefficient, is related to the particle diffusion coefficient D_{j} by

$$D_{\mathrm{f}} = (\partial\ln a_j/\partial\ln c_j)D_j \tag{59}$$

The term preceding D_j, known as the thermodynamic factor, can sometimes be large and enchance the current flux above that which would be expected from the concentration gradient alone (Steele and Riccardi, 1974). Under conditions approaching ideality, the thermodynamic factor becomes unity.

†The symbol $\bar{\mu}$ is used to distinguish chemical potential from μ, the mobility.

Oxides that are predominantly ionic or mixed conductors were shown to possess very high concentrations of ionic defects. Consequently, it is to be expected that the chemical potential of these species will be essentially independent of position within the material (with the possible exclusion of the variable-valence case at high P_{O_2} values). Thus the motion of such ionic species is influenced almost in its entirety by the condition of the internal electrostatic field, i.e.,

$$\iota_{ion} = -\sigma_{ion}(\partial\phi/\partial x) \tag{60}$$

The consequences this has on the transport of the relatively dilute electronic carriers can be appreciated by considering what occurs when electrodes that block ions are used. In accordance with Eq. (60) this requires that the internal electric field approach zero. Under these conditions, the flow of the minority electronic carriers must proceed solely by diffusion. The consequences of these observations will become clear when we consider the simultaneous transport of several carriers.

B. Galvanic Cell Methods for Determining Transference Numbers

We next consider the situation in which no external current of any form flows. Such open-circuit conditions occur, for example, in the operation of galvanic cells, as well as during the growth of oxide layers on metals. Open-circuit conditions require only that all partial currents sum to zero:

$$\sum_j \iota_j = 0 = \sum_j \frac{\sigma_j}{Z_j q} \frac{\partial \eta_j}{\partial x} \tag{61}$$

which when combined with Eq. (3) gives

$$\sum_j \frac{t_j}{Z_j} \frac{\partial \eta_j}{\partial x} = 0 \tag{62}$$

Since the potential drop across such a cell is equal to the corresponding difference in electrochemical potential for electrons (Weppner and Huggins, 1978), it is useful to extract this component from the summation $\sum_j(\partial\eta_j/\partial x)$. This may be accomplished by considering the equilibrium condition for the ionization of neutral species

$$\bar{\mu}^*_{j(\text{neutral})} = \eta_j^{Z+} + Z_j\eta_e \tag{63}$$

together with Eq. (62) to give

$$\frac{\partial \eta_e}{\partial x} = \sum_j \frac{t_j}{Z_j} \frac{\partial \bar{\mu}^*_j}{\partial x} \tag{64}$$

which when integrated between points I and II results in the measurable emf

$$E = -\frac{1}{q}\int_{\mathrm{I}}^{\mathrm{II}} d\eta_e = -\frac{1}{q}\int_{\mathrm{I}}^{\mathrm{II}} \sum_j \frac{t_j}{Z_j} d\bar{\mu}_j^* \tag{65}$$

For a binary oxide, M_xO_y, Eq. (65) takes the form

$$E = \frac{1}{2q}\int_{\mathrm{I}}^{\mathrm{II}} (t_M + t_o)\, d\bar{\mu}_o^* \tag{66}$$

After the application of the Gibbs–Duhem relation, Eq. (66) may with the aid of Eq. (54) be rewritten in terms of the externally controllable variable P_{O_2}

$$E = \frac{kT}{4q}\int_{\mathrm{I}}^{\mathrm{II}} t_{\mathrm{ion}}\, d\ln P_{O_2} \tag{67}$$

This reduces to the well-known Nernst emf

$$E_N = (kT/4q)\ln((P_{O_2}^{\mathrm{II}}/P_{O_2}^{\mathrm{I}})) \tag{68}$$

when the ionic transference number is unity. This chemically induced voltage E_N is of course the principle behind the operation of a number of electrochemical devices including fuel cells, batteries, and chemical sensors. We return to a discussion of such applications in the latter part of this chapter.

Such cells may also be used to determine t_{ion} and thus enable one to separate the ionic and electronic contributions to the total conductivity, namely, $\sigma_{\mathrm{ion}} = t_{\mathrm{ion}}\sigma_{\mathrm{tot}}$ and $\sigma_{\mathrm{elec}} = (1 - t_{\mathrm{ion}})\sigma_{\mathrm{tot}}$. Commonly, this is done by assuming t_{ion} in Eq. (67) to be nearly independent of position within the sample, resulting in an expression for t_{ion} given by

$$t_{\mathrm{ion}} = E/E_N \tag{69}$$

Although the above assumption is close to true when operating within the electrolytic domain away from the boundaries, it becomes progressively weaker as one approaches the electrolytic boundaries. Under these circumstances, a more appropriate expression for t_{ion} may be obtained by differentiating Eq. (67) with respect to the upper limit while holding $P_{O_2}^{\mathrm{I}}$ constant to obtain

$$t_{\mathrm{ion}}(P_{O_2}^{\mathrm{II}}) = \frac{4q}{kT}\left(\frac{\partial E}{\partial \ln P_{O_2}}\right)_{P_{O_2} = P_{O_2}^{\mathrm{II}}} \tag{70}$$

The above equation shows that the ionic transference number at any P_{O_2} can be obtained experimentally by evaluating the slope of the E-versus-$\ln P_{O_2}$ curve obtained at that P_{O_2}. This approach, taken most recently in evaluating ionic transference numbers in doped Al_2O_3 (Dutt *et al.*, 1975), MgO (Sempolinsky and Kingery, 1980), and ThO_2 (Fujimoto and Tuller, 1979) as a function of P_{O_2} is to be preferred over the previous approach which at best results only in approximate values for t_i.

One may relate these expressions to the known partial pressure dependence of the carriers [Eqs. (41)–(43), (46)], as first done by Schmalzried (1962), by noting that

$$\sigma_e = \sigma_n + \sigma_p = \frac{\sigma_{min}}{2}\left[\left(\frac{P_0}{P_{O_2}}\right)^{1/4} + \left(\frac{P_{O_2}}{P_0}\right)^{1/4}\right] \tag{71}$$

for the case of fixed-valence impurities. This may be recast in terms of t_{ion} by

$$t_{ion} = \frac{\sigma_{ion}}{\sigma_{ion} + \sigma_e} = \left\{1 + \frac{\sigma_{min}}{2\sigma_{ion}}\left[\left(\frac{P_0}{P_{O_2}}\right)^{1/4} + \left(\frac{P_{O_2}}{P_0}\right)^{1/4}\right]\right\}^{-1} \tag{72}$$

Since σ_{ion} is equal to σ_n and σ_p at P_n and P_p, respectively, one may solve for P_n and P_p in terms of σ_{min} and σ_{ion} and obtain

$$P_n = \left(\frac{\sigma_{min}}{2\sigma_{ion}}\right)^4 P_0, \qquad P_p = \left(\frac{2\sigma_{ion}}{\sigma_{min}}\right)^4 P_0 \tag{73}$$

Substituting these results into Eq. (72) gives

$$t_{ion} = \left[1 + \left(\frac{P_n}{P_{O_2}}\right)^{1/4} + \left(\frac{P_{O_2}}{P_p}\right)^{1/4}\right]^{-1} \tag{74}$$

which domonstrates the utility of the constants P_n and P_p (i.e., the position of the domain boundaries) in determining the ionic transference number at any oxygen partial pressure. A plot of t_{ion} versus $\log P_{O_2}$ (Fig. 15) shows the major features of interest including a $t_i = 1$ plateau at intermediate P_{O_2} values and that P_p and P_n are the high and low P_{O_2} values at which t_i drops

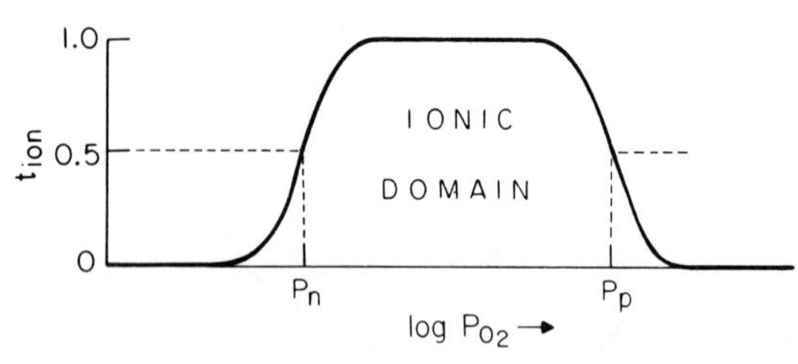

Fig. 15. The ionic transference number as a function of oxygen partial pressure in an oxide with fixed-valence impurities. The locations of P_n and P_p define the bounds of the ionic domain.

to a value of $\frac{1}{2}$. Integration of Eq. (67) with t_{ion} replaced by the above expression leads to

$$E = \frac{kT}{q}\left\{\ln\left[\frac{P_p^{1/4} + \left(P_{O_2}^{I}\right)^{1/4}}{P_p^{1/4} + \left(P_{O_2}^{II}\right)^{1/4}}\right] + \ln\left[\frac{P_n^{1/4} + \left(P_{O_2}^{II}\right)^{1/4}}{P_n^{1/4} + \left(P_{O_2}^{I}\right)^{1/4}}\right]\right\} \tag{75}$$

assuming $P_p \gg P_n$, a condition satisfied for most cases of interest.

Equation (75) had special utility when considered for a number of special limiting cases which include (Goto and Pluschkell, 1972)

(a) For $P_p \gg P_{O_2}^{II} > P_{O_2}^{I} \gg P_n$,

$$E = (kT/4q)\ln(P_{O_2}^{II}/P_{O_2}^{I}) = E_N \tag{76}$$

In this regime one obtains the Nernst emf, making this the optimal regime for use of the oxide as a solid electrolyte

(b) For $P_p \gg P_{O_2}^{II} > P_n \gg P_{O_2}^{I}$,

$$E = (kT/4q)\ln(P_{O_2}^{II}/P_n) \tag{77}$$

which says that, if the reference $P_{O_2}^{II}$ is maintained within the electrolytic domain but $P_{O_2}^{I}$ is chosen to be outside this domain, then the emf achieves a constant value depending only on the reference $P_{O_2}^{II}$ and P_n. This serves as a useful means for evaluating P_n.

(c) For $P_{O_2}^{II} > P_p \gg P_{O_2}^{I} > P_n$,

$$E = (kT/4q)\ln(P_p/P_{O_2}^{I}) \tag{78}$$

which essentially mirrors the results of Eq. (77) but at high P_{O_2} values. Other limiting cases are discussed by Goto and Pluschkell (1972).

Heyne and Beekmans (1971) have used the values of P_n and P_p obtained electrochemically to determine the band gap of stabilized zirconia. Rearrangement of the quantities in Eq. (73) shows that the intrinsic electronic conductivity is related to P_n and P_p by

$$\sigma_{\min} = 2(P_n/P_p)^{1/8}\sigma_{\text{ion}} \tag{79}$$

from which the band gap follows from Eq. (47), assuming a negligible temperature dependence for μ_n and μ_p.

C. Chemical Diffusion

We earlier observed that, even under open-circuit conditions, currents may continue to flow so long as they sum to zero [Eq. (61)]. Under such conditions the movement of particles, a consequence of the existence of composition gradients, occurs by so-called chemical diffusion. As previously noted, such conditions exist during composition reequilibration, gas permeation, and the like, and are characterized by the flux of neutral species. Heyne (1977), employing the ambipolar description, derived the following expression for the chemcial or ambipolar diffusion coefficient in mixed conductors:

$$D_a = t_e D_O + t_O D_e \tag{80}$$

in which D_O and D_e are the oxygen-ion and electronic diffusion coefficients, respectively. For predominantly ionic conductors, $t_O \gg t_e$, and consequently D_a becomes

$$D_a \sim t_O D_e \sim D_e \tag{81}$$

which is the same approxmate result obtained by Wagner (1936) in his "scaling theory" of metals. Thus, as anticipated earlier, the reequilibration kinetics of such materials may be used to obtain the diffusion coefficients or equivalently the mobilities of the minority electronic carriers. The two experimental approaches most commonly used to evaluate such small electronic contributions are the permeation and polarized cell techniques. In the following we treat these two approaches successively.

1. Permeation

If a fixed gradient of P_{O_2} is maintained across a specimen, a steady state flux of neutral oxygen from the high-P_{O_2} to the low-P_{O_2} side given by

$$J_O = -D_a \frac{\partial c_O}{\partial x} \tag{82}$$

will be observed. Taking the result of Eq. (81), $D_a \sim t_O D_e$, and assuming high ion defect levels and retention of local charge neutrality, this becomes after some manipulation in terms of a current flux

$$i = \frac{kT}{4q} \sigma_{ion} t_e \frac{\partial}{\partial x} \ln P_{O_2} \tag{83}$$

Integration over the sample thickness leads to

$$i = \left(\frac{kT}{4qL}\right) \sigma_{ion} \int_{I}^{II} t_e \, d \ln P_{O_2} \tag{84}$$

while differentiation of this expression with respect to the upper integration limit yields

$$\left(\frac{\partial i}{\partial \ln P_{O_2}}\right)_{P_{O_2} = P_{O_2}^{II}} = \left(\frac{kT}{4qL}\right) \sigma_{ion} t_e \tag{85}$$

In this way the electronic transference number t_e can be evaluated at any $P_{O_2}^{II}$ from the slope of the permeation current versus $\ln P_{O_2}$ at that P_{O_2}.

Taking into account the pressure dependence of σ_n and σ_p under conditions where $P_{O_2}^{II} \gg P_{O_2}^{I}$ Eq. (84) yields for the permeation current (Heyne, 1977)

$$i \approx \frac{kT}{qL}(\sigma_n^{I} + \sigma_p^{II}) \tag{86}$$

where σ_n^{I} and σ_p^{II} are the electron and hole conductivities at the two electodes. This result shows that the permeation rate is controlled by the larger of the two boundary conductivities and should therefore exhibit the same P_{O_2} dependence as that of the dominant electronic conductivity. Such results have been confirmed by a number of investigators using this technique to investigate hole conductivity at elevated P_{O_2} values in stabilized zirconia electrolytes. In a recent review of such results by Fouletier *et al.* (1976) one finds that $\sigma_h \approx \sigma_h^0 \exp(-2\ \mathrm{eV}/kT) P_{O_2}^{+1/4}$.

The individual contributions of carrier concentration and mobility to the electronic conductivity can be separated by recording the transient permeation current against time after an instantaneous change in P_{O_2} on the opposite side of the specimen. Since the diffusion of holes, for example, determines this process, the diffusion equation for holes

$$\partial p/\partial t = D_a(\partial^2 p/\partial x^2) \tag{87}$$

must be solved. Heyne and Beekmans (1971) show that $D_a(\sim D_h)$ is simply related to $\tau_{1/2}$, the time it takes the permeation rate to reach one-half its steady-state value, by

$$\tau_{1/2} = 0.1387(l^2/D_a) \tag{88}$$

where l is the thickness of the specimen. Such results when combined with values of σ obtained from steady state permeation measurements can be used to determine electronic carrier densities (Heyne and Beekmans, 1971).

Values for D_a may also be obtained by monitoring the oxidation–reduction kinetics of specimens, for example, by monitoring oxygen desorption after the P_{O_2} in the sample chamber is suddenly reduced. For short times, the total quantity of gas desorbed (expressed in terms of equivalent electrical charge) in time t is given by (Heyne and Beekmans, 1971)

$$Q_t = 2.26(D_a t)^{1/2} \tag{89}$$

2. *Polarized Cell Technique*

Partial electronic conductivities may also be determined by using ion-blocking electrodes to suppress ionic transport so that only electrons and holes can pass through the system. This may be accomplished by using a chemically inert material at one electrode so that no ions are supplied to the sample when a voltage is applied. This results in depletion of the mobile ions at that interface, with ultimately no ionic current flowing under steady state conditions. An electrode reversible to ionic species is used at the other end of the specimen, however, in order to fix the chemical potentials of the components and thus specify the state of the system.

Since the ionic current is virtually zero, at steady state $\partial\eta_{\mathrm{O}}/\partial x = 0$. Also, from Eq. (60), $\partial\phi/\partial x = 0$, so it follows from Eqs. (56) and (62) that the residual current is given by

$$i = (\sigma_{\mathrm{e}}/q)(\partial\bar{\mu}_{\mathrm{e}}/\partial x) = (\sigma_{\mathrm{e}}/2q)(\partial\bar{\mu}_{\mathrm{O}}^{*}/\partial x) \tag{90}$$

Integration over the sample thickness L gives

$$i = \frac{1}{2qL}\int_{\mathrm{I}}^{\mathrm{II}} \sigma_{\mathrm{e}}\, d\bar{\mu}_{\mathrm{O}}^{*} = \frac{kT}{4qL}\int_{\mathrm{I}}^{\mathrm{II}} \sigma_{\mathrm{e}}\, d\ln P_{\mathrm{O}_2} \tag{91}$$

While $P_{\mathrm{O}_2}^{\mathrm{I}}$ at the reversible electrode is fixed to the equilibrium P_{O_2} of the adjacent phase, $P_{\mathrm{O}_2}^{\mathrm{II}}$ at the blocked electrode is a function of the applied potential E given by

$$E = -\frac{1}{q}\int_{\mathrm{I}}^{\mathrm{II}} d\bar{\mu}_{\mathrm{e}} = -\frac{kT}{4q}\int_{\mathrm{I}}^{\mathrm{II}} d\ln P_{\mathrm{O}_2} = -\frac{kT}{4q}\ln\left[\frac{P_{\mathrm{O}_2}^{\mathrm{II}}}{P_{\mathrm{O}_2}^{\mathrm{I}}}\right] \tag{92}$$

Differentiating Eq. (91) with respect to the upper limit gives

$$\left(\frac{\partial i}{\partial \ln P_{\mathrm{O}_2}}\right)_{P_{\mathrm{O}_2} = P_{\mathrm{O}_2}^{\mathrm{II}}} = \frac{kT\sigma_{\mathrm{e}}^{\mathrm{II}}}{4qL} \tag{93}$$

which with Eq. (92) may also be written as

$$(\partial i_{\mathrm{e}}/\partial E)_E = -\sigma_{\mathrm{e}}^{\mathrm{II}}/L \tag{94}$$

This allows one to determine the electronic conductivity at a P_{O_2} corresponding to the applied voltage E from a plot of steady state polarization current as a function of applied voltage. By application of Eq. (54) we may rewrite Eq. (92) to give

$$E = -\frac{(\mu_{\mathrm{e}}^{\mathrm{II}} - \mu_{\mathrm{e}}^{\mathrm{I}})}{q} = \frac{-kT}{q}\ln\left(\frac{n^{\mathrm{II}}}{n^{\mathrm{I}}}\right) = \frac{kT}{q}\ln\left(\frac{p^{\mathrm{II}}}{p^{\mathrm{I}}}\right) \tag{95}$$

where n^{II} (p^{II}) and n^{I} (p^{I}) refer to the electron (hole) concentrations at the blocking and reversible electrodes, respectively. Consequently the electronic conductivity may be related to the applied potential by

$$\sigma_{\mathrm{e}} = \sigma_n + \sigma_p = \sigma_n^{\mathrm{I}}\exp(-qE/kT) + \sigma_p^{\mathrm{I}}\exp(qE/kT) \tag{96}$$

where σ_n^{I} and σ_p^{I} represent the electron and hole conductivities at the reference electrode. Substitution of Eq. (96) into (91), and assuming constant electron and hole mobilities, yields (Heyne, 1977)

$$i = (kT/qL)\{\sigma_p^{\mathrm{I}}[1 - \exp(-qE/kT)] + \sigma_n^{\mathrm{I}}[\exp(qE/kT) - 1]\} \quad (97a)$$

If one normalizes this relation by dividing through by $[1 - \exp(-qE/kT)]$, as first done by Patterson *et al.* (1967), one obtains

$$\frac{i}{1 - \exp(-qE/kT)} = \frac{kT}{qL}\left[\sigma_p^{\mathrm{I}} + \sigma_n^{\mathrm{I}}\exp(qE/kT)\right] \quad (97b)$$

which when plotted against $\exp(qE/kT)$ yields σ_n^{I} from the slope of the curve and σ_p^{I} from the intercept with the ordinate.

This approach was used by Patterson *et al.* (1967) to determine electron and hole conductivities in ZrO_2 plus 15 mole % CaO and ThO_2 plus 15 mole % $YO_{1.5}$. Difficulties were, however, encountered in obtaining perfect blocking conditions as a result of leakage conductance through the gas phase.

Recently Wang and Nowick (1980) demonstrated that the polarized cell technique need not be limited to conditions of complete ionic blocking at one electrode or complete reversibility at the other. They showed that the form of Eq. (97a) is maintained at sufficiently large potentials where E is replaced by the cathodic overpotential η_c. This substantially relieves the limitations on the application of this technique to oxides for which the ion blocking requirement is particularly difficult to obtain.

Transient measurements performed on such polarized cells may also be used to determine separately the mobilities of the minority electronic species. For example, if the externally applied voltage is removed after attaining a steady state, the concentation gradients given by Eq. (95) will decay until equilibrium concentrations, as determined by the reference electrode, are achieved throughout the specimen. These transients exhibit themselves as measurable voltage relaxations. Weppner (1976) has derived expressions that relate D_e and D_h to the time dependence of such voltage relaxations for a number of limiting conditions. Applying this technique to pellets of ZrO_2 plus 10 mole % Y_2O_3, he was able to obtain values for D_e and D_h in the temperature range of 700–900°C, as reproduced in Fig. 16.

A similar approach was utilized by Wang and Nowick (1980) to determine D_e in CeO_2 + 15 mole % CaO without, however, the restriction of a fully blocking electrode. The value $D_e = 10^{-4}$ cm^2/sec obtained in this way at 800°C was consistent with values determined by Tuller and Nowick (1977) from measurements of conductivity and thermoelectric power under equilibrium conditions.

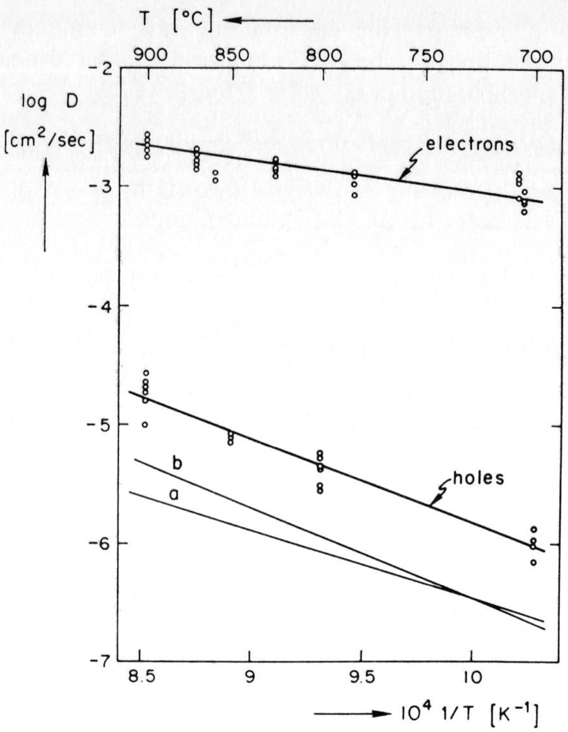

Fig. 16. Diffusion coefficients of electrons and holes in ZrO_2 plus 10 mole % Y_2O_3 obtained by a voltage relaxation technique. Curves a and b correspond to hole diffusion coefficients derived from permeation experiments (a) Heyne and Beekmans (1971); (b) Kitazawa and Coble (1974). (From Weppner, 1976.)

D. Other Techniques for Establishing Transference Numbers

1. Conductivity versus P_{O_2}

The total conductivity, as observed earlier, is composed of three terms:

$$\sigma_{tot} = \sigma_p + \sigma_{ion} + \sigma_n \tag{98}$$

which for the pure and fixed-valence cases becomes [Eqs. (40)–(42)]

$$\sigma_{tot} = \sigma_p^0 \exp(-E_p/kT) P_{O_2}^{+1/4} + \sigma_i^0 \exp(-E_i/kT) P_{O_2}^0 + \sigma_n^0 \exp(-E_n/kT) P_{O_2}^{-1/4} \tag{99}$$

If a σ_{tot}-versus-P_{O_2} isotherm includes a P_{O_2}-independent regime over a range of P_{O_2} values, it becomes a simple matter to evaluate the electron and hole contributions by subtracting the P_{O_2}-independent component from the total conductivity at each P_{O_2} of interest. This technique has been routinely used to separate P_{O_2}-independent ionic conductivities from P_{O_2}-dependent electron and hole conductivities in doped CeO_2 (Tuller *et al*, 1981; Tuller and Nowick, 1975c), ThO_2 (Iqbal and Baker, 1973; Bauerle, 1966; Lasker and Rapp, 1966; Maiti and Subbarao, 1976), cubic stabilized ZrO_2 (Patterson *et al.*, 1967), monoclinic ZrO_2 (Nasrallah and Douglass, 1974), Bi_2O_3 (Takahashi *et al*, 1977) and $LaAlO_3$ (Kilner *et al.*, 1978), as well as others. One should, however, keep in mind, when utilizing this approach, that situations can arise where the P_{O_2}-independent conductivity is electronic rather than ionic. This occurs, for example, when operating within region II of Fig. 3 or 5. A P_{O_2}-independent *p*-type conductivity has in fact been observed recently in Y_2O_3-doped UO_2 (Dudney *et al.*, 1981).

Complications in applying this approach also arise when dealing with oxides doped with variable-valence ions. Under these conditions, as observed earlier, σ_{ion} may become P_{O_2}-dependent (Fig. 6, curve 1). Under these circumstances, one may take advantage of the isothermal dependence of conductivity on N_t at constant P_{O_2}. This approach is discussed in the following section.

2. *Conductivity versus* $[N_M]_{tot}$

Just as the partial conductivities have characteristic P_{O_2} dependences [Eq. (99)], they also possess characteristic N_t dependences which can be used to distinguish them. For instance the intermediate region of Fig. 5 is characterized by the following relations:

$$n \propto N_t^{-1/3} P_{O_2}^{-1/6}, \qquad p \propto N_t^{1/3} P_{O_3}^{+1/6}$$
$$[V_O^{\cdot\cdot}] \propto N_t^{2/3} P_{O_2}^{-1/6}, \qquad [O_i''] \propto N_t^{-2/3} P_{O_2}^{+1/6} \tag{100}$$

Because of the identical P_{O_2} dependence exhibited by n and $[V_O^{\cdot\cdot}]$ and by p and $[O_i'']$, measurement of σ versus P_{O_2} would be insufficient to identify the more highly conducting species of each pair (compare curves 1 and 4 in Fig. 6, region III). If such measurements were performed on specimens with different dopant levels, however, it would be straightforward to identify the more highly conducting species. This procedure, recently applied to results obtained for Ce-doped ThO_2 (Fujimoto and Tuller, 1979), identified the $\sigma \propto P_{O_2}^{-1/6}$ component as being ionic rather than electronic (σ increased with Ce_t). This result was later confirmed with concentration cell measurements.

3. *Diffusion Measurements*

Combining Eq. (55) with (57) we obtain an expression relating the *j*th partial conductivity to its diffusion coefficient

$$\sigma_j = [c_j(Z_j q)^2 / kT] D_j \tag{101}$$

The transference number of a particle may therefore be determined by measuring its self-diffusion coefficient and comparing its equivalent conductivity [Eq. (101)] with that of the total conductivity. In this manner, Kingery *et al.* (1959) and Simpson and Carter (1966) confirmed that conduction in stabilized zirconia was, within experimental error, exclusively by oxygen ion transport.

4. Other Methods

A number of other techniques that make use of specialized instrumentation and/or are applicable over limited experimental conditions can sometimes be used to great advantage in separating the electronic and ionic contributions to the total conductivity.

The individual contributions of anions and cations may be evaluated by Faraday law experiments in which the passage of known quantities of charge through a solid are related to the changes in the masses of the anode and cathode. These so-called Tubandt- or Hittorf-type experiments are discussed in some detail by Weppner and Huggins (1978).

The study of thermoelectronic emission from oxides at elevated temperatures has been useful in identifying electronic contributions. The density of electrons emitted by a solid as a function of temperature T and work function ϕ is given by Richardson's law:

$$J = AT^2\exp(-\phi/kT) \tag{102}$$

which for the case of nondegenerate semiconductors takes the form (Rifflet *et al*, 1975)

$$J = BnT^{1/2}\exp(-\chi/kT) \tag{103}$$

where n is the electron concentration, χ the electron affinity, and B a constant.

Such measurements, performed on refractory oxides such as ZrO_2 and Y_2O_3 as a function of P_{O_2}, resulted in simple power law dependencies, i.e., $J \propto n \propto P_{O_2}^{-1/x}$, as expected from defect equilibria. Since the thermoelectronic flux is sensitive only to the electron density, it serves as a useful means of separating this contribution (even when small) from the overall conductivity, as well as distinguishing contributions of n and μ_n to σ_n. Experimental details are discussed by Rifflet *et al.* (1975).

Other more traditional means used in the semiconductor field for separating electrons and holes, such as thermoelectric power and Hall effect measurements, have been reviewed many times in the past (e.g., see Sze, 1969).

V. Mixed Conduction in Oxides–A Survey

Dependable quantitative data concerning mixed ionic and electronic conduction are available only for a limited number of oxides. In the following, emphasis is therefore focused on a selected number of oxides, primarly those crystallizing with the fluorite structure for which significant data have recently been reported.

A. Cubic—Fluorite Oxides

A number of well-known oxides crystallize with the cubic fluorite structure, including CeO_2, ThO_2, HfO_2, and ZrO_2, the latter two only if stabilized by the addition of sufficient levels of di- or trivalent impurities. Each of these systems can accommodate large additions of alkaline earth and rare earth oxides in solid solution. The di- and trivalent ions enter the fluorite lattice substitutionally at cation sites, thereby stimulating the formation of uncommonly high concentrations of oxygen vacancies approaching 10–15 mole %. These high levels of ionic defects, coupled with relatively high ion mobilities (Table I), combine to give high ionic conductivities ($\sim 10^{-2}$–10^{-1} $(\Omega\text{-cm})^{-1}$, 1273 K) and thus satisfy the first criteria outlined earlier for achieving mixed or predominantly ionic conductivity. Phase relations, electrical conductivity, and other properties of these oxides are reviewed by Etsell and Flengas (1970).

The band gaps of all these systems are reported to be $\sim$5 eV (Strehlow and Cook, 1973) and so possess low intrinsic concentrations of electrons and holes even at elevated temperatures (Table III). The conduction bands of these oxides appear to be of either a d or f nature, which favors low electron mobilities (μ_n). This seems to be supported by the data assembled in Table IV. The valence band, on the other hand, believed to be of oxygen 2p character, would be expected to support electron–hole mobilities (μ_p) of > 1 cm^2/V sec. While the data for ThO_2 support this expectation, μ_p values for HfO_2, UO_2, and ZrO_2 are substantially below this value. Heyne (1977) has suggested that these low values may not be intrinsic values but may rather be a consequence of substantial carrier trapping. In either case, the second criterion for mixed conduction requiring low intrinsic electronic carrier concentations coupled with reasonably low mobilities is also satisfied for these materials.

The final major criterion, i.e., stability against composition excursions in reducing or oxidizing environments, is what ultimately differentiates the above materials with respect to their electrical properties. In the following we review recent results obtained for these systems.

1. Cerium Dioxide

a. Undoped CeO_2. Cerium dioxide, CeO_{2-x}, because of its multivalent cation (Ce^{4+}, Ce^{3+}), deviates strongly from stoichiometry at elevated temperatures and reduced P_{O_2} to produce an oxygen-deficient *n*-type semiconductor. Some earlier studies (Blumenthal *et al.*, 1971) favored cerium interstitials as the predominant ionic defect, but x-ray (Ray *et al.*, 1975; Faber *et al.*, 1976a), neutron diffraction (Ray *et al.*, 1975; Faber *et al.*, 1976b), oxygen diffusion (Steele and Floyd, 1971), and more recent electric conductivity measurements (Tuller and Nowick, 1979; Naik and Tien, 1978) leave no doubt that oxygen vacancies predominate. Extensive measurements of $\sigma(T, P_{O_2})$ coupled with thermoelectric and thermogravimetric data showed that doubly ionized oxygen vacancies, the dominant ionic defect at small x ($x < 10^{-3}$), becomes displaced by singly ionized vacancies at larger x (Tuller and Nowick, 1979). The controlling defect reactions in CeO_{2-x}

$$O_O \rightarrow V_O^{\cdot\cdot} + 2e' + \tfrac{1}{2}O_2 \tag{104}$$

$$V_O^{\cdot} \rightarrow V_O^{\cdot\cdot} + e' \tag{105}$$

are characterized by equilibrium constants $K_1(T) = 1.2 \times 10^{74} \exp(-4.67 \pm 0.07\ \text{eV}/kT)\ \text{cm}^{-9}\ \text{atm}^{1/2}$ and $K_2(T) = 6.0 \times 10^{22} \exp(-0.56\ \text{eV}/kT)\ \text{cm}^{-3}$, respectively.

For very small x, CeO_{2-x} becomes a mixed conductor, as indicated by oxygen concentration cell (Van Handel and Blumenthal, 1974), electric conductivity, and thermoelectric power measurements (Tuller and Nowick, 1975b). Under these conditions the ionic conductivity is believed to be controlled by background impurities, in agreement with expectations based on Fig. 4. Recent conductivity measurements preformed at high P_{O_2} values (400 atm), on the other hand, suggest the onset of *p*-type conductivity (Baker *et al.*, 1977). These results seem suspect, given the predicted location of the *p*–*n* transition.

Measurements of electronic conductivity in combination with thermoelectric power (Tuller and Nowick, 1977) or thermogravimetric data (Naik and Tien, 1978; Blumenthal and Sharma, 1975) have confirmed that electron transport in CeO_{2-x} proceeds via a small polaron process [Eq. (13)]. The first approach presumably is more accurate, since thermoelectric power monitors electron carrier concentrations directly. All studies find μ_n ($\sim 10^{-2}\ \text{cm}^2/\text{V sec}$ at 1273 K) to be activated with hopping energies in the range of ~0.2–0.6 eV as x increases from 0.01 to 0.25. The variation in E_H has been related to the onset of defect ordering at larger values of x (Tuller and Nowick, 1977).

Comparing μ_n and $\mu(V_O^{\cdot\cdot})$ (Tables I and IV) one comes to the surprising conclusion that $\mu_n/\mu(V_O^{\cdot\cdot})$ is only on the order of 10, thus making CeO_2 a mixed conductor even at large deviations from stoichiometry.

b. Doped CeO_2. Addition of di- and trivalent cation impurities to CeO_2 serves to increase its ionic conductivity orders of magnitude as confirmed by electrical conductivity versus P_{O_2} (Blumenthal *et al.*, 1973; Tuller and Nowick, 1975a, c), oxygen diffusion (Steele and Floyd, 1971), and concentration cell measurements (Kudo and Obayashi, 1976). For dopant levels of $\gtrsim 1$ mole %, doped CeO_2 becomes a predominantly ionic conductor at high P_{O_2} values and temperatures below ~1000°C. Under these circumstances, the total conductivity is observed to be independent of P_{O_2} [Eq. (40)], indicating operation within the electrolytic zone (Fig. 3b, curve 1). A detailed discussion of the dependence of the ionic conductivity on type and concentration of dopant is given in Chapter 5 by Kilner and Steele and will not be repeated here.

At sufficiently reduced P_{O_2} values, a component of conductivity $\propto P_{O_2}^{-1/4}$ is always observed, indicating the appearance of *n*-type electronic conductivity (Tuller and Nowick, 1975c; Tuller *et al.*, 1981; Blumenthal and Garnier, 1976), confirming the predictions of Eq. (23). A typical plot showing both the P_{O_2}-independent region at high P_{O_2} and the $\sigma \propto P_{O_2}^{-1/4}$ region at low P_{O_2} is given in Fig. 17 for $(CeO_2)_{0.95}$ plus $(Y_2O_3)_{0.05}$. The

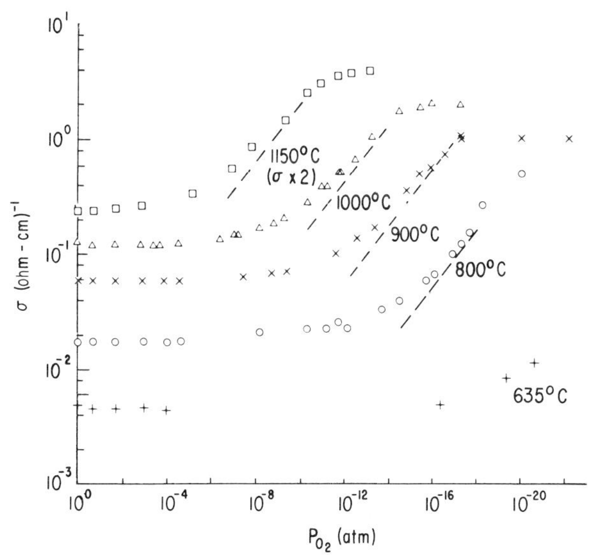

Fig. 17. Electrical conductivity as a function of P_{O_2} for a sample of $(CeO_2)_{0.95}$–$(Y_2O_3)_{0.05}$ at five different temperatures. The dashed lines indicate the electronic component of the conductivity σ_e. (From Tuller and Nowick, 1975c; reprinted by permission of the publisher, The Electrochemical Society, Inc.)

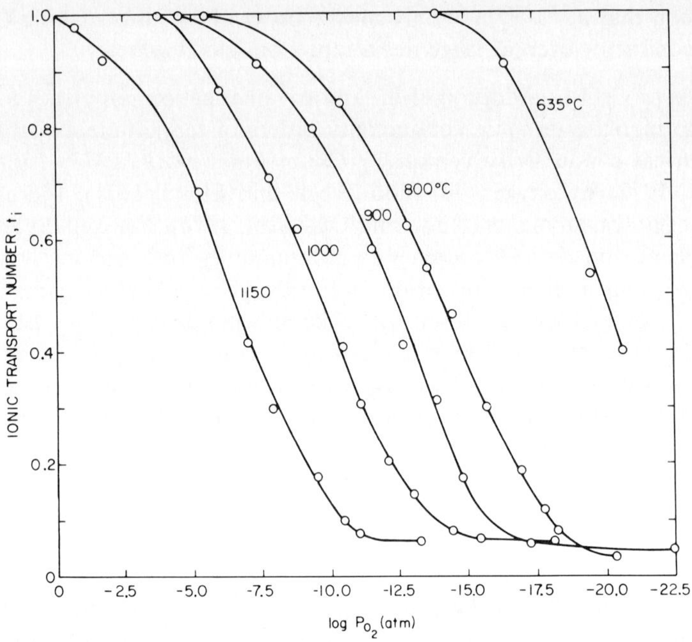

Fig. 18. The ionic transference number t_i as a function of P_{O_2}, as derived from Fig. 17 for various temperatures. (From Tuller and Nowick, 1975c; reprinted by permission of the publisher, The Electrochemical Society, Inc.)

electronic component is obtained by subtractlng the P_{O_2}-independent conductivity from the total conductivity as described in a previous section.

Given both σ_{ion} and σ_n at every temperature and P_{O_2} it becomes a simple matter to calculate $t_{ion}(T, P_{O_2})$. Such data, presented in Fig. 18, show that the electrolytic domain shrinks as the temperature is increased. This is as expected given the increasing tendency for Ce^{4+} to be reduced to Ce^{3+} at elevated temperatures. Values of P_n [i.e., P_{O_2} ($t_i = 0.5$)] are readily deduced from such curves and are shown plotted against $1/T$ in Fig. 19. The straight line drawn through these points defines the electrolytic domain boundary for $(CeO_2)_{0.95} + (Y_2O_3)_{0.05}$ under reducing conditions. From the slopes of this curve, $E_n - E_i$ was found to be equal to 1.61 eV [Eq. (45)] which, when combined with $E_i = 0.76$ eV from ($\sigma_{ion}T$ versus $1/T$), results in a value for $E_n = 2.37$ eV. Since $E_n = E_R/2 + E_H$ [Eqs. (13), (23), and (42)] and $E_H \sim 0.4$ eV (Tuller and Nowick, 1977), this gives $E_R \sim 4$ eV for the energy of reaction Eq. (19)]. The equivalent value obtained for E_R in pure CeO_2 (Tuller and Nowick, 1979) is ~ 4.7 eV, thus indicating a decrease in E_R with doping. Similar results have been observed for CeO_2 doped with Nd_2O_3, La_2O_3 (Tuller *et al.*, 1981), and Gd_2O_3 (Kudo and Obayashi, 1976).

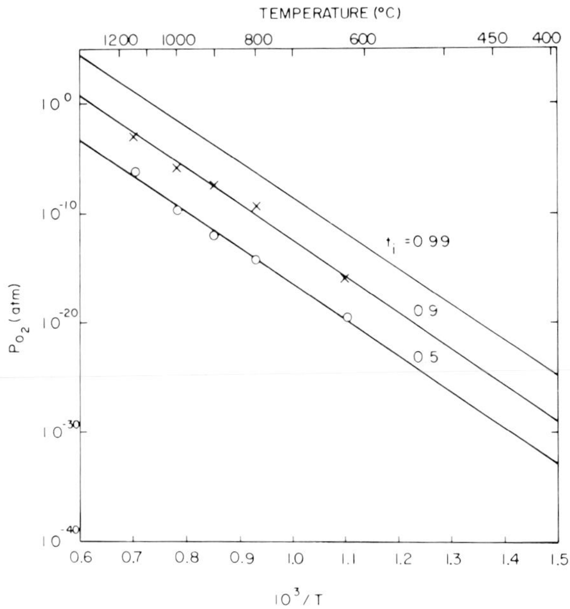

Fig. 19. Curves of P_{O_2} versus T^{-1} at fixed values of t_i for $(CeO_2)_{0.95}$–$(Y_2O_3)_{0.05}$. The upper curves, obtained by extrapolation, mark the boundary of the electrolytic domain ($t_i \geqslant 0.99$). (From Tuller and Nowick, 1975c; reprinted by permission of the publisher, The Electrochemical Society, Inc.)

Specimens doped with different impurities but at equivalent dopant levels (i.e., equal $[V_O^{\cdot\cdot}]$) do not result in equivalent electrolytic domains, as illustrated in Fig. 20. CeO_2 plus Y_2O_3 is observed to be least susceptible to reduction of the three and thus might be more desirable, for example, in a sensor application. Table VIII lists P_n (800°C) for a number of CeO_2-based

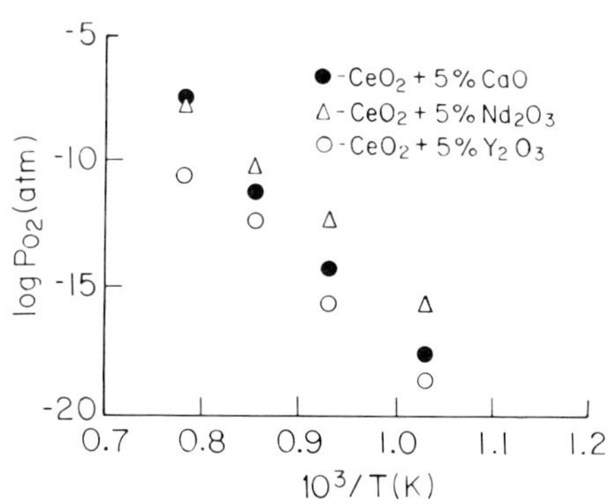

Fig. 20. Domain boundaries defined for $t_i = 0.5$ are plotted for a series of ceria solid solutions containing 5 mole % CaO, Nd_2O_3, and Y_2O_3, respectively. (From Tuller *et al.*, 1981).

TABLE VIII

Values of P_n for Compositions in the System $Ce_{1-x}N_xO^a_{2-x}$

	Composition, x	log P_n, 800°C
N = Gd	0.05	−14.6
	0.10	−16.0
	0.15	−16.1
	0.20	−16.2
	0.25	−14.0
N = Ca	0.01	−10.4
	0.05	−14.1
	0.12	−13.9
N = Nd	0.02	−12.7
	0.05	−12.5
	0.075	−14.5
N = Y	0.05	−15.4
N = La	0.022	−13.4
	0.07	−14.9
	0.15	−15.3

[a]Values of P_n for $Ce_{1-x}Gd_xO_{2-x}$ derived from data of Kudo and Obayashi (1976). The remainder from Tuller *et al.* (1981).

electrolytes. Although P_n is expected to decrease with increasing dopant level (Figs. 3a and 4), it is observed rather to saturate and at times even to decrease with increasing N_t. This is understandable if one takes into account that (a) σ_{ion} goes through a maximum with increasing N_t (see Chapter 5 by Kilner and Steele) and (b) E_R decreases with increasing N_t, resulting in a higher concentration of electrons than expected by simple extrapolation of the defect model (Tuller and Nowick, 1975; Tuller, 1978).

Finally, it may be observed that doping with pentavalent impurities, i.e., Ta and Nb, serves to increase the electronic conductivity σ_n in CeO_2 by orders of magnitude in air (Hosler and Frederikse, 1975; Naik and Tien, 1979), indicating that such impurities enter the lattice substitutionally.

2. *Thorium Dioxide*

a. Undoped ThO_2. Thorium dioxide, ThO_2, one of the most refractory oxides known, is always characterized by mixed conduction which takes the form

$$\sigma_{tot} = \sigma'_p P_{O_2}^{+1/4} + \sigma_{ion} \tag{106}$$

The source of ionic conductivity at intermediate P_{O_2} values is attributed to background impurities at concentrations of several hundred ppm (Bauerle, 1966). Values of t_i are readily obtained from σ-versus-P_{O_2} measurements in the manner previously described for doped CeO_2. The value of P_p has in this way been found to be nearly temperature-independent and equal to $\sim 10^{-5}$ atm (800–1200°C) (Etsell, 1972). Following Eq. (44), this result implies $E_i \approx E_p$.

Values for the apparent activation energy for hole conduction E_p have been reported to range from 0.85 eV (Hammou and Deportes, 1974) to 1.23 eV (Iqbal and Baker, 1973), while the energy for ionic conduction E_i varies from 0.89 eV (Bransky and Tallan, 1970) to $\sim$1.7 eV (Hammou and Deportes, 1974). In the latter study, $\log P_p$ shifted with temperature from ~ -8.5 (1013°C) to ~ -3.8 (1515°C), as might be expected given the large difference in E_i and E_p (0.85 eV) in this particular material. Clearly the exact nature of the impurities must be accounted for in explaining these differences.

For temperatures above $\sim$1300°C an n-type conductivity begins to appear at the lowest P_{O_2} values. This component, usually attributed to the onset of electron conduction, exhibits an apparent activation energy of $E_n \sim 3$ eV (Bransky and Tallen, 1970). Etsell (1972) points out that $E_n + E_p = E_g$ [Eq. (18)], assuming temperature-independent mobilities. Choosing $E_p = 1$ eV and $E_g = 5$ eV (from optical data) he calculates $E_n = 4$ eV. In searching for a reason for the discrepancy in the two values of E_n, it is interesting to note that Hammou and Deportes (1974) attribute a major part of their n-type conductivity ($\sigma \propto P_{O_2}^{-1/x}$) to ion transport. This would explain how an incorrect value for E_n could be extracted from the temperature-dependence of this component of the conductivity. The P_{O_2}-dependent ionic conductivity was attributed to the presence of background levels of variable-valence impurities. Assuming $E_n = 4$ eV, Etsell estimates the following relation for P_n in undoped ThO_2

$$\log P_n \sim (-60.5 \times 10^3)/T + 23.3 \qquad (1000\text{–}1600°C) \qquad (107)$$

which he finds in reasonably good agreement with limited σ-versus-P_{O_2} results.

In a recent study combining electrolysis with dilatometry, Duclot and Hammou (1978) measured the transport number for thorium ions t_{Th}. As expected, they found t_{Th} to be small ($\sim 10^{-6}$) and independent of P_{O_2}. Thorium ion conductivity exhibited an apparent activation energy of 2.3 eV and an equivalent diffusion coefficient of 10^{-16} cm^2/sec at $\sim$1000°C. Comparison with tracer diffusion data showed wide variations between results of different investigators, which were attributed to complications due to grain boundaries and surface treatment.

b. Doped ThO_2. ThO_2 is most commonly doped with Y_2O_3, which results in high concentrations of oxygen vacancies as confirmed by density measurements (Subbarao *et al.*, 1965). The electrical conductivity exhibits the same general character as that of undoped Y_2O_3 except for the much higher magnitude of the ionic conductivity.

To test the applicability of the defect relations derived on the basis of dilute solution approach, i.e., $p \propto [Y'_{Th}]^{1/2}P_{O_2}^{1/4}$ and $[V_O^{\cdot\cdot}] \propto [Y'_{Th}]$ (Table V), σ_p and σ_i were measured as a function of $[Y'_{Th}]$ and P_{O_2} by Lasker and Rapp (1966) and Hammou (1975). Hammou (1975) found $\sigma_i \propto [Y'_{Th}]$ for $0.05 < YO_{1.5} < 1$ mole % with a constant activation energy of $E_i = 1.1$ eV. Above this dopant range ($2 < YO_{1.5} < 21.5$ mole %) σ_i increased more slowly, ultimately passing through a maximum at $\sim$ 15 mole % $YO_{1.5}$. The hole conductivity maintained a $P_{O_2}^{+1/4}$ dependence throughout with $E_p \sim$ 0.85 eV up to 15 mole % $YO_{1.5}$. Lasker and Rapp (1966) found $\sigma_p \sim [Y'_{Th}]^{1/2}$ over a wide range of Y_2O_3 doping in agreement with predictions. Since σ_i increases more rapidly with $[Y'_{Th}]$ than σ_p, it is reasonable that P_p would increase with increasing $[Y'_{Th}]$. Again, since $E_p \sim E_i$, no strong dependence of P_p on temperature is expected.

Similar $\sigma(P_{O_2}, T)$ measurements have been performed more recently on ThO_2 doped with CaO (1–7 mole %) by Maiti and Subbarao (1976). In general, the results are similar, except for the fact that the magnitude of ionic conductivity is substantially lower in ThO_2 plus CaO than in ThO_2 plus Y_2O_3 with equivalent concentrations of oxygen vacancies. Similar results have been observed in ceria (Tuller and Nowick, 1975a, c) and in stabilized zirconia (Etsell and Flengas, 1970). Hole conductivity in ThO_2 plus CaO satisfied the relation $\sigma_p \propto P_{O_2}^{1/4}\exp(-1.1\ \text{eV}/kT)$ as above. Plots of total conductivity versus $1/T$ showed a three-slope behavior which is likely a result of polarization at the electrodes and grain boundaries rather than changes in lattice conductivity.

Taking the data of Lasker and Rapp (1966), Tuller (1978) calculated the position of the electrolytic domain boundary characterized by P_p as a function of $[Y'_{Th}]$ in YDT. The results presented in Fig. 21 show the domain boundary to shift to higher P_{O_2} with increasing yttria additions as expected. However, as in the case of ceria-based electrolytes discussed earlier, a saturation effect occurs at high dopant levels. Note also that the slope of the curve, which according to Eq. (44) corresponds to $4(E_i - E_p)$, decreases. Since E_p has been shown to remain relatively constant, this implies that E_i is decreasing with increasing additives.

In Fig. 22 similar data derived from the study by Maiti and Subbarao (1976) are compared with the yttria-doped data. The domain boundaries for the calcia doped materials are observed to lie at considerably lower P_{O_2} than for the equivalent yttria-doped material. This simply reflects the fact

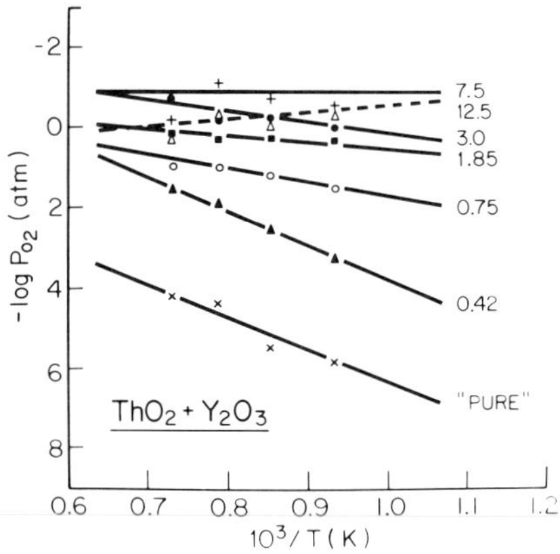

Fig. 21. Domain boundaries defined for $t_i = 0.5$ for a series of ThO_2 plus Y_2O_3 solid solutions. Data points derived from conductivity results of Lasker and Rapp (1966). Note the strong dependence on concentration, given in mole % for each curve.

that, while σ_p remains about the same in both materials, σ_{ion} in the former material is considerably lower in magnitude. The results of Fig. 22 therefore emphasize the importance of both the type and concentration of additive in determining the location of the electrolytic domain boundaries.

The desirability of thoria-based electrolytes is based on their utility at reduced P_{O_2} values, implying very low values for P_n. In this regime of P_{O_2}, oxygen concentration cell or polarization techniques are more readily applied than conductivity techniques. Etsell (1972), in reviewing techniques used to determine P_n, found that steady state concentration cell measure-

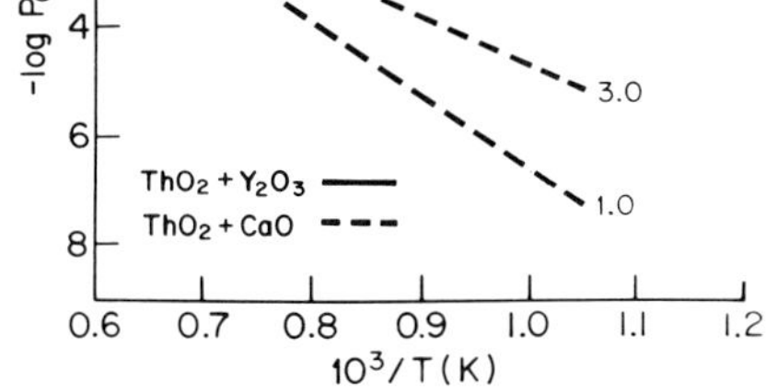

Fig. 22. Comparison of the location of domain boundaries ($t_i = 0.5$) for a number of CaO- and Y_2O_3-doped ThO_2 electrolytes with nearly equivalent dopant levels. In all cases the Y_2O_3-doped materials exhibit larger values of P_p.

ments operated at very low P_{O_2} values were susceptible to error due to oxygen transfer through the electrolyte either by permeation or via pore diffusion. Leakage problems associated with polarization measurements have been discussed previously. The most satisfactory approach in such cases appears to be via coulometric titration in which the anode of the electrolyte is temporarily reduced electrochemically to below P_n, followed by immediate measurement of the resultant open-circuit emf. The value of P_n may then be evaluated from Eq. (77) after ensuring that the measured emf is at its saturation value. Following this approach, the expression below was obtained for the domain boundary under reducing conditions

$$\log P_n = (-57.9 \times 10^3)/T + 12.4 \qquad (700\text{–}1600°\text{C}) \qquad (108)$$

Comparison of this equation with Eq. (45) and taking $E_i = 1.0$ eV yields $E_n = 3.9$ eV, which taken together with the reported values of $E_p \sim 1$ eV results in a band gap value of $E_g \sim 5$ eV.

A recent study on the electrical properties of ThO_2 doped with Ce (Fujimoto and Tuller, 1979) demonstrates the possible effects that multivalent ions may have on the defect equilibria of such oxides. The total electrical conductivity was found to exhibit what appeared to be a $p \rightarrow n$ transition at intermediate P_{O_2} values, which shifted to a higher P_{O_2} with increasing temperature (Fig. 7). The first inclination was to identify the $\sigma \propto P_{O_2}^{-1/x}$ component as being electronic. Increasing levels of Ce_t were, however, found to increase this component which according to Eq. (100) would identify it as ionic rather than electronic. This hypothesis was in fact confirmed by subsequent oxygen concentration cell measurements. The controlling defect relation given by

$$O_O + 2Ce^*_{Th} \rightarrow 2Ce'_{Th} + V_O^{\cdot\cdot} + \tfrac{1}{2}O_2 \qquad (109)$$

is characterized by an equilibrium constant $K_R = K_R^0 \exp(-E_R/kT)$. For conditions of $[Ce^*_{Th}] \sim Ce_t$, the vacancy concentration $[V_O^{\cdot\cdot}] \propto \exp(-E_R/3kT)Ce_t^{2/3}P_{O_2}^{-1/6}$. A plot of σ_{ion} versus $1/T$ yielded an apparent activation energy of ~2.9 eV which according to the above corresponds to $E_R/3 + E_m$. Taking $E_m = 1.2$ eV (Lasker and Rapp, 1966) this yields a value for $E_R \simeq 5.1$ eV. Hole conduction at a high P_{O_2} characterized by $E_p \sim 1$ eV remains nearly the same as in yttria- and calcia-doped ThO_2.

3. *Zirconium Dioxide*

a. Monoclinic and Tetragonal ZrO_2. Undoped monocylinic ZrO_2 is reportedly a semiconductor at temperatures above ~800°C (Vest *et al.*, 1964; Nasrallah and Douglass, 1974), which transforms from p to n type at $\sim 10^{-14}$ atm. Ionic conduction is, however, observed at 600°C and intermediate P_{O_2} values. Doping with Y_2O_3 (~1 mole %) increases the ionic component substantially and broadens the electrolytic domain. Nasrallah and Douglass propose a defect model which includes O_i'' at high P_{O_2} values and $V_O^{\cdot\cdot}$ at low P_{O_2} values as the predominant ionic defects.

Measurements of $\sigma(P_{O_2}, T)$ on undoped tetragonal zirconia (McClaine and Coppel, 1966) suggest mixed conduction for which ionic conduction appears to dominate below 1200°C. Above this temperature it becomes pregressively more semiconducting, with a $p \rightarrow n$ transition occurring at $\sim 10^{-5}$ atm.

b. Cubic ZrO_2. Calcia- and yttria-stabilized zirconia today remain the most popular solid oxide electrolytes. This is a consequence of their relatively high oxygen ion conductivity and low electronic conductivity at both high and low P_{O_2} values. Nevertheless electron conduction still limits their usefulness at elevated temperatures and reduced pressures, while hole conduction controls oxygen permeation at P_{O_2} values in the vicinity of 1 atm.

Hole conduction σ_p has been studied in stabilized zirconia by permeation, redox kinetics, dc polarization, and voltage relaxation techniques. Using permeation techniques on CSZ, Heyne and Beekmans (1971) obtained an expression for

$$\sigma_p = 6.6 \times 10^2 P_{O_2}^{1/4} \exp(-1.9 \text{ eV}/kT) \quad (\Omega\text{-cm})^{-1} \tag{110}$$

with an associated hole diffusion coefficient equal to $\sim 10^{-6}$ cm^2/V sec at 1273 K and an activation energy of 1.24 eV. By monitoring the voltage relaxations following removal of the polarizing voltage in a dc polarization cell, Weppner (1976) deduced both the electron and hole diffusion coefficients in ZrO_2 plus 10% Y_2O_3 as shown previously in Fig. 16. The electron mobility μ_n, with an activation energy of 0.56 ± 0.15 eV, is observed to be several orders of magnitude greater than the hole mobility, with an activation energy of 1.4 ± 0.3 eV. One suspects that such low electronic carrier mobilities, characterized by large activation energies, are not likely to be intrinsic values but are rather trap-limited, as pointed out by Heyne and Beekmans (1971).

Electron conduction σ_n in stabilized zirconia has been investigated by σ-versus-P_{O_2}, dc polarization, permeation, oxygen concentration, and electrolysis techniques. Employing concentration cell measurements of the type defined by Eq. (77), Etsell and Flengas (1972) obtained the following expression for the low-P_{O_2} domain boundary in ZrO_2 plus 10 mole % CaO:

$$\log P_n = (-54.5 \times 10^3)/T + 14.0 \tag{111}$$

Combining this with $E_i = 1.1$ eV and Eq. (45), they obtained a value for $E_n = 3.8$ eV. Assembling similar data from the literature, Heyne and

Beekmans (1971) deduced the following relation for σ_n

$$\sigma_n = 3.7 \times 10^6 P_{O_2}^{-1/4} \exp(-3.7\ \text{eV}/kT) \quad (\Omega\text{-cm})^{-1} \tag{112}$$

Fouletier and Kleitz (1978) recently used a novel technique to characterize σ_n in $(ZrO_2)_{0.87}$–$(Y_2O_2)_{0.12}$–$(CeO_2)_{0.01}$ as a function of oxygen deficiency. Reduction of the specimen was effected by electrolysis, while deviation from stoichiometry was monitored by an oxygen gauge. In this way they established that σ_n, at fixed deviation from stoichiometry, was activated with energies from 0.27–0.71 eV depending on the composition.

Combining the above expressions for σ_n and σ_p, Heyne and Beekmans (1971) derived the following relation for the $n \rightarrow p$ boundary

$$P_0 = 3.4 \times 10^7 \exp(-3.6\ \text{eV}/kT) \quad \text{atm} \tag{113}$$

while the corresponding conductivity at this boundary was

$$\sigma_{\min} = 2 \times 4.92 \times 10^4 \exp(-2.8\ \text{eV}/kT) \quad (\Omega\text{-cm})^{-1} \tag{114}$$

By equating this relation with Eq. (48), Heyne and Beekmans (1971) deduced a value for the band gap of 5.6 eV, a value in reasonable agreement with optical data (Strehlow and Cook, 1973). A serious difficulty with this approach arises, however, when the term $(\mu_n \mu_p)^{1/2}$ in Eq. (48) itself exhibits a strong dependence on temperature, as appears to be the case in stabilized zirconia. If we substitute the values of 0.56 and 1.4 eV obtained by Weppner for the activation energies of μ_n and μ_p, respectively, into Eq. (48), we come up with a considerably smaller value of ~3.95 eV for the band gap. This discrepancy deserves further attention.

Recent studies have been performed on the effect of mixed conduction due to additions of Ce and other multivalent ions to stabilized zirconia. These studies have been stimulated by the search for mixed conductors for magnetohydrodynamic (MHD) electrodes (Rossing *et al.*, 1975) or electrolyzers (Browall and Doremus, 1977). Although such additives are shown to increase electronic conductivity, the studies are insufficiently detailed to establish the defect equilibria in these systems.

B. Perovskites

Perovskite-type oxides, ABO_3 (where A is di- or trivalent and B is tetra- or trivalent), can often accommodate substantial concentrations of lower-valent impurity ions at either the A or B sites, resulting in the formation of substantial oxygen deficiency (Coates and McMillan, 1964). Takahashi and Iwahara (1971) have shown that such doping may lead to high ionic conductivities in the systems $CaTi_{1-x}Al_xO_{3-\alpha}$, $La_{1-x}Ca_xAlO_{2-\alpha}$, and

$SrTi_{1-x}Al_xO_{3-\alpha}$. However, all such oxides tested maintained predominant ionic conduction only over an intermediate range of P_{O_2} ($\sim 10^{-6}$–10^{-16} atm) as confirmed by $\sigma(P_{O_2})$ and concentration cell measurements. Similar results have more recently been reported for other perovskites (Browall and Muller, 1976; Kilner *et al.*, 1978), but inadequate data exist for establishing the position of the domains and their dependence on doping.

C. MgO

Although MgO has been known to be a mixed conductor at elevated temperatures for some time (Schmalzried, 1960), no consistent picture of the defect structure had been established. Very recently dilatocoulometric, concentration cell, and $\sigma(P_{O_2}, T)$ measurements have resulted in considerable new insight into the defect structure and transport properties of MgO. Duclot and Deportes (1980) measured the cationic transport number t_{Mg} in undoped single crystals of MgO by dilatocoulometry over a range of temperatures and P_{O_2} values. The measurements, which showed t_{Mg} to decrease from ~0.6 at 1100°C to ~0.2 at 1305°C, clearly support a mixed conduction model in which the mobile ion is Mg. A plot of $\sigma_{Mg} = t_{Mg}\sigma_{tot}$ versus $1/T$ gave an activation energy for ionic conduction of 2.2 eV. Furthermore, σ_{Mg} was observed to increase with increasing P_{O_2}, which Duclot and Deportes suggested might be due to the presence of multivalent Fe. Unfortunately, because the nature and concentration of the controlling impurities could not be established, further analysis was limited.

More recently a systematic study on $\sigma(P_{O_2}, T)$ and $t_i(P_{O_2}, T)$ was performed on MgO single crystals doped with controlled amounts of Al, Sc, and Fe (320–1500 ppm) (Sempolinsky and Kingery, 1980; Sempolinsky *et al.*, 1980). With such high levels of trivalent impurities, the conductvity was observed to be entirely ionic at intermediate P_{O_2} values for temperatures below 1400°C. A plot of σ_{ion} versus $[N^{\cdot}_{Mg}]$ showed a linear dependence, thereby allowing calculation of $\mu(V''_{Mg}) = \sigma_{ion}/[N^{\cdot}_{Mg}]q$ which was found to be equal to

$$\mu(V''_{Mg}) = (8800/T)\exp(-2.27 \text{ eV}/kT) \quad \text{cm}^2/\text{V sec}$$

Note the good agreement between this value of activation energy and that obtained by Duclot and Deportes (1980).

For temperatures above 1400°C the specimens became progressively more electronic, exhibiting p-type conduction at high P_{O_2} and n-type conduction at low P_{O_2} in general agreement with the predictions of Fig. 13. More specifically $\sigma_e = \sigma_{tot}(1 - t_{ion})$ was observed to pass through a $p \rightarrow n$ transition at $\sim 10^{-5}$ atm, with P_0 shifting weakly to lower P_{O_2} values with increasing T. Analysis of σ_{min} in terms of Eq. (47) resulted in values for the electronic carrier mobilities ($\mu_p \sim 5$ cm^2/V sec, $\mu_n \sim 15$ cm^2/V sec, 1673 K) and the band gap energy ($E_g = 6.8 \pm 0.5$ eV).

While the magnesium ion conductivity in Sc- and Al-doped MgO remained independent of P_{O_2}, it showed an increase with increasing P_{O_2} in the Fe-doped sample. Following an approach similar to that described in Section III.A.2 for variable-valence impurities, Sempolinski *et al.* (1980) derived the following defect relations for the oxidation of MgO:

$$\tfrac{1}{2}O_2 \rightarrow O_O + V''_{Mg} + 2h^{\cdot}: \quad K_{ox}(T)$$

$$K_{ox}(T) = 7 \times 10^{63} \exp(-6.26 \pm 0.8 \text{ eV})/kT \quad \text{cm}^{-9}\text{atm}^{-1/2}$$

and, for the ionization of Fe_{Mg},

$$Fe_{Mg} \rightarrow Fe^{\cdot}_{Mg} + e': \quad K_{Fe}(T)$$

$$K_{Fe}(T) = 1.7 \times 10^{21} \exp(-3.3 \pm 0.5 \text{ eV}/kT)$$

D. Al_2O_3

Mixed conduction in Al_2O_3 has been actively investigated by Kröger and co-workers over the past 5 yr. These studies have included measurements of $\sigma(P_{O_2})$, $t_i(P_{O_2})$, optical absorption and EPR on Al_2O_3 doped with Co (Dutt *et al.*, 1975) Ti, and Mg (Mohapatra and Kröger, 1977a, b), as well as undoped Al_2O_3 (Mohapatra *et al.*, 1979). As in the case of Ce-doped ThO_2, Al_2O_3 doped with variable-valence transition metal ions exhibited a minimum in the conductivity of which one side was ionic and one side electronic. Cobalt doping led to ionic conductivity (Al_i) at low P_{O_2} and *p*-type conductivity at high P_{O_2}, while Ti, which takes on a charge greater than 3+, resulted in ionic conductivity (V_{Al}) at high P_{O_2} and *n*-type conductivity at low P_{O_2}. The P_{O_2} dependence of the ionic conductivity was shown to be related to the change in valence of the impurity ion, as discussed in Section III.A.2.

Defect models were used to derive values for ionic and electronic mobilities and the appropriate equilibrium constants. Because of difficulties in establishing the concentration of active impurities (possible compensation effects were noted) further confirmation is desirable. A recent summary of some of these results (Tiku and Kröger, 1980) points out some of these difficulties.

E. Other Systems

Mixed conduction has been reported in many other systems including Sm_2O_3, Gd_2O_3, Dy_2O_3, (Tare and Schmalzried, 1964), Y_2O_3 (Schieltz *et al.*, 1971), La_2O_3 (Etsell and Flengas, 1969), and Ta_2O_5 (Stroud *et al.*, 1974). Based on the criteria established in the early part of this chapter, it can be expected that any oxide with a reasonably large band gap ($E_g \gtrsim 3$–4 eV) that is appropriately doped will exhibit mixed conduction over some temperature and P_{O_2} interval. In evaluating the transport properties of oxides, this important factor should always be kept in mind.

F. Applications

Mixed conductors play an important role in many practical applications. Since it is not the purpose of this chapter to treat such applications in an exhaustive manner, I will only mention several for which the requirements can be viewed in light of the concepts reviewed here.

The majority of applications take advantage of the high ionic conductivity that a number of oxides exhibit over extended operating conditions (i.e., electrolytic domain). Here they may be used as solid electrolytes in various types of electrochemical cells to generate electricity, electrolyze water, or measure oxygen activities in various media.

The conversion of chemical to electrical energy in high-temperature, solid electrolyte fuel cells and batteries [Eq. (67)] has a number of attractive features including (a) no Carnot cycle limitations, (b) improved kinetics at electrodes, (c) the use of nonnoble electrodes, and (d) potentially nonpolluting and noiseless operation. Wide electrolytic domains allow the use of highly reducing fuel gases, resulting in large emfs per cell. High ionic conductivities are necessary to minimize ohmic losses, while low electronic transference numbers t_e are necessary to maximize output voltage [Eq. (69)] and limit cell discharge [Eq. (83)]. Tannhauser (1978) and Ross and Benjamin (1976) calculate conversion efficiencies for cells exhibiting mixed conduction. Similar cells may also be run in reverse (externally applied potential) and operated as electrolyzers to generate H_2 fuel from water vapor.

Although stabilized zirconia is most actively considered in such applications, intensive efforts to identify more highly conducting oxides (e.g., doped CeO_2, perovskites, Bi_2O_3) at lower temperatures are now proceeding. Development of nonnoble metal electrodes compatible with the requirements of fuel cells has revolved around the possibility of developing oxide materials with high electronic conductivity for low resistance but which also exhibit reasonably high ion conductivity which enhances mass transfer to and away from the electrode–electolyte interface. Obayashi and Kudo (1977) review the materials problems to be faced in the development of such devices.

When operated in an open-circuit mode, such cells may be used to monitor oxygen activities in equilibrium with M/MO systems, dissolved in liquid metals, and in the gas phase and as such have been used extensively to determine standard free energies of formation of oxides, P_{O_2} values in equilibrium with nonstoichiometric phases, oxygen diffusion in melts, and other kinetic processes (Goto and Pluschkell, 1972; Worrell, 1977). Extensive use of such cells is now being envisioned in automobilies as automobile exhaust sensors. Such cells must operate at substantially reduced temperatures within less than ideal environments (Fleming, 1977). Although such cells are now manufactured exclusively from stabilized zirconia, efforts are being directed toward developing alternatives such as ceria-based cells (Dirstine *et al.*, 1979) and sensors where P_{O_2} excursions result in resistivity changes ($\sigma \propto P_{O_2}^{\pm 1/x}$) (Park and Logothetis, 1977). Heyne and Engelson (1977) have recently considered the factors that influence the response time of such solid electrolyte sensors.

Conversion of fossil fuel energy at high temperatures to electricity via MHD techniques has received a great deal of attention over the last 10 yr. Here electrodes are required that can operate at temperatures above 1500°C, possess conductivities above 1 $(\Omega\text{-cm})^{-1}$, and simultaneously are resistent to thermal shock, chemical attack, and erosion. Because of their stability and reasonably high conductivities, the cubic fluorite oxides including zirconia, ceria, and hafnia have received the most attention. For this application, high ionic conduction is detrimental, since it can lead to electrolysis of the electrodes under high current drain. Efforts are therefore being directed to enhance the electronic conducivity of these materials by doping with multivalent or pentavalent ions (Rossing *et al*, 1975). A review of materials design considerations in MHD reactors is given by Bowen and Rossing (1976).

Finally, increased attention is being focused on the role mixed conduction plays in the catalytic activity of different oxide-based catalysts. Studies have been reported for instance for CO oxidation on ZrO_2 and ThO_2 (Berkstresser *et al.*, 1974) and NO reduction on perovskite-like compounds derived from $LaMnO_3$ (Voorhoeve *et al.*, 1975). As noble metals become less available, considerably more research in this area can be expected.

Acknowledgment

The author thanks the Department of Energy (Contract No. EG77-S-02-4436) and the National Science Foundation (Contract No. DMR-78-262-6) for supporting the research programs that have contributed in great measure to the development of the material in this chapter.

References

Appel, J. (1968). *Solid State Phys.* **21**, 193.

Armstrong, R. D., Bulner, R. S., and Dickinson, T. (1973). *In* "Fast Ion Transport in Solids" (W. Van Gool, ed.), pp. 269–282. Am. Elsevier, New York.

Baker, E. H., Iqbal, M., and Knox, B. E. (1977). *J. Mater. Sci.* **12**, 305–310.
Bauerle, J. E. (1966). *J. Chem. Phys.* **45**, 4162–4166.
Berkstresser, G. W., Brook, R. J., and Whelan, J. M. (1974). *J. Mater. Sci.* **9**, 491–493.
Blumenthal, R. N., and Garnier, J. E. (1976). *J. Solid State Chem.*, **16**, 21–34.
Blumenthal, R. N., and Sharma, R. K. (1975). *J. Solid State Chem.* **13**, 360–364.
Blumenthal, R. N., Lee, P. W., and Panlener, R. J. (1971). *J. Electrochem. Soc.* **118**, 123–129.
Blumenthal, R. N., Brugner, F. S., and Garnier, J. E. (1973). *J. Electrochem. Soc.* **120**, 2130.
Bosman, A. J., and Van Daal, J. H. (1970), *Adv. Phys.* **19**, 1.
Bowen, H. K., and Rossing, B. R. (1976). *In* "Critical Problems in Energy Production" (C. Stein, ed.), pp. 311–334. Academic Press, New York.
Bransky, I., and Tallan, N. M. (1970). *J. Am. Ceram. Soc.* **53**, 625–629.
Bransky, I., and Tallan, N. M. (1971). *In* "Physics of Electronic Ceramics," Part A (L. L. Hench and D. B. Dove, eds.), pp. 67–98. Dekker, New York.
Brouwer, G. (1954). *Philips Res. Rep.* **9**, 366.
Browall, K. W., and Doremus, R. H. (1977). *J. Am. Ceram. Soc.* **60**, 262–267.
Browall, K. W., and Muller, O. (1976). *Mater. Res. Bull.* **11**, 1475–1482.
Cline, C. F., and Newkirk, H. W. (1968). *J. Chem. Phys.* **49**, 3496–3504.
Coates, R. V., and McMillan, J. W. (1964). *J. Appl. Chem.* **14**, 346–350.
Deputy, G. O., and Vest, R. W. (1978). *J. Am. Ceram. Soc.* **61**, 321–325.
Dirstine, R. T., Gentry, W. O., Blumenthal, R. N., and Hammetter, W. (1979). *Am. Ceram. Soc. Bull.* **58**, 778–783.
Duclot, M., and Deportes, C. (1980). *J. Solid State Chem.* **31**, 377–385.
Duclot, M., and Hammou, A. (1978). *Mater. Res. Bull.* **13**, 707–715.
Dudley, G. J., Steele, B. C. H., and Howe, A. T. (1976), *J. Solid State Chem.* **18**, 141–147.
Dudney, N. J., Coble, R. L., and Tuller, H. L. (1981). *J. Am. Ceram. Soc.* In press.
Dutt, B. V., Hurrell, J. P., and Kroger, F. A. (1975). *J. Am. Ceram. Soc.* **58**, 420–427.
Etsell, T. H. (1972). *Z. Naturforsch, Teil A* **27**, 1138–1149.
Etsell, T. H., and Flengas, S. N. (1969). *J. Electrochem. Soc.* **116**, 771–778.
Etsell, T. H., and Flengas, S. N. (1970). *Chem. Rev.* **70**, 339–376.
Etsell, T. H., and Flengas, S. N. (1972). *J. Electrochem. Soc.* **119**, 1–7.
Faber, J., Jr., Seitz, M. A., and Mueller, M. H. (1976a). *J. Phys. Chem. Solids* **37**, 903–907.
Faber, J., Jr., Seitz, M. A., and Mueller, M. H. (1976b). *J. Phys. Chem. Solids* **37**, 909–915.
Fleming, W. J. (1977). *J. Electrochem. Soc.* **124**, 21–28.
Fouletier, J., and Kleitz, M. (1978). *J. Electochem. Soc.* **125**, 752–755.
Fouletier, J., Fabry, P., and Kleitz, M. (1976). *J. Electrochem. Soc.* **123**, 204–213.
Fujimoto, H. H., and Tuller, H. L. (1979). *In* "Fast Ion Transport in Solids" (P. Vashishta, J. N. Mundy, and G. K. Shenoy, eds.), pp. 649–652. Elsevier/North-Holland, Amsterdam.
Goodenough, J. B. (1978). *In* "Solid Electrolytes" (P. Hagenmuller and W. Van Gool, eds.), pp. 393–415. Academic Press, New York.
Goto, K. S., and Pluschkell, W. (1972). *In* "Physics of Electrolytes" (J. Hladik, ed.), Vol. 2, pp. 540–622. Academic Press, New York.
Hammou, A. (1975). *J. Chim. Phys. Phys.-Chim. Biol.* **72**, 431–438.
Hammou, A., and Deportes, C. (1974). *J. Chim. Phys.* **7/8**, 1071–1080.
Heyne, L. (1977). *In* "Solid Electrolytes" (S. Geller, ed.), pp. 169–221. Springer-Verlag, Berlin and New York.
Heyne, L., and Beekmans, N. M. (1971). *Proc. Br. Ceram. Soc.* **19**, 229–263.
Heyne, L., and Engleson, D. (1977). *J. Electrochem. Soc.* **124**, 727–735.
Hosler, W. R., and Frederikse, H. P. R. (1975). *Proc. Int. Conf. MHD Electr. Power Gener., 6th, Washington, D.C.* **2**, pp. 67–76.
Huggins, R. A. (1978). *In* "Solid Electrolytes" (P. Hagenmuller and W. Van Gool, eds.), pp. 27–44. Academic Press, New York.

Hughes, R. C. (1979). *Phys. Rev. B* **19**, 5318–5328.
Ihrig, H., and Hennings, D. (1978). *Phys. Rev. B* **17**, 4593–4599.
Iqbal, M., and Baker, E. H. (1973). *High Temp.—High Pressures* **5**, 265–271.
Karim, D. P., and Aldred, A. T. (1979). *Phys. Rev. B* **20**, 2255–2263.
Kilner, J. A., Barrow, P., Brook, R. J., and Norgett, M. J. (1978). *J. Power Sources* **3**, 67–80.
Kim, K. K., Mundy, J. N., and Chen, W. K. (1979). *J. Phys. Chem Solids* **40**, 743–755.
Kingery, W. D., Pappis, J., Doty, M. E., and Hill, D. C. (1959). *J. Am. Ceram. Soc.* **42**, 393.
Kitazawa, K., and Coble, R. L. (1974). *J. Am. Ceram. Soc.* **57**, 360–363.
Kittel, C. K. (1976). "Introduction to Solid State Physics," 5th ed. Wiley, New York.
Kudo, T., and Obayashi, H. (1976). *J. Electrochem. Soc.* **123**, 415–419.
Lasker, M. F., and Rapp, R. A. (1966). *Z. Phys. Chem.* (*Frankfurt am Main*) **49**, 198–221.
McClaine, L. A., and Coppel, C. P. (1966). *J. Electrochem. Soc.* **113**, 80–85.
Maiti, H. S., and Subbarao, E. C. (1976). *J. Electrochem. Soc.* **123**, 1713–1718.
Metselaar, R., and Huyberts, M. A. H. (1977). *J. Solid State Chem.* **22**, 309–319.
Mohapatra, S. K., and Kröger, F. A. (1977a). *J. Am. Ceram. Soc.* **60**, 141–148.
Mohapatra, S. K., and Kröger, F. A. (1977b). *J. Am. Ceram. Soc.* **60**, 381–387.
Mohapatra, S. K., Tiku, S. K., and Kröger, F. A. (1979). *J. Am. Ceram. Soc.* **62**, 50–57.
Naik, I. K., and Tien, T. Y. (1978). *J. Phys. Chem. Solids* **39**, 311–315.
Naik, I. K., and Tien, T. Y. (1979). *J. Electrochem. Soc.* **126**, 562–566.
Nasrallah, M. M., and Douglass, D. L. (1974). *J. Electrochem. Soc.* **121**, 255–262.
Obayashi, H., and Kudo, T. (1977). *In* "Solid State Chemistry of Energy Conversion and Storage" (J. B. Goodenough and M. S. Whittingham, eds.), pp. 316–363. Am. Chem. Soc. Washington, D. C.
Park, K., and Logothetis, E. M. (1977). *J. Electrochem. Soc.* **124**, 1443–1446.
Patterson, J. W. (1971). *In* "Physics of Electronic Ceramics," Part A (L. L. Hench and D. B. Dove, eds.), pp. 131–154. Dekker, New York.
Patterson, J. W. (1974). *In* "Electrical Conductivity in Ceramics and Glass," Part B (N. M. Tallan, ed.), pp. 453–558. Dekker, New York.
Patterson, J. W., Bogren, E. C., and Rapp, R. A. (1967). *J. Electrochem. Soc.* **114**, 752–758.
Pearsall, T. P., and Lee, C. A. (1974). *Phys. Rev. B* **10**, 2190–2194.
Ray, S. P., Nowick, A. S., and Cox, D. E. (1975). *J. Solid State Chem.* **15**, 344.
Rifflet, J. C., Odier, P., Anthony, A. M., and Loup, J. P. (1975). *J. Am. Ceram. Soc.* **58**, 493–497.
Ross, P. H., and Benjamin, T. G. (1976). *J. Power Sources* **1**, 311.
Rossing, B. R., Cadoff, L. H., and Gupta, T. K. (1975). *Proc. Int. Conf. MHD Elect. Power. Gener. 6th, Washington, D.C.*, 105–117.
Schieltz, J., Patterson, J. W., and Wilder, D. R. (1971). *J. Electrochem. Soc.* **118**, 1140–1144.
Schmalzried, H. (1960). *J. Chem. Phys.* **33**, 940.
Schmalzried, H. (1962). *Z. Elektrochem.* **66**, 572–576.
Seltzer, M. S. (1976). *In* "Oxides and Oxide Films" (J. W. Diggle and A. K. Vijh, eds.), Vol. 4, pp. 1–98. Dekker, New York.
Sempolinsky, D. R., and Kingery, D. R. (1980). *J. Am. Ceram. Soc.* **63**, 664–668.
Sempolinsky, D. R., Kingery, D. W., and Tuller, H. L. (1980). *J. Am. Ceram. Soc.* **63**, 669–674.
Simpson, L. A., and Carter, R. E. (1966). *J. Am. Ceram. Soc.* **49**, 139.
Smith, A. W., Maszaros, F. W., and Amata, C. D. (1966). *J. Am. Ceram. Soc.* **49**, 240–244.
Steele, B. C. H., and Floyd, J. M. (1971). *Proc. Br. Ceram. Soc.* **19**, 55.
Steele, B. C. H., and Riccardi, C. C. (1974). *Proc. Int. Symp. React. Solids, 7th* pp. 123–135.
Stratton, T. G., Reed, D., and Tuller, H. L. (1980). *Proc. Int. Symp. Grain Boundary Phenom. Electron. Ceram., Am. Ceram. Soc.* (to be published).
Strehlow, W. H., and Cook, E. L. (1973). *J. Phys. Chem. Ref. Data* **2**, 163–200.

Stroud, J. E., Tripp, W. C., and Wimmer, J. M. (1974). *J. Am. Ceram. Soc.* **57**, 172–175.
Subbarao, P. H., Sutter, P. H., and Hrizo, J. (1965). *J. Am. Ceram. Soc.* **48**, 443–446.
Sze, S. M. (1969). "Physics of Semiconductor Devices." Wiley (Interscience), New York.
Takahashi, T., and Iwahara, H. (1971). *Energy Convers.* **11**, 105–111.
Takahashi, T., and Iwahara, H. (1978). *Mater. Res. Bull.* **13**, 1447–1453.
Takahashi, T., Iwahara, H., and Esaka, T. (1977). *J. Electrochem. Soc.* **124**, 1563–1569.
Tallan, N. M., Tripp, W. C., and Vest, R. W. (1967). *J. Am. Ceram. Soc.* **50**, 279.
Tannhauser, D. S. (1978). *J. Electrochem. Soc.* **125**, 1277–1282.
Tare, V. B., and Schmalzried, H. (1964). *Z. Phys. Chem.* (*Frankfurt am Main*) **43**, 30.
Tiku, S. K., and Kröger, F. A. (1980). *J. Am. Ceram. Soc.* **63**, 31–32.
Tuller, H. L. (1977). *Proc. Int. Conf. Defects Insulat. Cryst., Gatlinburg, Tenn.* NTIS, U. S. Dep. Commer. CONF-771002, pp. 444–445. (Extended Abstr.)
Tuller, H. L. (1978). *Proc. Workshop High Temp. Solid Oxide Fuel Cells, Brookhaven Lab., New York* pp. 104–113.
Tuller, H. L., and Nowick, A. S. (1975a). *In* "Mass Transport Phenomena in Ceramics" (A. R. Cooper and A. H. Heuer, eds.), pp. 177–185. Plenum, New York.
Tuller, H. L., and Nowick, A. S. (1975b). *J. Electrochem. Soc.* **122**, 836–838.
Tuller, H. L., and Nowick, A. S. (1975c). *J. Electrochem. Soc.* **122**, 255–259.
Tuller, H. L., and Nowick, A. S. (1977). *J. Phys. Chem. Solids* **38**, 859–867.
Tuller, H. L., and Nowick, A. S. (1979). *J. Electrochem. Soc.* **126**, 209–217.
Tuller, H. L., Tannhauser, D. S., and Calahora, Z. (1981). To be published.
Van Handel, G. J., and Blumenthal, R. N. (1974). *J. Electrochem. Soc.* **121**, 1198–1202.
Vest, R. W., Tallan, N. M., and Tripp, W. C. (1964). *J. Am. Ceram. Soc.* **47**, 635–640.
Voorhoeve, R. J. H., Remeika, J. P., Trumble, L. E., Cooper, A. S., Disalvo, F. J., and Gallagher, P. K. (1975). *J. Solid State Chem.* **14**, 395–406.
Wagner, C. (1936). *Z. Phys. Chem. Leipzig* **B32**, 447.
Wagner, J. B., Jr. (1979). *In* "Corrosion Chemistry" (G. R. Brubaker and P. B. P. Phipps, eds.), ACS Symposium Series, No. 89, pp. 76–95. Am Chem. Soc., Washington, D.C.
Wang, D. Y., and Nowick, A. S. (1980). *J. Electrochem. Soc.* **127**, 113–122.
Weppner, W. (1976). *Z. Naturforsch., Teil A* **31**, 1336–1343.
Weppner, W., and Huggins, R. A. (1978). *Annu. Rev. Mater. Sci.* **8**, 269–311.
Worrell, W. L. (1977). *In* "Solid Electrolytes" (S. Geller, ed.), pp. 143–168. Springer-Verlag, Berlin and New York.

7

*Structure, Defects, and Nonstoichiometry in Oxides: An Electron Microscopic View**

LeRoy Eyring

Department of Chemistry and
The Center for Solid State Science
Arizona State University
Tempe, Arizona

. . . each man called perfect that which he saw agree with the general idea he had formed of the thing in question, and called imperfect that which he saw agree less with his own preconceived type, even though it had evidently been completed in accordance with the idea of its artificer. This seems to be the only reason for calling natural phenomena, which indeed, are not made with human hands, perfect or imperfect: for men are wont to form general ideas of things natural, no less than

*Supported by the National Science Foundation under grants DMR77-08473 and DMR78-05722.

ISBN 0-12-655280-0

things artificial, and such ideas they hold as types, believing that Nature (who they think does nothing without an object) has them in view, and has set them as types before herself. Therefore when they behold something in Nature, which does not wholly conform to the preconceived type which they have formed of the thing in question, they say that Nature has fallen short or has blundered and has left her work incomplete. Thus we see that men are wont to style natural phenomena perfect or imperfect rather from their own prejudices than from true knowledge of what they pronounce upon . . . .

Spinoza, B. (1677). From Part IV of the "Ethics,"
translated from the Latin by R. H. M. Elwes (1883).

I. Introduction

Thus far the contributions to this volume have dealt primarily with the thermodynamic and dynamic characteristics of nonstoichiometric oxides (having variable stoichiometry). In these concluding chapters, structures that reflect nonstoichiometry are discussed. Here the unique contributions of electron microscopy (EM) to these studies is surveyed. It has only been within the past decade that enough information has come from EM on the subject of defects in solids, in the sense intended here, to be reviewed and discussed. For example, Amelinckx and Van Landuyt (1970), Ashbee (1970), and Hyde and Bursill (1970) discussed then-current observations in a conference on extended defects in nonmetallic solids in the spring of 1969. A little later, Tilley (1972), with insight gained from EM studies, discussed crystallographic shear (CS) in inorganic oxides resulting in ordered and disordered intermediate compounds. Hyde *et al.* (1974) treated relationships between solid structures and structural elements that provide a coherent transition between them. Considerable EM, some at high resolution, had already been done. At a colloquy on defects and transport in oxides, Anderson (1974) and Cowley and Iijima (1974) emphasized the contributions made by high-resolution transmission electron microscopy (HRTEM). Cowley (1976) described the imaging process in HRTEM and its application to inorganic materials. The structural chemistry of some complex oxides, many of which are the subject of this book, was selectively reviewed by Eyring and Tai (1976). In a short review, Iijima and Cowley (1977) discussed developments of the theory and application of EM and optical transforms to studies on order and disorder in complex materials. Anderson (1977) summarized the current understanding of stoichiometry, defects, and ordering and the exploitation of high-resolution electron microscopy (HREM) in their study. The direct imaging of atoms in crystals was the subject of the 47th Nobel Symposium (Kihlborg, 1978–1979). These proceedings give an account of the present state of HREM.

The term "nonstoichiometry" applies to exceptions to the central rules of stoichiometry, the law of constant composition, and the law of definite proportions fundamental to classical chemistry. It is now realized that all solid compounds, except molecular solids, have a composition that sensibly depends on the environment, i.e., the temperature, pressure, and activity of the components of the solid. The chemical makeup conforms to the law of constant composition only when the limits of analysis are sufficiently relaxed. Compounds of atoms of fixed single valence usually show a very small deviation from the law of definite proportions, but oxides of multivalent transition, lanthanide, and actinide elements frequently have metal-to-oxygen (M/O) ratios far from that of small whole numbers. They often have ratios quite near those of large whole numbers, such as 20/57, considered by Wadsley (1968) to represent "grotesque stoichiometry." There are, however, innumerable carefully prepared compounds whose atomic ratios remain irrational or can easily be made so.

Nonstoichiometry has great significance for structure. Crystal chemistry tended, on the basis of fixed contents of a unit cell, to require that solid compounds be stoichiometric. It is clear that these concepts have had to be modified with the appearance of HRTEM which provides information on local structural anomalies at the atomic level. We shall direct our attention to the compounds possessing composition-altering features in easily detectable amounts. The use of the word "structure" shall not always mean that the exact location of each atom is known. Rather, it will indicate that the atom positions are approximately those of some well-known crystal type with much more detail about local deviations from that structure. It is a more intimate view of atomic makeup than can be obtained by the averaging methods of x-ray and neutron diffraction.

The unique ability of HRTEM techniques to observe defects in solids and to follow their change with time is of enormous importance in science. It should be possible to determine mechanisms of solid state reactions directly in some cases and to suggest them in many others, including those occurring or being promoted at interfaces or surfaces.

The examples chosen in later sections will illustrate a large number of ways crystals adjust their composition in response to their environment. A nonstoichiometric crystal clearly has inhomogeneities above the atomic level. These could result from varied occupancy of sites preserving the framework of the basic structure, or aggregates of atoms with a compatible but distinctive arrangement could be present within the structure, giving texture to the material. The latter appears to be more common. Nonstoichiometry comes about through the creation of extended defects, first as isolated features coherently intergrown within the parent structure; then, as the concentration of these defects increases, they tend toward further aggregation or intermittent order, resulting in a new, ordered, intermediate structure. This extended defect may be simply a compatible variation, or it may be an element of another structure. The final ordered structure may be an intergrowth of both the old and new elements in some repetitive way. It may also consist of regions of the old structure regularly interrupted by a new feature. The extended defect frequently modifies substantially one of the atomic substructures; i.e., one sublattice may remain intact while the other is modified, often forming a planar fault.

In some cases, an increasing concentration of one type of atom causes the defect mode to evolve in a discontinuous way as the structure adapts to changing atomic ratios. We shall see this, for example, as the CS of WO_3 gives way to tunneled structures and then to block structures of different connectivities as the M/O ratio decreases. It appears that simple point defects that may be formed initially quickly aggregate into extended defects by elimination, as CS, or are preserved in an ordered way, as in the fluorite-related structures. When features such as tunnels occur to accommodate variable amounts of certain types of atoms, the question of the extent to which they are ordered in the tunnels is an interesting one just beginning to be answered.

The key to these complex structures appears to be an ability to form compatible interfaces between structural elements along crystallographic directions, thus providing easy transition between the textured elements of the composite structure. This is analogous to building up complex molecular structures such as proteins by utilizing compatible amino acid building blocks, in this case in three dimensions. When the elements of the two structures are intergrown in an ordered way, a new larger unit cell can be defined with an intermediate composition. The almost endless examples of homologous series of compounds are a result of this phenomenon, and several will be discussed in detail below. These systems may either be cation-rich or -poor relative to the parent structure, depending on the system, but as of now many more of the former are known than the latter.

II. The Techniques of High-Resolution Electron Microscopy

Electron microscopy is most powerful as a tool for structure determination when used in conjunction with the conventional techniques of x-ray and neutron diffraction. Nevertheless, even used alone, it can reveal the symmetry, lattice parameters, and sometimes the structure of crystalline materials. In principle, it is possible to perform direct-space crystallography with EM, but so far this has remained sparingly developed (Cowley, 1978). Structures have been convincingly proposed from HRTEM results when x rays have been unable to yield the answer (Summerville *et al.*, 1978) or have led to the wrong structure (Allpress *et al.*, 1973). In addition, we shall see that EM provides information about structure in very small crystal volumes, which cannot be obtained in any other way. The basis of nonstoichiometry is revealed precisely by the analysis of such small volumes in space.

The use of x-ray and neutron diffraction in the study of nonstoichiometry is the topic of Chapter 8. A comparison between x-ray and electron diffraction methods of crystal analysis indicates the areas of particular usefulness and weakness of each method. The wavelength of 100-kV electrons is 0.037 Å, compared to 1.54 Å for CuK_α x rays. The Ewald sphere for electrons is so large that when Bragg's law is satisfied the reciprocal lattice points of greatest interest fall almost in a plane. This makes possible the recording of large sections of the reciprocal lattice without the complicated camera machinations required for the x-ray precession camera.

The scattering factors for electrons are about 1000 times greater than for x rays. This makes possible the observation of diffraction effects from very small regions of space. Sharp diffraction patterns are obtained from an area 10^{-10} cm^2 and 10^{-6} cm thick in a few seconds, while crystals tenths of a centimeter on an edge commonly used for x-ray analysis require exposures of many hours. This ability to gain detailed structural information from about 10^4 unit cells rather than the 10^{18} required for x rays is frequently an advantage.

This greater interaction of electrons raises serious problems for the analysis of electron scattering since, in a thick specimen, an electron in transit may be scattered many times. For this reason, except for very thin specimens, where kinematic scattering theory may be successfully applied, dynamical theory must be used. Dynamical effects cannot presently be dealt with adequately for crystals with a thickness greater than about 100 Å. This limits direct-space crystallography by electron diffraction (Cowley, 1978).

Most of the results of EM studies on the structure of nonstiochiometric oxides have been obtained by conventional transmission techniques. HRTEM combines the ability to obtain selected-area electron diffraction from a region approximately 3×10^6 $Å^3$ with the ability to produce an image from about 30% of this area with a point-to-point resolution of $\leqslant 3.5$ Å in 100-kV machines and about 2 Å in 1-MeV microscopes, virtually at atomic resolution, and is improving progressively. Because of their charge, electrons may be focused and images magnified.

The theory of diffraction and imaging of electrons scattered from thin specimens is discussed at length by Cowley (1975, see esp. Chaps. 13, 17, and 18). Less detailed accounts may be found in Allpress (1972), Allpress and Sanders (1973), and Allpress *et al.* (1972). Programs for calculating images to be expected from thin crystals have been made available by O'Keefe (1973), Skarnulis *et al.* (1978), and O'Keefe *et al.* (1978).

Specimens from bulk samples can be prepared for examination in the microscope by grinding, crushing, or ion-beam thinning crystal fragments to obtain edges about 50 Å thick. These can be placed on a coated grid support on a tilting goniometer stage, allowing the aligned thin edges to be examined by electron diffraction and electron image formation without interference from the support. The diffraction pattern (spots, streaking, and diffuse scatter) yields information concerning the crystal structure, perfection, and whether more than one phase inhabits the exposed region. The high-resolution images are frequently taken as a through-focus series to be analyzed. In favorable cases, as illustrated below, the nature of small regions is revealed at the atomic level. In others, features at the level of coordination polyhedra or small defect clusters are clearly revealed.

The electron micrograph is an array of blobs of varying contrast arranged in a more-or-less regular way. In order for the image to be interpreted intuitively as a two-dimensional projection of the atomic potentials, the microscope must be well aligned, the beam must be along the crystal axis within 2 or 3 milliradians, the crystal must be very thin (less than 100 Å), and the Scherzer focus (Scherzer, 1949) must be achieved. Under other conditions, different contrast patterns are obtained, which must be interpreted with the aid of calculations. Programs are now available (O'Keefe *et al.*, 1978) based on the *n*-beam multislice theory of Cowley and Moodie (1957). Calculated images, which include through-focus and thickness effects, can match observed images in a convincing way in establishing the structure.

The many techniques associated with scanning transmission electron microscopy (STEM) can now be added to those of fixed-beam instruments (HRTEM) in the study of nonstoichiometry in crystalline materials. These methods depend on observing the effect of a small, bright electron beam as it sweeps the area of interest on a specimen using a television-type raster. The transmitted electron beam, backscattered electrons, emitted x rays, etc., can then be monitored to give structural and analytical information from very limited regions. If the scan of the beam is stopped, a convergent beam diffraction pattern can be obtained from a diameter of the specimen comparable to the resolution limit of the microscope (at present about 5 Å in diameter) (Cowley, 1978–1979). One of the greatest advantages is that, in this technique, the sample is usually outside the lens, making space available for instrumentation for special experiments. It also makes it easier to have a controlled-environment cell for the specimen.

So far, STEM methods have not been very fully developed for the study of crystalline oxides, but many desirable features of the method should be exploited. With a STEM attachment for currently available microscopes, it is possible to perform high-resolution studies at almost atomic resolution and to obtain x-ray analytical results for regions of a few hundred cubic angstroms at the same time. This will facilitate greatly the study of compositionally complex systems important in mineralogy and chemistry.

Radiation damage to specimens caused by the intense electron beam during imaging must always be kept in mind. In fact, this limits ultimate resolution studies to particularly stable materials. Organic molecules are virtually eliminated from consideration, as are many other heat-sensitive compounds such as nitrates and hydroxides. Oxides as a class tend to be refractory, hence among the most stable, in the electron beam. (An interesting observation is the great variation in stability to the beam of different crystals of the same preparation.) The range of materials that may be advantageously studied extends beyond those that apparently suffer no degradation in the beam required for their observation to those affected but where change is still within the bounds allowing informative study. Whereas some structural studies require complete stability, one often wishes to observe the radiation damage itself and the accompanying chemical and physical changes. Indeed, a future use of the HRTEM technique will involve observation of *in situ* reactions either in the vacuum of the microscope or in controlled-environment cells.

III. The Application of Electron Microscopy to the Study of Nonstoichiometry in Oxides

A. Fluorite-Related Higher Oxides of the Rare Earths

We begin this discussion with the fluorite-related oxides of the rare earths not because they were the first studied by this method, or are the best suited to the technique, but because they have played such an important part in the discussions of the preceding chapters of this volume. In fact, successful application of electron optical techniques to these oxides underlies the usefulness of the method, since they are not particularly suited to HRTEM. The unit cells are, for the most part, dumpy and rather small in all directions but not small enough to satisfy the requirements of the "thin phase object" approximation, hence not well suited in either theory or practice. This they have in common with a large number of interesting chemical systems. These oxide systems have recently been broadly reviewed by Eyring (1979).

Furthermore, patient studies employing x-ray and neutron diffraction have failed to determine either the unit cells or structural principle relating members of this complex system. A primary reason for this failure is that they consist of strongly absorbing heavy metal atoms and light oxygen atoms with a propensity for twinning in ways too difficult to treat. Kunzmann and Eyring (1975) were able, with the use of HRTEM, to obtain the unit cells of the members of the homologous series and to suggest the structural principle that relates them. The relationships between the unit cells are shown in Fig. 1. Table I lists the relationship between members of the homologous series and the fluorite structure. The homologous series of oxides R_nO_{2n-2} was shown actually to consist of two series, one when n is odd and the other, with polysynthetic twinning at the unit cell level, when n is even. Furthermore, it was possible to propose the structure of the $n = 11$ member of the odd series based on its lattice image (Fig. 2). Skarnulis *et al.* (1978) obtained images of the structural prototype Pr_7O_{12} and compared them with those calculated by the n-beam multislice method which they had reduced to a useable computer package.

The unit cell of R_7O_{12} was first proposed for Tb_7O_{12} by Baenziger *et al.* (1961) from x-ray powder diffraction. The structure of an isomorphous phase, UY_6O_{12}, was determined later by Bartram (1966). The structure of Pr_7O_{12} was determined by the Rietfeld method of total profile analysis of powder neutron diffraction curves by Von Dreele *et al.* (1975).

In the fluorite-related Pr_7O_{12} structure, all the metal atoms are either 6- or 7-coordinated. The 6-coordinated metal atoms all lie in {135} planes,

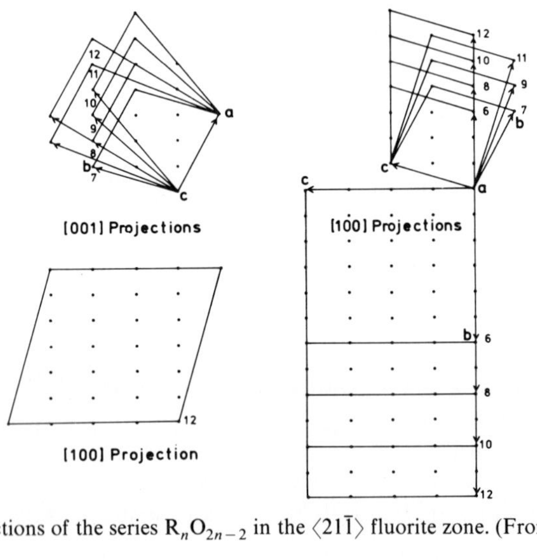

Fig. 1. Projections of the series R_nO_{2n-2} in the $\langle 21\bar{1} \rangle$ fluorite zone. (From Kunzmann and Eyring, 1975.)

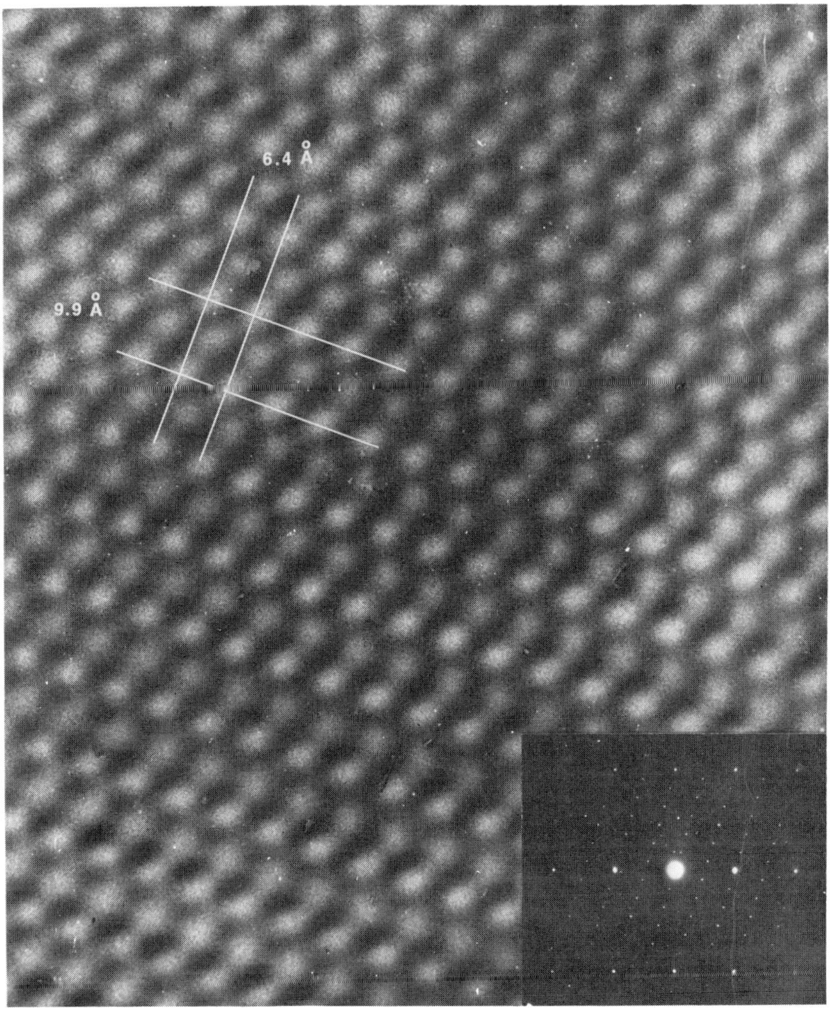

Fig. 2. Optically enhanced $\langle 21\bar{1}\rangle_F$ lattice image of $Tb_{11}O_{20}$ showing the vacancy arrangement. Magnification: 2×10^7. (From Kunzmann and Eyring, 1975.)

and each is surrounded by two oxygen vacancies and six 7-coordinated metal atoms along $[111]_F$ (all indices used in this section refer to the fluorite substructure). The six metal atoms, three above and three below, around the two oxygen vacancies complete what has been called the Bevan cluster found in the ternary $Zr_{10}Sc_4O_{26}$. The metal atoms have moved away from the relatively positive oxygen vacancy by about 0.01 Å. It is this (135) plane that is the structural feature relating the homologous series in binary rare

TABLE I

Structural Data on Intermediate Rare Earth Oxide Phases[a]

Composition	n (in R_nO_{2n-2})[b]	Symmetry	Lattice parameters	Relation to basic structure	Relative volume	No. of vacancies per unit cell
$RO_{1.500}$	4	$Ia3$	$a = 11.152$ Å ($PrO_{1.5}$) $a = 10.728$ Å ($TbO_{1.5}$)	$a = 2a_f$	8	16
$RO_{1.714}$	7	$R\bar{3}$	$a = 6.750$ Å, $\alpha = 99°23'$ ($PrO_{1.714}$) $a = 6.509$ Å, $\alpha = 99°21'$ ($TbO_{1.714}$)	$a = a_f + \frac{1}{2}b_f - \frac{1}{2}c_f$	$\frac{7}{4}$	2
$PrO_{1.778}$	9	Triclinic	$a = 6.5$ Å $b = 8.4$ Å $c = 6.5$ Å $\alpha = 97.3°$ $\beta = 99.6°$ $\gamma = 75.0°$	$a = a_f + \frac{1}{2}b_f - \frac{1}{2}c_f$ $b = \frac{3}{2}b_f + \frac{1}{2}c_f$ $c = \frac{1}{2}a_f - \frac{1}{2}b_f + c_f$	$\frac{9}{4}$	2
$PrO_{1.800}$	10	Pn	$a = 6.7$ Å $b = 19.3$ Å $c = 15.5$ Å $\beta = 125.2°$	$a = a_f + \frac{1}{2}b_f - \frac{1}{2}c_f$ $b = \frac{5}{2}(-b_f - c_f)$ $c = 2(-b_f + c_f)$	10	8

$TbO_{1.809}$	$10\frac{1}{3}$	Triclinic	$a = 13.8$ Å $b = 16.2$ Å $c = 12.1$ Å $\alpha = 107.4°$ $\beta = 100.1°$ $\gamma = 92.2°$	$a = -\frac{1}{2}a_f + \frac{5}{2}c_f$ $b = -2a_f - 2b_f - c_f$ $c = 2a_f - b_f$	$\frac{31}{2}$	12
$TbO_{1.818}$	11	Triclinic	$a = 6.5$ Å $b = 9.9$ Å $c = 6.5$ Å $\alpha = 90.0°$ $\beta = 99.6°$ $\gamma = 96.3°$	$a = a_f + \frac{1}{2}b_f - \frac{1}{2}c_f$ $b = -\frac{1}{2}a_f + \frac{3}{2}b_f + c_f$ $c = \frac{1}{2}a_f - \frac{1}{2}b_f + c_f$	$\frac{11}{4}$	2
$TbO_{1.833}$	12	*Pn*	$a = 6.7$ Å $b = 23.2$ Å $c = 15.5$ Å $\beta = 125.2°$	$a = a_f + \frac{1}{2}b_f - \frac{1}{2}c_f$ $b = 3(-b_f - c_f)$ $c = 2(-b_f + c_f)$	12	8
$PrO_{1.833}$	12	*Pn*	$a = 6.687$ Å $b = 11.602$ Å $c = 15.470$ Å $\beta = 125°15'$	$a = a_f + \frac{1}{2}b_f - \frac{1}{2}c_f$ $b = \frac{3}{2}(-b_f - c_f)$ $c = 2(-b_f + c_f)$	6	4
$RO_{2.000}$	∞	*Fm3m*	$a = 5.393$ Å (PrO_2) $a = 5.220$ Å ($TbO_{1.95}$)		1	0

[a] From Kunzmann and Eyring (1975).

[b] If a phase occurs in the PrO_x as well as in the TbO_x system R is used, otherwise the specific symbol is used.

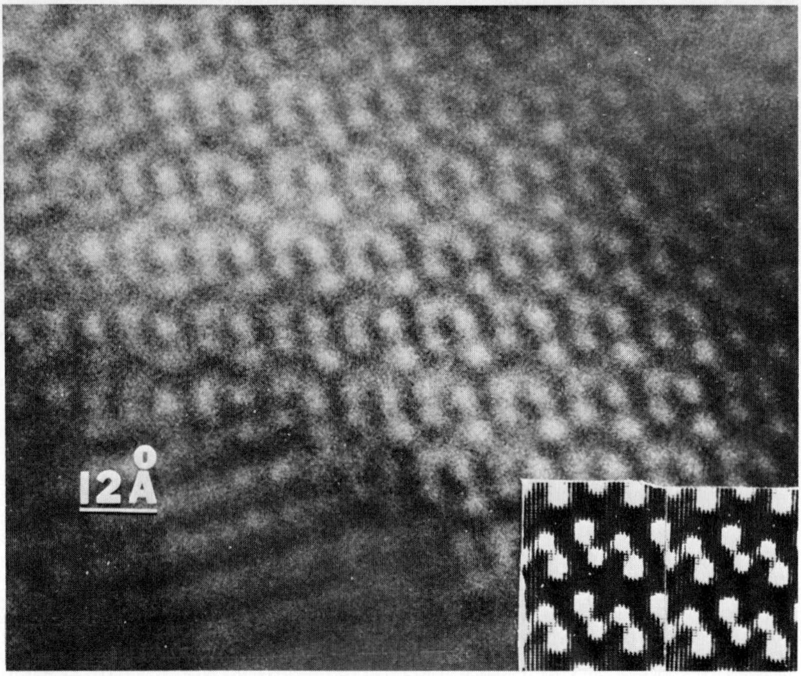

Fig. 3. Crystral structure $\langle 211 \rangle_F$ image of a thin β-phase crystal. Inset is a calculated image from the Pl structural model. (From Summerville *et al.*, 1978.)

earth oxides. It is not a shear plane, since there is no functional shear to remove the oxygen vacancies.

Summerville *et al.* (1978) were then able to propose a structure for $Pr_{24}O_{44}$, the $n = 12$ member of the series, on the basis of HRTEM images. Figure 3, which compares observed and calculated images, shows that the rows of oxygen vacancies appear as white dots. Tuenge and Eyring (1979) confirmed the proposed structure of Pr_9O_{16} as illustrated in Fig. 4. Based on these studies it is virtually estabished that the odd members of the series have $\{135\}$ planes that contain all the 6-coordinated rare earth atoms with two oxygen atoms missing from each coordination cube along [111]. The spacing of these planes determines the *b* axis and composition. In the case of the even members, the defect feature is a corrugation of $\{135\}$ planes which result from twinning at the unit cell level. Again, the even number that characterizes the series member is determined by the spacing of these planes. In some cases the corrugations are modified and the unit cell dimensions are doubled again, this time along the *b* axis. This provides an opportunity to display a large number of possible polymorphs, which is not missed by these remarkable oxides.

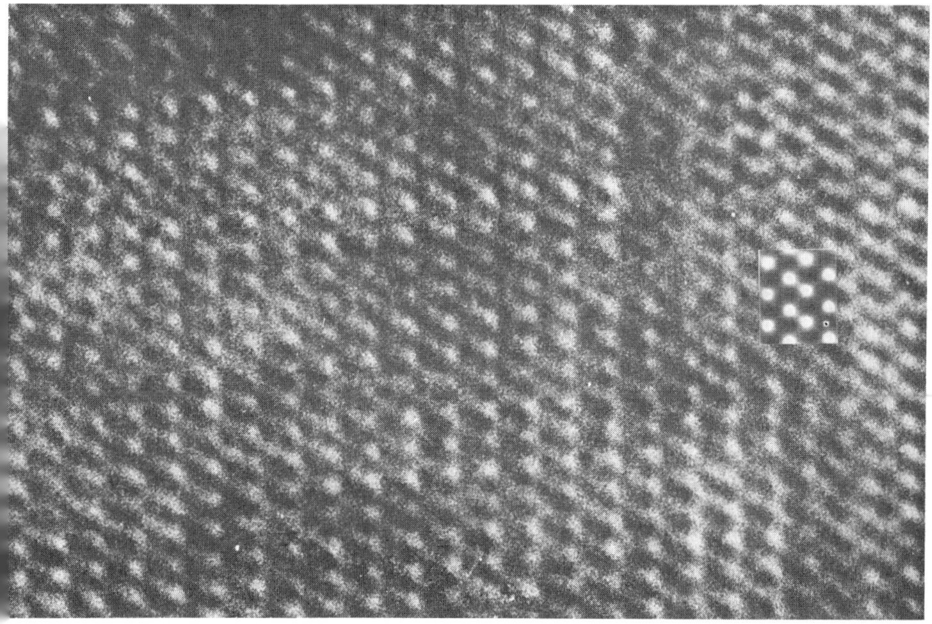

Fig. 4. An experimental image of a ζ phase in the $\langle 100 \rangle_9$ zone at a defocus of -900 Å. The inset is the calculated image. (From Tuenge and Eyring, 1979.)

In spite of a homologous compositional series for Ce, Pr, and Tb binary oxides, only the $n = 7$, R_7O_{12}, member is structurally isomorphous in the three series. For $n = 12$, on the other hand, $Pr_{12}O_{22}$ has been observed in two polymorphs—one is indicated as the primitive cell in Fig. 1 and the other, $Pr_{24}O_{44}$, is formed by twinning the primitive cell across (110)—whereas two entirely different polymorphs of $Tb_{12}O_{22}$ have been observed. One of these is similar to the twinned $Pr_{24}O_{44}$ except that the parallel *a* and *c* planes have been slipped relatively a distance of one metal atom. The other, with a unit cell content $Tb_{48}O_{88}$, is doubled again along the *b* axis as illustrated in Fig. 1. It is obvious that extremely slight differences in atomic properties result in different fluorite-related structures. This independent behavior appears to be maintained in ternary fluorite-related materials (see, e.g., Allpress and Rossell, 1975, to be discussed below).

A small compositional variation may occur within a single phase of the homologous series. Specimens prepared at composition almost always show a large amount of disorder in the microscope. This disorder sometimes consists of planar inclusions of other phases one or a few unit cells thick, and at other times contrast variation suggests much smaller regions of imperfection. The crystal defects and twinning that prevented structure determination by conventional methods are apparent in microscope images. In addition, these oxides are not very stable in the electron beam in the vacuum of the microscope and can easily be induced to lose oxygen and thereby be transformed to other members of the homologous series. This fortuitous circumstance makes possible observation of the *in situ* decomposition of these oxides discussed in Section IV.

B. Rare Earth Sesquioxides and Some Ternary Phases

Caro, Schiffmacher, and co-workers in France have studied rare earth sesquioxides and have given some attention to ternary phases (pseudobinary) such as the La_2O_3–CeO_2 system. Shiffmacher (1977) has reviewed much of this work. Rare earth sesquioxides have been of interest because of this group's interest in exploring the breadth of application of the idea that $(LnO)_n^{n+}$ sheets behave structurally like those of $(BiO)_n^{n+}$ which can sandwich almost any anionic group.

The cubic C type is fluorite-related and isotropic, whereas the A type is hexagonal and may convincingly be considered layers of $(LnO)_n^{n+}$ in which OLn_4 tetrahedra share edges to form a layer framework with oxygen deployed between the sheets to make Ln_2O_3. The B-type sesquioxide is similar to the A type, except that the sheets are shifted with respect to each other and the hexagonal symmetry changes to monoclinic.

A particularly interesting study on twinning was made in B-type samarium oxide prepared by the oxidation of thin metal films. In these materials there is a large population of defects of various types. Figure 5 shows the occurrence of a step along $[3\bar{1}3]$ in the (111) plane observed by Schiffmacher *et al.* (1976), with contrast variation along its length. Figure 6 shows the nature of the step. Sibieude *et al.* (1978) have published the results of a high-resolution microscopic study of the pseudobinary La_2O_3–CeO_2 system in the La_2O_3-rich region. In this material the two oxides grow in bands with the *c* axis of La_2O_3 along the threefold axis of cubic CeO_2. In the case of the La_2O_3–ThO_2 system, also prepared by Sibieude, the phases form a polytypic series with a dispersion of coherent layers (L and T) at the unit cell level (Van Tendeloo *et al.*, 1979). In Fig. 7 the contrast shift at the layers of thorium atoms is very striking. The technologically important stabilized cubic structures of zirconia and hafnia by calcia have not yet been imaged at the atomic level, but electron diffraction studies (Allpress and Rossell, 1975) have revealed ordered intermediate phases and a small domain size in cases of disordered intermediate compositions. For example, analysis of diffuse scatter in electron diffraction patterns suggested domains of the ϕ phase of 30 Å diameter coherently embedded in the cubic matrix in all possible orientations.

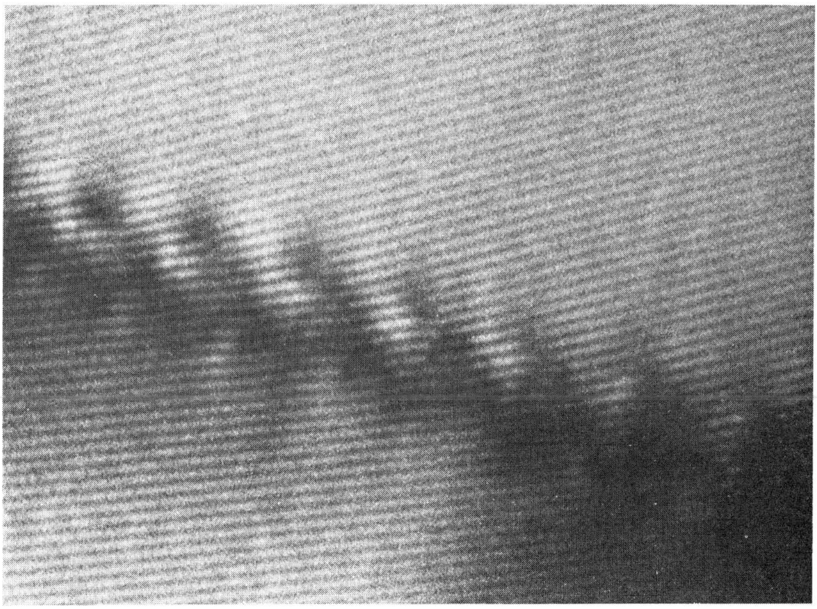

Fig. 5. A step along [3$\bar{1}$3] in the (111) plane of Sm_2O_3. (From Schiffmacher, 1977.)

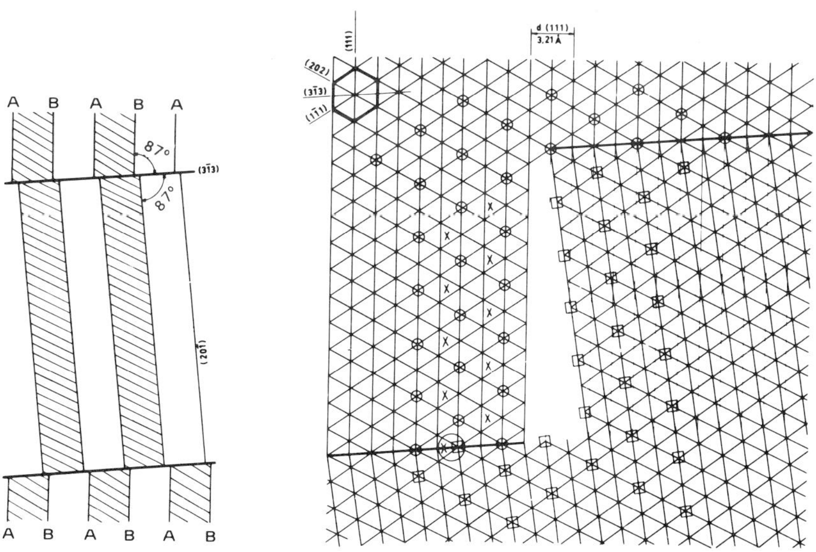

Fig. 6. A model of the step shown in Fig. 7 (From Schiffmacher, 1977.)

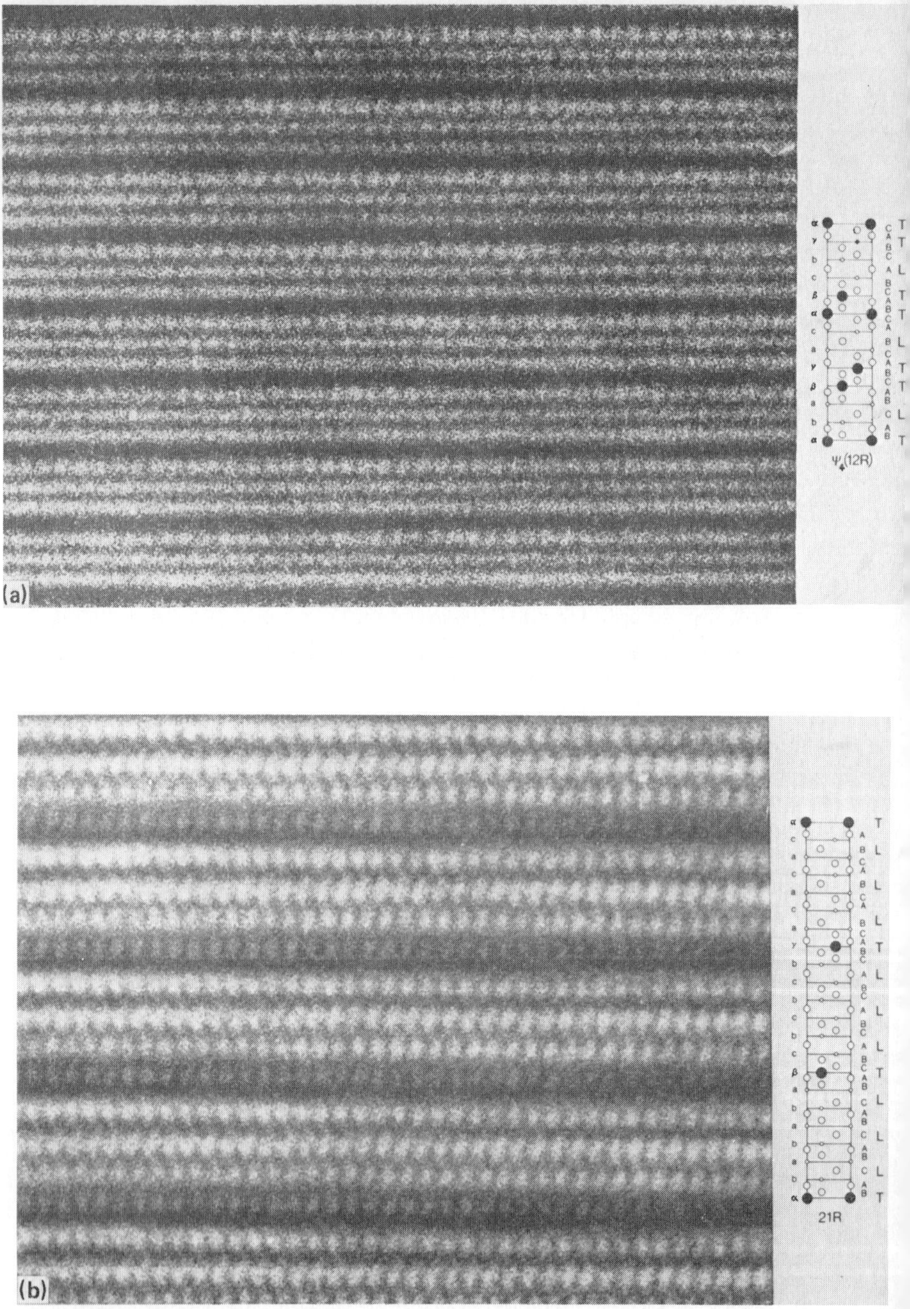

Fig. 7. Structure images of mixed lanthanum–thorium oxides. (a) Ψ_4 (12R); (b) 21R. (From Van Tendeloo *et al.*, 1979.)

C. Oxides with ReO_3-Related Structures

1. Shear Structures

Undoubtedly, the oxides most studied with the new electron microscopic techniques are those with ReO_3-related structures. This is the natural outcome of the developments of the 1950s and 1960s in which the idea of homologous series of oxide phases was presented (Magnéli, 1949) and the structural feature associated with ReO_3-related series was called CS by Wadsley (1955). An isolated or disordered CS structural defect is called a Wadsley defect (Andersson, 1970). As we shall see, it is this defect that is responsible for nonstoichiometry in a large number of transition metal oxides.

It was, in fact, the application of EM to ReO_3-related structures that began the flurry of examinations of the ultrastructure of high-temperature oxides (Allpress *et al.*, 1968) a dozen years ago. The attainable resolution of electron microscopes was greatly improved a decade ago, as demonstrated by Iijima (1971) with a specimen of $Ti_2Nb_{10}O_{29}$. As shown in Fig. 8, he obtained images of this ReO_3-related block structure at about 3-Å resolu-

Fig. 8. Two-dimensional lattice image of orthorhombic $Ti_2Nb_{10}O_{29}$ obtained with the electron beam incident parallel to the *b* axis. (From Iijima, 1971.)

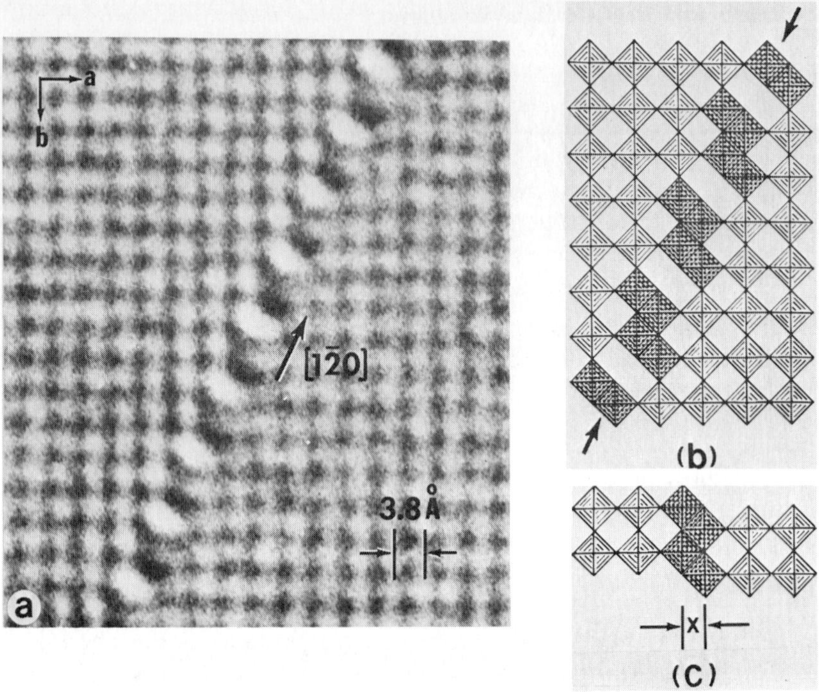

Fig. 9. (a) High-resolution electron micrograph of a WO_{3-x} crystal obtained with the electron beam incident parallel to the *c* axis. A prominent feature of the contrast shows the (210) CS plane across which square arrays of dark dots (indicating W atom positions) are shifted. The contrast distributions are in good accordance with the idealized model of the (210) CS plane (b). However, the image showed that edge-sharing octahedra (shaded) are considerably distorted, so that the distance *x* in (c) is much larger than that in the ideal model. (From Iijima, 1975.)

tion, which was enough to see an intuitive two-dimensional display of the square tunnels in the ReO_3-type structure (see Fig. 9b). These structures are ideal for HRTEM because of their large unit cell dimensions in the plane of the specimen and the short dimension of one octahedron along the beam direction.

It is easy to visualize CS in ReO_3-type structures. In the parent structure the octahedra formed by coordination of the metal atom with six oxygen atoms share all corners in a square network illustrated in one direction by the unperturbed regions in Fig. 9b. One can visualize this in three dimensions, since the cubic array would appear the same in all three directions. The checkerboard array represents columns of metal-filled octahedra sharing corners, the projected square outline of which is the maximum dimension. The light squares are empty tunnels in all three directions.

Imagine all the oxygen atoms in a (210) plane to be removed and the original oxygen atom array reestablished by joining the two parts. This shearing action of $\frac{1}{2}[1\bar{1}0]$ is a move in the direction and the distance of one edge of an octahedron. This results in the maintenance of the octahedral coordination of the metal atoms by introducing edge sharing, as illustrated in Fig. 9b and c, producing a Wadsley defect designated $(210)\frac{1}{2}[1\bar{1}0]$. The higher M/O ratio and the nonstoichiometry are maintained by creating a slab in the parent structure with shorter metal–metal distances.

The Wadsley defect has been imaged at a resolution of about 3.5 Å by Iijima (1975) in ReO_3-type $WO_{3-\delta}$. Figure 9a shows the altered contrast resulting from the edge sharing of tunnels and filled octahedra. In a thin crystal, such as was examined here, the image is an intuitive projection of the structure in the direction of the electron beam. There is a distortion of the connectivities at the shear plane such as to produce a greater distance x (Fig. 9c) in the real structure than in the idealized one shown.

Figure 10 shows a well-developed Wadsley defect in $WO_{3-\delta}$ flanked by regular contrast variation which, upon further electron beam heating, develops into additional shear planes parallel to the first at more-or-less

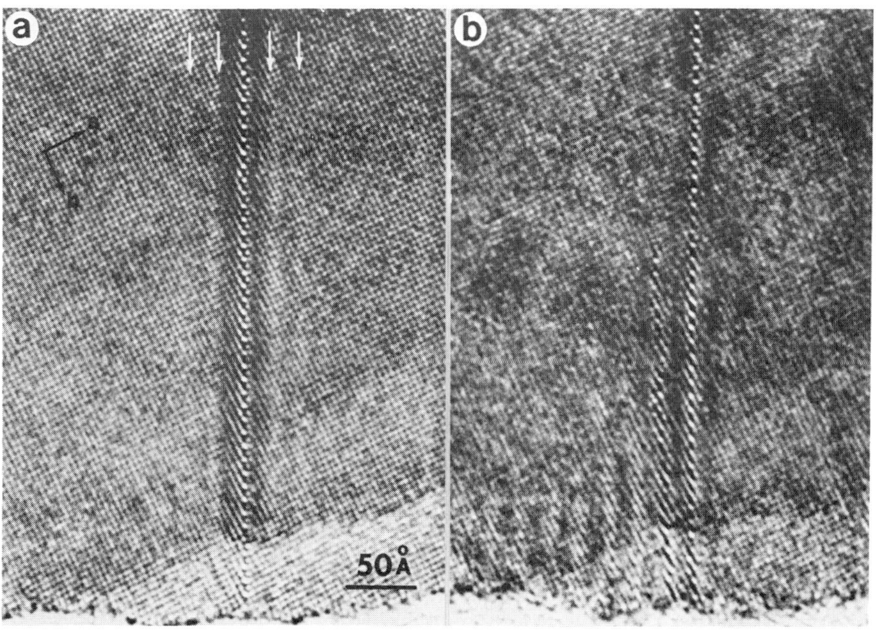

Fig. 10. (a) Oscillatory dark and light contrast bands running parallel to the CS plane on both sides. (b) With electron beam irradiation new CS planes develop with the same periodicity as that of the oscillatory contrast in (a). (From Iijima, 1975.)

constant intervals. This appearance of parallel lines of dark strain fields along [102] is indicative of early development of shear. This strain is relieved as the effective shear is completed. As the Wadsley defects increase in concentration, they tend to become parallel and to purge themselves of the kinks they frequently show at low concentrations. This behavior is consistent with the model of CS given by Anderson and Hyde (1967) in which oxygen vacancies aggregate on a disk, surrounded by partial dislocations, which expands and moves longitudinally.

Crystallographic shear may be based on other planes, as we shall see later, but the original material reduced at 1022°C displayed only (210) and $(2\bar{1}0)$ CS planes. No shear planes were observed in the [010] zone. During beam heating, however, CS occurs on (201) and $(20\bar{1})$. The distortion of the monoclinic WO_3 structures from cubic ReO_3 probably accounts for this anisotropic behavior.

Tilley and co-workers have studied $WO_{3-\delta}$, where δ varies over a considerable range. This was done by reducing WO_3 crystals and by substituting a large number of different altervalent metal atoms for tungsten at various concentrations. For example, Sundberg and Tilley (1974) reduced WO_3 to an approximate composition $WO_{2.91}$ at 950, 1000, and 1070°C. The oxygen loss resulted in {103} CS planes being produced at 1070°C, {102} CS at 950°C, and both {103} and {102} CS at 1000°C. The annealed crystals always had needle-shaped crystallites attached as a result of vapor growth. These, formed at any temperature, were members of the W_nO_{3n-2} homologous series based on {103} CS.

Sundberg and Tilley (1974) observed the growth of CS planes directly at high resolution. The structure of W_nO_{3n-2}, based on {103} CS, was modified by adding and extending the CS planes when it was irradiated in the electron beam. A dislocation mechanism was proposed for CS growth. There was, however, much evidence of change by other mechanisms. There is considerable similarity between these observations and those of Iijima's (1975) work mentioned above.

Pickering and Tilley (1976) examined more completely reduced $WO_{3-\delta}$ specimens in the electron microscope. Near WO_3 the homologous series W_nO_{3n-2} was observed, the lowest value of n being 16. At the other end of the compositional range the tunneled structure $W_{18}O_{49}$ existed, apparently without compositional variation. A new phase more closely related to tunnel structures than CS structures was observed between these two compositions. An interesting feature of a study on pure $WO_{3-\delta}$ by Allpress *et al.* (1971) was that the Wadsley defects collected in packets, producing regions of the crystal with very different compositions. Magnéli (1978) has summarized the electron microscopic results on reduced binary $WO_{3-\delta}$. He has emphasized the results of Sundberg (1976), Kihlborg and Sundberg

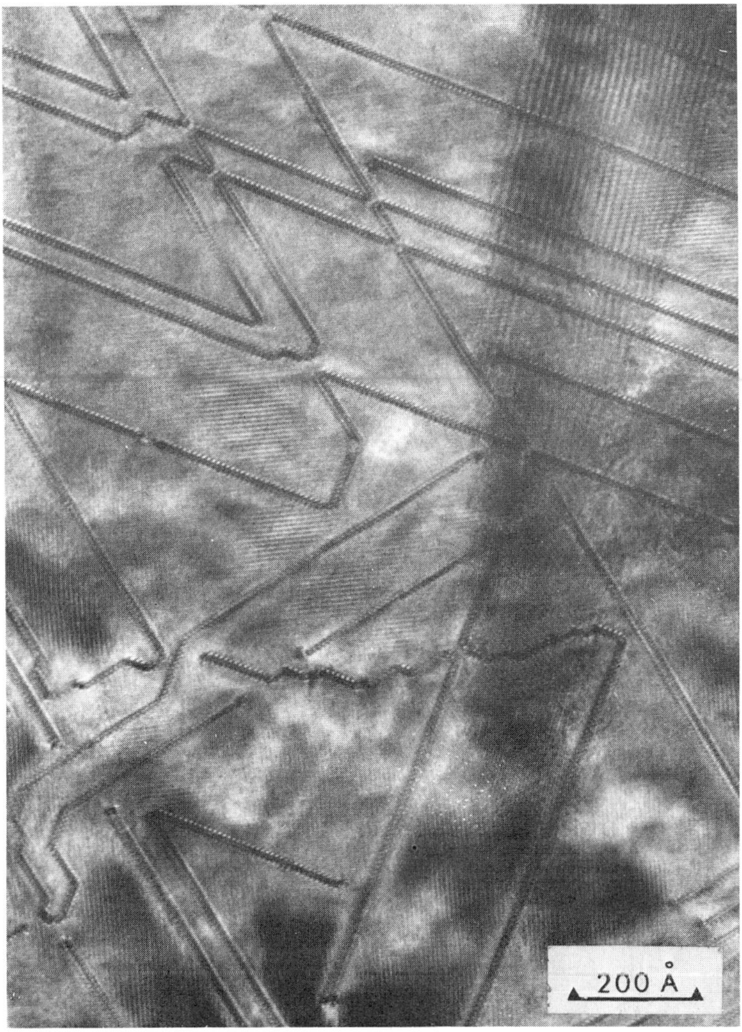

Fig. 11. Lattice image of a very slightly reduced tungsten trioxide crystal, WO_{3-x}, containing {201} CS planes in four different orientations. (From Kihlborg and Sundberg, 1976.)

(1976), and Sundberg and Tilley (1974), with the conclusion that, at elevated temperatures, as the oxygen content is reduced, {102}-type shear planes are introduced in all possible directions at irregular intervals. Figure 11 shows a region of $WO_{2.99}$ in which this situation obtains. As the defect concentration increases, the structure is modified by the suppression of {102} shear plane growth in all but one direction. Figure 12 depicts such a

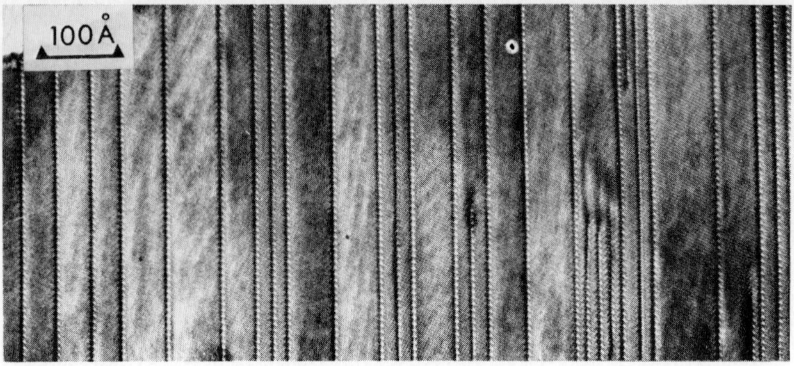

Fig. 12. Structure image of a $WO_{2.97}$ crystallite. Shear planes ending within the crystal are likely to have formed under the influence of the electron beam in the microscope. (From Magnéli, 1978.)

crystal of $WO_{2.97}$ in which only [102] CS occurs, but at very irregular intervals, preserving nonstoichiometry. At greater degrees of reduction {103} CS occurs, yielding planes of lower oxygen content at greater spacing. Crystallographic shear on {103} can result in ordered intermediate phases which are members of the homologous series W_nO_{3n-2} with regular

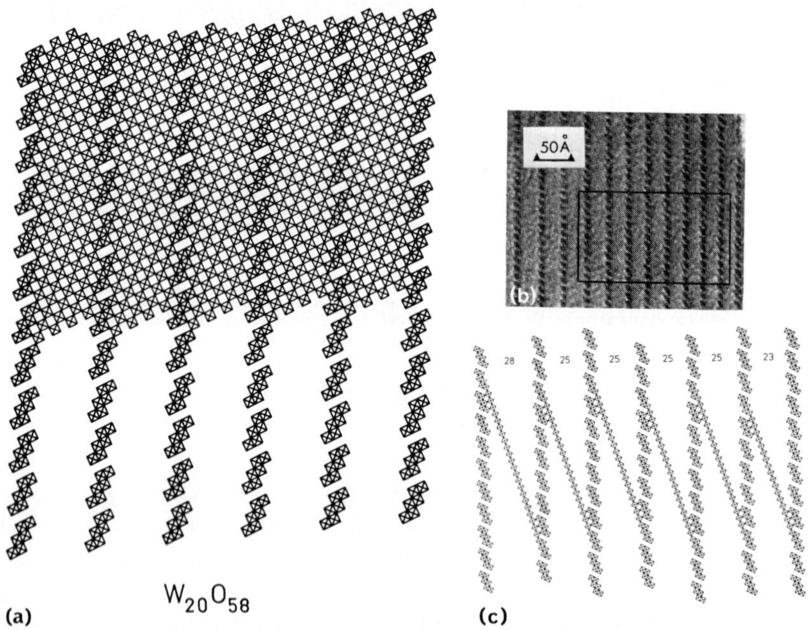

Fig. 13. (a) Structure of $W_{20}O_{58}$ as formed from WO$_6$ octahedra. (b, c) Structure of $W_{25}O_{73}$: (b) shows an electron optical image and (c) the interpretation in terms of WO_6 octahedra of the portion within the rectangle. (From Magnéli, 1978.)

spacings. However, phases that are incompletely annealed retain some variation in CS spacing as shown in Fig. 13.

Crystallographic shear structures for reduced WO_3 caused by the substitution of metal atoms of lower oxidation number have been investigated by Ekström and Tilley (1974, 1976), who found CS structures for Ti substitution but not for V. As a matter of fact, except for Ti, transition metals do not substitute in WO_3 to produce structures having CS. Ti substituting in WO_3 at up to 4 at. % produces CS phases based on {102}, {103}, and {001} planes. As the degree of reduction decreases {102} CS gives way to {103} and finally to that based on {001}. Intergrowth phases of {103} and {001} were observed, but no mixed phases.

2. *Tunnel and Column Structures Based on ReO_3-Type Coordination*

As the degree of reduction increases in WO_3-based materials, new defect structure types emerge. Elements of the first type are formed by a complicated cylindrical shear (Hyde and O'Keeffe, 1973), which preserves the corner sharing of the metal coordination octahedra while generating trigonal and pentagonal tunnels and, in some cases, pentagonal columns and hexagonal tunnels along the periphery. Among the pentagonal column (PC) structures that have been elucidated by EM are ternary phases with the Mo_5O_{14} structure, $\Theta_{Mo}(M)$. Figure 14 illustrates this structure. The binary molybdenum compound decomposes above 820 K (Kihlborg, 1959, 1963).

Ekström and Tilley (1976) have reported extensive studies in which elements were added to WO_3 to give an $\Theta_W(M)$ phase. Among the many that do so is iron. Figure 15 is a high-resolution electron micrograph of this ternary oxide. This micrograph shows a regular structure relatively free of defects of any type including occupation of the tunnels.

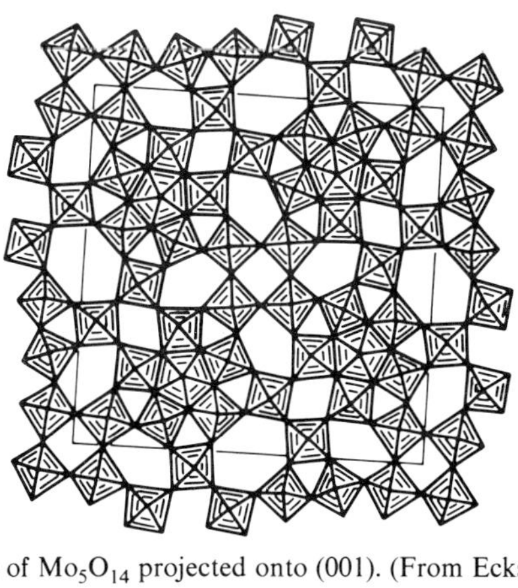

Fig. 14. The structure of Mo_5O_{14} projected onto (001). (From Eckström and Tilley, 1976.)

Fig. 15. High-resolution lattice image of the structure of an O_w(Fe) oxide viewed along [001] with the structure of Mo_5O_{14} superimposed, confirming the identity of the two structure types. (From Eckström and Tilley, 1976.)

As is clear from the results of studies on WO_{3-x} ($0.28 \leqslant x \leqslant 0.10$) by Pickering and Tilley (1976), at large values of x tunnel compounds are formed. This tendency toward tunnel formation is enhanced by the addition of altervalent cations of lower oxidation number. For example, Iijima and Allpress (1974a, b) performed electron microscopic studies on mixtures of WO_3 and Nb_2O_5 that retain 6-coordinated tungsten in corner-sharing octahedra. Functionally, four such strings in the ReO_3 structure shear cylindrically along c such that, where corner sharing is reestablished, triangular and pentagonal tunnels are created. The existence of these altervalent cations stabilizes the structure, and –Nb–O–Nb– strings are accommodated in the pentagonal tunnels in a variety of superstructures depending on the Nb concentration (hence the composition). Figure 16 illustrates the cylindrical shear required to form the tetragonal tungsten bronze (TTB) structure, as well as units characteristic of $4Nb_2O_5 \cdot 9WO_3$ and $2Nb_2O_5 \cdot 7WO_3$.

In a way analogous to the CS structures mentioned above, this element of cylindrical shear can occur isolated and disordered as a defect in the WO_3 structure resulting in nonstoichiometry, or can be ordered to form an

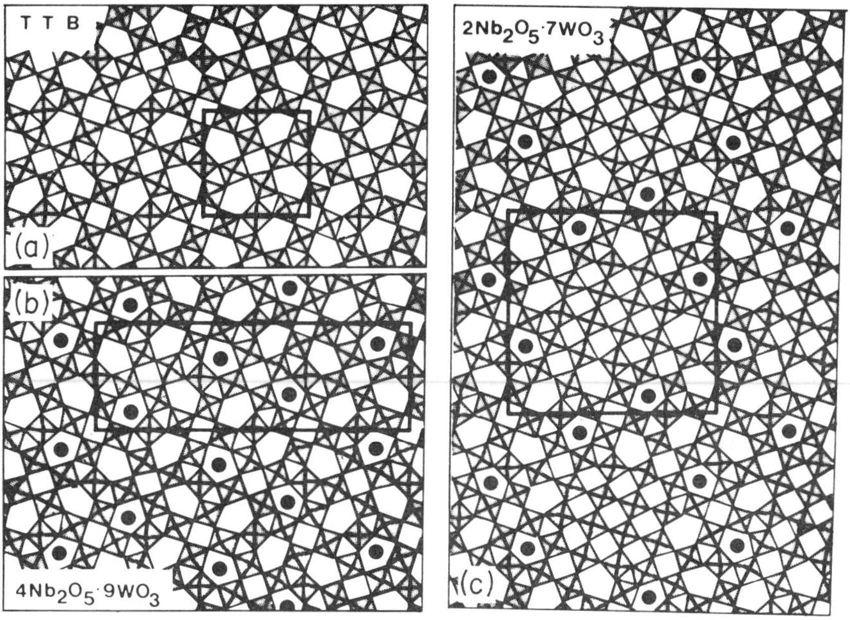

Fig. 16. Structural relationships among (a) TTB, (b) $4Nb_2O_5 \cdot 9WO_3$, (c) $2Nb_2O_5 \cdot 7WO_3$. (From Eyring and Tai, 1976.)

element of a new intermediate structure. Figure 17 shows the basic structure for $4Nb_2O_5 \cdot 9WO_3$ accommodating a modified defect region between two out-of-phase sections of the crystal. Figure 18 illustrates an analogous defect-stepped boundary in $2Nb_2O_5 \cdot 7WO_3$. In these cases there is a marked similarity between a two-dimensional model and the thin-crystal images, as it should be, since for such thin specimens an intuitive interpretation is usually valid. Because of this, it has been possible to clarify structures, see the nature of structural defects, and even find new structures by this technique (Allpress *et al.*, 1973).

3. *Intergrowth Tungsten Bronzes*

It has already been pointed out that, when certain large metal atoms are added to tungsten oxides, the connectivities of the coordination octahedra change in such a way as to provide large tunnels to accommodate them. In the case of the hexagonal tungsten bronzes (HTBs), these tunnels are themselves hexagons in cross section, formed by corner-sharing octahedra in a hexagonal array. This structure is stable over a range of composition A_xWO_3, where x exceeds a minimum large metal atom concentration, $0.13 \leqslant x \leqslant 0.33$. What happens if these concentrations fall below this

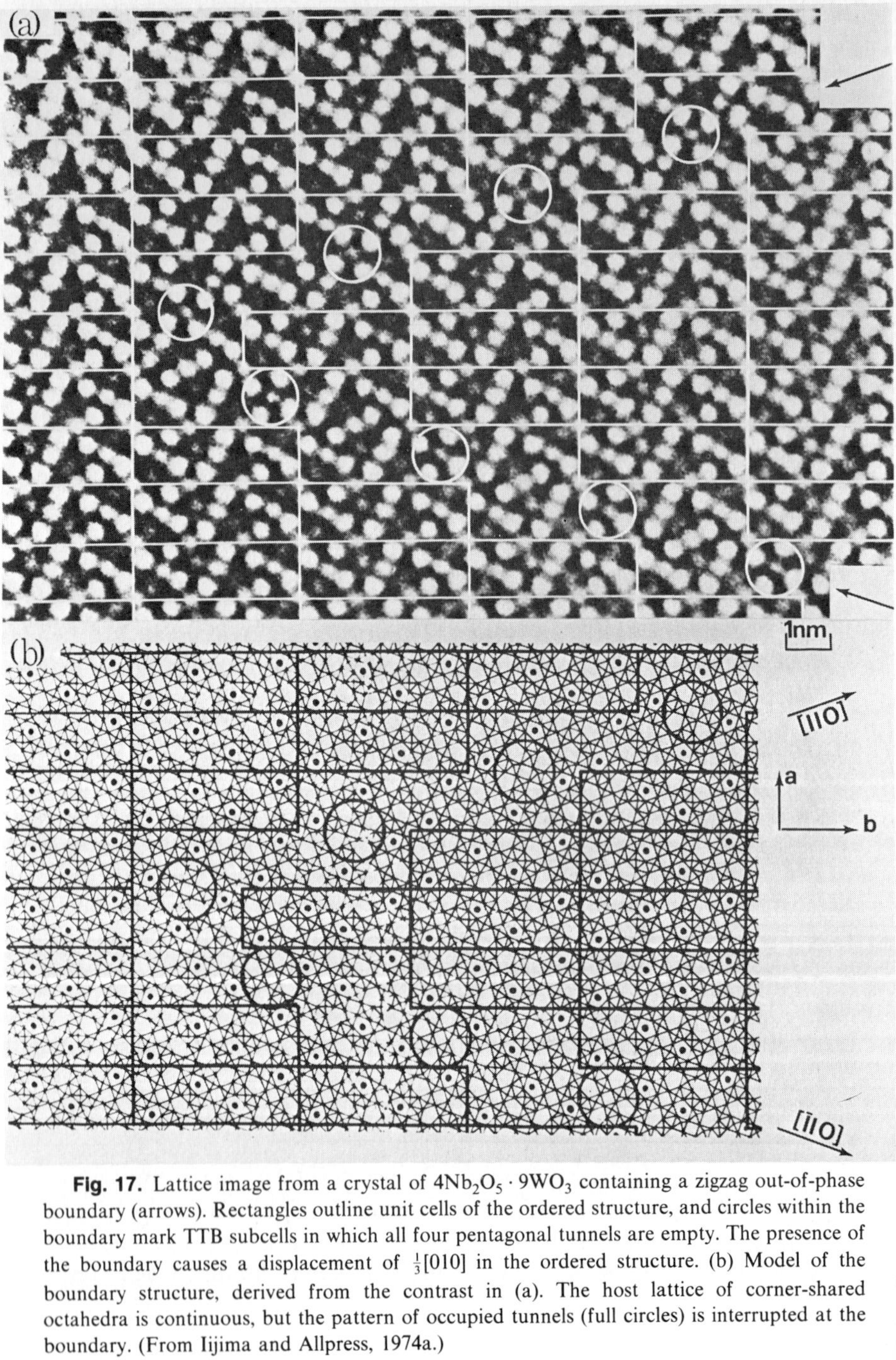

Fig. 17. Lattice image from a crystal of $4Nb_2O_5 \cdot 9WO_3$ containing a zigzag out-of-phase boundary (arrows). Rectangles outline unit cells of the ordered structure, and circles within the boundary mark TTB subcells in which all four pentagonal tunnels are empty. The presence of the boundary causes a displacement of $\frac{1}{3}[010]$ in the ordered structure. (b) Model of the boundary structure, derived from the contrast in (a). The host lattice of corner-shared octahedra is continuous, but the pattern of occupied tunnels (full circles) is interrupted at the boundary. (From Iijima and Allpress, 1974a.)

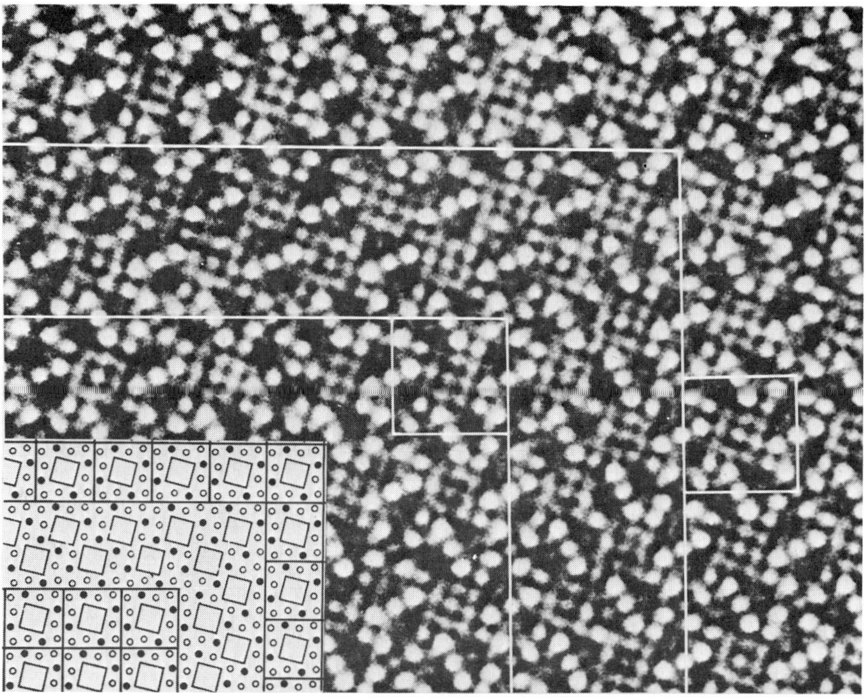

Fig. 18. Antiphase boundary in $2Nb_2O_5 \cdot 7WO_3$. The domains C_1 and C_2 are displaced by $\frac{1}{2}[110]$ with respect to each other, and the boundary has the same composition as the matrix. (From Iijima and Allpress, 1974a).

threshold? This question has been answered by Hussain and Kihlborg (1976), who have studied potassium bronzes and find that the lower concentration level for potassium in the HTB structure is 13 at. %. A new phase was discovered with $x \approx 0.10$ in A_xWO_3. It was studied utilizing EM, and a most interesting result obtained. When a lower concentration of alkali metal atoms is to be accommodated, a strikingly economical way is found. The connectivities of the coordination octahedra are modified only to the extent necessary, as shown in the model and electron micrograph in Figs. 19 and 20, respectively. Notice that hexagonal tunnels are supplied as needed and are concentrated in ordered planar prisms of appropriate thickness. These are intergrown with slabs of WO_3, which have been distorted as required to intergrow coherently with the slabs of HTB. This means tilting the octahedra about 15° while maintaining their corner sharing. Image calculations for this model are similar to those observed, strongly supporting the structure that was firmly established by x-ray analysis.

Electron micrographs also revealed (Fig. 21) that the width of slabs of

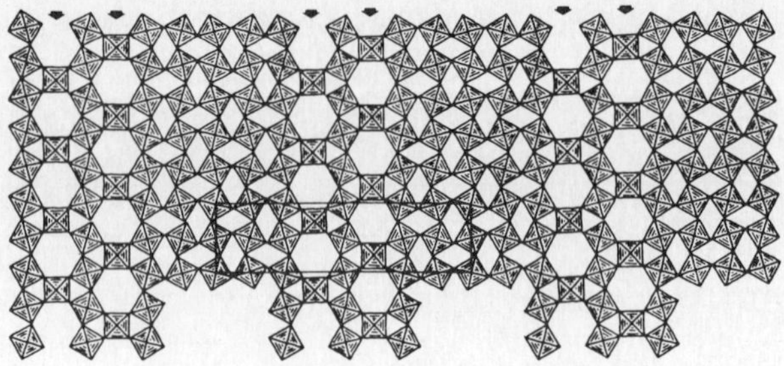

Fig. 19. Structure model derived from the image in Fig. 20. WO_6 octahedra are linked by corner sharing. The alkali metal atoms in the hexagonal tunnels are not indicated. The HTB elements have been extended downward for clarity. (From Hussain and Kihlborg, 1976.)

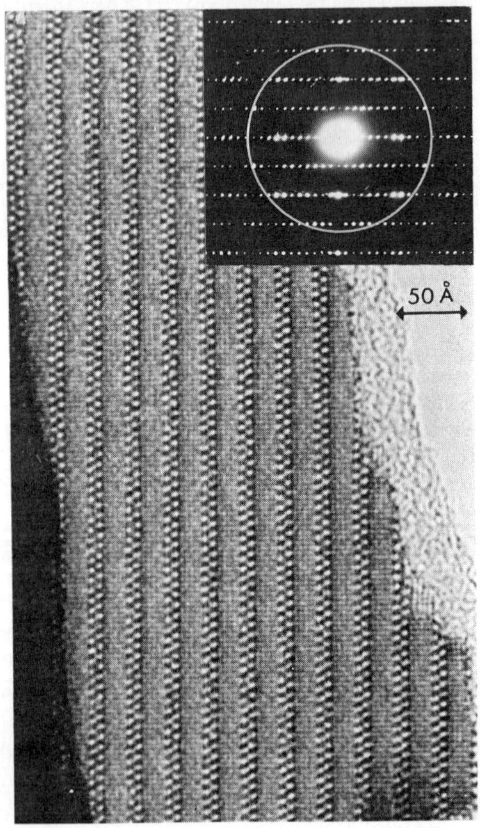

Fig. 20. Lattice image of a thin edge of a crystal of the (1, 5) ITB phase from a sample of composition $K_{0.10}WO_3$. A corresponding diffraction pattern is also shown, with the size of the objective aperature indicated. (From Hussain and Kihlborg, 1976.)

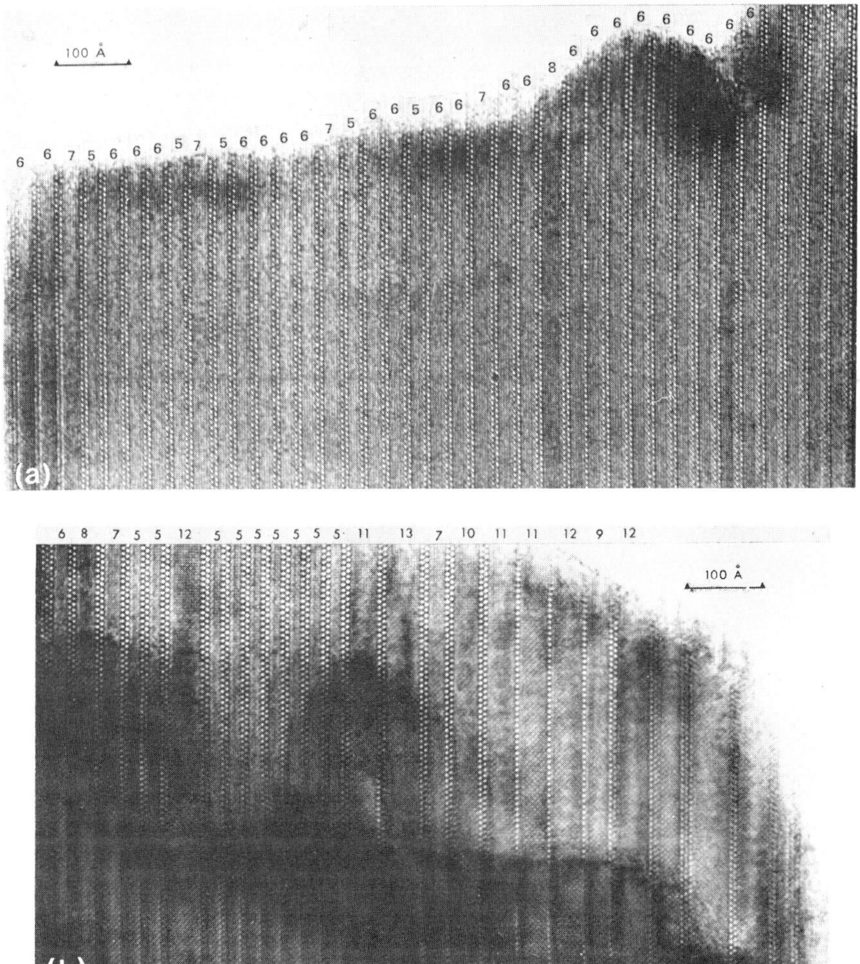

Fig. 21. Lattice images from Rb_xWO_3 crystals in a sample with $x = 0.03$. The widths of the WO_3 slabs in terms of n are given above each slab. (a) A moderately disordered crystal where the tunnel rows are double with only one exception (indicated by an arrow). (b) A highly disordered fragment with single, double, and triple tunnel rows and large variations in the distance between them. (From Hussain and Kihlborg, 1976.)

both the WO_3 and HTB structures could vary. This led to the proposal that new families of a homologous series could occur in which ordered but varied widths of each type of structure could be accommodated in different families. These are designated intergrowth tungsten bronzes (ITBs).

This phenomenon is not limited to alkali metal bronzes, but in studies by

Steadman *et al.* (1972) was shown to occur in Sn_xWO_3. Ekström and Tilley (1978) discussed the system Pb_xWO_3 ($0.01 \leqslant x \leqslant 0.28$) prepared between 973 and 1373 K. The high-temperature phase for $x > 0.18$ was of the TTB type. The low-temperature phases at $x < 0.18$ were ITB structures, and HRTEM revealed images that appeared to consist of WO_3 slabs separated by single rows of filled hexagonal tunnels. However, the actual nature of these parallel faults was not completely established. No attempt to calculate the images based on probable models was made.

4. *The Block Structures*

The first application of EM to complex oxide materials was its use by Allpress *et al.* (1968) in a study on Nb-W oxides. Lattice images showed planar faults to exist corresponding to shear planes from the intergrowth of one phase within another. The faults varied in width from less than one unit cell to several and resulted in nonperiodic structures responsible for composition variation. These results confirmed Wadsley's proposed cause of nonstoichiometry and promoted a new and powerful technique for the study of defects and solids.

With further reduction in ReO_3-related structures, new structure types based on CS on both (100) and (001) of ReO_3 result in columns of ReO_3-like material a few coordination octahedra wide in two dimensions ($m \times n$). These may be stacked together in different sizes at levels that differ by one-half an octahedron and are separated by shear planes in two directions at right angles to each other (Wadsley and Andersson, 1970). These columns are the building blocks of a large number of oxides with O/M ratios of about 2.5. Working out the details of these connectivities, and particularly the defects in these materials, has been one of the most notable successes of HRTEM. Since many of these block structure compounds are difficult if not impossible to prepare in a fully ordered and stoichiometric state, visualizing the details is of utmost value.

Nonstoichiometry in block structures is determined by point defects in the established structure, imperfect order, intergrowth of regions of different composition, or the capricious filling of rows of tetrahedral sites, which results when columns at the same level have no edge sharing. The observation of point defects has been reported by Iijima *et al.* (1973). These point defects are apparent in Fig. 22 where a black spot has replaced one of the white spots due to a square tunnel. Figure 23 indicates their interpretation of this contrast change. Here two extra oxygen atoms (open circles) are associated with metal atoms displaced $\frac{1}{2}\{001\}$ from their regular site in the columns outlined with faint lines. If oxygen is lost from the specimen by heating in the electron beam, these point defects could disappear upon the

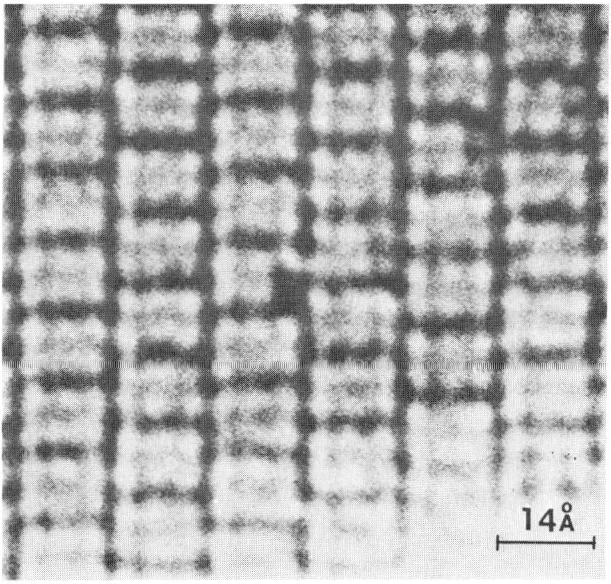

Fig. 22. Enlarged image of the black spot in an image of $Nb_{12}O_{29}$. The position of the spot corresponds to one of the 2 × 3 channels in a block. (From Iijima *et al.*, 1973.)

return of the displaced metal atoms to their regular sites. The concentration of defects decreases as the specimen is irradiated by electrons. This defect involves the shift of only a few atoms with a component perpendicular to the electron beam.

Iijima and Allpress (1973) illustrate several ways in which intergrowth of various kinds can occur in titanium-doped Nb_2O_5 and lead to nonstoichiometry. The composition of the crystal studied was between that of the established structures $TiNb_{24}O_{62}$ and $Ti_2Nb_{10}O_{29}$. The defects are described in terms of the matrix in which they occur. Figure 24 is a diagram of the $TiNb_{24}O_{62}$ structure. A micrograph of this structure with a calculated image inserted is shown in Fig. 25. Some defects in this structure that accommodate nonstoichiometry are shown in Fig. 26. Here elements of the $TiNb_{24}O_{62}$ structure formed by a 1 : 1 intergrowth are designated E ($M_{13}O_{33}$) and F ($M_{12}O_{29}$). The compositions of these regions of two fragments, which can be thought of as intergrowths of $TiNb_{24}O_{62}$ and $M_{12}O_{29}$ or $M_{13}O_{33}$ and $M_{12}O_{29}$, are $Ti_5Nb_{44}O_{120}$ and $Ti_4Nb_{58}O_{153}$, respectively. Two regions of $TiNb_{24}O_{62}$ were observed to be out of register, separated by a slice of H–Nb_2O_5. Such displacements are not always regular, as shown in Fig. 27, where localized insertions are observed accompanied by the required jogs in the structure.

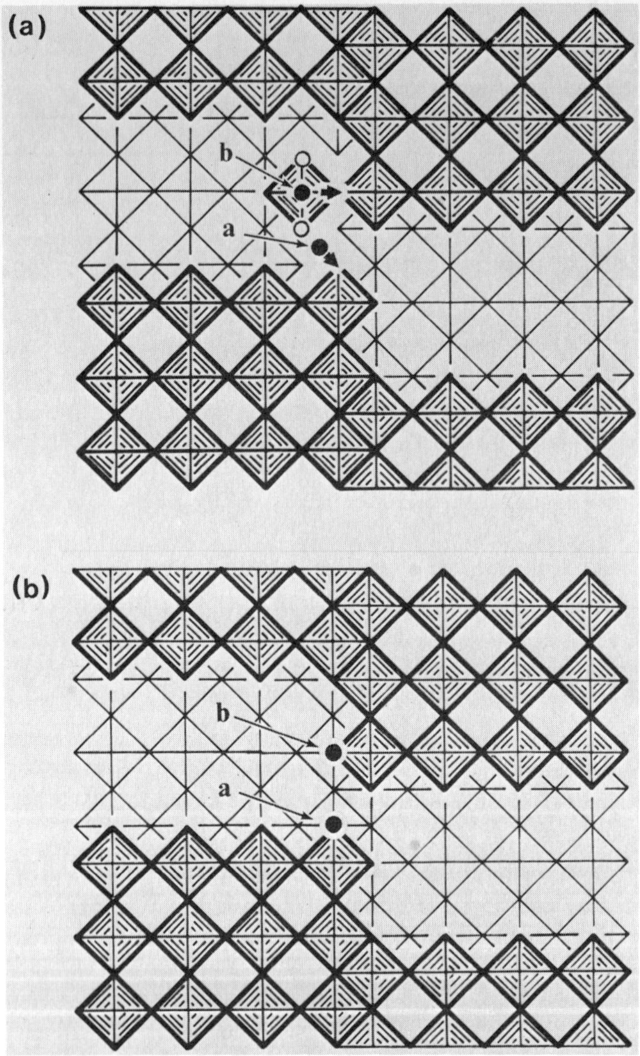

Fig. 23. (a) A proposed model for the point defect derived from the image shown in Fig. 22. The point defect is a complex composed of two interstitial metal atoms (indicated by a and b) and two oxygen atoms (open circles). (b) Perfect structure corresponding to the one in (a). (From Iijima *et al.*, 1973.)

The equilibrium state of nonstoichiometric and intergrown material certainly is in question. A detailed HRTEM study of twinning and other defects in Nb_2O_5 was made by Anderson *et al.* (1972). Blocks of M– or N–Nb_2O_5 and other defects observed inserted into the main phase, H–Nb_2O_5, were considered out-of-equilibrium structures. The types of defects

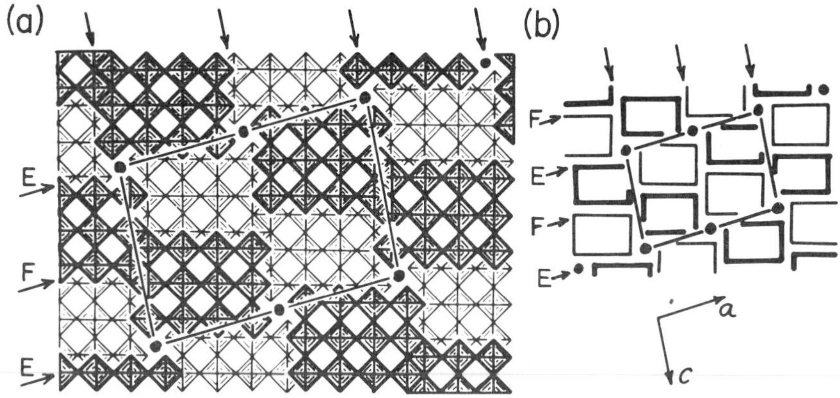

Fig. 24. Idealized representations of the structure of $TiNb_{24}O_{62}$. (a) Each hatched square represents an octahedron of 6 oxygen atoms surrounding a central metal atom, viewed down a body diagonal. The columns of corner-shared 4×3 octahedra share edges where they join. (b) An abstract of (a) where the 4×3 columns are rectangles and the tetrahedral sites are marked by a dark spot. (From Iijima and Allpress, 1973.)

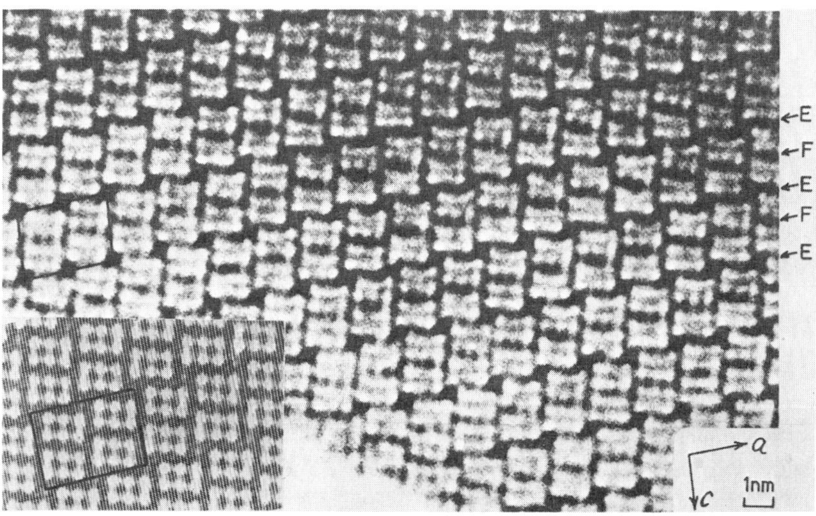

Fig. 25. Two-dimensional lattice image from a fragment of $TiNb_{24}O_{62}$, recorded at the same orientation as that of the model in Fig. 24. Some E- and F-type CS planes are identified. The inset shows the contrast computed for an 8-nm-thick crystal of $TiNb_{24}O_{62}$ imaged 50 nm underfocus using 65 contributing *h0l* beams. (From Iijima and Allpress, 1973.)

that occur in any particular solid are restricted, and the cataloging of these guides the discernment of mechanisms of solid state reactions. In many of the block structures with a composition near M_2O_5 there are what have been termed "tetrahedral columns," formed when the ReO_3-type blocks have no edge sharing at the same level. These are strings of two types of

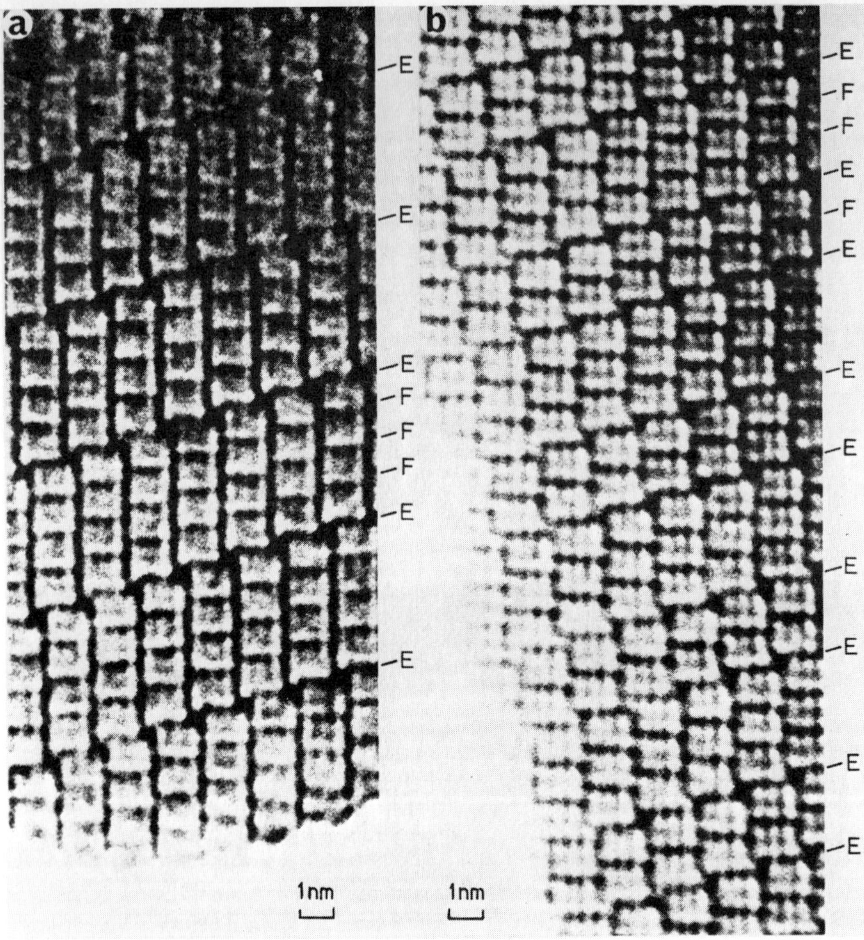

Fig. 26. Lattice images from fragments of nominal composition $TiO_2 \cdot 6Nb_2O_5$, showing several ordered intergrowths of the parent structures, $M_{13}O_{33}$ and $M_{12}O_{29}$, identified by their corresponding CS planes, labeled E and F, respectively. (a) Four periods of the sequence –E–3F–E– having the composition $Ti_5Nb_{44}O_{120}$. (b) Four periods of the sequence –E–F–E–2F–E– with the composition $Ti_4Nb_{58}O_{153}$. (From Iijima and Allpress, 1973.)

tetrahedral sites surrounded by a sheath of two types of octahedral sites. Many of these phases exhibit nonstoichiometry even when high-resolution images do not show any imperfections in the block structure. It is believed that this nonstoichiometry results from a partial filling of these tetrahedral columns.

Anderson *et al.* (1975) made a thorough study of the structure of $GeO_2 \cdot 9Nb_2O_5$, a nonstoichiometric oxide inexplicably isomorphous with

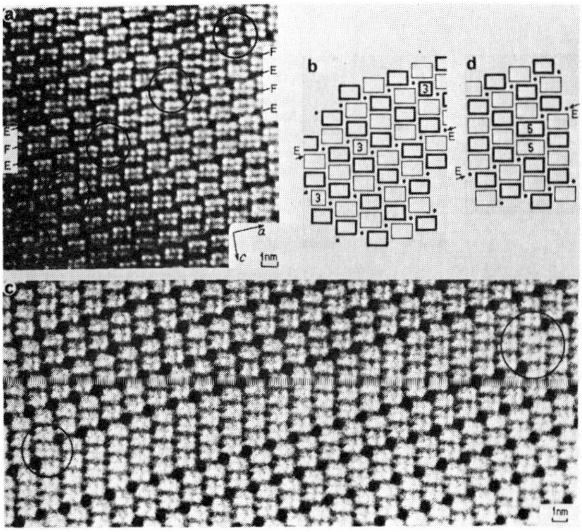

Fig. 27. (a) Lattice image showing the incorporation of five 3 × 3 blocks (circled) in a matrix of $TiNb_{24}O_{62}$. Their presence displaces the E and F planes, some of which are identified. (b) Idealized model, showing how the 3 × 3 blocks in the central part of (a) are built into the matrix. (c) Lattice image showing displacements along E planes containing tetrahedral sites (dark blobs), associated with the presence of 5 × 3 blocks (circled). (d) Model of the displacement caused by the double 5 × 3 block at the top right of (c). (From Iijima and Allpress, 1973.)

$P_2O_5 \cdot Nb_2O_5$. This latter structure consists of 3 × 3 blocks of the ReO_3 type edge-sharing as in other block structures, but not with blocks at the same level. There are two tetrahedral columns associated with the projection of the unit cell along the *b* axis. That there is an excess of metal atoms can be seen from the composition alone. The structure was determined by x ray, neutron total profile analysis, and HRTEM. All were consistent in placing the extra metal atoms in the tetrahedral columns, but it was uncertain how they should be distributed among the tetrahedral and octahedral sites. Based on chemical intuition, a distribution was proposed. The structure image suggested that the block complement of the structure was regular; hence the extra metal must be in the tetrahedral columns. The contrast at these locations varied from white spots on the thin edge to very dark spots in the thicker parts of the crystal. A pattern of column filling was suggested, but no attempt was made to determine whether or not it would fit the observed contrast.

Skarnulis *et al.* (1976) attempted a refinement of the structure of "$GeNb_9O_{25}$" by HRTEM. They considered a number of plausible models for site filling in the columns, calculated the image to be expected using the multislice method, and obtained somewhat surprising results. Placing metal atoms in the tetrahedral sites could not lead to dark contrast at these positions in the image. Black spots at the tetrahedral column positions could only be calculated when the extra metal atoms were placed at the octahedral sites, suggesting that the column sheath was much more fully occupied than suggested by Anderson *et al.* (1975). Figure 28 shows the favorable comparison and agreement between calculated and observed images. This makes clear that in some cases an intuitive interpretation of observed images may not be reliable even for thin crystals; hence calculations should be made. Also, a fit between observed and calculated images over a defocus range is desirable. Dark contrast at the tetrahedral sites is regularly seen on all but the thinnest edges in images of phases with tetrahedral columns. The occupation of these sites by atoms, particularly in the octahedral position, can indeed account for nonstoichiometry in otherwise perfect crystals. Skarnulis *et al.* (1976) further point out that electron bombardment of these phases causes fluctuations in the contrast at a particular tetrahedral site. They attribute this to a shift in the occupation of sites within the column (more atoms at tetrahedral sites tend to bleach the dark spot), and transport of metal atoms through the structure is not necessary. These results support the previous discussion of point defects in block structures (Iijima *et al.*, 1973).

D. Layer Compounds

1. Hexagonal Barium Ferrites

Verma and Tregunazat (1974) define polytypism as being characterized by crystallization of a solid into several chemically similar modifications. These polytypes or types differ only in the stacking arrangement of the close-packed planes. They are one-dimensional polymorphs, but their origin is different and the distinction appears to be useful. The most commonly known polytypes are those of SiC, where about 40 structures are known and more than 130 have been reported. The story is similar for CdI_2 and ZnS. The unit cells may be any length along the *c* axis ($c \cong 1500$ Å in the 594 R polytype of SiC, Honjo *et al.*, 1950). If the layers all have the same composition, the overall atomic ratios will remain unchanged by a stacking disorder, but if they are different, compositional ranges may be observed.

Hexagonal ferrites include some having technically important magnetic properties. They are layer compounds based on two types of close-packed oxygen atoms. One, the hexagonal M structure, has the composition $BaFe_{12}O_{19}$ and consists of two equal parts rotated 180° with respect to each other. The other fraction, called Y, is rhombohedral $Ba_2Me^{II}Fe_{12}O_{22}$

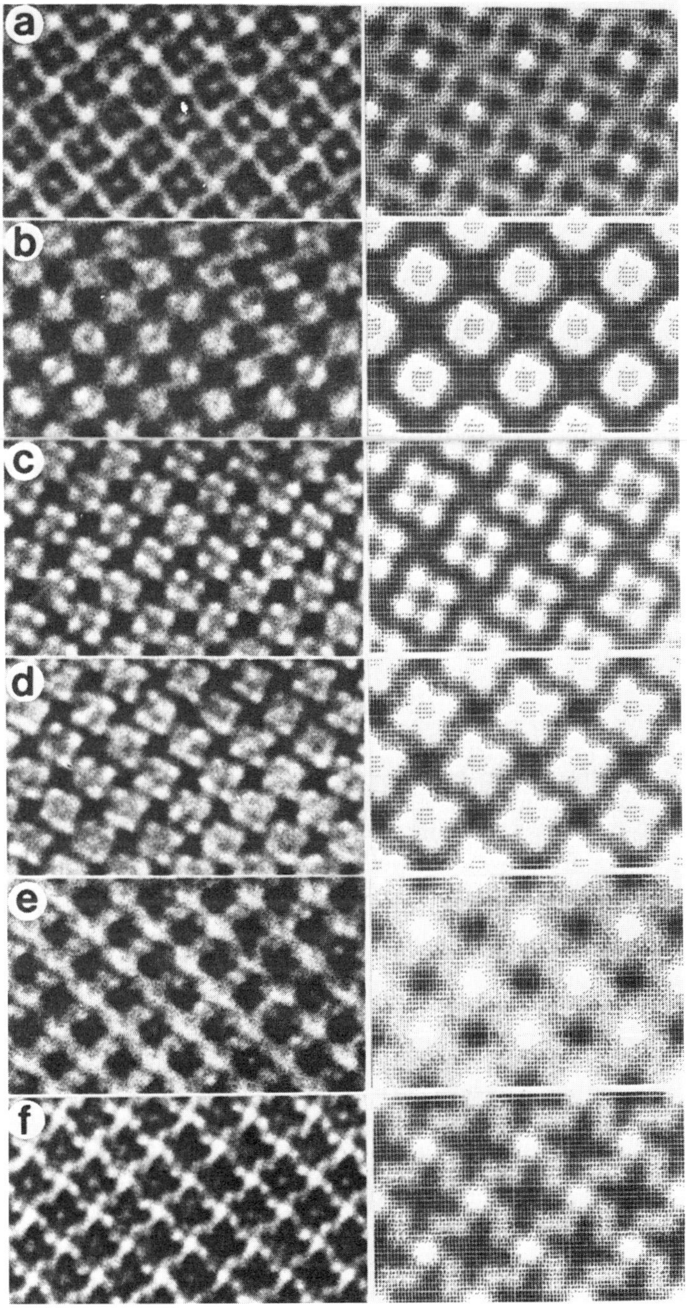

Fig. 28. Comparison of experimental through-focus images (left column) with calculated best-fit images (right column). Defocus values for experimental and theoretical images, respectively, are: (a) −100, −100 Å; (b) −450, −425 Å; (c) −700, −700 Å; (d) −950, −975 Å; (e) −1200, −1175 Å; (f) −1500, −1500 Å. Structural data used for the calculation are from the refined structure obtained by comparing the calculated images with the experimental images. The image in (c) is the optimum defocus image. (From Skarnulis *et al.*, 1976.)

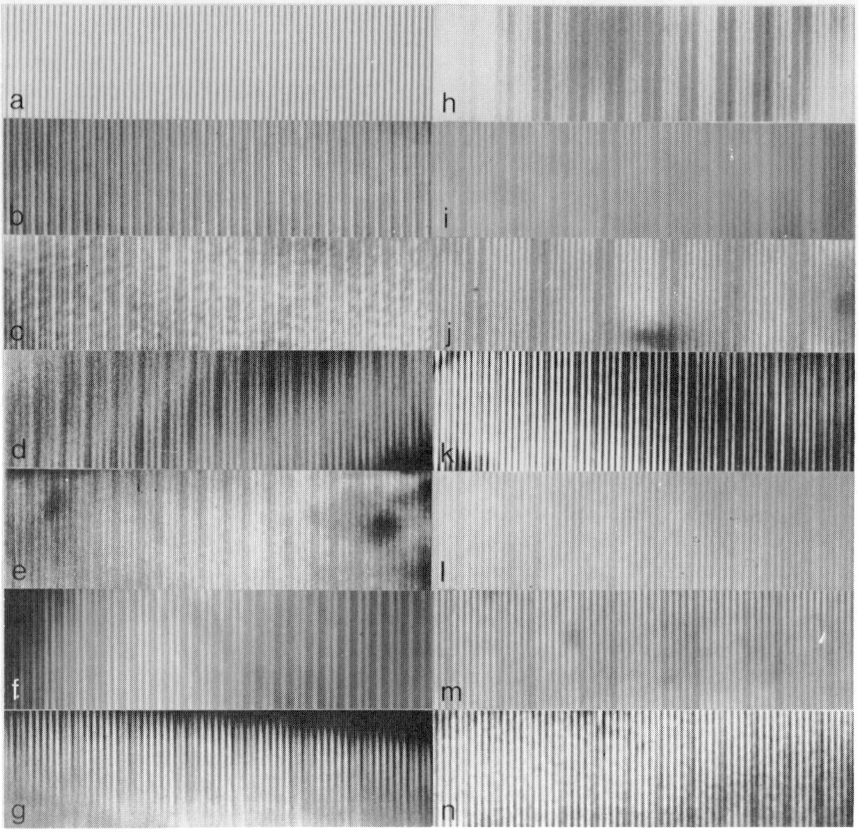

Fig. 29. A number of regular sequences. The characteristics of the different sequences are summarized in Table II. (From Van Landuyt *et al.*, 1974.)

(Me^{II} = Zn). In each block only two positions are used for the oxygen layer. Polytypes are formed by repeated stacking sequences of these coherent layers of M and Y, such as are shown in Fig. 29. Table II indicates the stacking sequences.

In these fringe images Van Landuyt *et al.* (1974) and their colleagues sorted out the sequences and even detected stacking errors among the faults. The same series of compounds was studied at higher resolution by McConnell *et al.* (1974) and discussed in detail by Anderson and Hutchinson (1975). The resolution was such that the fine detail of the image could be correlated with structural features within the layers themselves. Figure 30 is an enlarged region of the compound $MYMY_3$ on which a model of the structure is superimposed. There is a striking correlation of the pentagonal tunnels characteristic of the structure and the white dots of the image.

TABLE II

A Survey of Observed Structures[a]

M/Y ratio	Stacking sequence	Fig.
1 : ∞	Y	29a
2 : 1	MMY	29c
2 : 2	MYMY	29b
2 : 3	$MYMY_2$	29f
2 : 4	$MYMY_3$	29h
2 : 5	MY_2MY_3	29e
2 : 6	$MYMY_5$	29i
	MY_2MY_4	29k
	MY_3MY_3	29d
2 : 7	$MYMY_6$	29g
2 : 8	$MYMY_7$	29j
	MY_4MY_4	29n
2 : 10	MY_3MY_7	29l
2 : 11	MY_4MY_7	29m

[a] From Van Landuyt *et al.* (1974).

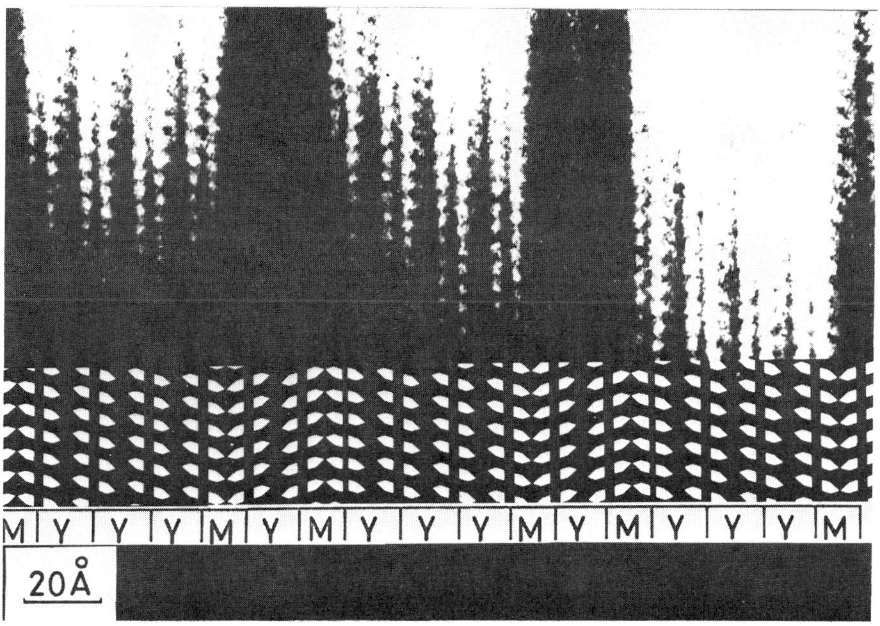

Fig. 30. Lattice image of $MYMY_3$ (greatly enlarged) superposed on a crude projection of charge density distribution. (From McConnell *et al.*, 1974.)

The unexplained dark contrast of the MYM groups correlates with the contrast difference in the fringe images (Fig. 29h). This illustrates the many levels of explicit information that can be obtained about complex systems.

2. Ferroelectric Bismuth Oxides

These oxides are another example of the formation of an endless number of possible compositions of a material made up of varied stacking arrangements of two different types of coherent layers. Here the layers consist of tetrahedrally coordinated Bi atoms in slabs of composition $(Bi_2O_2)^{2+}$, together with perovskite-like layers of composition $(A_{n-1}B_nO_{3n+1})^{2-}$ arranged in a charge-compensating manner. In the latter, n layers of corner-sharing B atoms in octahedral coordination enclose $n - 1$ layers of A atoms in 12-coordination (Hutchison *et al.*, 1977). The fact that these materials are ferroelectric and present the possibility of visualizing ferrolectric domains heightens our interest in them. Figure 31 reveals a dislocation core in $Ba_2Bi_4Ti_5O_{18}$ ($n = 5$) of the series. Two extra slabs have been inserted into the structure at the top of the figure. In view of the clarity of this image, careful measurements and models of the dislocation core at virtually atomic level were made. Domain walls in these ferroelectrics have also been

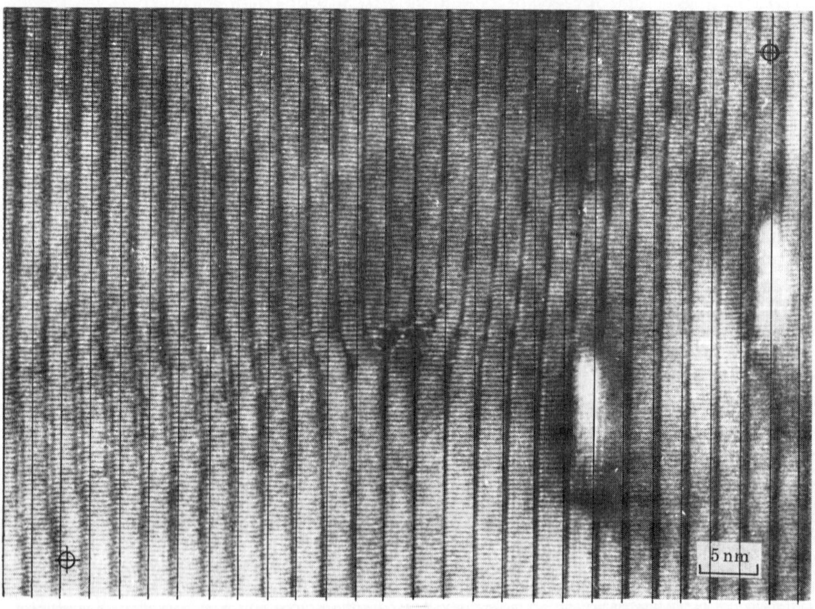

Fig. 31. Lattice image of $Ba_2Bi_4Ti_5O_{18}$ ($n = 5$) containing a dislocation. An unperturbed grid showing the positions of Bi_2O_2 layers in an unperturbed structure is superimposed for comparison. (From Hutchinson *et al.*, 1977.)

imaged. For a comparison between images of this material drawn to scale and taken on higher-voltage microscopes see Section III.G and Fig. 37.

Anderson (1973) has suggested that certain materials are capable of achieving fully ordered structures at every composition within a wide range, with no two-phase regions between. Many systems are intergrown on such a fine scale that they appear to have this ability, for example, polytypes with compositionally variable layers. There are other examples, of course, such as the swinging shear planes of the TiO_{2-x} system (Bursill and Hyde, 1972), to be mentioned later, vernier structures characteristic of systems like lanthanide florides, and incommensurate lattices. Many such systems have been observed with HRTEM in other connections and are included in this discussion.

E. Rutile-Based Oxides

Soon after the important observations by Allpress *et al.* (1968) of identifiable structural features in crystalline specimens of complex materials based on the ReO_3 substructure, a study of rutile-related phases was begun utilizing electron microscopic techniques. The results of these studies on the binary rutile system have been reported in many places, including a review by Bursill and Hyde (1972).

As in the WO_3-related phases, near the composition WO_3, the reduced phases accommodate oxygen deficiences by an increase in edge sharing of coordination octahedra along CS planes. The type of CS that occurs and the facility with which it is formed depend on the composition and on whether or not the specimen has been doped with another oxide. In the binary phases CS occurs along $(121)_r$ at high defect concentrations and is responsible for the early members of the homologous series Ti_nO_{2n-1} (all indices refer to the rutile subcell). At higher concentrations and, especially, with metal atoms like Cr^{3+} incorporated, the direction of shear changes according the equation $(hkl)_{cs} = p(121) + q(011)$, where p and q are integers and (011) is an antiphase boundary formed by the shearing operation $(011)\frac{1}{2}[0\bar{1}1]$. The extremes of the series observed are $(121)_{cs}$ and $(132)_{cs}$, the latter a result of equal amounts of (121) and (011) shear. In between are an unlimited number of possible combinations, many of which are observed. At low oxygen deficiency, $(132)_{cs}$ is preferred in the binary oxides at temperatures below 1300°C. The discovery of this phenomenon of swinging shear planes was important in establishing EM as a primary tool in unraveling the details of composition and structure.

Studies on doped rutile phases have been continued at higher resolution by Bursill and colleagues. For example, new types of CS have recently been observed in rutile specimens with about 15% Fe_2O_3, quenched from about 1500°C (Bursill *et al.*, 1976). In these specimens CS (020) $[0\frac{1}{2}\frac{1}{2}]_r$ is observed. Diffuse scatter indicates the ordering of metal atoms in the shear planes (Grey and Bursill, 1978). This gives families of CS planes with $(hkl)_{cs} = p(020) + q(011)$.

Bursill *et al.* (1978) have reported wavelike variations in CS in these same iron-doped rutile phases quenched from near the melting point. Three different types of such long-wavelength compositional fluctuations are observed. They are described as $k(101)$, $k(100)$, and $k(001)$ modulation structures, carrying the designation of the wave vectors in each case. Figure 32 illustrates the form of the modulation in $k(100)$ at modest resolution. The structural model assumes a cooperative reorientation of the CS indices from (hkl) in $\langle 111 \rangle$ zones to $\{Okl\}$ in $\langle 100 \rangle$ zones. A consideration of the spacings in the high-temperature phase that are directly related to the $\{121\}$ CS low-temperature phases suggests that changes in orientation occur through diffusional readjustment rather than dislocation mechanisms. It is suggested that $k(100)$ modulation structures are accounted for by

$$(021) = (121) + (\bar{1}21) \qquad \text{and} \qquad (02\bar{1}) = (12\bar{1}) + (\bar{1}2\bar{1})$$

These reactions combine to give

$$(020) = (021) + (02\bar{1})$$

Thus the low-temperature structures are transformed at high temperatures.

One of the most intriguing conclusions of this work is that the transverse composition waves can be predicted from Young's modulus plots by finding the path of steepest descent from low- to high-temperature structures. These transformations result from changes in elastic strain energy. The possibility is considered that at near melting temperatures (> 1550°C) these modulations become dynamic standing waves or even traveling waves. A diffusion mechanism has been found to be compatible, and orientations but not spacings would change cooperatively as observed.

Gibb and Anderson (1972) reported the homologous series $Ga_4Ti_{m-4}O_{2m-2}$ ($15 < m < 23$, m odd) in the gallia–rutile system. The structure of one of the members, $Ga_4Ti_{24}O_{48}$, was determined. The structural principle is that adjacent slabs of rutile are displaced by $[\frac{1}{2}\frac{1}{4}\frac{1}{2}]_r$ or as a $[\frac{1}{2}\frac{1}{4}\frac{1}{2}]\,(210)_r$ CS phase (Lloyd *et al.*, 1976). The rutile slabs are separated by layers of Ga_4O_6. The homologous series $Ga_4Ti_{m-4}O_{2m-2}$ is formed when a periodic operation occurs on $(210)_r$ planes. It was possible in this study to determine that the gallium atoms were ordered in the CS planes as columns of β-Ga_2O_3 separated by hexagonal tunnels, parallel to b, with an effective

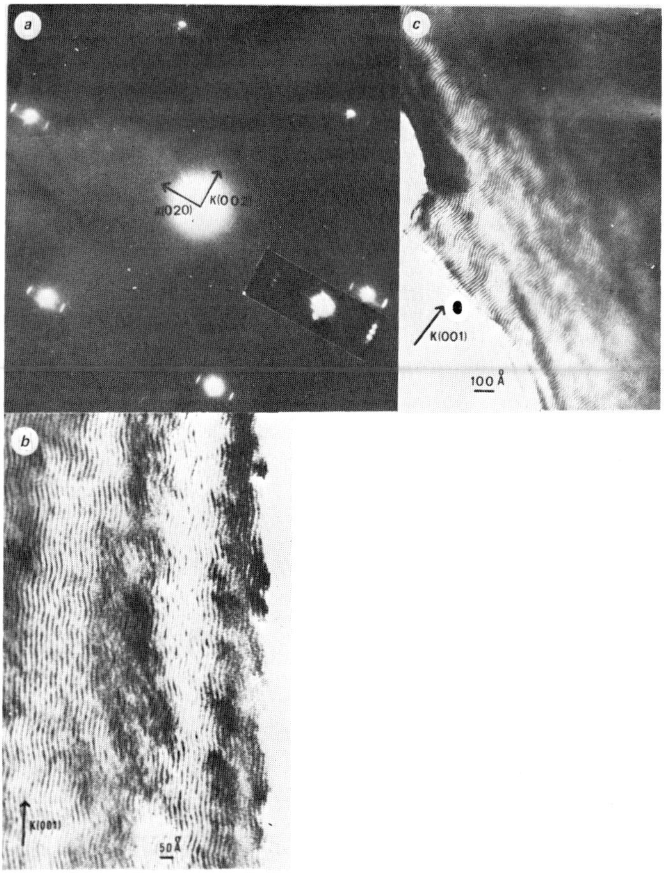

Fig. 32. (a) [100] zone diffraction pattern from (Ti, Fe)$O_{1.90}$ heated at 1500°C for 24 h. Note $k(001)$ modulation sidebands with a periodicity of approximately 250 $Å^{-1}$ (see inset). (b) High-resolution image corresponding to (a) showing a 23-Å-spacing wavelike CS structure. Note transverse modulation vector $k(001)$ with an amplitude of approximately 5 Å and a wavelength of 240 Å. (c) High-resolution image of the same preparation, showing perfect sinusoidal CS waves oscillating from $k(041)$ to $k(04\bar{1})$ with the mean plane (020). Transverse modulations have an amplitude of 25 Å and a wavelength of approximately 335 Å. (From Bursill *et al.*, 1978.)

diameter of 2.5 Å. Figure 33a is a direct image of the tunnels in the $Ga_4Ti_{33}O_{72}$ structure. These are modeled in Fig. 33b. In Fig. 33c, the calculated image is compared with that observed.

Kamiya and Tilley (1977) investigated the same system and found that at a low Ga content isolated $\{210\}$ faults occurred. These tended to cluster at higher concentrations and became essentially ordered arrays at still

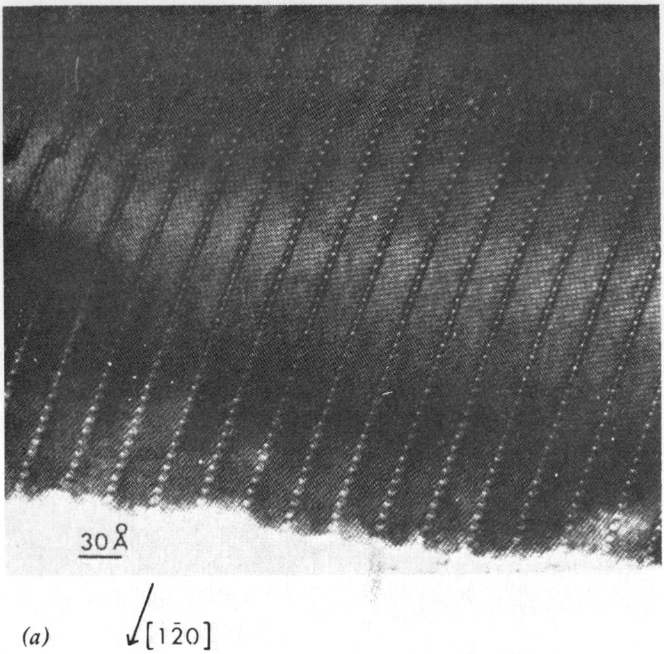

(a)

(b)

Fig. 33. (a) High-resolution image of $Ga_4Ti_{33}O_{72}$. Objective lens defocus ~ -1200 Å, objective aperture 60 μm, beam divergence 0.4×10^{-3} radians. Note the rows of white blobs, spacing 10.3 Å, and the 3.2-Å cross-grating fringes. (b) Crystal structure of $Ga_4Ti_{33}O_{72}$ [001] projection. Octahedra share edges, and tetrahedra share corners, parallel to the projection axis. β-Gallia-like columns are heavily outlined. Note the edge-shared octahedra and the tetrahedra.

higher concentrations.The (210) $[\frac{1}{2}\,\frac{1}{4}\,\frac{1}{2}]$ CS in gallia-doped rutile has been studied by Stone and Bursill (1977) at a 3.2-Å resolution. They have imaged isolated pairs of (210) boundaries with well-defined spacings of 4.8 Å with the stoichiometry Ga_4TiO_8. These give way to regularly spaced single boundaries up to 35 Å apart, yielding members of the series

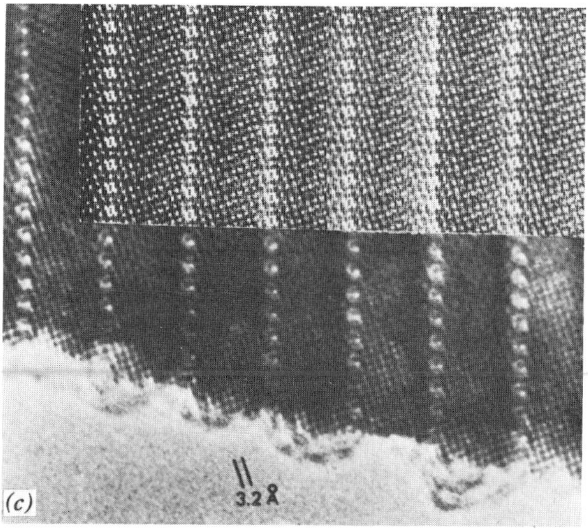

Fig. 33. (*Continued.*) (c) Computer-simulated image of Fig. 33a $C_8 = 2.4$ mm, thermal spread of 0.7 V, divergence of 0.4×10^{-3} radians. (From Stone and Bursill, 1977.)

$Ga_4Ti_{m-4}O_{2m-2}$. This rather complete microscopic study illustrates the occurrence of the isolated defect pair Ga_4TiO_8 (Fig. 34), a defect pair within an ordered member of the homologous series, and a defect pair of composition $Ga_4Ti_3O_{12}$. These observations have allowed a mechanism of Ga concentration in the shear plane and the growth of the Ga_4TiO_8 pair to be proposed.

F. High-Resolution Electron Microscopy of Silicates

The application of HRTEM to mineralogy has been rewarding. It has been extremely valuable as a tool in probing the basis of nonstoichiometry in natural minerals. Only a few examples shall be given here. A further guide to the literature is provided by Buseck and Iijima (1974), Thomas and Jefferson (1978), and more recently by Thomas *et al.* (1978–1979). We select as an example the occurrence of a form of polytypism where successive layers vary in composition.

Veblen *et al.* (1977) describe many silicate minerals that can be viewed as being made up of chains of SiO_4 tetrahedra linked by corner sharing, consisting of single, double, or triple chains. Minerals consisting of any particular chain type may exist, as well as others with regular mixtures such as polytypes. In addition, irregular faults of any variety may appear. Figure 35 shows ordered pyriboles made up, respectively, of double, triple, and

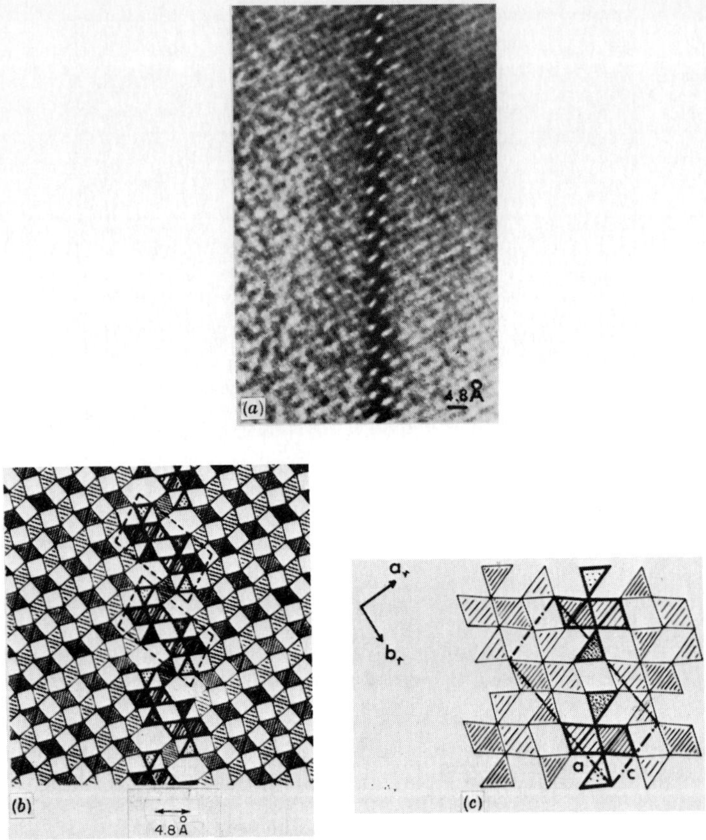

Fig. 34. (a) Rutile containing (210) defect pairs, separation 4.8 Å. Electron optical conditions same as for Fig. 33a. (b) Structural model for Ga_4TiO_8 defect pair. β-gallia-like elements are heavily outlined. (c) $[010]_\beta$ projection of β-gallia. (From Stone and Bursill, 1977.)

double-triple chains (1 : 1 intergrowth). Figure 36 illustrates the occurrence of disordered or long-period structures.

The successes in mineralogical study will be further amplified with the development of microanalytical analyses in conjunction with high resolution. These techniques, which analyze 10^7 Å^3 at present, are being improved. In the near future, the association of chemical composition with the structural features in these mineral specimens observed at about a 3-Å resolution should be possible. The minerals are made in nature under conditions, in environments, and from elemental sources that can only be surmised. These analytical techniques are therefore necessary. They will also be of enormous value in the solid state chemistry of multicomponent systems which are beginning to receive greater attention.

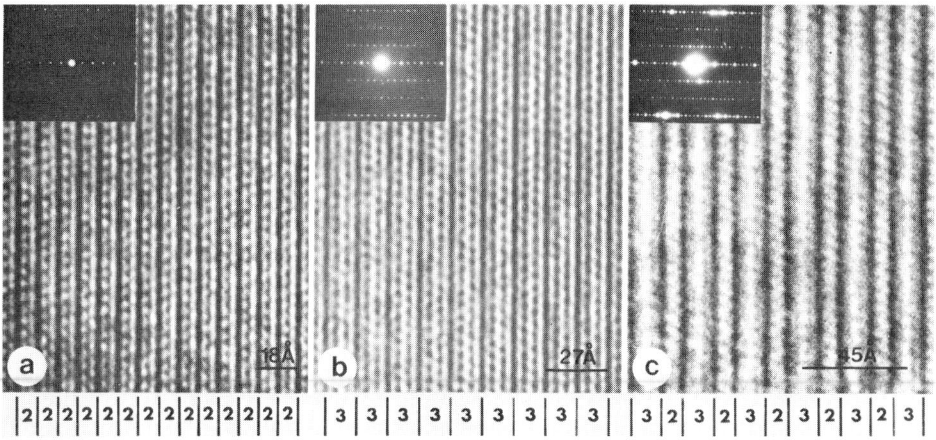

Fig. 35. HRTEM *a*-axis images and electron diffraction patterns of ordered pyriboles: (a) anthophyllite (double chains), (b) jimthompsonite (triple chains), and (c) chesterite (alternating double and triple chains). Dark fringes mark the chain edges; variation in detailed contrast results from variations in microscope operating conditions. The axial directions *b* and *c* are the same for all the minerals, and all subsequent figures are in the same orientation. (From Veblen *et al.*, *Science*, **198**, No. 4315, 359–365 (1977). Copyright 1977 by the American Association for the Advancement of Science.)

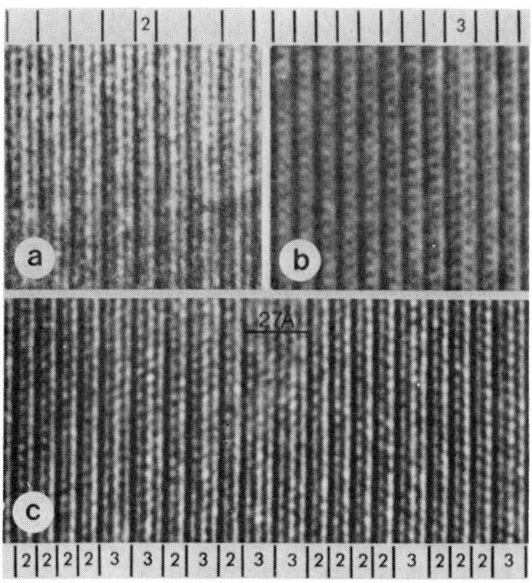

Fig. 36. (a) Isolated double chain (2) in jimthompsonite. (b) Isolated triple chain (3) in anthophyllite. (c) Disordered sequence of double and triple chains (2 and 3, respectively.). This can best be viewed at a low angle, parallel to the chains. (From Veblen *et al.*, *Science*, **198**, No. 4315, 359–365 (1977). Copyright 1977 by The American Association for the Advancement of Science.)

G. *High-Voltage, High-Resolution Transmission Electron Microscopy*

Most of the so-called HRTEM discussed so far in this chapter has been carried out with 100-kV microscopes with a resolution only occasionally better than 3.5 Å point-to-point. At this resolution it is difficult to be certain of atomic positions, but arrangements of coordination polyhedra such as the MO_6 octahedra are clearly seen. It is desirable in most studies to have atomic resolution revealing structural details not now visualized in the nonmetallic systems or ordered alloys illustrated so far. With atomic resolution, one could extend these studies to yield structural detail of point defects and close-packed systems without superstructure.

The limit of resolution is given by the

$$d_{\min} = 0.6 C_s^{1/4} \lambda^{3/4}$$

where $d_{\min}$ is the limit of resolution, C_s is the spherical aberration constant, and λ is the wavelength. Obviously, one of the ways to achieve better resolution is to reduce the electron wavelength. A new generation of microscopes is just beginning to be used that operate at 200 kV, and very impressive results have already been published. Using the JEM200CX, Hutchison (1978–1979) has imaged $Ba_2Bi_4Ti_5O_{18}$, which he had imaged

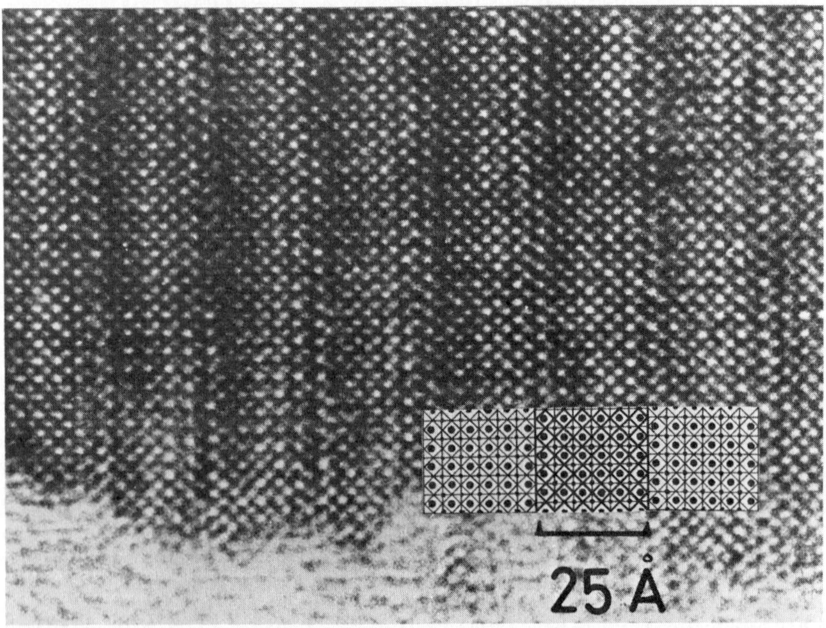

Fig. 37. ⟨110⟩ lattice image of $Ba_2Bi_4Ti_5O_{18}$, with an idealized projection superimposed on the image. (From Hutchison, 1978–1979.)

Fig. 38. A 1-MV crystal structure image of $9Nb_2O_5 \cdot 8WO_3$. (a) At low magnification. (*Continued*)

Fig. 38. (*Continued*) (b) At high magnification. MO_6 octahedra share corners to form 5×5 blocks, where M is a cation. The crystal belongs to the tetragonal system with the lattice parameters $a = 26.3$ and $c = 3.81$ Å.

previously at lower resolution (Hutchinson *et al.*, 1977). Figure 37 shows the image obtained. The inset, a drawing of the structure, shows that agreement is clearly obtained with both the A- and B-cation positions. This image should be compared to that in Fig. 31, which is of the same crystal taken on a 100 kV microscope. This great improvement in resolution from 3.5 to 2.45 Å is due not only to the higher voltage but also to the greater stability of the electronics.

Horiuchi *et al.* (1978) have described the design and performance of a high-voltage, high-resolution transmission electron microscope operating at

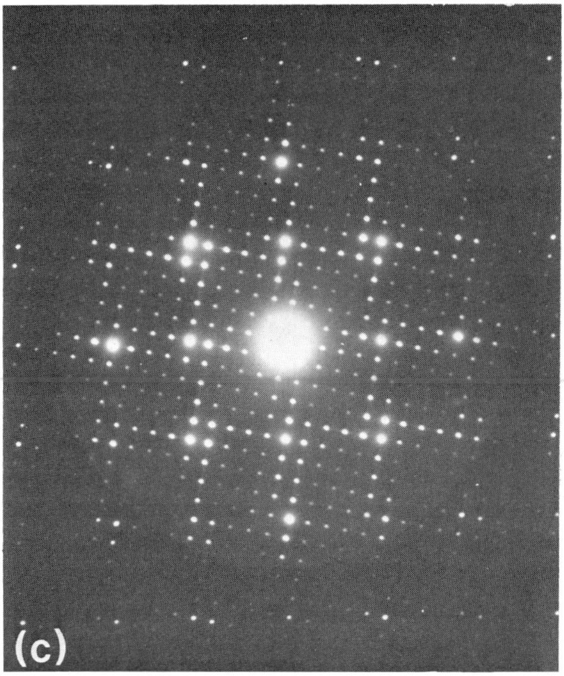

Fig. 38. (*Continued*) (c) The diffraction pattern corresponding to the image. The encircled area corresponds to the bore of the objective aperature. (From Horiuchi, 1978–1979.)

1000 kV with a point-to-point resolution of 2 Å. Figure 38 (Horiuchi, 1978–1979) is a structural image of $9Nb_2O_5 \cdot 8WO_3$ obtained with this instrument. Not only are the 3.8 and 1.9 Å holes visible, but one begins to see the occupation of the octahedral sites in the tetrahedral columns. This improvement in detail is clearly shown in the comparison of calculated and observed images by O'Keefe *et al.* (1978).

The success of this high-voltage microscope is further illustrated in the imaging of the mixed-layer compound $Bi_7Ti_4NbO_{21}$, revealing the 1 : 1 intergrowth of $Bi_4Ti_3O_{12}$- and Bi_3TiNbO_9-like layers. Figure 39 (Horiuchi *et al.*, 1977) shows the image of the structure juxtaposed on a two-dimensional representation. This is an example of the tool being used to obtain direct-space structures. As one other example, Bando *et al.* (1979) have determined the structure of $Na_2Ti_9O_{19}$ by direct imaging and resolution of the Ti and Na sites. The nonatitanate framework is of a new type clearly distinguished from others already known. Figure 40 shows the structure in projection perpendicular to [010]. Two kinds of sheets (B and D) are drawn to compare the proposed structure. Polytypism is to be expected from such a B-D system and is clearly indicated as possible in certain defect regions

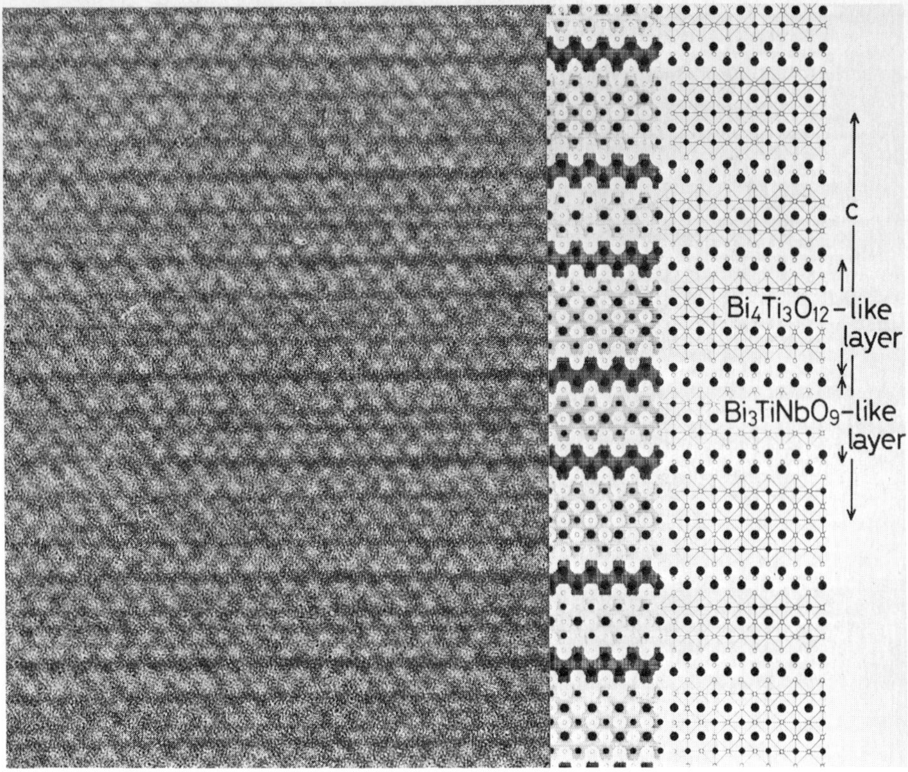

Fig. 39. Interpretation of the image of $Bi_7Ti_4NbO_{21}$. (a) An enlarged image. (b) The correspondence between the image contrast and the crystal structure. (c) The prototype structure of $Bi_7Ti_4NbO_{21}$ viewed along [110]. Large, dark circles indicate Bi ions, small circles Ti or Nb ions, and open circles O ions; $c = 58.1$ Å. (From Horiuchi *et al.*, 1977.)

and in other samples. It appears that direct-space crystallography is here at last, and that every variation, defect, and nuance will soon be visualized at the atomic level. We conclude our discussion of examples that illustrate the power of this technique on this optimistic note.

IV. Application of HRTEM to Chemical Dynamics

Up to this point the use of HRTEM in revealing structures and the defects that lead to nonstoichiometry has been emphasized. We now turn to an account of a number of cases where the method has been used to answer questions about how changes in solids occur.

Anderson (1972, 1978–1979) has outlined attempts to observe solid state reactions utilizing HRTEM and has laid down some general principles for

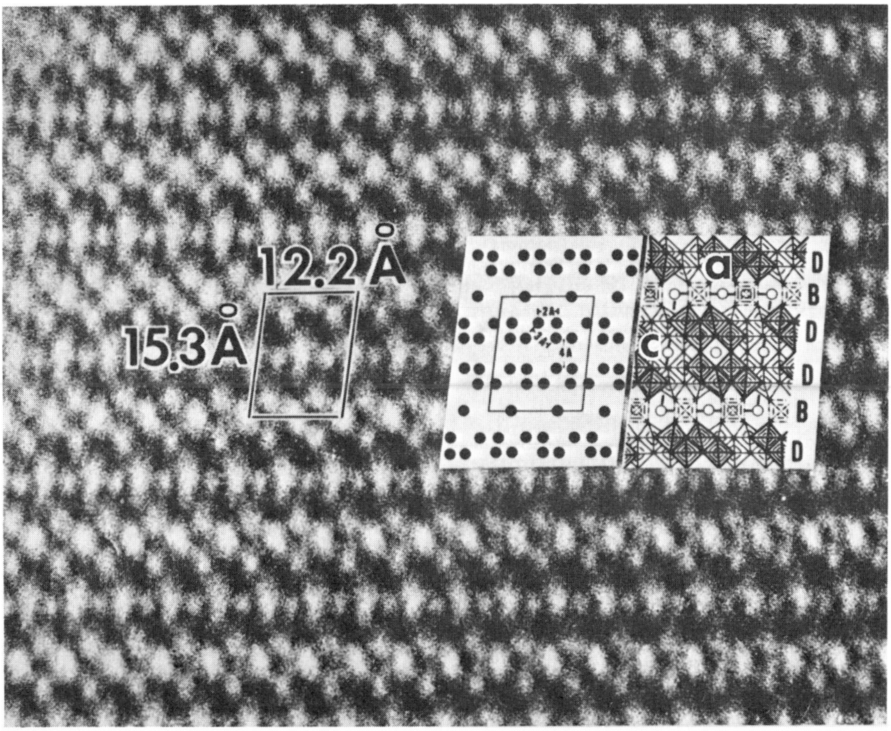

Fig. 40. A structure image of $Na_2Ti_9O_{19}$ (1-MV high-resolution electron microscope). The incident electron beam is parallel to the [010] direction. The two types of sheets are indicated as B and D. The schematic arrangement of the dark dots is shown as solid circles in the inset, in which the separations of the adjoining dark dots are indicated. The structure model is also shown as an inset. The heavy squares indicate TiO_6 octahedra at $y = 0$, and the light ones TiO_6 octahedra at $y = \frac{1}{2}$. The dashed squares show bridging TiO_5 octahedra at $y = \frac{1}{4}$ and $\frac{3}{4}$. The open circles show Na atoms. (Bando *et al.*, 1979.)

such changes. Since reactions in solids must involve local changes in atomic arrangement and since HREM is the best method for observing these local events at the atomic level, it should be clear that HREM is the method of choice in seeking to understand the atomic mechanism of solid state reactions.

Indirect methods involving observation of, say, the oxygen pressure dependence of composition, conductivity, etc., leave ambiguities which may persist for decades. HREM, while not a panacea, will be a powerful ally of the more conventional methods. For one thing, it can eliminate most possibilities summarily by simply observing the types of defects involved. Indeed, the cataloging of defects in materials undergoing reactive change, even if one cannot actually see their evolution, will be of enormous aid.

It has even been possible to make low-resolution movies of phase transformations, crystal growth, and other reactions. For example, Van Landuyt (1977) has discussed the use of EM in the study of phase transitions in solids. Obtaining movies at high resolution will require very elaborate techniques which have not yet been well developed. Rapid reactions will be particularly difficult to observe at the very low illumination levels available at high resolution. Nevertheless, successive images of the same area of a crystal after changes have occurred (usually from beam irradiation) permit a piecing together of the reaction path in some cases. This is one of the most important uses of HREM in chemistry and related materials sciences, and much more will be done with it in the future.

Several investigations have gone beyond simply cataloging defects in materials and have attempted to understand the mechanism of solid state changes. For example, Bursill (1969) observed the development of a phase having well-developed CS planes in $MO_{3-\delta}$ from patterns that appeared to have small domains yielding to a cross-hatched pattern which steadily evolved. Iijima (1975) observed a similar development of CS from faint beginnings in the $WO_{3-\delta}$ region.

Browne and Anderson (1974) investigated reaction mechanisms involved in the oxidation of $Nb_{12}O_{29}$. They found that the two forms of $Nb_{12}O_{29}$ (monoclinic and orthorhombic) oxidized at low temperature to produce two new forms of Nb_2O_5. Both electron diffraction and images show the development of a superstructure in $Nb_{12}O_{29}$. At this stage, oxidation of the monoclinic form was complete at $NbO_{2.500}$, and then considerable rearrangement of the substance at constant composition followed. A similar sequence of changes occurred when the orthorhombic $Nb_{12}O_{29}$ was oxidized. At high temperatures the reaction mechanism and, indeed, the product (H–Nb_2O_5) is quite different. A detailed analysis of the low-temperature reactions revealed homogeneous localized changes as diffusing oxygen arrived. The rate-determining step might be the metal atom jump. These atomic changes are coordinated to produce a superstructure with a rather large unit cell as a first stage. Although detailed atomic level displacements and locations cannot be followed, the general features of the reaction can be worked out.

Melts of $TiNb_{38}O_{97}$ were splat-cooled by Anothony *et al.* (1977) and examined by HRTEM. The material was crystalline, even after cooling at an estimated $10^5\,°C\,sec^{-1}$. They found, besides severely disordered regions, highly ordered block structures. The results were interpreted intuitively to consist of two overlapping structures in many cases.

Anderson *et al.* (1977) analyzed several defects and dislocations in Nb_2O_5 and reduced Nb_2O_5. In the analysis, mapping and modeling were done in a quite literal way to correspond to the observed images. It was possible to obtain convincing interpretations of inserted structure, twinning, and dislocations, and the Burgers vector was determined. It is suggested that this sort of topological treatment can be extended to other crystal types, e.g., those in $(Bi_2O_2)\ Ba_{n-1}Ti_nO_{3n+1}$ (Hutchinson *et al.*, 1977).

A study on reactions of the type $A(s) \rightarrow B(s) + C(s)$, where B(s) is capable of intergrowth with A(s) but C(s) is not, was undertaken by Obayashi and Anderson (1976). Reversible phase reactions in the Nb_2O_5–WO_3 system were used; for example, disproportion of the $(5 \times 4)_1$ block structure of $8Nb_2O_5 \cdot 5WO_3$ into coherent $(4 \times 4)_1$ blocks of $7Nb_2O_5 \cdot 3WO_3$, and that of $9Nb_2O_5 \cdot 8WO_3$ into coherent $8Nb_2O_5 \cdot 5WO_3$ or $(5 \times 5)_1$ blocks into $(5 \times 4)_1$. The homogeneous, coherent reaction occurs block by block and can be followed in some detail. The incoherent phase, in this case a pentagonal tunnel structure, must be formed at the surface by removal of the leftover Nb_2O_5 and WO_3.

Hutchinson and Anderson (1977) studied the final decomposition of $GaNbO_4$ to R–Nb_2O_5 and Ga_2O_3 (implied). During the reaction a number of new block structures, members of the homologous series $Me_{2n}O_{5n-2}$, were observed. The mechanism proposed for the reaction requires the cooperative displacement of a pair of cations with oxygen atoms by $\frac{1}{2} a_R$ $\langle 110 \rangle_R$ along an octahedral edge of the ReO_3 structure. This must be coupled with atom diffusion and place exchange, processes that are fast at high temperatures. Such localized changes in order have been observed after intense beam heating of $GaNbO_4$. Further irradiation transforms the structure to members of a new series, $Me_{2n}O_{5n-2}$.

In many oxide systems care must be exercised to avoid beam heating and radiation damage to the specimen under HRTEM examination. Rare earth oxides are such systems. The higher oxides of Pr and Tb are easily reduced to their sesquioxides in the microscope. This is an advantage in their study, since both prepared static states and transformations can be watched. As examples of this, let us observe the reduction reactions

$$TbO_2 \rightarrow TbO_{2-\delta}\ \text{(disordered)} \rightarrow Tb_nO_{2n-2}\quad \text{(ordered phases)}$$

$$Pr_{10}O_{18} \rightarrow Pr_{12}O_{22} + Pr_{17}O_{30}\quad (1:1\ Pr_8O_{14} + Pr_9O_{16})$$

and an actual sequence in the reduction

$$Pr_9O_{16} \rightarrow Pr_7O_{12}$$

In Fig. 41 (Eyring and Tuenge, unpublished results) a sequence is displayed in which the diffraction patterns on the right were taken on either side of the images on the left. The first diffraction pattern shows no superstructure and only slight diffuse scatter, but the image taken immedately afterward shows considerable ordering as indicated by the presence of regions of lattice fringes. The next diffraction pattern shows distinct extra reflections and more diffuse scatter. Easily identified regions of one of the known

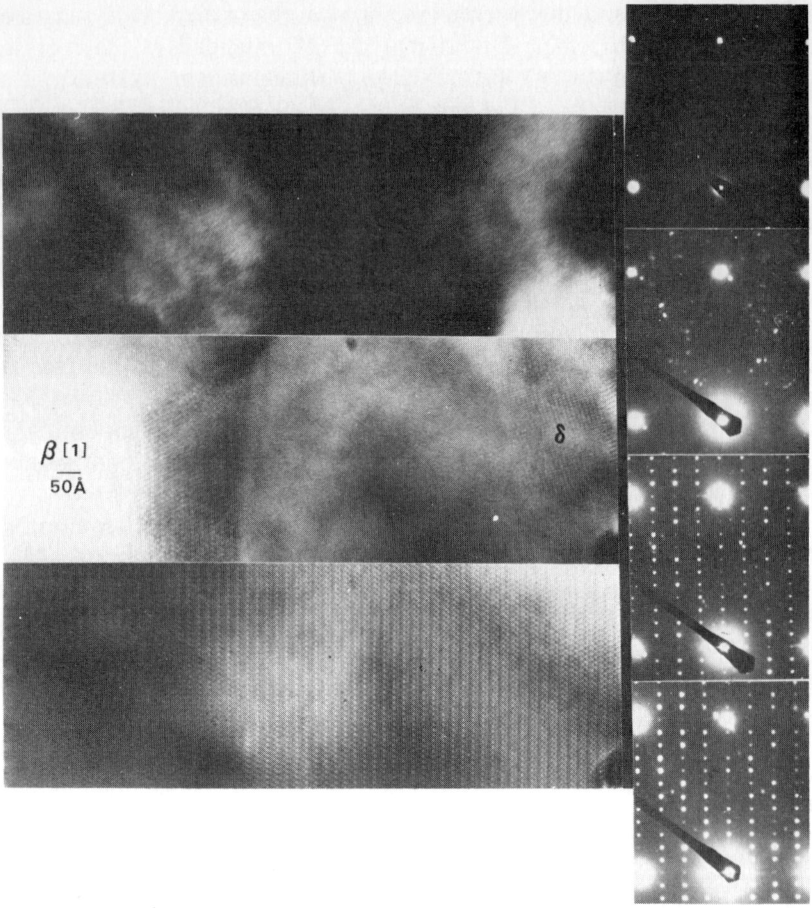

Fig. 41. The progressive ordering of sections of a crystal of TbO_x initially disordered and finally fully ordered. The images on the left were taken sequentially between the diffraction patterns on the right. (From Eyring and Tuenge, unpublished results.)

ordered phases of the $n = 12$ phase is observed to be progressively formed from the left. This process of ordering is continued by growth of the $n = 12$ phase as order fills the entire region.

A specimen of $Pr_{10}O_{18}$ was heated in the electron beam (Eyring and Tuenge, unpublished results). It was observed to disproportionate into phases based on simple $\{135\}$ defect planes, as indicated by the appearance of linear contrast changes in these directions. Later, the same region was observed to have become ordered into a large area of the triclinic primitive cell of $n = 12$ and a 1:1 intergrowth of Pr_8O_{14} and Pr_9O_{16} (Fig. 42). Changes of this kind are very rapid because the metal atoms maintain their

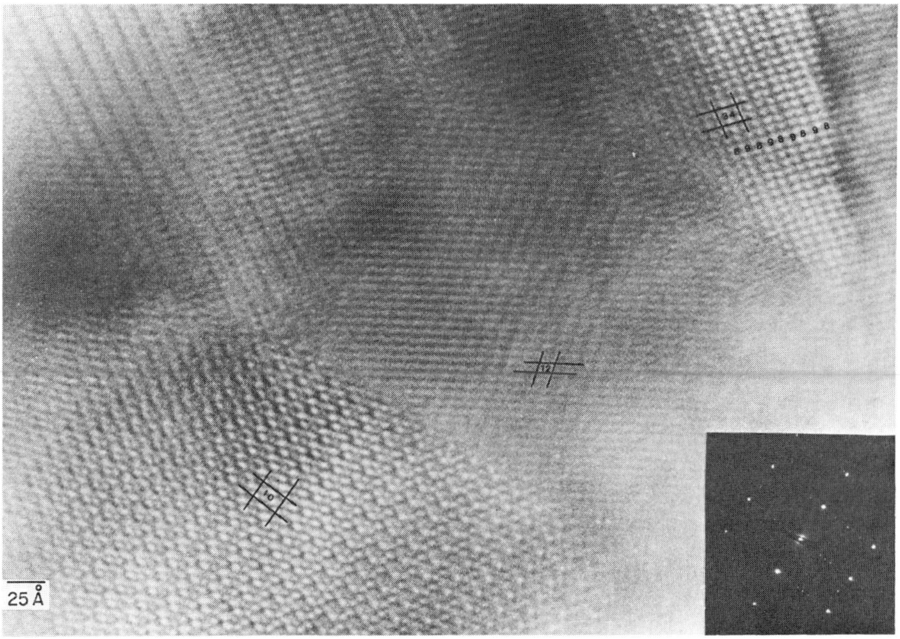

Fig. 42. A crystal, originally $Pr_{10}O_{18}$, undergoing disproportionation. The $Pr_{10}O_{18}$ phase persists in the bottom center, but the remainder of the micrograph shows the triclinic $n = 12$ phase and a 1 : 1 growth of $n = 9$ and 8, as marked. (From Eyring and Tuenge, unpublished results.)

fluorite positions while the oxygen atoms are extremely labile and easily assume new ordered positions.

The phase reaction $Pr_9O_{16} \rightarrow Pr_7O_{12}$ has received considerable attention. Irregular intergrowths of the two structures appear as reduction proceeds. Figure 43 shows this, with two regions of the reactive specimen modeled. A region of $n = 9$ exists on one side and a region of $n = 7$ on the other (the straight lines indicate the directions of the b axis), with the banded irregular intergrowth in between.

A very striking event was recorded in Fig. 44 (Eyring and Tuenge, unpublished results). At the top, a blade of Pr_7O_{12} appears to have grown from left to right. The image in the middle shows that many other blades have nucleated on the left and are moving to the right. Finally, at the bottom is seen the nearly completely banded development from nuclei developed to the left of the image. These three images were taken in succession, a minute or so apart. Notice especially the pointed character of the growing tips, which must involve a high density of misplaced atoms and vacancies. The reaction product achieves a certain width and propagates

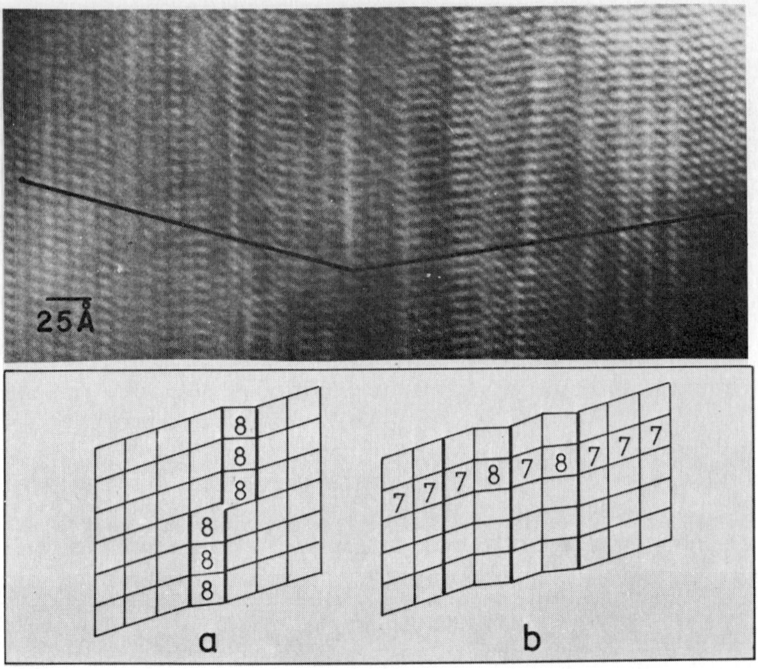

Fig. 43. A region of transition between Pr_9O_{16} on the left to Pr_7O_{12} on the right. The (135) plane is vertical. Disordered intergrowth occurs across the common *ac* plane, and the *b* directions are marked. The (a) and (b) regions are modeled. (From Eyring and Tuenge, unpublished results.)

without further lateral extension. This width gives perfect intergrowth of Pr_7O_{12} in a crystal of Pr_9O_{16}; i.e., seven rows of Pr_9O_{16} unit cells are replaced by nine rows of Pr_7O_{12}.

V. Concluding Remarks

Our understanding of the nature of nonstoichiometry in oxides has increased dramatically during the past 10 yr. The primary reason for this is the development and application of HRTEM. By this means structures and their defects have been directly visualized at the level of coordination polyhedra with their implied atom positions. Not only have defects that cause nonstoichiometry been clearly seen in many cases but, when they have not, the possibilities have been drastically reduced so that well-designed experiments can be more critically applied.

The state of the art is improving rapidly. The first of the relatively inexpensive next generation of microscopes with point-to-point resolution of about 2.5 Å are already being employed with great success. They yield

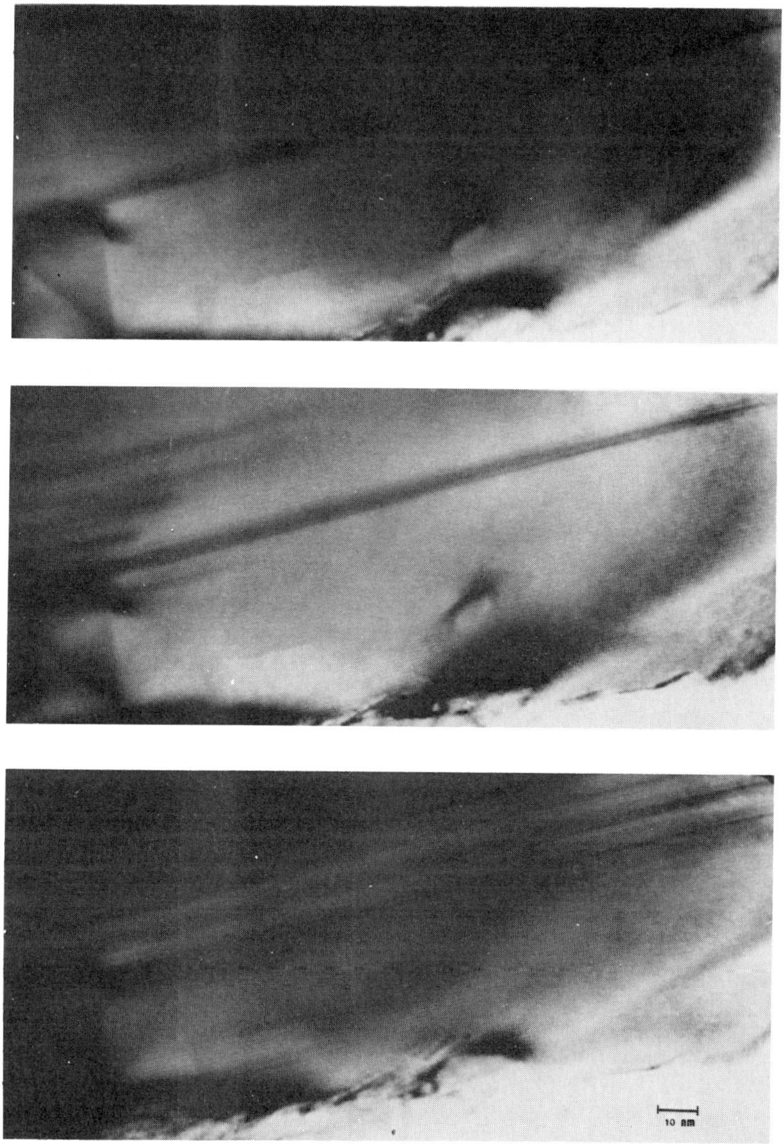

Fig. 44. Progressive reduction of Pr_9O_{16}. The layer of dark contrast in the top image consists of a layer of nine unit cells of Pr_7O_{12} replacing seven of Pr_9O_{16} in an undisturbed single crystal. Similar layers are propagated to the right from nuclei on the left, finally giving a banded appearance. The total time interval of the images is about 5 min. (From Eyring and Tuenge, unpublished results.).

images at atomic resolution, as we have seen above. In addition, nanoanalytical electron microscopy (NAEM) is about to be achieved, which will give compositional information at near the resolution limit of the microscope. The combination of HRTEM and NAEM will then be the ultimate chemical tool. These developments bring most materials, including close-packed structures, within the scrutiny of these powerful techniques.

Even with these prospects in mind, perhaps the greatest contribution of these methods will not be the elucidation of structure, even defect structure as such, but the ability to see directly the sequence of chemical change. In this way the mechanisms of chemical reaction in materials may be accurately discerned, whether they be desirable or undesirable, permitting an appropriate response.

References

Allpress, J. G. (1972). *Natl. Bur. Stand.* (*U.S.*), *Spec. Publ.* No. 364, 87–111.

Allpress, J. G., and Rosell, H. J. (1975). *J. Solid State Chem.* **15**, 68–78.

Allpress, J. G., and Sanders, J. V. (1973). *J. Appl. Crystallogr.* **6**, 165–190.

Allpress, J. G., Sanders, J. V., and Wadsley, A. D. (1968). *Phys. Status. Solidi* **25**, 541–550.

Allpress, J. G., Tilley, R. J. D., and Sienko, M. J. (1971). *J. Solid State Chem.* **3**, 440–451.

Allpress, J. G., Hewat, E. A., Moodig, A. F., and Sanders, J. V. (1972). *Acta Crystallogr., Sect. A* **28**, 528–536.

Allpress, J. G., Iijima, S., Roth, R. S., and Stephenson, N. C. (1973). *J. Solid State Chem.* **7**, 89–93.

Amelinckx, S., and Van Landuyt, J. (1970). *In* "The Chemistry of Extended Defects in Non-Metallic Solids" (L. Eyring and M. O'Keeffe, eds.), pp. 295–322. North-Holland Publ., Amsterdam.

Anderson, J. S. (1972). *In* "Reactivity of Solids." Proceedings of the 7th International Symposium on the Reactivity of Solids (J. S. Anderson, M. W. Roberts, and F. S. Stone, eds.), pp. 1–22. Chapman & Hall, London.

Anderson, J. S. (1973). *J.C.S. Dalton Trans.* 1107–1115.

Anderson, J. S. (1974). *In* "Defects and Transport in Oxides" (M. S. Seltzer and R. I. Jaffee, eds.), pp. 25–54, 260–266. Plenum, New York.

Anderson, J. S. (1977). *J. Phys.* (*Paris*) **38**(12), Suppl., 17–27.

Anderson, J. S. (1978–1979). *Chem. Scr.* **14**, 129–139.

Anderson, J. S., and Hutchison, J. L. (1975). *Contemp. Phys.* **16**, 443–467.

Anderson, J. S., and Hyde, B. G. (1967). *J. Phys. Chem. Solids* **28**, 1393–1408.

Anderson, J. S., Browne, J. M., and Hutchison, J. L. (1972). *J. Solid State Chem.* **5**, 419–431.

Anderson, J. S., Bevan, D. J. M., Cheetham, A. K., Von Dreele, R. B., Hutchison, J. L., and Strähle, J. (1975). *Proc. R. Soc. London, Ser. A* **346**, 139–156.

Anderson, J. S., Hutchison, J. L., and Lincoln, F. J. (1977). *Proc. R. Soc. London, Ser. A* **352**, 303–323.

Andersson, S. (1970). *In* "The Chemistry of Extended Defects in Non-Metallic Solids" (L. Eyring and M. O'Keeffe, eds.), pp. vi–vii. North-Holland Publ., Amsterdam.

Anthony, A.-M., Anderson, J. S., and Hutchison, J. L. (1977). *J. Solid State Chem.* **21**, 233–241.

Ashbee, K. H. G. (1970). *In* "The Chemistry of Extended Defects in Non-Metallic Solids" (L. Eyring and M. O'Keeffe, eds.), pp. 323–346. North-Holland Publ., Amsterdam.

Baenziger, N. C., Eick, H. A., Schuldt, H. S., and Erying, L. (1961). *J. Am. Chem. Soc.* **83**, 2219–2223.

Bando, Y., Watanabe, M., and Sekikawa, Y. (1979). *Acta Crystallogr., Sect. B* **35**, 1541–1545.

Bartram, S. F. (1966). *Inorg. Chem.*. **5**, 749–754.

Browne, J. M., and Anderson, J. S. (1974). *Proc. R. Soc. London, Ser. A* **339**, 463–482.

Bursill, L. A. (1969). *Proc. R. Soc. London, Ser. A* **311**, 267–290.

Bursill, L. A., and Hyde, B. G. (1972). *Prog. Solid State Chem.* **7**, 177–251.

Bursill, L. A., Grey, I. E., and Lloyd, D. J. (1976). *J. Solid State Chem.* **16**, 331–347.

Bursill, L. A., Netherway, D. J., and Grey, I. E. (1978). *Nature (London)* **272**, 405–410.

Buseck, P. R., and Iijima, S. (1974). *Am. Mineral.* **59**, 1–21.

Cowley, J. M. (1975). "Diffraction Physics," North-Holland Publ., Amsterdam.

Cowley, J. M. (1976). *Annu. Rev. Mater. Sci.* **6**, 53–81.

Cowley, J. M. (1978). *Inst. Phys. Conf. Ser.* No. 41, 156–166.

Cowley, J. M. (1978–1979). *Chem. Scr.* **14**, 33–38.

Cowley, J. M., and Iijima, S. (1974). *In* "Defects and Transport in Oxides" (M. S. Seltzer and R. I. Jaffee, eds.), pp. 205–224. Plenum, New York.

Cowley, J. M., and Moodie, A. F. (1957). *Acta Crystallogr.* **10**, 609–619.

Ekström, T., and Tilley, R. J. D. (1974). *Mater. Res. Bull.* **9**, 705–714.

Ekström, T., and Tilley, R. J. D. (1976). *J. Solid State Chem.* **19**, 125–133.

Ekström, T., and Tilley, R. J. D. (1978). *J. Solid State Chem.* **24**, 209–218.

Eyring, L. (1979). *In* "Handbook on the Physics and Chemistry of the Rare Earths" (K. A. Gschneidner, Jr. and L. Eyring, eds.), Vol. 3, pp. 337–399. North-Holland Publ., Amsterdam.

Eyring, L., and Tai, L.-T. (1976). *In* "Treatise on Solid State Chemistry" (N. B. Hannay, ed.), Vol. 3, pp. 167–252. Plenum, New York.

Eyring, L., and Tuenge, R. T. (Unpublished results).

Gibb, R. M., and Anderson, J. S. (1972). *J. Solid State Chem.* **5**, 212–225.

Grey, I. E., and Bursill, L. A. (1978). *Acta Crystallogr., Sect. B* **34**, 2412–2424.

Honjo, G., Miyake, S., and Tomita, T. (1950). *Acta Crystallogr.* **3**, 396–397.

Horiuchi, S. (1978–1979). *Chem. Scr.* **14**, 75–81.

Horiuchi, S., Kikuchi, T., and Goto, M. (1977). *Acta Crystallogr., Sect. A* **33**, 701–703.

Horiuchi, S., Matsui, Y., Bando, Y., Katsuta, T., and Matsui, I. (1978). *J. Electron Microsc.* **27**, 39–48.

Hussain, A., and Kihlborg, L. (1976). *Acta Crystallogr., Sect. A* **32**, 551–557.

Hutchison, J. L. (1978–1979). *Chem. Scr.* **14**, 181–186.

Hutchison, J. L., and Anderson, J. S. (1977). *J. Solid State Chem.* **22**, 135–144.

Hutchison, J. L., Anderson, J. S., and Rao, C. N. R. (1977). *Proc. R. Soc. London, Ser. A* **355**, 301–312.

Hyde, B. G., and Bursill, L. A. (1970). *In* "The Chemistry of Extended Defects in Non-Metallic Solids" (L. Eyring and M. O'Keeffe, eds.), pp. 347–378. North-Holland Publ., Amsterdam.

Hyde, B. G., and O'Keeffe, M. (1973). *Acta Crystallogr., Sect. A* **29**, 243–248.

Hyde, B. G., Bagshaw, A. N., Andersson, S., and O'Keeffe, M. (1974). *Annu. Rev. Mater. Sci.* **4**, 43–92..

Iijima, S. (1971). *J. Appl. Phys.* **42**, 5891–5893.

Iijima, S. (1975). *J. Solid State Chem.* **14**, 52–65.

Iijima, S., and Allpress, J. G. (1973). *J. Solid State Chem.* **7**, 94–105.

Iijima, S., and Allpress, J. G. (1974a). *Acta Crystallogr., Sect. A* **30**, 22–29.

Iijima, S., and Allpress, J. G. (1974b). *Acta Crystallogr., Sect. A* **30**, 29–36.

Iijima, S., and Cowley, J. M. (1977). *J. Phys. (Paris)* **38**(12), Suppl., 135–144.

Iijima, S., Kimura, S., and Goto, M. (1973). *Acta Crystallogr., Sect. A* **29**, 632–636.

Kamiya, S., and Tilley, R. J. D. (1977). *J. Solid State Chem.* **22**, 205–216.
Kihlborg, L. (1959). *Acta Chem. Scand.* **13**, 959–962.
Kihlborg, L. (1963). *Ark. Kemi* **21**, 427–437.
Kihlborg, L. (1978–1979). *Proc. Nobel Symp., 47th, Direct Imag. Atoms, Lidingö, Sweden, 1979*; *Chem. Scr.* **14**, 7–295.
Kihlborg, L., and Sundberg, M. (1976). *Siemens Rev.* **43**, 3–8.
Kunzmann, P., and Eyring, L. (1975). *J. Solid State Chem.* **14**, 229–237.
Lloyd, D. J., Grey, I. E., and Bursill, L. A. (1976). *Acta Crystallogr., Sect. B* **32**, 1756–1761.
McConnell, J. D. M., Hutchison, J. L., and Anderson, J. S. (1974). *Proc. R. Soc. London, Ser. A* **339**, 1–12.
Magnéli, A. (1949). *Ark. Kemi* **1**, 513–523.
Magnéli, A. (1978). *Pure Appl. Chem.* **50**, 1261–1271.
Obayashi, H., and Anderson, J. S. (1976). *J. Solid State Chem.* **19**, 331–351.
O'Keefe, M. A. (1973). *Acta Crystallogr., Sect. A* **29**, 389–401.
O'Keefe, M. A., Buseck, P. R., and Iijima, S. (1978). *Nature (London)* **274**, 322–324.
Pickering, R., and Tilley, R. J. D. (1976). *J. Solid State Chem.* **16**, 247–255.
Scherzer, D. (1949). *J. Appl. Phys.* **20**, 20–29.
Schiffmacher, G. (1977). *J. Microsc. Spectrosc. Electron.* **2**, 503–510.
Schiffmacher, G., Caro, P. E., and Boulesteix, C. (1976). *Phys. Status Solidi* **33**, K9–K13.
Sibieude, F., Schiffmacher, G., and Caro, P. (1978). *J. Solid State Chem.* **23**, 361–367.
Skarnulis, A. J., Iijima, S., and Cowley, J. M. (1976). *Acta Crystallogr., Sect. A* **32**, 799–805.
Skarnulis, A. J., Summerville, E., and Eyring, L. (1978). *J. Solid State Chem.* **23**, 59–71.
Steadman, R., Tilley, R. J. D., and McColm, I. J. (1972). *J. Solid State Chem.* **4**, 199–208.
Stone, G. G., and Bursill, L. A. (1977). *Philos. Mag.* **35**, 1397–1412.
Summerville, E., Tuenge, R. T., and Eyring, L. (1978). *J. Solid State Chem.* **24**, 21–31.
Sundberg, M. (1976). *Acta Crystallogr., Sect. B* **32**, 2144–2149.
Sundberg, M., and Tilley, R. J. D. (1974). *J. Solid State Chem.* **11**, 150–160.
Thomas, J. M., and Jefferson, D. A. (1978). *Endeavour (New Series)* **2**(3), 127–136.
Thomas, J. M., Jefferson, D. A., Mallinson, L. G., Smith, D. J., and Crawford, E. S. (1978–1979). *Chem. Scr.* **14**, 167–179.
Tilley, R. J. D. (1972). *In* "Solid State Chemistry" (H. F. Emeleus and L. E. J. Roberts, eds.), MTP International Review of Science, Inorganic Chemistry, Series One, Vol. 10, pp. 279–313. Butterworth, London.
Tuenge, R. T., and Eyring, L. (1979). *J. Solid State Chem.* **29**, 165–179.
Van Landuyt, J. (1977). *J. Phys. (Paris)* **38**(12), Suppl., 5–16.
Van Landuyt, J., Amelinckx, S., Kohn, J. A., and Eckart, D. W. (1974). *J. Solid State Chem.* **9**, 103–119.
Van Tendeloo, G., Van Dyck, D., Van Landuyt, J., and Amelinckx, S. (1979). *J. Solid State Chem.* **27**, 55–70.
Veblen, D. R., Buseck, P. R., and Burnham, C. W. (1977). *Science* **198**, 359–365.
Verma, A. R., and Tregunazat, G. C. (1974). *In* "Solid State Chemistry" (C. N. R. Rao, ed.), pp. 51–94. Dekker, New York.
Von Dreele, R. B., Eyring, L., Bowman, A. L., and Yarnell, J. L. (1975). *Acta Crystallogr., Sect. B* **31**, 971–974.
Wadsley, A. D. (1955). *Rev. Pure Appl. Chem.* **5**, 165–193.
Wadsley, A. D. (1968). *In* "Approach to Chemistry." Proceedings of the 7th Summer School for Senior Teachers of Chemistry (R. Foon and D. P. Graddon, eds.). Univ. of New South Wales Press, Sydney.
Wadsley, A. D., and Andersson, S. (1970). *In* "Perspectives in Structural Chemistry" (J. D. Dunitz and J. A. Ibers, eds.), Vol. 3, pp. 1–58. Wiley, New York.

8

Structural Studies on Nonstoichiometric Oxides Using X-Ray and Neutron Diffraction

A. K. Cheetham
Chemical Crystallography Laboratory
University of Oxford
Oxford, England

I. Introduction

Our present understanding of the physical and chemical properties of crystalline solids relies heavily upon a knowledge of structure at the unit cell level, and diffraction studies using x rays, neutrons, and electrons play a central role in contemporary studies on solid materials. In this chapter, the contribution of x-ray and neutron diffraction to the elucidation of structure in nonstoichiometric oxides is reviewed. The discussion will not be restricted entirely to nonstoichiometric phases but will also embrace the

ISBN 0-12-655280-0

structures of relevant ordered phases and other systems in which series of closely related ordered structures are observed.

In Section II the information that can be obtained from x-ray and neutron diffraction measurements is compared with that which is accessible in the electron microscope, and Section III examines the special advantages of neutron diffraction. Nonstoichiometric binary oxides are reviewed in Section IV with emphasis on compounds that adopt the sodium chloride and fluorite structures. Crystallographic shear planes in structures based upon the rutile and rhenium trioxide types are discussed in Section V, and compounds related to Nb_2O_5 are considered in Section VI. Finally, Section VII deals with some interesting compounds whose ordered structures adapt continuously to changes in composition. The review is not comprehensive and does not, for example, cover tungsten bronzes and related compounds, which have been reviewed recently by Hagenmuller (1973) and Dickens and Wiseman (1975).

II. Comparative Roles of Diffraction Methods and Electron Microscopy

Until approximately 1970, most of the structural data on metal oxides emanated from x-ray and neutron diffraction studies, but with the advent of powerful, high-resolution imaging techniques, electron microscopy is now playing a significant role in the determination of structure at the atomic level. It is important that the complementarity of these different methods be understood.

In a conventional diffraction experiment using x rays or neutrons, the intensities of a large number of Bragg reflections are measured. With the use of standard crystallographic procedures, the intensity data are then manipulated to determine the structure that, for a fully ordered material, will uniquely define the contents of the unit cell. For a nonstoichiometric compound, however, the unit cell contents vary from one part of the crystal to another and the unit cell contents derived in this manner will represent a weighted average of all those present in the crystal (see, e.g., Willis, 1970). The disorder present in the sample will also lead to diffuse scattering of radiation into regions between the Bragg reflections. This diffuse scattering contains information about the clustering or short-range ordering of defects, but the data collection procedure in x-ray and neutron diffraction normally discriminates against the diffuse scattering. Information about defect structure can be extracted from diffuse scattering, and the technique will be illustrated in Section IV.

When the electron microscope is used to produce a high-resolution lattice image, both the Bragg and the diffuse scattering present in the diffraction pattern contribute to the final image, which contains contrast features reflecting local inhomogeneities in the crystal. This enables us to see the stacking faults, dislocations, and antiphase boundaries directly and, under favorable conditions, even more localized defects may be apparent (see Chapter 7). The advantages of electron microscopy are dramatically underlined by results for materials in which two closely related phases may be coherently intergrown in a random manner. This occurs in metastable, low-temperature phases of ferrous oxide where orthorhombic and tetragonal structures are intergrown. Both x-ray and electron diffraction photographs reveal apparently cubic symmetry, but the lattice image shows clearly the extent and complexity of the disorder (Andersson and Sletnes, 1977).

Any attempt to determine the x-ray crystal structure of a material that is disordered in the manner described above would probably lead to ambiguous conclusions, but it must be recognized that many nonstoichiometric materials are not intergrowth structures of this type. Neutron and x-ray studies remain the only techniques with which we can define accurately the coordinates of metal and oxygen atoms in crystals and determine quantitatively the occupancy numbers of partially occupied sites. Furthermore, it is important to note that the electron microscope produces a projection of the crystal structure along a chosen direction and though some materials, for example niobium oxides, give projections that are very informative, in other compounds the projections may be quite intractable. Clearly the different structural techniques have complementary qualities, and the most powerful approach to structural problems in the defect solid state is one that harnesses the advantages of more than one method.

III. X Rays or Neutrons?

The inconvenience and expense of using neutrons will always ensure that x rays are used in the vast majority of structure determinations. Nevertheless, in this chapter the results obtained using x rays and neutrons are reviewed, and it is therefore appropriate to consider briefly the relative advantages of the two methods.

The main advantage of neutrons to the crystallographer is that they are generally more sensitive to the positions of light atoms (e.g., hydrogen, carbon, oxygen) because the neutron-scattering power does not show a strong dependence on atomic number (Fig. 1). This clearly confers some advantage in the study of oxides, but it must be said that single-crystal x-ray methods are normally quite adequate for the location of oxygen in the presence of very heavy metal atoms. For example, in recent experiments there was no difficulty in locating oxygen in the structures of some

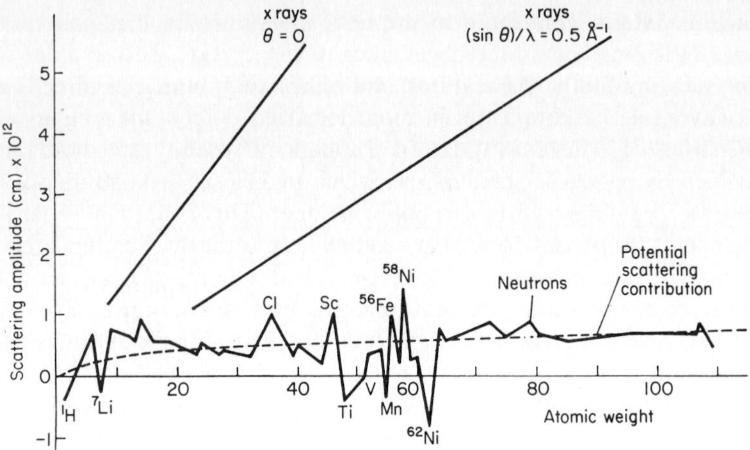

Fig. 1. The variation in neutron-scattering amplitude with atomic weight. (After Bacon, 1952.)

bismuth-rhenium oxides (Rae-Smith and Cheetham, 1979). However, nonstoichiometric oxides may contain partially filled oxygen sites for which the added sensitivity of neutrons is clearly desirable in the determination of precise coordinates and occupancy numbers.

The other main application of neutron diffraction, apart from its use in magnetic studies, is in systems where we need to distinguish between elements that are adjacent, or nearly so, in the periodic table. Such elements are usually indistinguishable with x rays, but in some cases neutrons offer a high degree of contrast because neutron scattering is a nuclear property and may vary dramatically from one element to another. It can be seen in Fig. 1 that some elements have a negative scattering amplitude, corresponding to an anomalous phase shift on scattering. Thus, for example, we have a very sensitive probe for the distribution of mangangese and iron in a mixed oxide or alloy containing these elements; the neutron-scattering lengths of Mn and Fe are -0.35 and $+0.95$ ($\times 10^{-14}$ m), respectively (Bacon, 1975). This principle has been used in a wide variety of systems including studies on the degree of inversion in spinels such as $MgAl_2O_4$ (Bacon, 1952) and $NiCo_2O_4$ (Battle *et al.*, 1979).

The low absorption cross sections of most elements for neutrons (Bacon, 1975) are useful when the sample has to be contained in a controlled environment, such as high pressure or high temperature. This has led to a number of applications with metal oxides, particularly using powders since it is then not necessary to move the specimen during data collection. High-temperature neutron studies have the additional advantage that attenuation of the scattered intensity by thermal motion is not enhanced by a dependence of the scattering power on sin θ/λ; with neutrons, the nuclear scattering is isotropic.

Finally it should be noted that powders, despite their obvious limitations, have been widely used in structural studies on nonstoichiometric compounds, especially with neutrons (Cheetham, 1972). There are several reasons for this. At the practical level, it is frequently difficult to produce single crystals of the desired composition, especially crystals large enough for neutron diffraction. Second, nonstoichiometry is frequently found in high-symmetry crystal systems that are relatively amenable to study using powders. Third, the collection of diffraction data as a function of temperature is clearly more facile with powders for the reasons given above. Last, the advent of total profile analysis using neutrons (Rietveld, 1969) or x rays (see, e.g., Malmros and Thomas, 1977) has clearly extended the scope of powder studies to more complex crystallographic systems (Cheetham and Taylor, 1977), and it can reasonably be expected that studies in this area will expand.

IV. Nonstoichiometric Binary Oxides

Wide ranges of nonstoichiometry are most frequently encountered in high-temperature phases, and it is not surprising to find that many compounds in this category adopt structures of high symmetry. In the case of oxides, two common stucture types, rock salt and fluorite, are most prevalent among binary nonstoichiometric phases and here the information obtained from diffraction studies on these classical disordered systems is discussed.

A. Phases with the Rock-Salt Structure

With the exception of CuO, in which Cu^{2+} exhibits a Jahn–Teller distortion (Åsbrink and Norrby, 1970) and CrO, which is not well documented, the monoxides of the first transition series adopt the NaCl structure. TiO is a metallic conductor (Denker, 1964, 1966) and shows distinct similarities to VO which is a semimetal (Loehman *et al.*, 1969). MnO, FeO, CoO, and NiO are semiconductors.

1. Titanium Monoxide, TiO

The Ti–O phase diagram has been studied by several workers (see, e.g., Porter and Roy, 1964; Wahlbeck and Gilles, 1966), and the region around TiO is shown in Fig. 2. A particularly interesting feature of TiO is that its

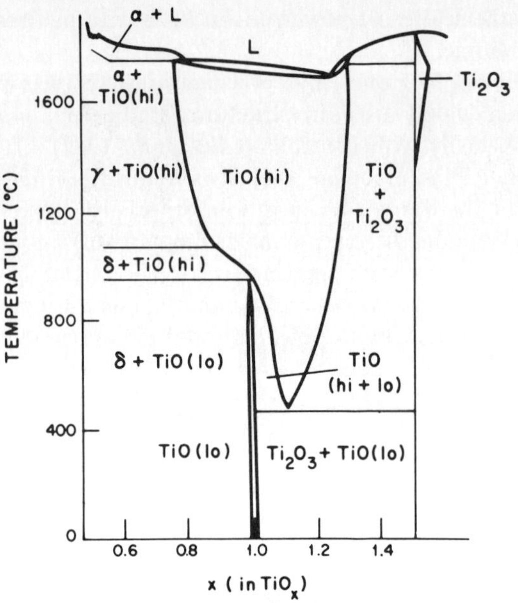

Fig. 2. The titanium–oxygen phase diagram in the vicinity of TiO. (After Banus and Reed, 1970.)

range of nonstoichiometry extends on both sides of the stoichiometric composition. TiO accommodates vacancies in both the cation and anion sublattices, and at the ideal composition, $Ti_{1.0}O_{1.0}$, 15% of the titanium and oxygen sites are unoccupied (Banus and Reed, 1970; Taylor and Doyle, 1970). Below 1200 K, there is an ordered phase with a narrow range of nonstoichiometry around TiO (Watanabe *et al.*, 1967, 1968), and the crystal structure of this ordered phase has been determined by single-crystal x-ray methods (Terauchi *et al.*, 1978). The structure is monoclinic, B2/m, and the titanium and oxygen vacancies are ordered in every third (110) plane in the cubic subcell (Fig. 3). The ordering is incomplete, with more titanium than oxygen vacancies concentrated in each plane. No evidence has been found for the presence of tetrahedrally coordinated interstitial metal atoms. Terauchi *et al.* proposed a model for the defect interactions in TiO; terms describing vacancy–vacancy, vacancy–electron, and vacancy–strain interactions were included. Terauchi and Cohen (1979) have also examined the diffuse scattering of x rays from a single crystal held at 1323 K and shown that local order resembling that of the low-temperature phase is retained at high temperatures.

Two other ordered phases have been reported in the literature. Samples of composition $TiO_{1.25}$ become ordered after prolonged annealing at 1070 K; the crystal symmetry of the ordered phase is tetragonal (Watanabe *et*

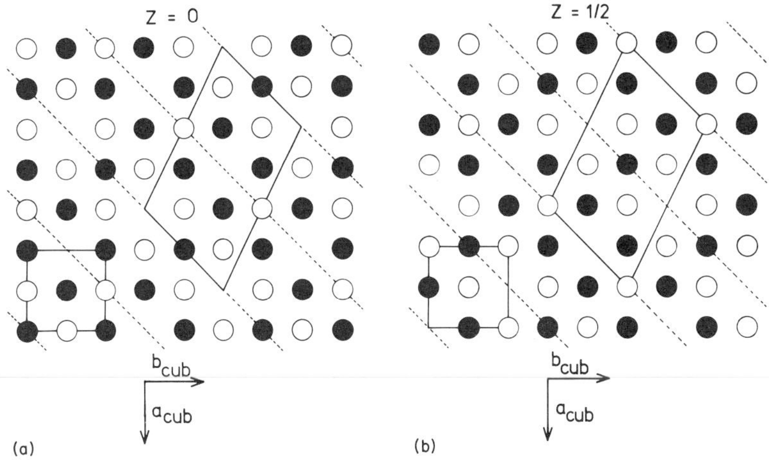

Fig. 3. The idealized structure of ordered TiO showing sections at (a) $z = 0$ and (b) $z = \frac{1}{2}$. Titanium and oxygen atoms are indicated by solid and open circles, respectively. The relationship to the rock-salt structure is shown.

al., 1968). An orthorhombic ordered phase has been obtained in the range $TiO_{0.7}$ to $TiO_{0.9}$ in samples cooled rapidly from the melt and annealed briefly at low temperatures, and in thin foils (Watanabe *et al.*, 1966); this transition structure is ordered in a manner closely related to that of the monoclinic phase.

It is also interesting to note that the number of vacancies in TiO can be reduced by the application of pressure (Taylor and Doyle, 1970). The complete removal of vacancies can be achieved by a careful selection of temperature and pressure; both the lattice parameter and density are increased by increasing pressure.

2. *Vanadium Monoxide, VO*

VO is similar to TiO in that its range of nonstoichiometry extends almost symmetrically on both sides of the stoichiometric composition from approximately $VO_{0.75}$ to $VO_{1.30}$ (Banus and Reed, 1970). The defect mechanism is again primarily metal and oxygen vacancies. At the oxygen-rich phase limit, the anion sublattice is almost complete, but at the vanadium-rich end, approximately 13% of the cation sites remain vacant. Like TiO, the stoichiometric compound has 15% vacancies in each sublattice. The behavior under high pressure is also similar to that of TiO (Taylor and Doyle, 1970).

An ordered structure has been observed at temperatures below 1073 K in the composition range $VO_{1.2}$ to $VO_{1.3}$ (Andersson and Gjønnes, 1970; Bell and Lewis, 1971). The composition of this phase is believed to be $V_{52}O_{64}$ (i.e., $VO_{1.23}$) and, although the structure has not been fully refined, it is thought to contain an ordered array of clusters in which tetrahedral vanadium interstitials are associated with four octahedral vanadium vacancies. The presence of interstitial vanadium in the disordered phase has been reported by Høier and Andersson (1974) and Morinaga and Cohen (1976). Consequently, the cluster proposed for the ordered phase is believed to be present at high temperatures in the disordered phase. At $VO_{1.25}$, where we can assume that the oxygen sublattice is filled, 3% of the vanadium atoms are found at tetrahedral sites; the structural formula may then be written $V^{oct}_{0.776}V^{tet}_{0.024}O$. It is interesting to note that tetrahedral interstitials are also found in ferrous oxide (Section IV.A.4), but the concentration of interstitials is very much smaller in VO. With a ratio of nearly 10 octahedral vacancies per interstitial, it is clear that the majority of cation vacancies are not present in the immediate proximity of interstitials.

The defect structures of TiO and VO show marked similarities, and they are clearly controlled to a large extent by metallic bonding, a feature that is absent from the later monoxides. It is not surprising, then, to find that metallic NbO adopts the rock-salt structure and has an ordered arrangement of vacancies in both sublattices (Bowman *et al.*, 1966). The niobium and oxygen atoms at 0, 0, 0 and $\frac{1}{2}, \frac{1}{2}, \frac{1}{2}$, respectively, are vacant, resulting in a primitive rather than a face-centered cell (Fig. 4). The coordination geometry of both cation and anions is square planar, and the niobium–niobium distance is 2.98 Å. The structure can also be depicted as a framework of Nb_6 octahedra, similar to those found in the lower halides of niobium. Unlike the oxides TiO and VO, the range of homogeneity of NbO is narrow ($0.982 < x < 1.008$ in NbO_x).

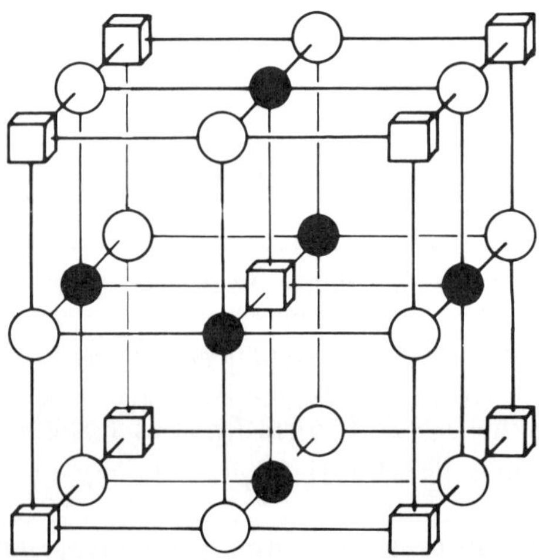

Fig. 4. The structure of NbO. Solid circles, niobium atoms; open circles, oxygen atoms; small cubes, vacant sites.

3. Manganese Oxide, MnO

At room temperature, MnO is a line phase. If it is prepared by low-temperature methods, for example, the thermal decomposition of manganous oxalate under vacuum at 750 K, it is produced in a finely divided form which oxidizes rapidly in air. The product of oxidation is a metastable phase of approximate composition $MnO_{1.13}$ whose defect structure has not yet been established. Samples of MnO prepared under high-temperature conditions, such as the reduction of Mn_3O_4 with hydrogen at 120 K, are indefinitely stable at room temperature.

Deviations from the ideal stoichiometry are of the type $MnO_{1+\delta}$, and such deviations (< 2%) are only observed at temperatures above approximately 1400 K (Fender and Riley, 1970). The partial oxidation of MnO must be accompanied by the formation of Mn^{3+} with its high-spin d^4 configuration. The Jahn–Teller distortion required by this ion could only be accommodated in the rock-salt structure with a certain amount of strain, and this probably accounts for the narrow range of homogeneity. No high-temperature diffraction studies on the defect structure of MnO have been reported.

Substantial deviations from the monoxide stoichiometry are found in solid solutions of MnO with FeO (Tret'yakov *et al.*, 1965), and the defect structure (Hope and Cheetham, 1980) is then similar to that found in FeO itself (Section IV.A.4). It is also worth noting that MnO becomes ordered antiferromagnetically on cooling below 120 K and that the transition is accompanied by a rhombohedral distortion of the structure (Roth, 1958).

4. Ferrous Oxide, FeO

Ferrous oxide is a unique member of the present group. It is a high-temperature phase which disproportionates into iron and Fe_3O_4 on cooling below 840 K and, rather surprisingly, the stoichiometric composition lies outside the phase range at normal pressures. The phase equilibria have been studied by several workers (Darken and Gurry, 1946; Vallet and Raccah, 1964; Fender and Riley 1969), and a typical phase diagram is shown in Fig. 5. The stoichiometric composition has been prepared at 1040 K under a pressure of 36 kbars by Katsura *et al.* (1967).

An interesting aspect of the thermodynamic measurements reported by Vallet and Raccah (1964) and Fender and Riley (1969) is that they both suggest that the ferrous oxide phase is subdivided into three regions (labeled W_1, W_2, and W_3 in Fig. 5) which are separated by second- or higher-order phase transitions (see Chapter 1). However, precise lattice parameter measurements by Hayakawa *et al.* (1972) failed to detect any variations at these boundaries, and none of the diffraction experiments described below has produced any supportive evidence.

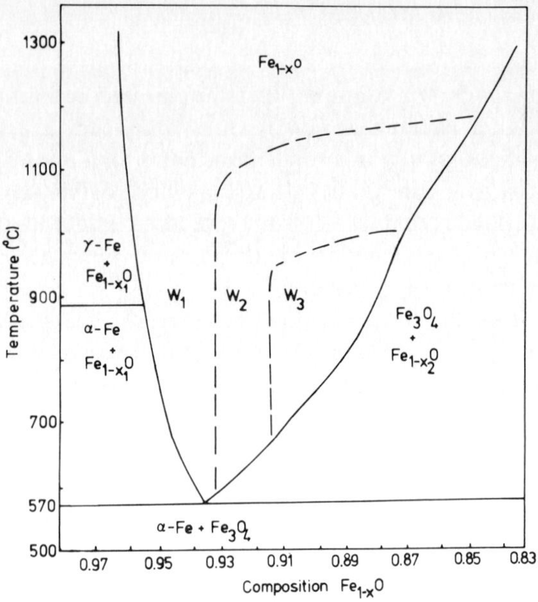

Fig. 5. The ferrous oxide phase diagram.

It can be seen from Fig. 5 that the ferrous oxide phase is grossly iron-deficient and is normally written as $Fe_{1-x}O$. Although early x-ray powder measurements (Jette and Foote, 1933) had indicated a simple iron vacancy defect mechanism, more recent evidence overwhelmingly favors a cluster model. The first indication of this was the detection of tetrahedral interstitials by powder neutron diffraction measurements on samples quenched to room temperature (Roth, 1960). A simple interionic distance calculation shows that such interstitials must always be surrounded by a tetrahedron of octahedral iron vacancies. It is reasonable to assume that the interstitials are Fe^{3+}, and this is confirmed by Mössbauer measurements (Greenwood and Howe, 1972) and theoretical work (Catlow and Fender, 1975).

With the structure represented as $Fe^{oct}_{1-x-y}Fe^{tet}_{y}O$, attention has focused on the ratio of octahedral vacancies $(x + y)$ to tetrahedral interstitials (y). If it is assumed that all the vacancies are associated with interstitials, then the vacancy/interstitial ratio will give a measure of the cluster size. Thus an isolated cluster would have a ratio of 4.0, whereas two such clusters sharing a common corner would have a ratio of 3.5 (Fig. 6). Roth's original experiments yielded a ratio of 2.0, but more recent results suggest a higher value. Koch and Cohen (1969) examined a metastable ordered phase by single-crystal x-ray methods and derived an extended cluster containing 13 vacancies and 4 interstitials (ratio 3.25) as shown in Fig. 26, Chapter 1.

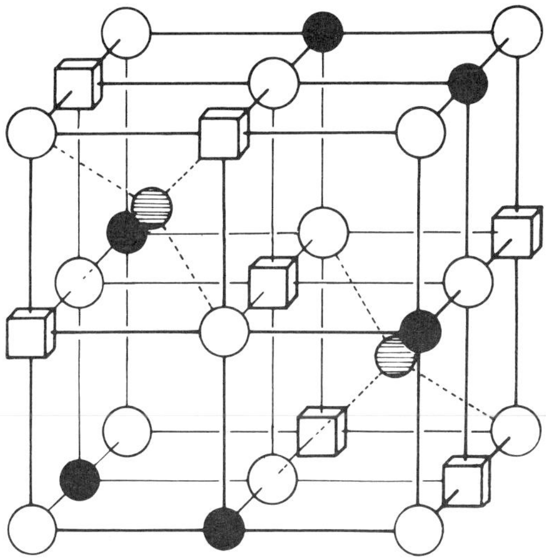

Fig. 6. $Fe_{1-x}O$: two tetrahedral clusters sharing a common corner to give an octahedral vacancy/tetrahedral interstitials ratio of 7 : 2. Solid circles, octahedral iron atoms; hatched circles, tetrahedral iron atoms; open circles, oxygen atoms; small cubes, vacant sites.

Their metastable phase was prepared by quenching a crystal from the equilibrium region to room temperature, although measurements at high temperatures showed that long-range order did not persist under equilibrium conditions.

Powder neutron diffraction studies at high temperatures (Cheetham *et al.*, 1971b) gave ratios in the range 2.8–4.0, but more extensive measurements by this method (Gavarri *et al.*, 1979) gave a ratio of 2.4 which was insensitive to both temperature and composition. A ratio of about 2.6 is indicated by recent room temperature neutron studies on quenched powders (Battle and Cheetham, 1979) and by Seebeck measurements (Gleitzer, 1980). Calculations by Catlow and Fender (1975) are in accord with these recent determinations and point to a condensed cluster model in which tetrahedra share common edges rather than corners. The clusters shown in Figs. 7a and b have ratios of 3.0 and 2.67, respectively.

Metastable phases of the type examined by Koch and Cohen have been studied by electron diffraction and high-resolution electron microscopy (Iijima, 1974; Andersson and Sletnes, 1977), and some interesting results have emerged. Several low-symmetry superstructures have been identified, and it is probable that the cubic superstructure reported by Koch and Cohen is in fact a complex intergrowth of phases with lower symmetry.

A further feature of ferrous oxide that deserves comment is its magnetic properties. Samples that have been quenched to room temperature become antiferromagnetically ordered when cooled below approximately 200 K

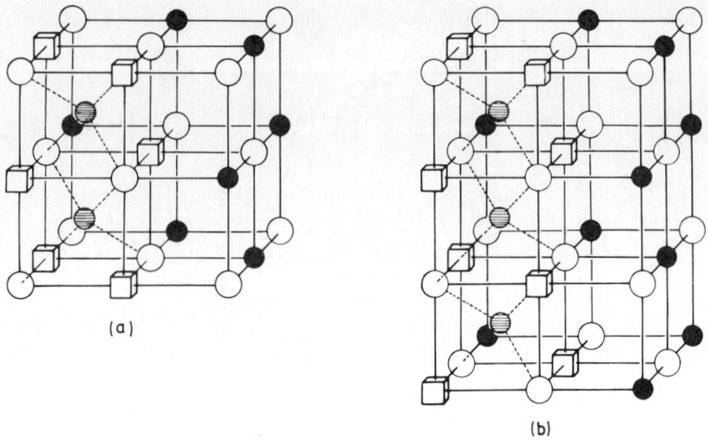

Fig. 7. $Fe_{1-x}O$: tetrahedral clusters sharing common edges. (a) A 6 : 2 cluster; (b) a 8 : 3 cluster. See Fig. 6 for atom types.

(Shull *et al.*, 1951). Susceptibility measurements (Koch and Fine, 1967) show that T_N increases with increasing defect concentration, even though the magnetic scattering of neutrons seems to decrease substantially. Battle and Cheetham (1979) have accounted for these observations with a model in which the ions within each cluster are antiferromagnetically aligned but do not contribute to the magnetic Bragg scattering because the clusters are not arranged in a regular, periodic manner. The magnetic scattering results are also compatible with clusters as shown in Fig. 7.

5. *Cobalt Oxide, CoO, and Nickel Oxide, NiO*

Unlike ferrous oxide, CoO and NiO show only small deviations from their ideal compositions, below the level that would be detectable by diffraction methods. Magnetic susceptibility and neutron diffraction studies have shown that both oxides order antiferromagnetically with T_N values of 289 and 523 K, respectively (Foex, 1948). The magnetic ordering is apparently accompanied by a lowering of symmetry to tetragonal for CoO (Shull *et al.*, 1951), although more recent x-ray measurements (Saito *et al.*, 1966) have revealed a small rhombohedral component of the distortion so that the true crystal symmetry at low temperatures is monoclinic.

It is worth noting that a second form of CoO has been reported by Ok and Mullen (1968, 1969), CoO(II) also has the rock-salt structure, but density measurements indicate approximately 25% vacancies in each sublattice.

Unlike the other monoxides, NiO is antiferromagnetic at room temperature and the magnetic structure is similar to that of MnO, except that the resultant rhombohedral distortion is very much smaller. NiO can take up to 40 mole % Li_2O into solid solution, but the monoxide stoichiometry is retained by the formation of corresponding amounts of Ni^{3+}: $Li_x^+Ni_{1-2x}^{2+}Ni_x^{3+}O$. The rhombohedral distortion diminishes as the Li content increases, and for $x > 0.12$ the symmetry is cubic (Toussaint and Vos, 1968).

B. Phases with the Fluorite Structure

Dioxides with the fluorite structure are found primarily with lanthanide and actinide metals. Of the lanthanide elements, only cerium, praseodymium, and terbium readily adopt the IV oxidation state. Their dioxides are well documented, and each can accommodate a substantial concentration of oxygen vacancies. In the actinide series, fluorite dioxides are formed by the elements thorium through californium and the ability of the fluorite structure to carry large deviations from its ideal stoichiometry is seen here with UO_2 which readily becomes oxygen-rich, and PuO_2, AmO_2, and CfO_2 which can be oxygen-deficient. It seems likely that the latter systems are qualitatively similar to lanthanide oxides.

1. Anion-Deficient Fluorites

The phase diagram for the Pr–O system is shown in Fig. 8 (Hyde *et al.*, 1966). While the other anion-deficient systems may differ in detail, this example will serve to illustrate the general properties of these systems. Above 875 K, there is a nonstoichiometric phase (α) extending from PrO_2 to approximately $PrO_{1.75}$. There have been no high-temperature diffraction studies on this particular system, but analogy with the ternary systems described below suggests that the crystal defects are incorporated simply as anion vacancies. If the relaxation effects around the defects are small, and this is believed to be the case, then the diffraction experiment reveals only the number of oxygen vacancies. This is of limited value since we are primarily interested in the distribution of the defects which can only then be obtained from an examination of the diffuse scattering or from techniques that monitor the local environment such as extended x-ray absorption fine structure (EXAFS) and resonance methods.

At temperatures below 875 K, vacancy ordering takes place with the formation of a homologous series of phases Pr_nO_{2n-2}, where $n = 7$, 9, 10, 11, and 12. The structure of one of these ordered phases, Pr_7O_{12}, has been examined by both x-ray (Eyring and Baenziger, 1962) and neutron diffraction (Von Dreele *et al.*, 1975); it appears to be isostructural with UY_6O_{12} (Bartram, 1966). An element of this ordered derivative is shown in Fig. 9,

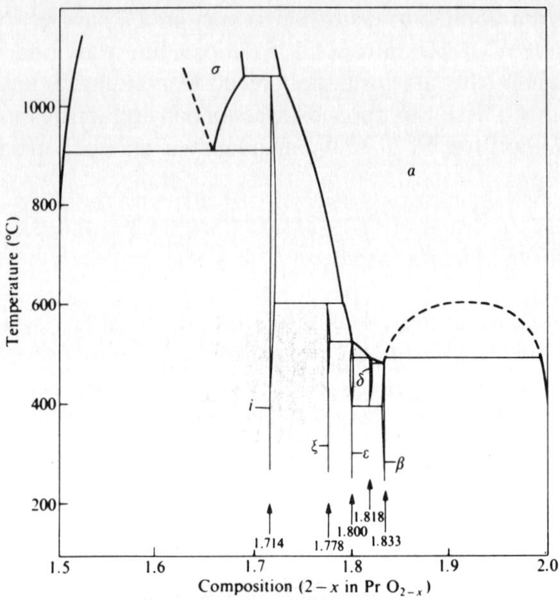

Fig. 8. PrO_{2-x} phase diagram. (After Cheetham, 1972.)

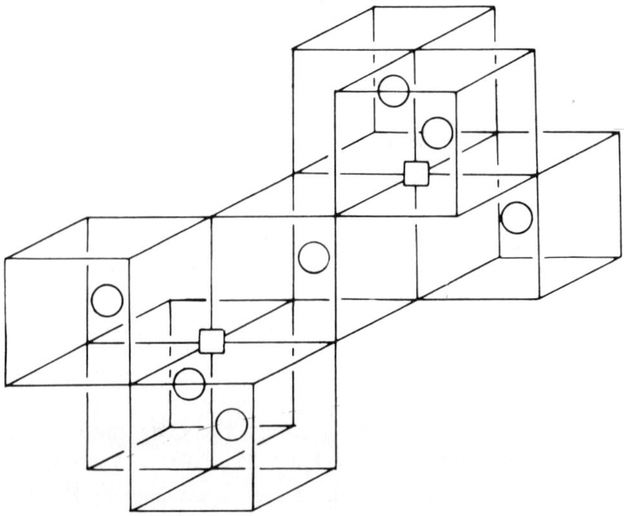

Fig. 9. A section of the ordered Pr_7O_{12} structure. Oxygen atoms are at the corners of the large cubes, small cubes represent oxygen vacancies along ⟨111⟩, and circles respresent cations at 6- and 7-coordinated sites.

and it is widely held that this unit, with oxygen vacancies in chains along ⟨111⟩ of the fluorite cell, represents the basic building block for this family of phases. However, confirmation of this through structure determinations for other members of the series has not yet been achieved in binary systems. For example, attempts to grow homogeneous single crystals of $Pr_{12}O_{22}$ were fraught with difficulties, and determination of the structure posed insurmountable problems (Lowenstein *et al.*, 1972). More success has been achieved using electron microscopy (Summerville *et al.*, 1978), and the results are discussed in Chapter 7 of this book.

Ternary oxides with anion-deficient fluorite structures, including ThO_2–Y_2O_3, ZrO_2–Y_2O_3, and ZrO_2–CaO, have been extensively studied. ZrO_2–CaO is, of course, widely known as cubic stabilized zirconia and is used as a solid electrolyte in high-temperature fuel cells. ZrO_2 only adopts the fluorite structure at very high temperatures (Cohen and Schaner, 1963; Smith and Cline, 1962), but the cubic phase is extended to lower temperatures by the introduction of rare earth sesquioxides or CaO. Both non-stoichiometric and ordered phases are again found, and here we illustrate the results by reference to ThO_2–Y_2O_3, ZrO_2–CaO, and ZrO_2–Sc_2O_3.

Steele (1971) examined thoria and zirconia solid solutions using powder neutron diffraction and the diffuse scattering of long-wavelength neutrons. The diffraction measurements on ThO_2–Y_2O_3 showed a random distribution of thorium and yttrium with vacancies in the oxygen sublattice, but no atomic displacements or new sites were observed. An identical model was proposed by Subbarao *et al.* (1965) on the basis of precise lattice parameter and density measurements. The diffuse scattering measurements of Steele are also consistent with a random distribution of defects. In the system CeO_2–Y_2O_3, diffraction measurements again indicated a random model, but the diffuse scattering studies revealed a significant degree of short-range ordering. The incoherent scattering is compatible with a model in which pairs of oxygen vacancies, lying along ⟨111⟩, are associated with charge-compensating Y^{3+} cations. The similarity between this model and the structure of Pr_7O_{12} (Fig. 9) is very clear.

Bragg measurements on zirconia solid solutions reveal a more complex behavior with significant relaxation effects around the defects. In ZrO_2–CaO, for example, although lattice parameter–density studies (Tien and Subbarao, 1963) again indicate a simple oxygen vacancy model, structure refinements based on powder neutron diffraction data reveal unexpected displacements of oxygen atoms from their normal positions (Carter and Roth, 1968). At lower temperatures, these displacements assume a regular arrangement, presumably in association with ordered vacancies, and superlattice reflections are observed. More detailed neutron studies by Steele and Fender (1974) suggest that only oxygen atoms adjacent to anion vacancies are displaced. These displacements appear to be characteristic of ZrO_2 solid solutions and are presumably a result of the small size of the Zr^{4+} cation.

The crystal structures of two ordered phases in the ZrO_2–Sc_2O_3 system have been determined by Thornber *et al.* (1968): $Zr_3Sc_4O_{12}$ (the δ phase) and $Zr_{10}Sc_4O_{26}$ (the γ phase). Both structures are based on a rhombohedrally distored fluorite cell. The δ phase is similar to Pr_7O_{12}, with chains of vacancies lying along $\langle 111 \rangle$, and in the γ phase the oxygen vacancies occur in pairs across the body diagonal of alternate MO_8 cubes along the threefold axis. The confirmation of this unit in both ZrO_2–Sc_2O_3 and CeO_2–Y_2O_3 lends support to the hypothesis that it is the fundamental building block of anion-deficient fluorite systems.

2. *Anion-Excess Fluorites*

The incorporation of excess anions into the fluorite structure is a common feature in fluorides, for example, EuF_{2+x} (Catalano *et al.*, 1969) and a wide range of ternary solid solutions of the type MF_2–LnF_3 (M = alkaline earth; Ln = lanthanide, yttrium, or bismuth) (Ippolitov *et al.*, 1967). In binary oxides, however, this behavior seems restricted to the system UO_{2+x}. The U–O phase diagram (Fig. 10) has been studied in great detail (Keller, 1973) and reveals that, above approximately 675 K, UO_2 can accommodate further oxygen with a phase limit of $UO_{2.20}$ at 1400°C.

The structure of this nonstoichiometric phase was studied by single-crystal neutron diffraction at 800°C by Willis (1963). He found evidence for two types of interstitials, one (O′) displaced from the $\frac{1}{2},\frac{1}{2},\frac{1}{2}$ position of

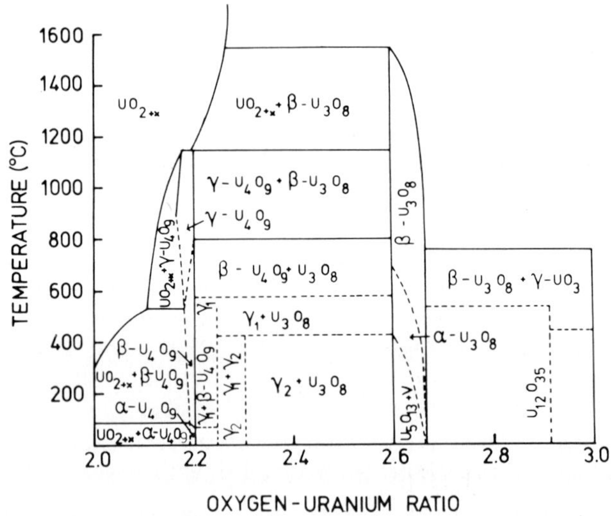

Fig. 10. The UO_{2+x} phase diagram. (After Keller, 1973.)

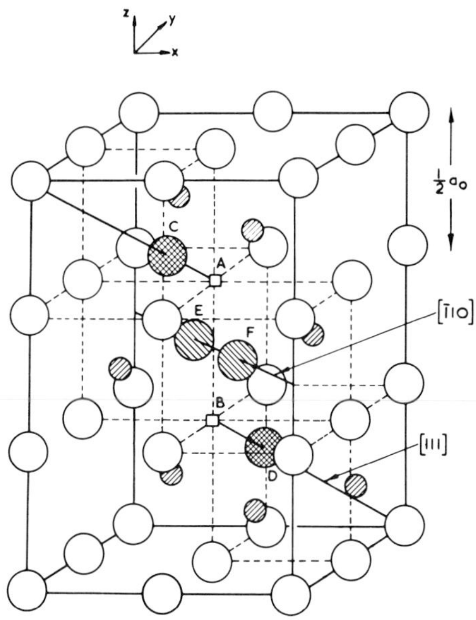

Fig. 11. The 2 : 2 : 2 cluster in UO_{2+x}. Large open circles, oxygen atoms; large hatched circles, O′ and O″ interstitials; small circles, uranium atoms.

the fluorite along ⟨110⟩ and the other (O″) displaced along ⟨111⟩. A significant concentration of anion vacancies was also observed. On the basis of the measured occupancy numbers, Willis proposed a defect cluster containing two O′ interstitials, two O″ interestitials, and two normal oxygen vacancies; this is commonly known as the 2 : 2 : 2 or Willis cluster (Fig. 11). There is also a phase with the composition U_4O_9 which appears to be an ordered derivative of UO_{2+x}; it is certainly related to fluorite, and refinements of the average unit cell using only fluorite reflections confirm that it contains the same structural features as UO_{2+x} (Willis, 1964). The structure of U_4O_9 contains the key to this system, but it has not yet been satisfactorily determined. More recent neutron experiments (Cheetham *et al.*, 1971a; Steele *et al.*, 1972) have shown that CaF_2–YF_3 is very similar to UO_{2+x}, and theoretical studies appear to confirm the validity of the 2 : 2 : 2 cluster. Spectroscopic measurements (Tallant and Wright, 1975; Baker and Chaplin, 1971) also indicate low symmetry in the vicinity of the defects.

One rather puzzling aspect of the 2:2:2 cluster is the very short oxygen–oxygen contact (approximately 2.1 Å) between the O′ interstitials. Recent structure determinations of ordered fluorides including KY_3F_{10} (Pierce and Hong, 1974) reveal a model that does not contain this short anion–anion contact, although a relationship to the Willis cluster can be seen, and it is my view that, in the near future, new clusters will replace the currently accepted models for anion-excess fluorites.

V. Crystallographic Shear Structures

In contrast to the compounds described above, there is a class of oxides in which deviations from ideal stoichiometry are accommodated not by vacancies, interstitials or defect clusters, but by the formation of so-called crystallographic shear (CS) planes. The principle is best illustrated by means of an example. The reduction of MoO_3 at high temperatures leads to the formation of a series of ordered phases of which one member, Mo_8O_{23}, is shown in Fig. 12. The structure is based on that of ReO_3, in which MO_6 octahedra are linked in three dimensions by corner sharing. The deficiency of oxygen in reduced MoO_3 is incorporated by the formation of planes of edge-sharing octahedra. These shear planes adopt a specific orientation, in this case {120}, and are regularly spaced. Other members of the series, which can be described by the general formula Mo_nO_{3n-1}, contain CS planes in the same orientation but with different spacings.

The shear phases are not strictly nonstoichiometric in either a structural or a thermodynamic sense, and each member of the series has a well-defined ordered structure with a specific, albeit obscure, stoichiometry. However, since the formation of CS planes is a common mechanism by which simple compounds may deviate from their ideal composition, shear structures are normally considered under the broad heading of non-stoichiometry. This behavior is a characteristic of the systems TiO_{2-x},

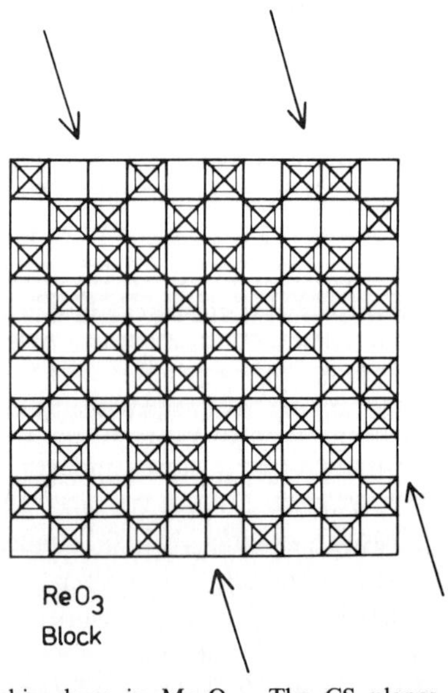

Fig. 12. Crystallographic shear in Mo_8O_{23}. The CS planes, across which the MoO_6 octahedra share edges, are indicated by arrows.

VO_{2-x}, CrO_{2-x}, MoO_{3-x}, and WO_{3-x}. All these oxides contain cations in high, formal oxidation states, and it is probable that the driving force for the formation of CS planes is the need to maintain octahedral coordination around the cation. By comparison, the formation of oxygen vacancies would be accompanied by a lowering of the cation coordination number from 6 to 5 and the breaking of a metal–oxygen bond which has a high degree of covalency. Binary fluorides with rutile (MF_2) or ReO_3 (MF_3) structures are not known to form CS planes, but these compounds of course contain cations with lower charges.

There has been considerable interest in the mechanism by which CS planes are formed (Gado, 1965; Andersson and Wadsley, 1966; Anderson and Hyde, 1967), but most of the direct evidence for or against the alternative mechanisms has been obtained by high-resolution electron microscopy. The nature of the interactions between the planes has also been a subject of debate and is discussed in Chapter 2.

Much of the recent structural work on these systems has been carried out by electron microscopy because the shear structures are essentially two-dimensional in nature and usually yield high-resolution images which can be interpreted in a simple manner. This area is taken up in some detail in Chapter 7. However, our present understanding of the structural principles of shear phases stems from a series of x-ray structure determinations by Magnéli, and in recognition of his unique contribution to this field, these compounds are now commonly referred to as Magnéli phases. Although most of our detailed structural knowledge has emerged from x-ray studies, neutrons too have played a role, particularly in the determination of cation distributions in ternary oxides. In the following sections, the crystallographic properties of shear structure based on ReO_3 and rutile are discussed in more detail.

A. MoO_3 and WO_3

An early investiation of the WO_{3-x} system (Glemser and Sauer, 1943) indicated three nonstoichiometric phases: α-WO_x ($2.98 < x < 3.0$), β-WO_x ($2.83 < x < 2.90$), and γ-WO_x ($2.66 < x < 2.72$). In contrast, Magnéli (1946) found only stoichiometric compounds: WO_3 and the reduced phases $W_{20}O_{58}$ and $W_{18}O_{49}$. His structural study (Magnéli, 1950) confirmed this view and, together with concurrent studies on reduced MoO_3 phases, revealed the new concept of crystallographic shear.

In $W_{20}O_{58}$, the CS planes are oriented along $\{130\}$ in the ReO_3 framework, and subsequent structural studies in the composition range WO_3 to $WO_{2.90}$ have revealed the members $n = 20$, 39, 40, and 50 in a homologous

TABLE I

Phases in the Homologous Series M_nO_{3n-2} That Have Been Characterized by Single-Crystal X-Ray Methods

Composition	Formula	n	Reference
$MO_{2.90}$	$W_{20}O_{58}$	20	Magnéli (1950)
$MO_{2.949}$	$W_{35}Ta_4O_{115}$	39	Gado *et al.* (1965)
$MO_{2.95}$	$W_{40}O_{118}$	40	Gado and Magnéli (1965)
$MO_{2.96}$	$W_{50}O_{148}$	50	Gebert and Ackermann (1966)

series W_nO_{3n-2} (Table I). More recent electron microscopy studies have demonstrated the ease with which neighboring members of this series may intergrow, and it is hardly surprising that homogeneous single crystals suitable for x-ray structure determination are difficult to obtain. It should be emphasized again, therefore, that the structural principles in these systems, which form the basis of the interpretation of recent electron optical studies, hinge entirely on these single-crystal x-ray studies.

MoO_3 itself does not adopt the ReO_3 structure but prefers a layer structure containing double layers of MoO_6 octahedra sharing corners and edges (Kihlborg, 1963a). In an earlier study, Glemser and Lutz (1947) found two nonstoichiometric phases on reduction, α-MoO_x ($2.95 < x < 3.0$) and β-MoO_x ($2.65 < x < 2.75$), but, as in the tungsten system, Magnéli (1946) found only stoichiometric phases: MoO_3, Mo_9O_{26}, Mo_8O_{23}, and Mo_4O_{11}. Single-crystal x-ray studies (Magnéli, 1948) have revealed that Mo_9O_{26} and Mo_8O_{23} belong to a family of shear structures based on ReO_3. Here the orientation of the CS planes is $\{120\}$ and the phases belong to a homologous series, Mo_nO_{3n-1}; the $n = 10$ and 11 members are known in the mixed oxides $(Mo_{0.85}W_{0.15})_{10}O_{29}$ and $(Mo_{0.5}W_{0.5})_{11}O_{32}$ (Blomberg *et al.*, 1953).

It is known that the MoO_{3-x} system accommodates CS planes in an ReO_3-based structure only above approximately 975 K, and below this temperature, the oxygen deficiency is incorporated by means of crystallographic shear within the MoO_3 layer structure; a typical example is shown in Fig. 13. Once more, a homologous series is generated, Mo_nO_{3n-2}, but it is based upon a $\{311\}$ CS plane in MoO_3. Several single-crystal studies have been reported (Kihlborg, 1963b).

B. Rutile-Based Shear Structures

The rutile phases TiO_2, VO_2, and CrO_2 also form CS planes on reduction. In the rutile structure, chains in which MO_6 octahedra share edges are linked by corner sharing and the oxygen deficit is sustained by the

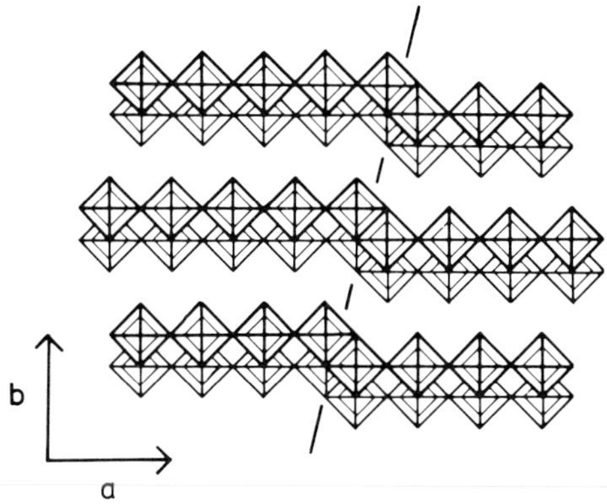

Fig. 13. Crystallographic shear in the layer structure of MoO_3.

introduction of face-sharing octahedra. This leads to some very short metal–metal contacts in the idealized structures, and there has been much interest in the role of the lower-valent cations in these systems. One relevant question is: Do the V^{3+} ions in VO_{2-x} form metal–metal bonds across the shear plane?

The existence of a series of phases in the TiO_{2-x} system was discovered by Andersson and Magnéli (1956) who found a homologous series, Ti_nO_{2n-1}, with $n = 4$ to 10 in the composition range $TiO_{1.75}$ to $TiO_{1.90}$. A single-crystal x-ray study on Ti_5O_9 (Andersson, 1960) revealed regularly spaced CS planes along $\{1\bar{2}1\}_{TiO_2}$, and for other members of the series the shear plane orientation remains the same. A similar series with the general formula V_nO_{2n-1} was subsequently found in the vanadium–oxygen system (Andersson *et al.*, 1957).

An interesting aspect of reduced titanium and vanadium oxides is that they undergo metal→semiconductor transitions on cooling below room temperature (Bartholomew and Frankl, 1969). Marezio and co-workers (Marezio *et al.*, 1973a, b, 1974, 1977) have carried out careful single-crystal x-ray studies on oxides in these systems at a range of temperatures, both above and below the phase transitions, and their experiments are summarized in Table II. The Ti_nO_{2n-1} phases exhibit two closely spaced transitions. In Ti_4O_7, for example, the Ti^{3+} and Ti^{4+} cations are localized in the low-temperature phase, and pairing of the Ti^{3+} ions in the chains along the pseudo rutile *c* axis is observed (Fig. 14). The formation of discreet metal–metal bonds is also found in several stoichiometric dioxides, for example VO_2, NbO_2, and MoO_2, and is often accompanied by a metal →semiconductor transition. Marezio's results indicate that metal–metal bonding takes place between MO_6 octahedra sharing edges within the rutile layers rather than across the shear planes where the octahedra share faces.

TABLE II

Single-Crystal X-Ray Studies on Ti_nO_{2n-1}, V_nO_{2n-1}, and Related Phases

Compound	Transition temperature (K)	Reference
Ti_4O_7	150, 125	Marezio *et al.* (1973a)
Ti_5O_9	140, 130	Marezio *et al.* (1977)
Ti_4O_7 (containing 0.25% V)	150, 125	Hodeau and Marezio (1979)
V_4O_7	250	Marezio *et al.* (1973b)
		Horiuchi *et al.* (1972)
		Hodeau and Marezio (1978)
V_5O_9	135	Marezio *et al.* (1974)

This is a surprising result and shows that the cation distribution is not determined by coulombic factors. A displacement of Ti^{4+} toward oxygen is also observed in the low-temperature phase, suggesting that the transition has an antiferroelectric component. The trivalent and tetravalent cations are randomly distributed in the high-temperature, metallic phase. The intermediate phase has also been examined, but the structure determination does not reveal the origin of the double transition. There is no apparent pairing of the Ti^{3+} ions in this phase. The structural properties of the three phases of Ti_5O_9 are very similar, but conductivity measurements show that these phases are all semiconducting, presumably because the composition ($TiO_{1.80}$) approaches that of the insulating end member, TiO_2. The behavior of Ti_5O_9 precludes the thesis that the phase transitions in these compounds are driven entirely by a change from metallic conductivity to semiconductivity.

The vanadium phases V_4O_7 and V_5O_9 exhibit single transitions that coincide with marked changes in conductivity. In V_4O_7, some V^{3+}–V^{4+} ordering is apparent in the metallic phase and ordering appears to be complete by the transition temperature. Some V–V pairing is apparent in the high-temperature phase, and it increases gradually on cooling to low temperatures. For example, the V(1)–V(1) distances are 2.793, 2.687, and 2.669 Å at 295, 200, and 120 K, respectively. Charge ordering takes place at the metal conductivity–semiconductivity transition in V_5O_9, but there is no evidence in this case for V–V pairing below the transition temperature.

The distribution of trivalent and tetravalent cations in these structures is determined from the average metal–oxygen distance at each site. For example, since M^{3+} is larger than M^{4+}, it is assumed to occupy the sites with the larger M–O contacts. The MO_6 octahedra in these structures are grossly distorted, and the M–O contacts around each cation show a wide variation. One cannot therefore be certain that this method will define the cation oxidation state unambiguously. A more definitive cation distribution

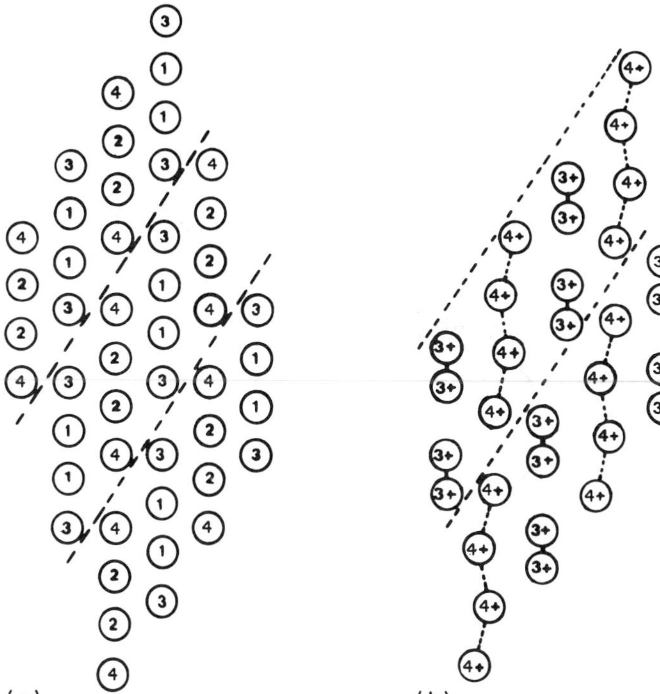

(a) (b)

Fig. 14. The structure of Ti_4O_7 showing only titanium atoms. (a) In the high-temperature phase the cations form chains M3–M1–M1–M3 and M4–M2–M2–M4 along the pseudo rutile direction [001]. (b) In the low-temperature phase Ti^{3+} and Ti^{4+} segregate in alternate chains along [001]. The shear planes are indicated by broken lines.

can be obtained in ternary systems, and an example will be described in Section VI.

A second series of shear structures has been observed in the Ti–O system in the composition range $TiO_{1.94}$ to $TiO_{2.00}$. In this new series, the orientation of the CS planes is $\{1\bar{3}2\}_{TiO_2}$, and the general formula is again Ti_nO_{2n-1}, with $15 < n < 36$ (Bursill and Hyde, 1971). These phases were observed by electron microscopy, and it has not proved possible to obtain monophasic crystals suitable for x-ray studies.

VI. Block Structures

Single-crystal x-ray studies have made a vital contribution to our understanding of the niobium oxides and related systems. Early thermodynamic and powder x-ray diffraction studies by Brauer (1941) revealed three phases of Nb_2O_5 itself and suggested a range of nonstoichiometry for the high-

Fig. 15. The idealized structure of PNb_9O_{25}. The MO_6 octahedra with thick and thin outlines have M at $z = 0$ and $z = \frac{1}{2}$, respectively. Solid circles represent tetrahedral tunnel sites. (After Cheetham and Taylor, 1977.)

temperature modification between $NbO_{2.40}$ and $NbO_{2.50}$. The x-ray studies of Wadsley, Andersson, Gruehn, and others have since revealed a kaleidoscope of ordered phases based on the shear structure principle. Here there are *two* sets of parallel CS planes separating columns or blocks of the ReO_3 structure. These compounds are normally referred to as block structures, and a typical example, PNb_9O_{25}, is shown in Fig. 15. Blocks of ReO_3, 3×3 octahedra in cross section and infinite along the direction of projection, are linked by edge-sharing octahedra. In this example, phosphorus atoms occupy the tetrahedral sites at the intersection of four blocks. Table III lists the binary and ternary niobium oxides and oxofluorides that have been shown by single-crystal x-ray methods to adopt block structures. The most striking feature of these compounds is the manner in which the block structures adapt to subtle changes in stoichiometry, and this can be controlled by altering the size of the ReO_3 block, its mode of linkage, or both. The structural chemistry of these materials has been reviewed extensively by Wadsley and Andersson (1969), so in this chapter it is only appropriate to describe some more recent structural contributions.

In the period 1960–1970, the underlying principles of the block structures were established by x-ray methods. These included not only the basic structure types, but confirmation that regular intergrowth of the simpler phases could lead to new stoichiometric compounds. For example, Andersson *et al.* (1966) showed that the structure of $W_4Nb_{26}O_{77}$ corresponds to an ordered intergrowth of $WNb_{12}O_{33}$ and $W_3Nb_{14}O_{44}$, and Wadsley and

TABLE III

Block Structures Based upon Niobium Oxides: Single-Crystal Structure Determinations Using X Rays

Anion/cation ratio	Unit cell formula	Compound	Reference(s)
2.4167	$M_{12}X_{29}$	$Nb_{12}O_{29}$	Norin (1963, 1966)
2.4167	$M_{12}X_{29}$	$Ti_2Nb_{10}O_{29}$	Wadsley (1961)
2.4667	$M_{15}X_{37}$	$Nb_{14}MgO_{35}F_2$	Lundberg (1970)
2.4681	$M_{47}X_{116}$	$Nb_{47}O_{116}$	Gruehn and Norin (1969)
2.4800	$M_{25}X_{62}$	$TiNb_{24}O_{62}$	Roth and Wadsley (1965a)
2.4906	$M_{53}X_{132}$	$Nb_{53}O_{132}$	Gruehn and Norin (1967)
2.5000	$M_{28}X_{70}$	H-Nb_2O_5	Gatehouse and Wadsley (1964)
2.5000	$M_{16}X_{40}$	N-Nb_2O_5	Andersson (1967)
2.5000	$M_{10}X_{25}$	PNb_9O_{25}	Roth *et al.* (1965)
2.5161	$M_{31}X_{78}$	$Nb_{31}O_{77}F$	Åström (1966)
2.5294	$M_{34}X_{86}$	$Nb_{17}O_{42}F$	Åström (1966)
2.5385	$M_{13}X_{33}$	$WNb_{12}O_{33}$	Roth and Wadsley (1965b)
2.5625	$M_{16}X_{41}$	$Nb_{15}MoO_{40}F$	Galy and Andersson (1968)
2.5667	$M_{30}X_{77}$	$W_4Nb_{26}O_{77}$	Andersson *et al.* (1966)
2.5882	$M_{17}X_{44}$	$W_3Nb_{14}O_{44}$	Roth and Wadsley (1965c)
2.6190	$M_{21}X_{55}$	$W_5Nb_{16}O_{55}$	Roth and Wadsley (1965b)
2.6538	$M_{26}O_{69}$	$W_8Nb_{18}O_{69}$	Roth and Wadsley (1965c)

Andersson (1970) have discussed other niobium oxides that can be described in a similar manner. The 1970s have seen many advances in this field, mainly using electron microscopy. Like the shear phases, the block structures are simple in projection and yield images that are normally straightforward to interpret. Careful studies have revealed domains of new intergrowth structures that cannot be obtained as homogeneous phases, but conventional diffraction methods have continued to play a role, and some recent experiments on germanium– and titanium–niobium oxides are described below.

The story of germanium–niobium oxide, $GeO_2 \cdot 9Nb_2O_5$, is particularly apposite in the context of this chapter, since it highlights one of the advantages of single-crystal diffraction techniques over electron optical methods. Electron micrographs and x-ray powder patterns of this material show very clearly that it adopts essentially the same structure as PNb_9O_{25} (Fig. 15) even though it has a different composition. This is surprising in view of the ability of block structures, emphasized in Table III, to adapt to small changes in composition. Electron microscopy studies do not reveal how the deficiency of oxygen compared with the ideal composition, $M_{10}O_{25}$, is accommodated, whether by oxygen vacancies or metal interstitials. In a very careful study using x rays and neutrons, Anderson *et al.* (1975) discovered a new octahedral site in the channels occupied by the tetrahedral interstitials. The occupancy of the new site is only approximately 10%, and it is hardly surprising that it could only be detected in Fourier maps calculated from single-crystal x-ray data. The occupation of both octahedral and tetrahedral sites in the channels described above permits the incorporation of excess cations, allowing deviations from the ideal $M_{10}O_{25}$ stoichiometry. It has been suggested by Anderson *et al.* (1973) that this mechanism establishes a basis for metal-excess block structures, which can be extended from the particular example of $GeO_2 \cdot 9Nb_2O_5$ to the other apparently nonstoichiometric niobium oxides reviewed by Gruehn (1972).

An important area in which x-ray and neutron diffraction have made unique contributions is in the determination of cation distributions in ternary niobium oxides. Electron microscopy is not yet sufficiently sensitive to detect the subtle differences in contrast that might be expected to arise from nonrandom cation distributions within the block, but neutron diffraction is a particularly sensitive probe for the distribution of cations in titanium–niobium oxides since the neutron-scattering lengths of titanium and niobium are different in sign ($b_{Ti} = -0.36 \times 10^{-14}$ m; $b_{Nb} = +0.691 \times 10^{-14}$ m). The crystal structures of $TiNb_2O_7$ and $Ti_2Nb_{10}O_{29}$ were originally determined by Wadsley (1961) using x-rays, but no ordering of the Ti^{4+} and Nb^{5+} cations was detected. In a powder neutron diffraction study of these compounds using profile analysis methods, Cheetham and Von Dreele (1974) were able to determine the percentage of titanium at each cation site with high precision. The idealized structure of $TiNb_2O_7$ is shown in Fig. 16. Blocks 3×3 octahedra in cross section form ribbons by edge sharing with identical blocks at the same level, $z = 0$; these ribbons also share octahedral edges with similar ribbons at $z = \frac{1}{2}$. The observed

Fig. 16. The idealized structure of $TiNb_2O_7$. The MO_6 octahedra with thick and thin outlines have M at $z = 0$ and $z = \frac{1}{2}$, respectively. Metal sites M1 to M5 are labeled. (After Cheetham and Taylor, 1977.)

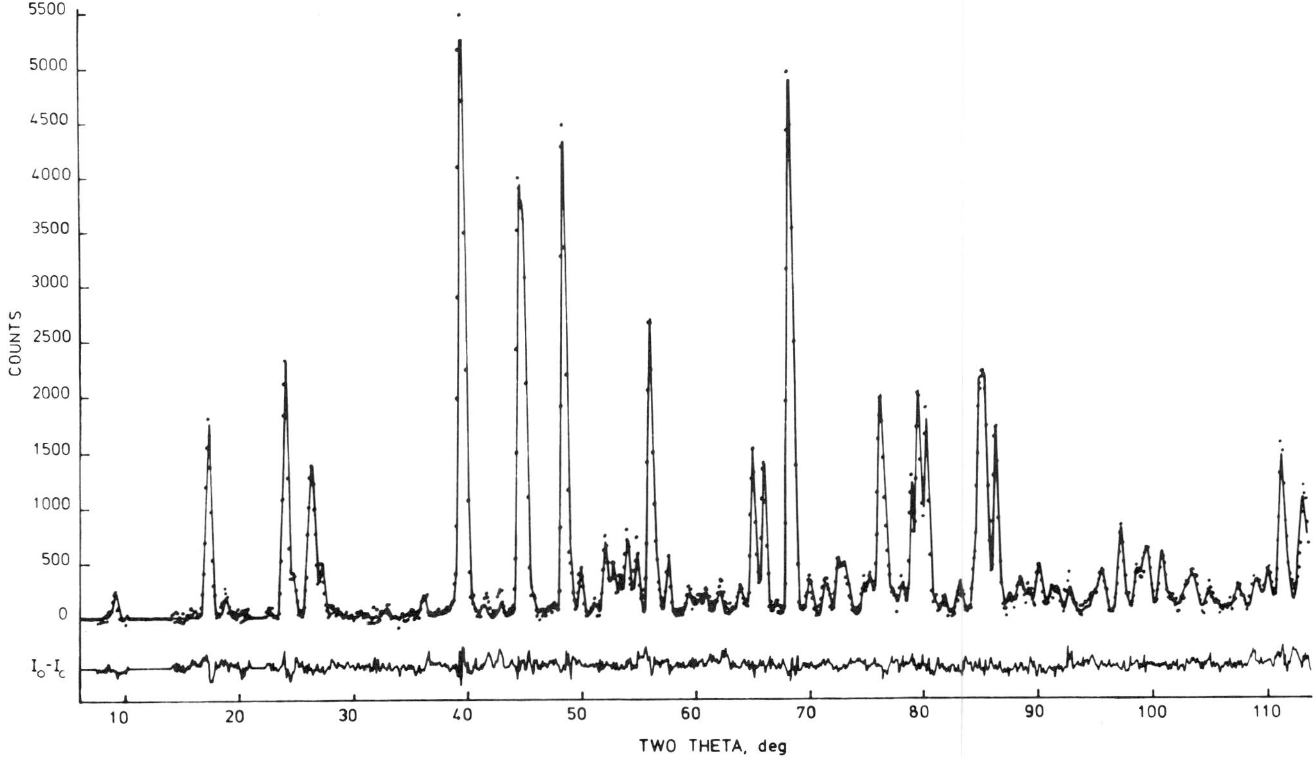

Fig. 17. Observed (dots) and calculated (smooth curve) powder neutron diffraction profiles of $TiNb_2O_7$. A difference curve is also shown. Six hundred and three allowed reflections contribute to the profile. (After Cheetham and Taylor, 1977.)

TABLE IV

Charge Distribution and Site Potentials in $TiNb_2O_7$

Atom	Titanium (%)	Average site charge	Site potential: Obs. dist.	Site potential: Random dist.
M1	33.8 (1.1)	+ 4.66	− 3.25	− 3.20
M2	64.5 (1.3)	+ 4.35	− 3.16	− 3.19
M3	26.0 (1.3)	+ 4.74	− 3.44	− 3.35
M4	14.0 (1.9)	+ 4.86	− 3.99	− 4.04
M5	20.6 (1.4)	+ 4.79	− 3.56	− 3.59

and calculated neutron profiles are shown in Fig. 17, and the percentage of titanium at each metal site is given in Table IV.

The lowest concentration of Ti^{4+} is found at site M4 in the center of the block, in an octahedron that shares only corners, as in ReO_3. The highest concentration of Ti^{4+} appears at site M2 in the corner of the block, a site that participates in both sets of CS planes. The coulombic potential at each site was calculated using both the observed cation distribution and a random model (Table IV), and it was found that the ranking order of titanium occupancies could be reproduced by this simple ionic model. A similar result was obtained for $Ti_2Nb_{10}O_{29}$. These results, like those of Marezio on Ti_nO_{2n-1}, demonstrate that cation ordering is an important feature in shear structure compounds and must contribute significantly to the lattice energies and heats of formation.

VII. Infinitely Adaptive Structures

In Sections V and VI, the architecture of materials in which variations in composition are achieved by changing the spacing between regular shear planes is described. This behavior leads to homologous series of shear structures which are close in composition, but biphasic regions in which two adjacent members coexist can still be observed in some instances. Alternatively, domains of the adjacent phases may intergrow coherently in a disordered manner. The 1970s have seen the discovery of a new mechanism for accommodating changes in stoichiometry whereby the structure adapts uniquely to any particular composition; these compounds are known as "infinitely adaptive structures" (Anderson, 1973). Every composition becomes ordered into a superstructure that is characteristic of that composition only, so that a *continuum* of closely spaced structures is generated without intervening two-phase regions. Such a system exhibits pseudobivariant behavior, so that it cannot be diagnosed by thermody-

TABLE V

Systems in Which Infinitely Adaptive Structures Are Observed

System	Reference
Silicides, etc., based on $TiSi_2$ structure	Nowotny (1970)
Bi–Te (50–57% Te)	Brebrick (1970)
$Ba_{1+x}Fe_2S_4$ $(0.062 < x < 0.143)$	Grey (1974)
$Y(O/F)_{2+x}$ $(2.13 < x < 2.22)$	Mann and Bevan (1972)
Ta_2O_5–WO_3	Roth *et al.* (1970); Stephenson and Roth (1971a, b, c, d)
TiO_2–Cr_2O_3	Bursill *et al.* (1971)

namic means, and our present knowledge in this area has been obtained almost entirely by structural methods. The range of materials that exhibit this phenomenon embraces alloys, chalcogenides, oxofluorides, and oxides (Table V), but the systems of particular relevance to the present chapter are Ta_2O_5–WO_3, $Y(O/F)_{2+x}$, and TiO_2–Cr_2O_3.

The existence of a continuum of structures in the composition range Ta_2O_5 to $11Ta_2O_5 \cdot 4WO_3$ was first reported by Roth *et al.* (1970). The basic structure is related to that of U_3O_8 (Andresen, 1958), in which an orthorhombic unit cell is built up from edge-sharing pentagonal bipyramids and distorted octahedra. The superstructures in the system described by Roth *et al.* are also orthorhombic, with approximately constant a and c lattice parameters ($\sim$6.2 and $\sim$3.9 Å: respectively) but with a b parameter that is a multiple of $b' \sim 3.65$ Å. The structural basis for the continuum of phases was revealed in a series of single-crystal x-ray studies by Stephenson and Roth (1971a, b, c, d). These authors proposed a number of basic superstructures with $b = mb'$, where $m = 5$, 8, and 11; the units with $m = 5$ and 8 are shown in Fig. 18. Stephenson and Roth determined the structure of L-Ta_2O_5, which corresponds to the compound with $m = 11$, and that of $15Ta_2O_5 \cdot 2WO_3$ with $m = 8$. Between Ta_2O_5 and $15Ta_2O_5 \cdot 2WO_3$, a continuum of structures is generated by the ordered intergrowth of the $m = 8$ and $m = 11$ units; for example, the structure of $37Ta_2O_5 \cdot 2WO_3$ corresponds to $m = 19$, i.e., an intergrowth of one unit with $m = 8$ and one with $m = 11$. The region between $15Ta_2O_5 \cdot 2WO_3$ and $11Ta_2O_5 \cdot 4WO_3$, the phase boundary, is based on ordered intergrowth of $m = 8$ and $m = 5$ units, and the structure of the limiting composition, $11Ta_2O_5 \cdot 4WO_3$, corresponds to $m = 13$; i.e., $1 \times (m = 8) + 1 \times (m = 5)$. On the basis of this structural principle, hundreds of compounds can be formulated in the narrow composition range Ta_2O_5 ($MO_{2.5}$) to $11Ta_2O_5 \cdot 4WO_3$ ($MO_{2.577}$).

(a)

(b)

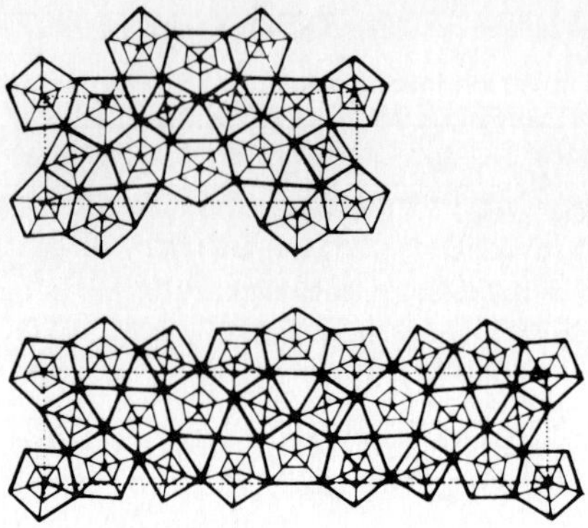

Fig. 18. The basic superstructure units with $m = 5$ (a) and $m = 8$ (b) in the Ta_2O_5–WO_3 system..

A similar principle has been elucidated by Bevan and co-workers in the $Y(O/F)_{2+x}$ system, $2.13 < x < 2.22$. In this case the basic superstructure units are layers of fluorite subcells, n fluorite cells in thickness, between which are accommodated excess fluoride ions. Five superstructures with $n = 4$, 5, 6, 7, and 8 are needed to account for the continuum of compounds in this system; these have compositions given by the general formula $Y_nO_{n-1}F_{n+2}$. Single-crystal structure determinations have been reported for $n = 7$ (Bevan and Mann, 1975) and, more recently, $n = 5$ and 6 (Mohyla, 1979). The members with $n = 4$ and 8 lie outside the stability range of the continuum, but they are found intergrown with the members $m = 5$ and 7, respectively. A sample of composition $Y(O/F)_{2.187}$ has a repeating unit, containing a sequence of 57 fluorite cells, made up of two layers with $n = 6$ and nine with $n = 5$ to give a lattice parameter $b \sim 317$ Å. A remarkable feature of these systems is their ability to sustain long-range order over such large distances.

An interesting variation on the theme of infinite adaptivity is found in the rutile shear system, TiO_2–Cr_2O_3. In the analogous system TiO_{2-x} (Section V) there are two homologous series of shear phases, one with a shear plane orientation $\{1\bar{2}1\}$ in the range $TiO_{1.75}$ to $TiO_{1.90}$ and another with $\{1\bar{3}2\}$ CS planes between $TiO_{1.937}$ and $TiO_{2.00}$. The behavior within the intermediate range of composition was established by electon microscopy in TiO_2–Cr_2O_3 by Bursill *et al.* (1971). As the composition is changed from $MO_{1.90}$ to $MO_{1.937}$, the CS planes pivot around from $\{1\bar{2}1\}$ to $\{1\bar{3}2\}$, traversing every intermediate orientation in the $[1\bar{1}1]$ zone. Gibb and Anderson (1972) have proposed a mechanism for this remarkable transformation. There can be no doubt that future structural studies will show that the phenomenon of infinite adaptivity is widespread in many areas of solid state chemistry.

VIII. Conclusion

From the crystallographer's point of view, nonstoichiometric compounds present an exciting challenge. Simple defective oxides have high symmetry, and their structures can normally be determined without difficulty. Here the challenge is found in extracting useful information from an average unit cell, and it has been seen that this normally requires the careful use of other chemical information. When the defects in high-symmetry crystals become ordered, for example, in Pr_7O_{12}, a superlattice is observed. In this situation, the phasing of the superstructure reflections presents a complex crystallographic problem, and it is not surprising that many well-known ordered phases such as U_4O_9 and Pr_6O_{11} have yet to be solved. The rewards for success, however, are considerable, since the ordered structures normally contain the key to understanding their disordered parents. A further difficulty is presented by ordered phases, whether derived from high-symmetry precursors or belonging to homologous series of shear or block structures; this is the problem of crystal growth. The preparation of homogeneous single crystals is very exacting and has proved to be a stumbling block in many systems. In the absence of single crystals, powders may be used, but if homogeneous polycrystalline samples prove illusive, then the electron microscope affords the only alternative approach.

Despite these difficulties, it has been seen that x-ray and neutron diffraction have been prominent in developing our understanding of nonstoichiometry. The complementarity with electrons is clearly illustrated in the shear and block structure systems; x rays have been used to elucidate the basic crystal architecture, and electron microscopy studies have utilized this as a basis for probing partial order and coherent intergrowth. Neutron and x-ray diffraction remain the only techniques that can yield precise and comprehensive information about interatomic distances and angles, and it can be confidently expected that they will continue to play a useful and unique role in the field of nonstoichiometry.

References

Anderson, J. S. (1973). *J.C.S. Dalton Trans.* 1107–1115.
Anderson, J. S., and Hyde, B. G. (1967). *J. Phys. Chem. Solids* **28**, 1393–1408.
Anderson, J. S., Browne, J. M., Cheetham, A. K., Von Dreele, R. B., Hutchison, J. L., Lincoln, F. J., Bevan, D. J. M., and Strähle, J. (1973). *Nature (London)* **243**, 81–83.
Anderson, J. S., Bevan, D. J. M., Cheetham, A. K., Von Dreele, R. B., Hutchison, J. L., and Strähle, J. (1975). *Proc. R. Soc. London, Ser. A* **346**, 139–156.
Andersson, B., and Gjønnes, J. (1970). *Acta Chem. Scand.* **24**, 2250–2253.
Andersson, B., and Sletnes, J. O. (1977). *Acta Crystallogr., Sect. A* **33**, 268–276.
Andersson, S. (1960). *Acta Chem. Scand.* **14**, 1161–1172.

Andersson, S. (1967). *Z. Anorg. Allg. Chem.* **351**, 106–112.
Andersson, S., and Magnéli, A. (1956). *Naturwissenschaften* **43**, 495–496.
Andersson, S., and Wadsley, A. D. (1966). *Nature (London)* **211**, 581–583.
Andersson, S., Collen, B., Kuylenstierna, U., and Magnéli, A. (1957). *Acta Chem. Scand.* **11**, 1641–1652.
Andersson, S., Mumme, W. G., and Wadsley, A. D. (1966). *Acta Crystallogr.* **21**, 802–808.
Andresen, A. F. (1958). *Acta Crystallogr.* **11**, 612–614.
Åsbrink, S., and Norrby, L. J. (1970). *Acta Crystallogr., Sect. B* **26**, 8–15.
Åström, A. (1966). *Acta Chem. Scand.* **20**, 969–982.
Bacon, G. E. (1952). *Acta Crystallogr.* **5**, 684–686.
Bacon, G. E. (1975). "Neutron Diffraction." Oxford Univ. Press, London and New York.
Baker, J. M., and Chaplin, D. M. (1979). *J. Phys. C* **12**, L129–L131.
Banus, M. D., and Reed, T. B. (1970). *In* "Chemistry of Extended Defects in Non-Metallic Systems" (L. Eyring and M. O'Keefe, eds.), pp. 488–522. North-Holland Publ., Amsterdam.
Bartholomew, R. F., and Frankl, D. R. (1969). *Phys. Rev.* **187**, 828–833.
Bartram, S. F. (1966). *Inorg. Chem.* **5**, 749–754.
Battle, P. D., and Cheetham, A. K. (1979). *J. Phys. C* **12**, 337–345.
Battle, P. D., Cheetham, A. K., and Goodenough, J. B. (1979). *Mater. Res. Bull.* **14**, 1013–1024.
Bell, P. S., and Lewis, M. H. (1971). *Phys. Status Solidi A* **7**, 431–439.
Bevan, D. J. M., and Mann, A. W. (1975). *Acta Crystallogr., Sect. B* **31**, 1406–1411.
Blomberg, B., Kihlborg, L., and Magnéli, A. (1953). *Ark. Kemi* **6**, 133–138.
Bowman, A. L., Wallace, T. C., Yarnell, J. L., and Wenzel, R. G. (1966). *Acta Crystallogr.* **21**, 843.
Brauer, G. (1941). *Z. Anorg. Allg. Chem.* **248**, 1–31.
Brebrick, R. F. (1970). *In* "Chemistry of Extended Defects in Non-Metallic Systems" (L. Eyring and M. O'Keefe, eds.), pp. 183–197. North-Holland Publ., Amsterdam.
Bursill, L. A., and Hyde, B. G. (1971). *Acta Crystallogr., Sect. B* **27**, 210–215.
Bursill, L. A., Hyde, B. G., and Philp, D. K. (1971). *Philos. Mag.* **23**, 1501–1513.
Carter, R. E., and Roth, W. L. (1968). *In* "E.M.F. Measurements in High Temperature Systems" (C. B. Alcock, ed.), pp. 125–144. Inst. Mining and Metallurgy, London.
Catalano, E., Bedford, R. G., Silviera, V. G., and Wickman, V. H. (1969). *J. Phys. Chem. Solids* **30**, 1613–1627.
Catlow, C. R. A., and Fender, B. E. F. (1975). *J. Phys. C* **8**, 3267–3279.
Cheetham, A. K. (1972). *In* "Chemical Applications of Thermal Neutron Scattering" (B. T. M. Willis, ed.), pp. 225–249. Oxford Univ. Press, London and New York.
Cheetham, A. K., and Taylor, J. C. (1977). *J. Solid State Chem.* **21**, 253–275.
Cheetham, A. K., and Von Dreele, R. B. (1974). *Proc. R. Soc. London, Ser. A* **338**, 311–326.
Cheetham, A. K., Fender, B. E. F., and Cooper, M. J. (1971a). *J. Phys. C* **4**, 3107–3121.
Cheetham, A. K., Fender, B. E. F., and Taylor, R. I. (1971b). *J. Phys. C* **4**, 2160–2165.
Cohen, I., and Schaner, B. E. (1963). *J. Nucl. Mater.* **9**, 18–24.
Darken, L. S., and Gurry, R. W. (1946). *J. Am. Chem. Soc.* **68**, 798–816.
Denker, S. P. (1964). *J. Phys. Chem. Solids* **25**, 1397–1405.
Denker, S. P. (1966). *J. Appl. Phys.* **37**, 142–149.
Dickens, P. G., and Wiseman, P. J. (1975). MTP International Review of Science (M. J. Emeléus, ed.), Inorganic Chemistry, Series Two, Vol. 10, pp. 211–254. Butterworth, London.
Eyring, L., and Baenziger, N. C. (1962). *J. Appl. Phys.* **33**, Suppl., 428–433.
Fender, B. E. F., and Riley, F. D. (1969). *J. Phys. Chem. Solids* **30**, 793–798.

Fender, B. E. F., and Riley, F. D. (1970). *In* "The Chemistry of Extended Defects in Non-Metallic Systems" (L. Eyring and M. O'Keefe, eds.), pp. 54–61. North-Holland Publ., Amsterdam.
Foex, M. (1948). *C. R. Acad. Sci.* **227**, 193–194.
Gado, P. (1965). *Acta Phys. Hung.* **18**, 111–117.
Gado, P., and Magnéli, A. (1965). *Acta Chem. Scand.* **19**, 1514–1515.
Gado, P., Holmberg, B., and Magnéli, A. (1965). *Acta Chem. Scand.* **19**, 2010–2011.
Galy, J., and Andersson, S. (1968). *Acta Crystallogr., Sect. B* **24**, 1027–1031.
Gatehouse, B. M., and Wadsley, A. D. (1964). *Acta Crystallogr.* **17**, 1545–1554.
Gavarri, J.-R., Carel, C., and Wiegel, D. (1979). *J. Solid State Chem.* **29**, 81–95.
Gebert, E., and Ackermann, R. J. (1966). *Inorg. Chem.* **5**, 136–142.
Gibb, R. M., and Anderson, J. S. (1972). *J. Solid State Chem.* **5**, 212–225.
Gleitzer, C. (1980). Personal communication.
Glemser, O., and Lutz, G. (1947). *Naturwissenschaften* **34**, 215.
Glemser, O., and Sauer, H. (1943). *Z. Anorg. Allg. Chem.* **252**, 144–159.
Greenwood, N. N., and Howe, A. T. (1972). *J.C.S. Dalton Trans.* 110–116.
Grey, I. E. (1974). *J. Solid State Chem.* **11**, 128–134.
Gruehn, R. (1972). *Natl. Bur. Stand.* (*U.S.*), *Spec. Publ.* No. 364.
Gruehn, R., and Norin, R. (1967). *Z. Anorg. Allg. Chem.* **355**, 176–181.
Gruehn, R., and Norin, R. (1969). *Z. Anorg. Allg. Chem.* **367**, 209–218.
Hagenmuller, P. (1973). *In* "Comprehensive Inorganic Chemistry" (A. F. Trotman-Dickenson, ed.), Vol. 4, pp. 541–605. Pergamon, Oxford.
Hayakawa, M., Cohen, J. B., and Reed, T. B. (1972). *J. Am. Ceram. Soc.* **55**, 160–164.
Hodeau, J. L., and Marezio, M. (1978). *J. Solid State Chem.* **23**, 253–263.
Hodeau, J. L., and Marezio, M. (1979). *J. Solid State Chem.* **29**, 47–62.
Høier, R., and Andersson, B. (1974). *Acta Crystallogr., Sect. A* **30**, 93–95.
Hope, D. A. O., and Cheetham, A. K. (1980). To be published.
Horiuchi, H., Tokonami, M., Morimoto, N., and Nagasawa, K. (1972). *Acta Crystallogr., Sec. B* **28**, 1404–1410.
Hyde, B. G., Bevan, D. J. M., and Eyring, L. (1966). *Philos. Trans. R. Soc. London, Ser. A* **259**, 583–614.
Iijima, S. (1974). *Proc. Int. Conf. Real Atoms Real Crys., Melbourne*, p. 135.
Ippolitov, E. G., Garashina, L. S., and Maklasklov, A. G. (1967). *Inorg. Mater.* (*USSR*) **3**, 59–62.
Jette, E. R., and Foote, F. (1933). *J. Chem. Phys.* **1**, 29–36.
Katsura, T., Iwasaki, B., Kimura, S., and Akimoto, S. (1967). *J. Chem. Phys.* **47**, 4559–4560.
Keller, C. (1973). *In* "Comprehensive Inorganic Chemistry" (A. F. Trotman-Dickenson, ed.), Vol. 5, pp. 219–276. Pergamon, Oxford.
Kihlborg, L. (1963a). *Ark. Kemi* **21**, 357–364.
Kihlborg, L. (1963b). *Ark. Kemi* **21**, 443–460, 471–495.
Koch, F., and Cohen, J. B. (1969). *Acta Crystallogr., Sect. B* **25**, 275–287.
Koch, F. B., and Fine, M. E. (1967). *J. Appl. Phys.* **38**, 1470–1471.
Loehman, R. E., Rao, C. N. R., and Honig, J. M. (1969). *J. Phys. Chem.* **73**, 1781–1784.
Lowenstein, M. Z., Kihlborg, L., Lau, K. H., Haschke, J. M., and Eyring, L. (1972). *Natl. Bur. Stan.* (*U.S.*), *Spec. Publ.* No. 364, 343–350.
Lundberg, M. (1970). *J. Solid State Chem.* **1**, 463–468.
Magnéli, A. (1946). *Ark. Kemi, Mineral. Geol.* **24A**, No. 2.
Magnéli, A. (1948). *Acta Chem. Scand.* **2**, 501–517.
Magnéli, A. (1950). *Ark. Kemi* **1**, 223–230, 513–523.
Malmros, G., and Thomas, J. O. (1977). *J. Appl. Crystallogr.* **10**, 7–11.

Mann, A. W., and Bevan, D. J. M. (1972). *J. Solid State Chem.* **5**, 410–418.
Marezio, M., McWhan, D. B., Dernier, P. D., and Remeika, J. P. (1973a). *J. Solid State Chem.* **6**, 213–221.
Marezio, M., McWhan, D. B., Dernier, P. D., and Remeika, J. P. (1973b). *J. Solid State Chem.* **6**, 419–429.
Marezio, M., Dernier, P. D., McWhan, D. B., and Kachi, S. (1974). *J. Solid State Chem.* **11**, 301–313.
Marezio, M., Tranqui, D., Lakkis, S., and Schlenker, C. (1977). *Phys. Rev. B* **16**, 2811–2821.
Mohyla, J. (1979). Ph.D. Thesis, Flinders Univ. of South Australia, Bedford Park.
Morinaga, M., and Cohen, J. B. (1976). *Acta Crystallogr., Sect. A* **32**, 387–395.
Norin, R. (1963). *Acta Chem. Scand.* **17**, 1391–1404.
Norin, R. (1966). *Acta Chem. Scand.* **20**, 871–880.
Nowotny, H. (1970). *In* "Chemistry of Extended Defects in Non-Metallic Systems" (L. Eyring and M. O'Keefe, eds.), pp. 223–237. North-Holland Publ., Amsterdam.
Ok, H. N., and Mullen, J. G. (1968). *Phys. Rev.* **168**, 550–562, 563–574.
Ok, H. N., and Mullen, J. G. (1969). *Phys. Rev.* **181**, 986.
Pierce, J. W., and Hong, H. Y.-P. (1974). *Proc. Rare Earth Res. Conf., 10th* **1**, 527–535.
Porter, V., and Roy, R. (1964). *Bull. Am. Ceram. Soc.* **43**, 263.
Rae-Smith, A. R., and Cheetham, A. K. (1979). *J. Solid State Chem.* **30**, 345–352.
Rietveld, H. M. (1969). *J. Appl. Crystallogr.* **2**, 65–71.
Roth, R. S., and Wadsley, A. D. (1965a). *Acta Crystallogr.* **18**, 724–730.
Roth, R. S., and Wadsley, A. D. (1965b). *Acta Crystallogr.* **19**, 32–38.
Roth, R. S., and Wadsley, A. D. (1965c). *Acta Crystallogr.* **19**, 38–47.
Roth, R. S., Wadsley, A. D., and Andersson, S. (1965). *Acta Crystallogr.* **18**, 643–647.
Roth, R. S., Waring, J. L., and Parker, H. S. (1970). *J. Solid State Chem.* **2**, 445–461.
Roth, W. L. (1958). *Phys. Rev.* **110**, 1333–1341; **111**, 772–781.
Roth, W. L. (1960). *Acta Crystallogr.* **13**, 140–149.
Saito, S., Nakahigoshi, K., and Shimomura, Y. (1966). *J. Phys. Soc. Jpn.* **21**, 850–860.
Shull, C. G., Strauser, W. A., and Wollan, E. O. (1951). *Phys. Rev.* **83**, 333–345.
Smith, D. K., and Cline, C. F. (1962). *J. Am. Ceram. Soc.* **45**, 249–250.
Steele, D. (1971). D. Phil. Thesis, Univ. of Oxford, Oxford.
Steele, D., and Fender, B. E. F. (1974). *J. Phys. C* **7**, 1–11.
Steele, D., Childs, P. E., and Fender, B. E. F. (1972). *J. Phys. C* **5**, 2677–2688.
Stephenson, N. C., and Roth, R. S. (1971a) *Acta Crystallogr., Sect. B* **27**, 1010–1017.
Stephenson, N. C., and Roth, R. S. (1971b). *Acta Crystallogr., Sect. B* **27**, 1018–1024.
Stephenson, N. C., and Roth, R. S. (1971c). *Acta Crystallogr., Sect. B* **27**, 1031–1036.
Stephenson, N. C., and Roth, R. S. (1971d). *Acta Crystallogr., Sect. B* **27**, 1037–1044.
Subbarao, E. C., Sutter, P. H., and Hrizo, J. (1965). *J. Am. Ceram. Soc.* **48**, 443–446.
Summerville, E., Tuenge, R. T., and Eyring, L. (1978). *J. Solid State Chem.* **24**, 21–31.
Tallant, D. R., and Wright, J. C. (1975). *J. Chem. Phys.* **63**, 2074–2085.
Taylor, A., and Doyle, N. J. (1970). *In* "The Chemistry of Extended Defects in Non-Metallic Solids" (L. Eyring and M. O'Keefe, eds.), pp. 523–540. North-Holland Publ., Amsterdam.
Terauchi, H., and Cohen, J. B. (1979). *Acta Crystallogr., Sect. A* **35**, 646–652.
Terauchi, H., Cohen, J. B., and Reed, T. B. (1978). *Acta Crystallogr., Sec. A* **34**, 556–561.
Thornber, M. R., Bevan, D. J. M., and Graham, J. (1968). *Acta Crystallogr., Sect. B* **24**, 1183–1190.
Tien, T. V., and Subbarao, E. C. (1963). *J. Chem. Phys.* **39**, 1041–1047.
Toussaint, C. J., and Vos, G. (1968). *J. Appl. Crystallogr.* **1**, 187–189.
Tret'yakov, Y. D., Saksonov, Y. G., and Gordeev, I. V. (1965). *Inorg. Mater. (USSR)* **1**, 382–386.

Vallet, P., and Raccah, P. (1964). *C. R. Acad. Sci.* **258**, 3679–3682.

Von Dreele, R. B., Eyring, L., Bowman, A. L., and Yarnell, J. L. (1975). *Acta Crystallogr., Sect. B* **31**, 971–974.

Wadsley, A. D. (1961). *Acta Crystallogr.* **14**, 660–664, 664–670.

Wadsley, A. D., and Andersson, S. (1970). *Perspect. Struct. Chem.* **3**, 1–58.

Wahlbeck, P. G., and Gilles, P. W. (1966). *J. Am. Ceram. Soc.* **49**, 180–183.

Watanabe, D., Castles, J. R., Jostsons, A., and Malin, A. S. (1966). *Nature (London)* **210**, 934–936.

Watanabe, D., Castles, J. R., Jostons A., and Malin, A. S. (1967). *Acta Crystallogr.* **23**, 307–313.

Watanabe, D., Terasaki, O., Jostsons, A., and Castles, J. R. (1968). *J. Phys. Soc. Jpn.* **25**, 292.

Willis, B. T. M. (1963). *Nature (London)* **197**, 755–756.

Willis, B. T. M. (1964). *J. Phys. (Paris)* **25**, 431–439.

Willis, B. T. M. (1970). *In* "Chemistry of Extended Defects in Non-Metallic Systems" (L. Eyring and M. O'Keefe, eds.), pp. 272–293. North-Holland Publ., Amsterdam.

Index